Springer Optimization and Its Applications

Volume 224

Aims and Scope

Optimization has continued to expand in all directions at an astonishing rate. New algorithmic and theoretical techniques are continually developing and the diffusion into other disciplines is proceeding at a rapid pace, with a spot light on machine learning, artificial intelligence, and quantum computing. Our knowledge of all aspects of the field has grown even more profound. At the same time, one of the most striking trends in optimization is the constantly increasing emphasis on the interdisciplinary nature of the field. Optimization has been a basic tool in areas not limited to applied mathematics, engineering, medicine, economics, computer science, operations research, and other sciences.

The series **Springer Optimization and Its Applications (SOIA)** aims to publish state-of-the-art expository works (monographs, contributed volumes, textbooks, handbooks) that focus on theory, methods, and applications of optimization. Topics covered include, but are not limited to, nonlinear optimization, combinatorial optimization, continuous optimization, stochastic optimization, Bayesian optimization, optimal control, discrete optimization, multi-objective optimization, and more. New to the series portfolio include Works at the intersection of optimization and machine learning, artificial intelligence, and quantum computing.

Volumes from this series are indexed by Web of Science, zbMATH, Mathematical Reviews, and SCOPUS.

Marija P. Stanić • Miloljub Albijanić •
Dragan Djurčić • Miodrag M. Spalević
Editors

Analysis, Approximation, Optimization: Computation and Applications

In Honor of Gradimir V. Milovanović on the Occasion of His 75th Birthday

Springer

Editors
Marija P. Stanić
Faculty of Science, Department of Mathematics and Informatics
University of Kragujevac
Kragujevac, Serbia

Dragan Djurčić
Faculty of Technical Sciences Čačak
University of Kragujevac
Čačak, Serbia

Miloljub Albijanić
FEFA Faculty
Metropolitan University
Belgrade, Serbia

Miodrag M. Spalević
Faculty of Mechanical Engineering, Department of Mathematics
University of Beograd
Belgrade, Serbia

ISSN 1931-6828 ISSN 1931-6836 (electronic)
Springer Optimization and Its Applications
ISBN 978-3-031-85745-4 ISBN 978-3-031-85743-0 (eBook)
https://doi.org/10.1007/978-3-031-85743-0

This Springer imprint is published by the registered company Springer Nature Switzerland AG
The registered company address is: Gewerbestrasse 11, 6330 Cham, Switzerland

This volume is dedicated to Academician Professor Gradimir V. Milovanović on the occasion of his 75th birthday

Academician Gradimir V. Milovanović

Preface

The book "Analysis, Approximation, Optimization: Computation and Applications" is dedicated to Academician Gradimir V. Milovanović on the occasion of his 75th anniversary and contains recent results in the fields of approximation theory, numerical analysis, mathematical analysis, optimization theory, and various applications of an interdisciplinary character.

This book contains chapters written by authors from 10 countries. Most of these results were presented in person during an International Conference "Analysis, Approximations and Applications" (AAA2023), organized by Faculty of Science, University of Kragujevac in Vrnjačka Banja, Serbia (June 21–24, 2023), in honor of the 75th birthday of Academician Gradimir V. Milovanović.

The book consists of the introductory text and 21 chapters.

In the introductory text under the title "The Life Path and Scientific Career of Academician Professor Gradimir V. Milovanović" the life path and rich scientific career of Academician Gradimir V. Milovanović is highlighted by his former PhD students and permanent collaborators Miodrag M. Spalević and Marija P. Stanić.

Chapters appear in alphabetical order with respect to the last name of the first-named author.

The shift-preserving operator $L : Vs \to Vs$ and the range operator R_s and their relationship, where V_s is a shift-invariant subspace of Sobolev space $Hs\ (Rn)$, $s \in R$, is studied by A. Aksentijević, S. Aleksić, and S. Pilipović.

M. Albijanić, M.M. Žižović, D. Pamučar, and M.R. Žižović present a new concept of criterion dominance in multicriteria analysis, which is applicable in situations of reduced satisfaction, normal satisfaction, and increased satisfaction of the dominant criterion.

H. Almutairi, D.Lj. Djukić, R.M. Mutavdžić Djukić, M.S. Pranić, L. Reichel, and M.M. Spalević propose to use the difference between enhanced averaged rules, introduced by M.S. Pranić, L. Reichel in 2015, and the associated Gauss rule to estimate the quadrature error in the latter.

On two lattices with interlacing nodes discrete measures for which a point weight is determined by the product of two classical Meixner weight functions are considered by A.I. Aptekarev, A.V. Dyachenko, and V.G. Lysov. For this system they prove perfectness, write the difference Pearson equation, and obtain a generalization of the Rodrigues formula for all the indices.

I. Aranđelović prove equivalence of triangle inequality and reverse triangle inequality on class of pseudometric spaces.

Some characteristics of the functional transformation of the Index function which is defined on the class of O—regularly varying function in the sense of Karamata are studied by D. Fatić and D.Djurčić.

J.C. García-Ardila and F. Marcellán consider Gaussian quadrature rules based on orthogonal polynomials associated with a weight function $w(x) = x^{\alpha} e^{-x}, \alpha > -1$, supported on an interval $(0, z)$, $z > 0$.

A.R Hayotov and S.S. Babaev derive and analyze an optimal quadrature formula in the Hilbert space $W_2^{(2,1)}(t, 1)$.

S. Ivković studies the dynamics of the adjoint of a weighted composition operator, gives necessary and sufficient conditions for this adjoint operator to be topologically transitive on the space of Radon measures on a locally compact Hausdorff space, and provides sufficient conditions for this operator to be chaotic.

D.R. Jandrlić, A.V. Pejčev, and M.M. Spalević present a survey of recent findings of authors concerning error estimates in Gaussian-type quadrature formulas for analytic functions defined on confocal ellipses.

Kurepa's hypothesis on the left factorial is discussed by Ž. Mijajlović.

R. Nikolić, V. Timotić, and D. Djurčić present results related to the mutual relationship between generalized inverse relations, weak asymptotic equivalence relations, and strong asymptotic equivalence relations in certain classes of functions studied in Karamata's theory of regular variability.

The existence of optimal solutions of a system of differential equations involving (j, φ)-Hilfer fractional derivative is studied by P.R. Patle, M. Gabeleh, D. Rabed Al-Zuhairi, and V. Rakočević.

An efficient iterative method of order six for the inclusion of the inverse of a given regular matrix is presented by M.D. Petković and M.S. Petković.

The process of constructing the optimal quadrature formula in the Hilbert space $\widetilde{W}_2^{(2,1)}(0, 1]$ of complex-valued periodic functions for the numerical calculation of Fourier coefficients is studied by K. Shadimetov, A. Hayotov, and U. Khayriev.

By using a modulus and non-negative summability matrix E. Savas introduces the notion of deferred almost sequences spaces, establishes some inclusion relations among these sequence spaces, and considers certain results on deferred almost sequences spaces.

A comprehensive analysis of the main topics concerning combinatorial numbers and polynomials with their generating functions, which have been studied up to the present and with future investigations, is given by Y. Simsek.

The convergence of difference schemes for parabolic interface problems with time-dependent coefficients are investigated by B. Sredojević, D. Bojović, and Z. Milovanović Jeknić.

T.V. Tomović Mladenović and M.P. Stanić give a survey of recent results on quadrature rules for trigonometric polynomials.

Ş. Yüzbaşı and B. Çetin offer a new and straightforward method for single degree of freedom system in terms of Boubaker polynomials.

M.R. Žižović, D. Pamučar, M.M. Žižović, and M. Albijanić provide a methodology for multicriteria analysis, in which the alternative evaluation function is multiplicative.

This book is intended for researchers and students of mathematics and other computational and applied sciences. As students, colleagues, and friends of Academician Gradimir V. Milovanović, we are honored by the opportunity to prepare this book.

Kragujevac, Serbia — Marija P. Stanić
Belgrade, Serbia — Miloljub Albijanić
Vrnjačka Banja, Serbia — Dragan Djurčić
Belgrade, Serbia — Miodrag M. Spalević
November 2024

Acknowledgments

We express our warmest thanks to all of the contributors of this volume and to the referees for their careful reading of the chapters.

Also, the editors would like to thank the younger colleagues Tatjana Tomović Mladenović, Aleksandra Milosavljević, Nevena Petrović, and Danica Fatić, who helped in the organization of the Conference as well as in the preparation of this volume.

Finally, the editors would like to express their deepest gratitude to Academician Gradimir V. Milovanović for the exceptional cooperation they have had with him so far, for all the help and support he has provided them, and for all the advice they have received from him.

Contents

Contributors

D. R. Abed Al-Zuhairi University of Diyala, College of Basic Education, Department of Mathematics, Diyala, Iraq

Aleksandar Aksentijević University of Kragujevac, Faculty of Science, Department of Mathematics and Informatics, Kragujevac, Serbia

Miloljub Albijanić University Metropolitan, FEFA Faculty, Belgrade, Serbia

Suzana Aleksić University of Kragujevac, Faculty of Science, Department of Mathematics and Informatics, Kragujevac, Serbia

Hanan Almutairi Department of Mathematical Sciences, Kent State University, Kent, OH, USA

Alexander I. Aptekarev Keldysh Institute of Applied Mathematics (RAS), Moscow, Russian Federation

Ivan D. Aranđelović University of Belgrade, Faculty of Mechanical Engineering, Belgrade, Serbia

Samandar S. Babaev V.I. Romanovsky Institute of Mathematics, Tashkent, Uzbekistan
Tashkent International University, Tashkent, Uzbekistan
Bukhara State University, Bukhara, Uzbekistan

Dejan Bojović University of Kragujevac, Faculty of Science, Department of Mathematics and Informatics, Kragujevac, Serbia

Beyza Çetin Missouri University Science and Technology, Department of Mathematics, Rolla, MO, USA

Dušan Lj. Djukić University of Belgrade, Faculty of Mechanical Engineering, Department of Mathematics, Belgrade, Serbia

Dragan Djurčić University of Kragujevac, Faculty of Technical Science, Department of Mathematics, Čačak, Serbia

Alexander V. Dyachenko Keldysh Institute of Applied Mathematics (RAS), Moscow, Russian Federation

Danica Fatić University of Kragujevac, Faculty of Technical Science, Department of Mathematics, Čačak, Serbia

Moosa Gabeleh Ayatollah Boroujerdi University, Department of Mathematics, Boroujerd, Iran

Juan C. García-Ardila Universidad Politécnica de Madrid, Departamento de Matemática Aplicada a la Ingeniería Industrial, Madrid, Spain

Abdullo R. Hayotov V.I. Romanovsky Institute of Mathematics, Tashkent, Uzbekistan
Central Asian University, Tashkent, Uzbekistan

Stefan Ivković Mathematical Institute of the Serbian Academy of Sciences and Arts, Belgrade, Serbia

Davorka R. Jandrlić University of Belgrade, Faculty of Mechanical Engineering, Department of Mathematics, Belgrade, Serbia

Umedjon Khayriev Bukhara State University, Bukhara, Uzbekistan
V.I. Romanovskiy Institute of Mathematics, Uzbekistan Academy of Sciences, Tashkent, Uzbekistan

Zorica Milovanović Jeknić University "Union-Nikola Tesla", Faculty of Construction Management, Belgrade, Serbia

Rada M. Mutavdžić Djukić University of Belgrade, Faculty of Mechanical Engineering, Department of Mathematics, Belgrade, Serbia

Vladimir G. Lysov Keldysh Institute of Applied Mathematics (RAS), Moscow, Russian Federation

Francisco Marcellán Universidad Carlos III de Madrid, Departamento de Matemáticas, Leganés, Spain

Žarko Mijajlović University of Belgrade, Faculty of Mathematics, Belgrade, Serbia

Rale M. Nikolić University of Defence, Military Academy, Department of Mathematics and Natural Sciences, Belgrade, Serbia

Dragan Pamučar University of Belgrade, Faculty of Organizational Sciences, Belgrade, Serbia

Pradip R. Patle VIT-AP University, School of Advanced Sciences, Department of Mathematics, Amravati, India

Aleksandar V. Pejčev University of Belgrade, Faculty of Mechanical Engineering, Department of Mathematics, Belgrade, Serbia

Marko D. Petković University of Niš, Faculty of Science and Mathematics, Niš, Serbia

Miodrag S. Petković University of Niš, Faculty of Electronic Engineering, Niš, Serbia

Stevan Pilipović University of Novi Sad, Faculty of Sciences, Department of Mathematics and Informatics, Novi Sad, Serbia

Miroslav S. Pranić University of Banja Luka, Faculty of Science, Department of Mathematics and Informatics, Banja Luka, Bosnia and Herzegovina

Vladimir Rakočević University of Niš, Faculty of Sciences and Mathematics, Department of Mathematics, Niš, Serbia

Lothar Reichel Kent State University, Department of Mathematical Sciences, Kent, OH, USA

Ekrem Savas Uşak University, Department of Mathematics, Uşak, Turkey

Kholmat Shadimetov Tashkent State Transport University, Tashkent, Uzbekistan
V.I. Romanovskiy Institute of Mathematics, Uzbekistan Academy of Sciences, Tashkent, Uzbekistan

Yilmaz Simsek University of Akdeniz, Faculty of Science, Department of Mathematics, Antalya, Turkey

Miodrag M. Spalević University of Belgrade, Faculty of Mechanical Engineering, Department of Mathematics, Belgrade, Serbia

Bratislav Sredojević University of Kragujevac, Faculty of Mechanical and Civil Engineering in Kraljevo, Kraljevo, Serbia

Marija P. Stanić Department of Mathematics and Informatics, Faculty of Science, University of Kragujevac, Kragujevac, Serbia

Valentina Timotić University of East Sarajevo, Faculty of Philosophy, Department of Mathematics, Pale, Bosnia and Herzegovina

Tatjana V. Tomović Mladenović University of Kragujevac, Faculty of Science, Department of Mathematics and Informatics, Kragujevac, Serbia

Şuayip Ş. Yüzbaşı Bartin University, Department of Mathematics, Bartin, Turkey

Miodrag M. Žižović Economics Institute, Belgrade, Serbia

Mališa R. Žižović University of Kragujevac, Faculty of Technical Sciences in Čačak, Čačak, Serbia

Part I
Introduction

Chapter 1
The Life Path and Scientific Career of Academician Professor Gradimir V. Milovanović

Miodrag M. Spalević and Marija P. Stanić

Dedicated to Academician Gradimir V. Milovanović
on the occasion of his 75th birthday

1.1 Introduction

In this chapter, the life path and rich scientific career of the world-known numerician and mathematician, scientist, professor, and Serbian Academician Gradimir V. Milovanović will be highlighted by his former PhD students and permanent collaborators.

Professor Milovanović obtained important results in different fields of Numerical Analysis, Approximation Theory, and Special Functions. The fields in which Academician Milovanović made the most important contributions are polynomials, orthogonal systems and polynomials, interpolation and quadrature processes, and approximations by polynomials and splines.

This chapter is a somewhat changed version of the lecture presented by the authors at the International Mathematical Conference "Analysis, Approximations and Applications" (AAA2023), held in Vrnjačka Banja (Serbia), from 21–23 June 2023 (Fig 1.1 - AAA2023 conference picture). The conference was dedicated to the 75th anniversary of Prof. Gradimir V. Milovanović (henceforth GVM for shortness), one of the best mathematicians that Serbia has ever had.

M. M. Spalević
Department of Mathematics, Faculty of Mechanical Engineering, University of Belgrade, Belgrade, Serbia
e-mail: mspalevic@mas.bg.ac.rs

M. P. Stanić (✉)
Department of Mathematics and Informatics, Faculty of Science, University of Kragujevac, Kragujevac, Serbia
e-mail: marija.stanic@pmf.kg.ac.rs

M. P. Stanić et al. (eds.), *Analysis, Approximation, Optimization: Computation and Applications*, Springer Optimization and Its Applications 224,
https://doi.org/10.1007/978-3-031-85743-0_1

Fig. 1.1 AAA2023 conference picture

A detailed account of GVM mathematical work, on the occasion of his 60th birthday, was featured by Ivić [13], Gautschi [6], and Rassias [26] in the book: "APPROXIMATION AND COMPUTATION: IN HONOR OF GRADIMIR V. MILOVANOVIĆ, Springer Optimization and Its Applications 42, W. Gautschi et al. (eds.), DOI 10.1007/978-1-4419-6594-3 16, Springer Science+Business Media, New York, LLC 2011," as well as on the occasion of his 70th birthday, by Rassias and Spalević [27, 28] in special issues of the journal *Applicable Analysis and Discrete Mathematics*.

This chapter is organized as follows. In the next section, the biography of GVM is given. In Sect. 1.3, we have highlighted his rich scientific career. In short, we point out his scientific results highlighting some of the most significant phases in his scientific career. In particular, we refer to some of his articles that initiated completely new directions of research. Finally, in Sect. 1.4, we give some personal remarks.

1.2 The Biography of GVM

GVM was born in Zorunovac (village near Knjaževac in Eastern Serbia) on January 2, 1948, to father Vukašin and mother Vukadinka (born Savić). He went to high school in the town of Knjaževac, where he successfully took part in mathematical

Fig. 1.2 GVM's family—a picture from a family lunch at the "Kalenic" restaurant in Belgrade

competitions. He is married to Dobrila, with whom he has a daughter, Irena, who lives with her husband Vladica Mitrović in Switzerland. Irena graduated from the Faculty of Medicine and specialized in microbiology. She works at Cantonal Hospital Lucerne (Medical Microbiology), Luzern, Switzerland. Her husband Vladica is a software engineer (he previously worked in the ABB corporation and now in his own company) (Fig. 1.2 - GVM's family picture).

In 1966, GVM began studies at the Section of Electronics of the Technical Faculty of Niš (Serbia), which soon became independent, as the Faculty of Electronics. From the beginning of his student days, GVM showed great interest for computer science and numerical mathematics. GVM obtained his bachelor's degree in 1971, master's degree in 1974 (at University of Niš), and his Ph. D. degree in 1976 (at University of Niš)—in both cases, his advisor was Prof. Dr. Dragoslav S. Mitrinović.

He became an assistant at the University of Niš in 1971, and he passed all the titles until he became a full professor in 1986 at the Faculty of Electronic Engineering. He was employed there until July 2008. During the period when he worked at the University of Niš, he served as the Head of the Department of Mathematics at the Faculty of Electronic Engineering (1983–2002), Vice Rector for Science and International Collaboration of the University of Niš (1989–1991), the Dean of the Faculty of Electronic Engineering (2002–2004), and the Rector of the University of Niš (2004–2006).

In 2008, GVM moved to Belgrade, where he was Dean of the Faculty of Computer Science at Megatrend University (2008–2011), and after that, he obtained a position of Research Professor at the Mathematical Institute of the Serbian Academy of Sciences and Arts (SASA) until his retirement in 2014. Retirement did not stop his fruitful scientific work, but he continued scientific activities through SASA.

In 2006, he was elected as a Corresponding Member of the SASA and, in 2012, became a Full Member. In addition, GVM is a member of other professional associations such as Full member of the Academy of Engineering Sciences of Serbia (2012 – ; corresponding member 2007–2012), Vijñāna Parishad (Science Academy) of India (2016 –), Scientific Society of Serbia, Mathematical Society of Serbia, Yugoslav Society for Applied and Industrial Mathematics (YUPIM), American Mathematical Society (AMS), Society for Industrial and Applied Mathematics (SIAM), SIAM Activity Group on Orthogonal Polynomials and Special Functions, and Gesellschaft für Angewandte Mathematik und Mechanik (GAMM).

During his rich university career, GVM taught a large number of different courses in mathematics at all levels of study from undergraduate to PhD level. These include numerical analysis, approximation theory, real analysis, linear algebra, special functions, mathematical programming, differential equations etc. This shows his wide mathematical culture and his many mathematical interests. He published nine books for students in Serbian language. He was a visiting professor at Purdue University (USA), Université de Pau (France), Università di Basilicata (Potenza, Italy), Université des Sciences et de la Technologie Houari Boumédiène, Faculté des Mathématiques, Algeria (January/February 2016: Approximation Theory Course at Doctoral Studies), etc.

According to The Mathematics Genealogy Project online database, Gradimir Milovanović supervised 14 PhD students, and he has 41 descendants.[1] Some of his students such as Predrag Stanimirović, Milan Kovačević, Marija P. Stanić, Miodrag M. Spalević, Nenad Cakić, Aleksandar Cvetković have become well-established mathematicians of their own. GVM supervised 16 master theses, too. He participated in the work of commissions for doctorate theses and promotion of professors in many countries (France, Italy, Cyprus, Australia, and India).

In addition to positions at universities, GVM held other important professional positions in the scientific community of Serbia. He was the President of the National Council for Scientific and Technological Development of the Republic of Serbia (from April 2006 to March 2010), President of the Scientific Committee for Mathematics, Computer Sciences and Mechanics of Serbian Ministry of Science and Technological Development (from October 2010 to October 2016), Scientific Secretary of Department of Mathematics, Physics and Geo-Sciences, SASA (from March 2016 to March 2021), the Vice-president of the Scientific Society of Serbia (from November 2002 to May 2016), President of the Scientific Council of the Mathematical Institute of SASA (from July 1997 to October 2010), and a member

[1] This information was downloaded from the site https://www.mathgenealogy.org on October 31, 2024.

of the Managing Board of the Mathematical Society of Serbia (from February 2003 to February 2006).

GVM participated in talks, including a large number of plenary and invited lectures, at numerous international conferences, congresses, symposia, workshops both in Serbia and around the world. Also, he organized several international mathematical conferences. We point out that he organized an international mathematical conference dedicated to the memory of his advisor Prof. Dragoslav S. Mitrinović in Niš in 1996. Also, appreciating his exceptional contribution, colleagues and students of GVM organized several conferences dedicated to him such as Approximation & Computation (2008, Niš, Serbia) on the occasion of his 60th anniversary; The Mediterranean International Conference of Pure &Applied Mathematics and Related Areas, (2018, Antalya, Turkey) on the occasion of his 70th anniversary; Analysis, Approximation, and Applications (2023, Vrnjačka Banja, Serbia) on the occasion of his 75th anniversary. In 2015 he received the "City of Belgrade" Award for Achievements for the Year 2014 in the Field: Natural and Technical Sciences, and in 2022 he received VPI Golden Jubilee Award, Vijñāna Parishad of India.

GVM was the founder of a scientific journal "Facta Universitatis: Series Mathematics and Informatics" at the University of Niš and its editor from the foundation 1986 to 2008. He has been served many scientific journals: Publication Institut Mathématique (Beograd) (N.S.) [Editor-in-Chief, 2011 –], Bulletin, Classe des Sciences Mathématiques et Naturelles, Sciences mathématiques, SASA, Belgrade [Editor-in-Chief, 2015 –], Optimization Letters. Springer [Associate Editor, 2014 –], Journal of Computational and Applied Mathematics [Guest Editor], Applicable Analysis and Discrete Mathematics (Formerly: Univ. Beograd. Publ. Elektrotehn. Fak. Ser. Mat.) [Editorial Advisory Board], Demonstratio Mathematica [Editorial Advisory Board], FILOMAT [Section Editor], Armenian Journal of Mathematics [Member of Editorial Board], Kragujevac Journal of Mathematics [Member of Editorial Board], Journal of Numerical Analysis and Approximation (Formerly: Revue d'Analyse Numérique et de Théorie de l'Approximation) [Editor], Studia Universitatis Babeş-Bolyai Mathematica [Member of Editorial Board], Yugoslav Journal of Operations Research [Member of Editorial Board], Novi Sad Journal of Mathematics [Member of Editorial Board], Applied Mathematics and Computation. Elsevier [Associate Editor, 2009 – 2019], Journal of Inequalities and Applications. Springer [Editor-in-Chief, 2016 – 2019], Annals of Functional Analysis. Springer [Associate Editor, 2010 – 2023], Banach Journal of Mathematical Analysis [Editorial Advisory Board, 2007 – 2010].

GVM is one of the most prolific Serbian mathematicians. So far, he has published more than 480 publications (9 monographs, 338 research papers in international reviewed journals, 38 chapters in monographs, and 82 in the proceedings of international and domestic conferences). He published papers in renowned mathematical journals such as "SIAM Journal on Scientific Computing," "IMA Journal of Numerical Analysis, "Numerische Mathematik," "Mathematics of Computation," "Journal of Computational and Applied Mathematics," "Journal of Mathematical Analysis and Applications," "Proceedings of the American Mathematical Society," "Constructive Approximation," and many others. Here, we have to point out three impressive monographs "*Topics in Polynomials: Extremal Problems, Inequalities, Zeros*" (written jointly with D.S. Mitrinović and Th.M. Rassias) [25], "*Interpolation process - Basic theory and applications*" (written jointly with G. Mastroianni) [14], and "*Extremal Problems and Inequalities of Markov-Bernstein Type for Algebraic Polynomials*" (written jointly with R.B. Gardner and N.K. Govil) [5]. The monograph [25] called by many the "Bible of Polynomials" has been cited over 1050 times, while the monograph [14] has been cited over 460 times so far. In a short time, the monograph [5] already has over 40 citations.

GVM was the leader of seven scientific projects. In recent years, he has been working on an individual project # Φ-96 (Serbian Academy of Sciences and Arts): "Interpolation and Quadrature Processes Based on Orthogonality Theory".

1.3 Scientific Career of GVM

During extremely fruitful scientific career of GVM, which spanned more than half a century, he worked successfully in various areas of mathematics. On the website

http://www.mi.sanu.ac.rs/~gvm/

one can find his complete bibliography. GVM collaborated with a large number of mathematicians around the world.

As we have already said, a detailed overview of his mathematical contribution is given in the papers [6, 13, 26–28]. We will not repeat here what was stated in those papers. Also, we will not list all the areas of mathematics in which GVM worked, nor all his collaborators, but we want to highlight, in our opinion, the most important phases in his work, marked by collaborations with certain people, as well as to point out the publications that had the most significant impact on the development of new directions of research.

1.3.1 Research Under the Influence of Professor Dragoslav S. Mitrinović

GVM started his research career working with Professor Dragoslav S. Mitrinović. In that initial period of his career, he primarily focused on proving different types of inequalities (integral inequalities, inequality of Ostrowski type, Wirtinger's type inequalities, etc.), then on iterative processes, functional equations, and extremal properties of polynomials. He published his first scientific paper [30] as well as his first book chapter [29] before defending his master's thesis. By the end of 1976, in which he defended his PhD dissertation, he had already published 12 scientific papers.

Fig. 1.3 Th.M. Rassias, D.S. Mitrinović, and GVM

In early period of his scientific career, the majority of his articles were published in Univ. Beograd. Publ. Elektrotehn. Fak. Ser. Mat. Fiz. (this journal continued as Applicable Analysis and Discrete Mathematics from 2007).

In that period, he also published three exceptional books [15–17] in the Serbian language, which have been used and studied by generations of students in Serbia, as well as in the former Yugoslavia, since then. Here, we want to point out that GVM was the author of the first complete text book in the field of numerical analysis in the Serbian language.

His first article in the field of numerical integration [24], an area that will largely be the focus of his scientific career, was also published in that period. Interesting coincidence is that article [24] was dedicated to Prof. Mitrinović on the occasion of his 75th birthday.

The crown of the collaboration with Prof. Mitrinović is the publication of exceptional monographs [25] in 1994 (Fig. 1.3).

Regarding his scientific work, GVM said that in the period up to 1990, full of energy, he expanded his knowledge the most and advanced his scientific and university career.

It is important to emphasize that GVM later continued to engage in research related to various types of inequalities and extremal properties of polynomials. That research led to the publication of another monograph in 2022 [5].

Fig. 1.4 GVM and W. Gautschi—Sicilia, Excursion during OPSFA3, 1990

1.3.2 Collaboration with Walter Gautschi

One of the most important moments in GVM's career was the beginning of collaboration with Professor Walter Gautschi in 1983. That collaboration had a great influence on his scientific work and further orientation (Fig. 1.4 - picture of GVM and Walter Gautschi from 1990).

GVM started to deal with the Constructive Orthogonality Theory. This opened up space for extensive numerical and symbolic work with orthogonal systems and their various applications such as:

- construction of new classes of (non-classical) orthogonal polynomials;
- development of different type of orthogonalities (s and σ-orthogonality; orthogonality on various curves in the complex plane; orthogonality on radial rays; orthogonality of Sobolev type; multiple orthogonality, etc.);
- applications in different fields of (numerical) analysis (numerical integration; applications in integral equations, differential equations, partial differential equations; interpolation processes, special functions, etc.);
- approximation theory;
- summation of slowly converging series;
- numerical integration of high oscillatory functions, etc.

The first joint paper with Walter Gautschi [7] was published in 1985. Their cooperation has not been interrupted to this day.

Walter Gautschi and GVM published 13 joint papers in the best journals for Numerical Analysis and Approximation Theory (e.g., Mathematics of Computation, Journal of Approximation Theory, Constructive Approximation, Numerische Mathematik, BIT, Journal of Computational and Applied Mathematics, ETNA, Numerical Algorithms). It can be said that the beginning of cooperation with Walter Gautschi had a great influence on GVM starting to deal with the orthogonality theory and its application in approximation theory and numerical integration, which are the areas he mostly dealt with in his scientific career.

Their cooperation until 2011 is described in detail in Gautschi's article [6]. Here, we mention only the main topics of their common research in that period: Gaussian quadrature involving Einstein and Fermi functions, summation processes, polynomials orthogonal on the semicircle, spline approximations, the numerical evaluation of singular integrals, s-orthogonality and construction of Gauss-Turán type quadrature rules. After that, they published four joint papers [9–12].

In [9], they developed procedures based on moments for computing the three-term recurrence relations for orthogonal polynomials relative to the Binet, generalized Binet, squared Binet, and related subrange weight functions. Also, they established monotonicity properties for the zeros of the respective orthogonal polynomials. In [10], they obtained in explicit forms, analytically as well as symbolically using Mathematica, the orthogonal polynomials and their three-term recurrence relation for generalized Marchenko–Pastur measure. Also, they provided the corresponding Matlab software. Article [11] was devoted to polynomials orthogonal with respect to the Prudnikov, the generalized Prudnikov, and Prudnikov-type weight functions and their symmetric extensions. Moment-based methods were used to generate the corresponding three-term recurrence relations. All developed procedures were implemented in MATLAB software. Finally, in [12], they found explicit expressions for the derivatives of the Ramanujan integral, proved complete monotonicity of the Ramanujan integral, and stated another useful integral representation of its $k-$th derivative.

W. Gautschi was a plenary speaker at the conference dedicated to GVM's 60th birthday (Fig. 1.5), while GVM visited Purdue University in 2018 when Gautschi's 90th birthday was celebrated (Fig. 1.6).

Professor Gautschi was the special guest at the conference AAA2023 organized last year dedicated to GVM, where he gave a lecture under the title *The Ramanujan integral and its derivatives: Computation and analysis*. This is also the title of their last published joint article [12] so far (Fig. 1.7).

1.3.3 Collaboration with Italian Mathematicians

GVM had collaboration with several Italian mathematicians: Luigi Gatteschi, Giovanni Monegato (Turin); Marco Frontini, Franca Calio (Milan); Laura Gori, Paulo Ricci (Roma); Mario Rosario Ocorssio (Naples); Francesco Altomare, Nicola Mastronardi (Bari), etc. (Fig. 1.8).

GVM had the most intensive collaboration with Giuseppe Mastroianni (Potenza) and his collaborators Donatella Ocorssio, Incoronata Notarangelo, Maria Carmela

Fig. 1.5 W. Gautschi during the lecture at the conference APP&COM dedicated to GVM's 60th birthday (Niš, Serbia, 2008)

de Bonnis, Maria Gracia Russo, Luisa Fermo. Collaboration with G. Mastroianni was intensified in the last decades of the 20th century and continues.

G. Mastroianni and GVM worked jointly on the following topics:

- weighted integration of periodic function on the real line;
- polynomial approximation;
- numerical methods for integral equations;
- interpolation processes;
- Gaussian quadrature rules related to the exponential weights, etc.

Probably the most significant result of their collaboration is the comprehensive monograph [14]. The description of that monograph was given in [13] (Fig. 1.9).

1.3.4 Collaboration with PhD Students

As already mentioned in the Sect. 1.2, GVM supervised 14 PhD students. Throughout his career, he collaborated with his Phd students as well as with his descendants. He had the most intensive and permanent ollaboration with the authors of this article and Aleksandar Cvetković (Fig. 1.10 - picture of A. Cvetković, W. Gautschi, and M. P. Stanić from 2007). Unfortunately, the extremely fruitful collaboration with A. Cvetković was stopped by his untimely death in 2022.

Fig. 1.6 W. Gautschi and GVM—Department of Computer Science, Purdue University, 2018

Fig. 1.7 W. Gautschi during the lecture at the conference AAA2023 dedicated to GVM's 75th birthday

GVM and A. Cvetković published more than 45 joint papers. Their joint work on the *Wolfram Mathematica* program package "*OrthogonalPolynomials*" [3] certainly enabled one of the key moments for the intensification of further research in the field of orthogonality theory and its application in various areas of mathematics, but also in other sciences (primarily in engineerings and computer sciences). That software is used by a large number of researchers dealing with orthogonal systems.

The most important areas of joint work between GVM and A. Cvetković were:

- numerical integration of functions with logarithmic end point singularity;

Fig. 1.8 G. Mastroianni, GVM, and G. Monegato—FAAT Conference, Acquafredda di Maratea, Italy, 2009

Fig. 1.9 G. Mastroianni and GVM—NAAT Conference, Cluj-Napoca, Romania (left), and cover of the joint monograph (right)

Fig. 1.10 A. Cvetković, W. Gautschi, and M. P. Stanić - ETH, Zurich, 2007

- numerical construction of orthogonal polynomials and Gaussian-type quadrature rules;
- orthogonal polynomials and Gaussian quadrature rules related to oscillatory complex weight functions;
- Gaussian interval quadrature rules for some classical weight functions;
- Gaussian type quadrature rules for Müntz systems;
- trigonometric orthogonal systems and quadrature rules;
- nonstandard Gaussian quadrature rules based on operator values;
- Gaussian quadrature rules using function derivatives, etc.

In terms of time, the longest permanent cooperation among his PhD students GVM has with Miodrag M. Spalević (Fig. 1.11). Also, it is a collaboration that includes the largest number of younger researcher, PhD students, and descendants of M. M. Spalević: Miroslav Pranić, Aleksandar Pejčev, Jelena Tomanović, Dušan Djukić, and Rada Mutavdžić Djukić.

GVM and M. M. Spalević published more than 25 joint papers in the following topics:

- power orthogonality (s and σ-orthogonal polynomials);
- quadrature rules with multiple nodes (of Gauss–Turán and Chakalov–Popoviciu type);
- error bounds for Gauss–Turán, Gauss–Turán–Radau, and Gauss–Turán–Lobatto quadratures for analytic functions;
- Kronrod extensions of Gaussian quadratures with multiple nodes;

Fig. 1.11 GVM and M. M. Spalević - ACTA Conference, SASA, Belgrade, 2017

- error bounds of some Gauss–Turán-Kronrod quadratures for analytic function;
- Kronrod extensions with multiple nodes of quadrature formulas for Fourier coefficients;
- quadratures with multiple nodes for Fourier–Chebyshev coefficients;
- moment-preserving spline approximation etc.

A monograph under the title "Quadrature Formulae With Multiple Nodes", written by G.V. Milovanović, M.M. Spalević, and M.S. Pranić will be published soon.

Finally, we will give a brief overview of the cooperation of GVM and Marija P. Stanić (Fig. 1.12) and her former and current PhD students and colalborators Tatjana Tomović Mladenović, Nevena Petrović, Aleksandar Jovanović, and Aleksandra Milosavljević.

GVM and M. P. Stanić published more than 20 joint articles in the following topics:

- multiple orthogonal algebraic polynomials and the optimal sets of quadrature rules in Borges' sense;
- multiple orthogonal polynomials on the semicircle and the corresponding optimal set of quadrature rules;
- trigonometric orthogonal systems and quadrature rules of Gaussian type;
- s and $\sigma-$ orthogonal trigonometric polynomials and the corresponding quadrature rules with multiple nodes;
- numerical integration of highly oscillating functions;
- multiple orthogonal trigonometric polynomials, etc.

Fig. 1.12 GVM, Th.M. Rassias, T. Igić, M. P. Stanić, GVM's wife, and Rassias' wife - Athens, Greece, 2012

1.3.5 GVM's Publications That Initiated New Directions of Research

Here we would like to highlight several very important GVM's publications, because they initiated completely new directions of research.

Article [18], although published in the Conference Proceedings, initiated the development of s-orthogonality and quadratures with multiple nodes. That article led to an invitation for GVM to write a review paper for Elsevier's series "Numerical Analysis 2000, Vol. V, Quadrature and Orthogonal Polynomials (W. Gautschi, F. Marcellan, and L. Reichel, eds.)." Such article was also published in J. Comput. Appl. Math. (see [22]).

Contour integration method for integration of highly oscillatory functions from [21] is used in the articles of many authors.

Orthogonal polynomials on the radial rays in the complex plane were introduced in [20]. In [4], orthogonal polynomials on the radial rays were particularly highlighted as an important case that has received attention in the literature. Also, an extended Dunkl oscillator model based on the generalized Hermitian polynomials on radial rays from GVM's papers was discussed very recently by Bouzeffour in [1, 2].

One of the papers that had a great impact on the work of large number of researchers is [19], where a method based on the use of Gaussian quadrature was proposed for summarizing slowly converging series.

The orthogonality on the semicircle with respect to non-Hermitian inner product introduced in joint work with W. Gautschi in [8] also opened up a whole range of new research studies and generalizations. This is also the case with the recently published extension to the orthogonal Laurent polynomials [23].

1.4 Final Personal Remarks

In view of everything that has been said about Academician Gradimir V. Milovanović and in view of our experience of collaboration with him, we want to say that we had a great privilege to be his PhD students and his permanent collaborators. We look forward to future joint research with him.

Also, we had a great privilege to organize the Conference dedicated to Academician Gradimir V. Milovanović on the occasion of his 75th birthday and in particular a great privilege to participate at the Conference talking among eminent world mathematicians about his life and scientific career.

We wish professor Gradimir good health and many more significant scientific results.

References

1. Bouzeffour, F. The extended Dunkl oscillator and the generalized Hermite polynomials on the radial lines. J. Nonlinear Math. Phys. **31**(56), 13 pp. (2024)
2. Bouzeffour, F. Supersymmetric Quesne-Dunkl quantum mechanics on radial lines. Symmetry **16**(11), 1508 (2024)
3. Cvetković, A.S., Milovanović, G.V.: The Mathematica Package "OrthogonalPolynomials". Facta Univ. Ser. Math. Inform. **19**, 17–36 (2004)
4. Deaño, A., Huybrechs, D., Iserles, A.: Computing Highly Oscillatory Integrals, X+180 pp. Society for Industrial and Applied Mathematics (SIAM), Philadelphia (2018)
5. Gardner, R.B., Govil, N.K., Milovanović, G.V.: Extremal Problems and Inequalities of Markov-Bernstein Type for Algebraic Polynomials. Mathematical Analysis and Its Applications, XVIII+424 pp. Elsevier/Academic Press, London (2022)
6. Gautschi, W.: My collaboration with Gradimir V. Milovanović. In: Gautschi, W., Mastroianni, G., Rassias, Th.M. (eds.) Approximation and Computation - In Honor of Gradimir V. Milovanović. Springer Optimization and Its Applications, vol. 42, pp. 33–43. Springer, Berlin (2011)
7. Gautschi, W., Milovanović, G.V.: Gaussian quadrature involving Einstein and Fermi functions with an application to summation of series. Math. Comp. **44**, 177–190 (1985)
8. Gautschi, W., Milovanović, G.V.: Polynomials orthogonal on the semicircle. J. Approx. Theory **46**, 230–250 (1986)
9. Gautschi, W., Milovanović, G.V.: Binet-type polynomials and their zeros. Electron. Trans. Numer. Anal. **50**, 52–70 (2018)
10. Gautschi, W., Milovanović, G.V.: Orthogonal polynomials relative to a generalized Marchenko-Pastur probability measure. Numer. Algorithms **88**(3), 1233–1249 (2021)
11. Gautschi, W., Milovanović, G.V.: Orthogonal polynomials relative to weight functions of Prudnikov type. Numer. Algorithms **90**, 263–270 (2022)

12. Gautschi, W., Milovanović, G.V.: A Ramanujan integral and its derivatives: Computation and analysis. Math. Comp. **93**(347), 1297–1308 (2024)
13. Ivić, A.: The scientific work of Gradimir V. Milovanović. In: Gautschi, W., Mastroianni, G., Rassias, Th.M. (eds.) Approximation and Computation - In Honor of Gradimir V. Milovanović. Springer Optimization and Its Applications, vol. 42, pp. 3–32. Springer, Berlin (2011)
14. Mastroianni, G., Milovanović, G.V.: Interpolation Processes - Basic Theory and Applications. Springer Monographs in Mathematics, XIV+444 pp. Springer, Berlin–Heidelberg (2008)
15. Milovanović, G.V.: Numerical Analysis I, Naučna knjiga, Belgrade, 1985 (Second Edition 1988, Third Edition 1991), VIII+435 pp. (in Serbian).
16. Milovanović, G.V.: Numerical Analysis II, Naučna knjiga, Belgrade, 1985 (Second Edition 1988, Third Edition 1991), VIII+210 pp. (in Serbian).
17. Milovanović, G.V.: Numerical Analysis III, Naučna knjiga, Belgrade, 1988 (Second Edition 1991), VIII + 204 pp. (in Serbian).
18. Milovanović, G.V.: Construction of s-orthogonal polynomials and Turán quadrature formulae. In: Numerical Methods and Approximation Theory III (Niš, 1987), pp. 311–328. University of Niš, Niš, (1988)
19. Milovanović, G.V.: Summation of series and Gaussian quadratures. In: Zahar, R.V.M. (ed.) Approximation and Computation, pp. 459–475, ISNM Vol. 119. Birkhäuser, Basel-Boston-Berlin (1994)
20. Milovanović, G.V.: A class of orthogonal polynomials on the radial rays in the complex plane. J. Math. Anal. Appl. **206**, 121–139 (1997)
21. Milovanović, G.V.: Numerical calculation of integrals involving oscillatory and singular kernels and some applications of quadratures. Comput. Math. Appl. **36**, 19–39 (1998)
22. Milovanović, G.V.: Quadrature with multiple nodes, power orthogonality, and moment-preserving spline approximation. J. Comput. Appl. Math. **127**, 267–286 (2001)
23. Milovanović, G.V.: Orthogonality on the semicircle: old and new results. Electron. Trans. Numer. Anal. **59**, 99–115 (2023)
24. Milovanović, G.V., Djordjević, R. Ž.: A generalization of modified Birkhoff-Young quadrature formula. Univ. Beograd. Publ. Elektrotehn. Fak. Ser. Mat. Fiz. No **735**(762), 130–134 (1982)
25. Milovanović, G.V., Mitrinović, D.S., Rassias, Th.M.: Topics in Polynomials: Extremal Problems. Inequalities, Zeros, XIV+822 pp. World Scientific, Singapore, New Jersey, London, Hong Kong (1994)
26. Rassias, Th.M.: On some major trends in mathematics. In: Gautschi, W., Mastroianni, G., Rassias, Th.M. (eds.) Approximation and Computation - In Honor of Gradimir V. Milovanović. Springer Optimization and Its Applications, vol. 42, pp. 45–49. Springer, Berlin (2011)
27. Rassias, Th.M., Spalević, M.: Gradimir V. Milovanović – a master in approximation and computation. Part I. Appl. Anal. Discrete Math. **13**(3), 643–648 (2019)
28. Rassias, Th.M., Spalević, M.M.: Gradimir V. Milovanović – a master in approximation and computation. Part II. Appl. Anal. Discrete Math. **14**(3), 528–559 (2020)
29. Ristić, M.M., Milovanović, G.V., Madić, P.: Fizičko-matematičeskaja formulirovka kinetiki spekanija, In: An Essay on the Generalization of the Sintering Theory, pp. 237–252. International Team for Studying Sintering, Beograd (in Russian) (1973)
30. Tošić, D.Dj., Milovanović, G.V.: An application of Newton's method to simultaneous determination of zeros of a polynomial. Univ. Beograd. Publ. Elektrotehn. Fak. Ser. Mat. Fiz. No **412**(460), 175–177 (1973)

Part II
Contributed Chapters

Chapter 2
Shift-Invariant Subspaces of Sobolev Spaces and Shift-Preserving Operators

Aleksandar Aksentijević, Suzana Aleksić, and Stevan Pilipović

Dedicated to Academician Gradimir V. Milovanović
on the occasion of his 75th birthday

2.1 Introduction

This chapter has the characteristics of a review paper in which results of Bownik [13] and Aguilera et al. [4, 5] are transferred to Sobolev spaces; thus, it also contains the original results. If after some assertion is written ([...], Theorem/Lemma/Proposition ...), it means that the corresponding result is proved in quoted paper for the L^2-case.

In the first part of this chapter, we follow the results of Bownik [13] (based on [19, 20, 22, 25]) for shift-preserving operator defined on shift-invariant subspaces of $L^2(\mathbb{R}^n)$ and give the corresponding results for the shift-preserving operator defined on V_s, where V_s is a shift-invariant subspace of Sobolev space $H^s := H^s(\mathbb{R}^n)$, $s \in \mathbb{R}$, building on our results [6]. By studying the shift-preserving operator defined on V_s, we give some characteristics of the dimension function. We show that every shift-preserving operator $L : V_s \to H^s$ can be represented by the corresponding range operator R_s. Using a range operator approach, for a given frame with bounds A and B, we obtain elements of the dual frame with bounds B^{-1} and A^{-1}.

In the second part of this chapter, we follow the results of [5] (based on [4, 7, 14]), and thus, the definitions of s-eigenvalue, s-eigenspace and s-diagonalization of the

---

A. Aksentijević · S. Aleksić (✉)
Department of Mathematics and Informatics, Faculty of Science, University of Kragujevac, Kragujevac, Serbia
e-mail: aksentijevic@kg.ac.rs; suzana.aleksic@pmf.kg.ac.rs

S. Pilipović
Department of Mathematics and Informatics, Faculty of Sciences, University of Novi Sad, Novi Sad, Serbia
e-mail: stevan.pilipovic@dmi.uns.ac.rs

M. P. Stanić et al. (eds.), *Analysis, Approximation, Optimization: Computation and Applications*, Springer Optimization and Its Applications 224,
https://doi.org/10.1007/978-3-031-85743-0_2

shift-preserving operator L are adapted to our spaces. We prove that every normal shift-preserving operator $L : V_s \to V_s$ is s-diagonalizable. Moreover, if the operator L is s-diagonalizable, then the corresponding range operator is diagonalizable. Finally, we give conditions under which the collection $\{L^j\phi_i : \phi_i \in V_s, i = 1, \dots, m, j = 0, \dots, d-1\}$ is a frame generator for shift-invariant space V_s generated by d functions $\varphi_1, \dots, \varphi_d \in H^s$. This problem is known as Dynamical Sampling. The dynamical sampling problem has got a lot of attention in the last few years. For a recent contribution to the theory, we refer the reader to [1–3, 7–10, 16–18, 23]. Applications can be found in [11, 27].

This chapter is organized as follows. After the introduction section where background information is provided, in Sect. 2.2, we reintroduce notation and assertions in much the same way as in [6]. The shift-preserving operator and the range operator are studied in Sect. 2.3. In Sect. 2.4, we prove that the frame operator has the corresponding range operator associated with the dual Gramian, and we give a result about the dual frame. Section 2.5 gives a presentation of results announced in [6]. The results proved in Sect. 2.6 will be used in Sect. 2.7. Finally, in Sect. 2.7, we find conditions on the shift-preserving operator and a finite set of functions of V_s so that the iterations of the operator on the functions produce a frame generator set of V_s.

2.2 Notation and Basic Assertions

We follow the notation of [6]. Let s be a real number, $\mu_s(\cdot) := (1 + |\cdot|^2)^{s/2}$ and $\mathbb{T}^n := [0,1)^n$, where $n \in \mathbb{N}$. We denote the translation of a function f for $k \in \mathbb{Z}^n$ by $T_k f(\cdot) = f(\cdot - k)$. By $\widehat{f}$ or $\mathcal{F}f$, we denote the Fourier transform of an integrable function f defined by $\mathcal{F}f(t) = \widehat{f}(t) = \int_{\mathbb{R}^n} f(x)\mathrm{e}^{-2\pi\sqrt{-1}\langle x,t\rangle}\,\mathrm{d}x$, $t \in \mathbb{R}^n$, where $\langle x, t\rangle = \sum_{i=1}^n x_i t_i$. Note that $\mathcal{F}^{-1} f(t) = \widehat{f}(-t)$. Recall,

$$(c_k)_{k\in\mathbb{Z}^n} \in \ell_s^2(\mathbb{Z}^n) \quad \text{if and only if} \quad \sum_{k\in\mathbb{Z}^n} |c_k|^2 \mu_s^2(k) < +\infty,$$

$$f \in H^s \quad \text{if and only if} \quad f \in \mathcal{S}'(\mathbb{R}^n) \text{ and } \int_{\mathbb{R}^n} |\widehat{f}(t)|^2 \mu_s^2(t)\,\mathrm{d}t < +\infty,$$

with the scalar products $\langle (c_k)_{k\in\mathbb{Z}^n}, (d_k)_{k\in\mathbb{Z}^n}\rangle_{\ell_s^2} = \sum_{k\in\mathbb{Z}^n} c_k \overline{d}_k \mu_s^2(k)$ for $\ell_s^2(\mathbb{Z}^n)$ and $\langle f, g\rangle_{H^s} = \int_{\mathbb{R}^n} \widehat{f}(t)\overline{\widehat{g}}(t)\mu_s^2(t)\,\mathrm{d}t$ for H^s. To shorten notation, we write ℓ_s^2 instead of $\ell_s^2(\mathbb{Z}^n)$. We will denote by $\{e_k : k \in \mathbb{Z}^n\}$, ($e_k = (0, \dots, 0, 1, 0, \dots)$, where the kth component of e_k equals 1) the standard basis of ℓ_s^2. For any measurable set A, we denote by $|A|$ the measure of the set A. Note, $L_s^2 := L_s^2(\mathbb{R}^n) = \mathcal{F}(H^s)$, i.e. $f \in L_s^2$ if and only if $\widehat{f} \in H^s$, $s \in \mathbb{R}$. By $H\left(\mathbb{T}^n, \ell_s^2\right)$, we denote the Hilbert space, which consists of all vector-valued measurable functions $F : \mathbb{T}^n \to \ell_s^2$ such that $\int_{\mathbb{T}^n} \|F(t)\|_{\ell_s^2}^2\,\mathrm{d}t < +\infty$, with scalar product $\langle F_1, F_2\rangle_{H(\mathbb{T}^n,\ell_s^2)} =$

$\int_{\mathbb{T}^n}\langle F_1(t), F_2(t)\rangle_{\ell^2_s}\,\mathrm{d}t$. By I is denoted the finite set $\{1, 2, \dots, r\}$ or $\mathbb{N}$. Next, $E_s(\mathcal{A}_{I,s}) = \{T_k\varphi : \varphi \in \mathcal{A}_{I,s}, k \in \mathbb{Z}^n\} \subset H^s$, where $\mathcal{A}_{I,s} = \{\varphi_i : \varphi_i \in H^s, i \in I\}$. If $I = \{1, \dots, r\}$, we will use notation $\mathcal{A}_{r,s}$ instead of notation $\mathcal{A}_{I,s}$.

Recall [6, 13], a closed subspace $V_s \subset H^s$ is a shift-invariant space if $T_k\varphi \in V_s$ for every $\varphi \in V_s$, for any $k \in \mathbb{Z}^n$. The space $S_s(\mathcal{A}_{I,s}) = \overline{\operatorname{span}}(E_s(\mathcal{A}_{I,s}))$ is a shift-invariant space generated by $\mathcal{A}_{I,s}$. It is called a principal shift-invariant (PSI) space if $S_s(\mathcal{A}_{I,s}) = S_s(\varphi)$. If $S_s(\mathcal{A}_{I,s}) = S_s(\varphi_1, \varphi_2, \dots, \varphi_r)$, we say that it is a finitely generated shift-invariant (FSI) space.

We recall some assertions from [6].

Lemma 2.1 ([6]) *A mapping $\mathcal{T}_s : H^s \to H(\mathbb{T}^n, \ell^2_s)$ defined by*

$$\mathcal{T}_s\varphi(t) = \left(\frac{\widehat{\psi}(t+k)}{\mu_s(k)}\right)_{k\in\mathbb{Z}^n}, \qquad t \in \mathbb{T}^n,$$

where $\varphi \in H^s$ and $\left(1 - \frac{\Delta}{4\pi^2}\right)^{s/2}\varphi = \psi \in L^2(\mathbb{R}^n)$ (Δ is the Laplacian), is an isometric isomorphism and

$$\mathcal{T}_s T_j\varphi(t) = \mathrm{e}^{-2\pi\sqrt{-1}\langle t,j\rangle}\mathcal{T}_s\varphi(t), \quad j \in \mathbb{Z}^n,\ t \in \mathbb{T}^n.$$

We introduce several notions, following [6]. A range function is a mapping

$$J_s : \mathbb{T}^n \to \left\{\text{closed subspaces of } \ell^2_s\right\}.$$

Next, J_s is measurable if for each $a, b \in \ell^2_s$, the scalar function $t \mapsto \langle P_{J_s}(t)a, b\rangle_{\ell^2_s}$ is measurable, where $P_{J_s}(t) : \ell^2_s \to J_s(t)$, $t \in \mathbb{T}^n$, are associated orthogonal projections. For a function J_s, let us define the space

$$M_{J_s} = \{F \in H(\mathbb{T}^n, \ell^2_s) : F(t) \in J_s(t) \text{ for a.e. } t \in \mathbb{T}^n\},$$

which is a closed subspace of $H(\mathbb{T}^n, \ell^2_s)$.

Theorem 2.1 ([6]) *Let J_s be a measurable range function. A closed subspace $V_s \subset H^s$ is shift-invariant if and only if*

$$V_s = \left\{\varphi \in H^s : \mathcal{T}_s\varphi(t) \in J_s(t) \textit{ for a.e. } t \in \mathbb{T}^n\right\}.$$

The correspondence between V_s and J_s is one-to-one under the convention that the range functions are identified if they are equal a.e. Furthermore, if $V_s = S_s(\mathcal{A}_{I,s})$ for some $\mathcal{A}_{I,s} \subset H^s$, then $J_s(t) = \overline{\operatorname{span}}\{\mathcal{T}_s\varphi(t) : \varphi \in \mathcal{A}_{I,s}\}$.

Thus, every measurable range function J_s determines a shift-invariant space V_s by $V_s = \mathcal{T}_s^{-1}M_{J_s}$, and vice versa, i.e., for every shift-invariant space $V_s \subset H^s$, there is a measurable range function J_s such that $\mathcal{T}_s V_s = M_{J_s}$.

The definition of the dimension function $\dim_{V_s} : \mathbb{T}^n \to \mathbb{N} \cup \{0, +\infty\}$ of V_s is adapted from [6], and it is defined by $\dim_{V_s}(t) = \dim J_s(t)$, where $V_s = \mathcal{T}_s^{-1} M_{J_s}$. The spectrum of V_s is defined by $\sigma_{V_s} := \{t \in \mathbb{T}^n : J_s(t) \neq \{\mathbf{0}\}\}$.

Recall that $E_s(\mathcal{A}_{I,s})$ is a Bessel family for $S_s(\mathcal{A}_{I,s})$ with bound $B < +\infty$ if

$$\sum_{(k,i)\in\mathbb{Z}^n\times I} |\langle f, T_k\varphi_i\rangle_{H^s}|^2 \leq B\|f\|_{H^s}^2, \quad f \in \operatorname{span}(E_s(\mathcal{A}_{I,s})).$$

A Bessel family $E_s(\mathcal{A}_{I,s})$ is called a frame for $S_s(\mathcal{A}_{I,s})$ with frame bounds A, B, if additionally there exists $A > 0$ such that

$$A\|f\|_{H^s}^2 \leq \sum_{(k,i)\in\mathbb{Z}^n\times I} |\langle f, T_k\varphi_i\rangle_{H^s}|^2, \quad f \in \operatorname{span}(E_s(\mathcal{A}_{I,s})).$$

A frame $E_s(\mathcal{A}_{I,s})$ is a Parseval frame if $A = B = 1$, and it is a fundamental frame if it is complete in $S_s(\mathcal{A}_{I,s})$, i.e. $\operatorname{span}(E_s(\mathcal{A}_{I,s}))$ is dense in $S_s(\mathcal{A}_{I,s})$.

We call $E_s(\mathcal{A}_{r,s})$ a Riesz family for $S_s(\mathcal{A}_{r,s})$ with bounds $A, B > 0$ if

$$A\|c\|_{(\ell_s^2)^r} \leq \left\| \sum_{i=1}^{r} \sum_{k\in\mathbb{Z}^n} c_k^i T_k\varphi_i \right\|_{H^s} \leq B\|c\|_{(\ell_s^2)^r},$$

hold for all (finitely supported) sequences $c = (c_k^i)_{k\in\mathbb{Z}^n, 1\leqslant i\leqslant r}$. If a Riesz family $E_s(\mathcal{A}_{r,s})$ is complete in $S_s(\mathcal{A}_{r,s})$, we say $E_s(\mathcal{A}_{r,s})$ forms a Riesz basis. The best general references here are [15] and [21].

Theorem 2.2 ([6]) *$E_s(\mathcal{A}_{I,s})$ is a frame or a Riesz basis for $S_s(\mathcal{A}_{I,s})$ with bounds A, B or a Bessel family with bound B, for every $s \in \mathbb{R}$ (equivalently, for some $s \in \mathbb{R}$) if and only if $\{\mathcal{T}_s\varphi(t) : \varphi \in \mathcal{A}_{I,s}\} \subset \ell_s^2$ is a frame or a Riesz basis for $J_s(t)$ with bounds A, B or a Bessel family with bound B for a.e. $t \in \mathbb{T}^n$, for every $s \in \mathbb{R}$ (equivalently for some $s \in \mathbb{R}$), respectively. Moreover, $E_s(\mathcal{A}_{I,s})$ is a fundamental frame for every $s \in \mathbb{R}$ if and only if $\{\mathcal{T}_s\varphi(t) : \varphi \in \mathcal{A}_{I,s}\} \subset \ell_s^2$ is a fundamental frame for a.e. $t \in \mathbb{T}^n$, for every $s \in \mathbb{R}$.*

We say that $\varphi_0 \in S_s(\varphi)$ is a tight-frame generator (or quasi-orthogonal generator) of $S_s(\varphi)$ if for every $f \in S_s(\varphi)$ holds $\|f\|_{H^s}^2 = \sum_{k\in\mathbb{Z}^n} |\langle T_k\varphi_0, f\rangle_{H^s}|^2$. If φ_0 is a tight-frame generator of $S_s(\varphi)$, then $\|\mathcal{T}_s\varphi_0(t)\|_{\ell_s^2} = \mathbf{1}_{\sigma_{S_s(\varphi)}}(t)$ for a.e. $t \in \mathbb{T}^n$, and vice versa (see [6] for more details).

The next theorem is given in [6] without proof, since it is similar to the proof of Theorem 3.3 in [13]. For the sake of completeness, we give here the proof. Unless otherwise stated, in the remainder of the chapter, we assume $\oplus$ to be an orthogonal sum and write $Z = X \oplus Y$ or $X = Z \ominus Y$.

Theorem 2.3 ([6], The Decomposition Theorem) *Suppose that V_s is a shift-invariant subspace of H^s. Then, V_s can be decomposed as an orthogonal sum*

$$V_s = \bigoplus_{i\in\mathbb{N}} V_s^i, \tag{2.1}$$

where V_s^i, $i \in \mathbb{N}$, are principal shift-invariant spaces with tight-frame generators φ_i, $i \in \mathbb{N}$, and $\sigma_{V_s^{i+1}} \subset \sigma_{V_s^i}$, for all $i \in \mathbb{N}$. Moreover, $\dim_{V_s^i}(t) = \|\mathcal{T}_s\varphi_i(t)\|_{\ell_s^2}$, $i \in \mathbb{N}$, and

$$\dim_{V_s}(t) = \sum_{i\in\mathbb{N}} \|\mathcal{T}_s\varphi_i(t)\|_{\ell_s^2}, \quad \text{for a.e. } t \in \mathbb{T}^n. \tag{2.2}$$

Proof Let $Y \subset H^s$ be an arbitrary shift-invariant space with the range function J_s and associated projections P_{J_s}. We will consider the function

$$F \in \mathcal{T}_s Y$$

defined as follows. If $Y = \{\mathbf{0}\}$, then $\mathcal{T}_s Y = \{\mathbf{0}\}$. Otherwise, we define $a_k \in H(\mathbb{T}^n, \ell_s^2)$ by

$$a_k(t) := \begin{cases} \dfrac{P_{J_s}(t)e_{\pi(k)}}{\|P_{J_s}(t)e_{\pi(k)}\|_{\ell_s^2}}, & t \in A_k, \\ 0, & \text{otherwise}, \end{cases}$$

where $\pi : \mathbb{N} \to \mathbb{Z}^n$ is bijection and $A_k := \{t \in \mathbb{T}^n : P_{J_s}(t)e_{\pi(k)} \neq \mathbf{0}\}$, $k \in \mathbb{N}$. Let $(B_k)_{k\in\mathbb{N}}$ be a sequence of sets such that

$$B_1 := A_1, \qquad B_{k+1} := A_{k+1} \setminus \bigcup_{j=1}^{k} A_j, \ k \in \mathbb{N},$$

and define

$$F := \sum_{k\in\mathbb{N}} a_k \mathbf{1}_{B_k}.$$

Then, since $\sigma_Y = \bigcup_{k\in\mathbb{N}} A_k$, we have $F(t) \in J_s(t)$ and $\|F(t)\|_{\ell_s^2} = \mathbf{1}_{\sigma_Y}(t)$ for a.e. $t \in \mathbb{T}^n$. Thus, it is clear that $\varphi := \mathcal{T}_s^{-1} F$ is a quasi-orthogonal generator of $S_s(\varphi) \subset Y$, and $\sigma_{S_s(\varphi)} = \sigma_Y$.

It is easy seen that

$$\mathcal{T}_s(Y \ominus S_s(\varphi)) = \left\{Q \in H(\mathbb{T}^n, \ell_s^2) : Q(t) \in J_s(t), \langle F(t), Q(t)\rangle_{\ell_s^2} = 0 \text{ a.e. } t \in \mathbb{T}^n\right\}.$$

If $Q \in \mathcal{T}_s(Y \ominus S_s(\varphi))$, then

$$\langle Q(t), \mathrm{e}_{\pi(k)} \rangle_{\ell_s^2} = \langle Q(t), P_{J_s}(t) e_{\pi(k)} \rangle_{\ell_s^2} = 0, \text{ for a.e. } t \in \mathbb{T}^n, \ k = 1, \ldots, k_0, \tag{2.3}$$

where $k_0 = \min\{k \in \mathbb{N} : |A_k| \neq 0\}$.

Next, by induction on k, we will construct a sequence $(\varphi_k)_{k \in \mathbb{N}}$ of quasi-orthogonal generators. Let $\varphi_1 := \mathcal{T}_s^{-1} F(V_s)$, and suppose that we have already constructed functions $\varphi_1, \ldots, \varphi_k$, for some $k \in \mathbb{N}$, with properties:

(1) the function $\varphi_i \in V_s$ is a quasi-orthogonal generator of $V_s^i = S_s(\varphi_i)$, for $i = 1, \ldots, k$;
(2) $V_s^i \perp V_s^j$ for $i \neq j$;
(3) for $Q \in \mathcal{T}_s W_s^k$, where

$$W_s^k = V_s \ominus \left(\bigoplus_{i=1}^{k} V_s^i \right), \tag{2.4}$$

we have $\langle Q(t), e_{\pi(i)} \rangle_{\ell_s^2} = 0$ for all $i = 1, \ldots, k$ and a.e. $t \in \mathbb{T}^n$.

Set $\varphi_{k+1} := \mathcal{T}_s^{-1} F(W_s^k)$. Now, the set $\{\varphi_1, \ldots, \varphi_{k+1}\}$ satisfies the properties (1) $-$ (3). Namely, (1) follows from the fact that $\|\mathcal{T}_s \varphi_{k+1}(t)\|_{\ell_s^2} = \mathbf{1}_{\sigma_{W_s^k}}(t)$; (2) is a consequence of the fact that $V_s^{k+1} \subset W_s^k$ and the equality (2.4) and, finally, (2.3) implies (3).

Take $Q \in \mathcal{T}_s\big(\bigcap_{i=1}^{+\infty} W_s^i\big)$. Then, from (3), we deduce that $\langle Q(t), e_{\pi(i)} \rangle_{\ell_s^2} = 0$ for a.e. $t \in \mathbb{T}^n$ and all $i \in \mathbb{N}$. Hence, $\bigcap_{i=1}^{+\infty} W_s^i = \{\mathbf{0}\}$. Thus, (2.1) follows. Next, since $W_s^{i+1} \subset W_s^i$, we conclude that $\sigma_{V_s^{i+1}} = \sigma_{W_s^{i+1}} \subset \sigma_{W_s^i} = \sigma_{V_s^i}$. Finally, (2.2) follows from (2.1). □

Proposition 2.1 ([4], Lemma 2.7) *Let J_{V_s} and J_{U_s} be the range functions for the shift-invariant subspaces V_s, U_s of H^s, respectively.*

(1) The orthogonal complement of V_s, denoted by $V_s^{\perp}$, is a shift-invariant space and $J_{V_s^{\perp}}(t) = (J_{V_s}(t))^{\perp}$ for a.e. $t \in \mathbb{T}^n$.
(2) If $J_{V_s}(t) = J_{U_s}(t)$ for a.e. $t \in \mathbb{T}^n$, then $V_s = U_s$.
(3) The space $V_s \cap U_s$ is a shift-invariant space and $J_{V_s \cap U_s}(t) = J_{V_s}(t) \cap J_{U_s}(t)$, for a.e. $t \in \mathbb{T}^n$.

Proof

(1) Let $f \in V_s^{\perp}$. Since $f \notin V_s$, we have $T_k f \notin V_s$ for every $k \in \mathbb{Z}^n$. Consequently, $T_k f \in V_s^{\perp}$, for every $k \in \mathbb{Z}^n$, i.e. $V_s^{\perp}$ is a shift-invariant space. Furthermore, for every $F \in H(\mathbb{T}^n, \ell_s^2)$, we have

$$F(t) \notin J_{V_s}(t) \text{ if and only if } F(t) \in (J_{V_s}(t))^{\perp}, \quad \text{for a.e. } t \in \mathbb{T}^n.$$

By Theorem 2.1, we get

$$F(t) \notin J_{V_s}(t) \text{ if and only if } \mathcal{T}_s^{-1}F \notin V_s$$
$$\text{if and only if } F(t) \in J_{V_s^{\perp}}(t), \quad \text{for a.e. } t \in \mathbb{T}^n.$$

Hence, $J_{V_s^{\perp}}(t) = (J_{V_s}(t))^{\perp}$ for a.e. $t \in \mathbb{T}^n$.

(2) If $J_{V_s}(t) = J_{U_s}(t)$ for a.e. $t \in \mathbb{T}^n$, then $M_{J_{V_s}} = M_{J_{U_s}}$. Since $\mathcal{T}_s V_s = M_{J_{V_s}}$ and $\mathcal{T}_s U_s = M_{J_{U_s}}$, we get $V_s = U_s$.

(3) Obviously, $V_s \cap U_s$ is a shift-invariant space. By Theorem 2.1, for $F \in H(\mathbb{T}^n, \ell_s^2)$, we have

$$F(t) \in J_{V_s \cap U_s}(t) \text{ for a.e. } t \in \mathbb{T}^n$$
$$\text{if and only if } \mathcal{T}_s^{-1}F \in V_s \cap U_s$$
$$\text{if and only if } \mathcal{T}_s^{-1}F \in V_s \text{ and } \mathcal{T}_s^{-1}F \in U_s$$
$$\text{if and only if } F(t) \in J_{V_s}(t) \text{ and } F(t) \in J_{U_s}(t) \text{ for a.e. } t \in \mathbb{T}^n$$
$$\text{if and only if } F(t) \in J_{V_s}(t) \cap J_{U_s}(t) \text{ for a.e. } t \in \mathbb{T}^n.$$

Thus, $J_{V_s \cap U_s}(t) = J_{V_s}(t) \cap J_{U_s}(t)$, for a.e. $t \in \mathbb{T}^n$.

□

Proposition 2.2 ([4], Lemma 2.9) *Let $V_s \subset H^s$ be a shift-invariant space with corresponding range function J_s such that $\dim J_s(t) < +\infty$ for a.e. $t \in \mathbb{T}^n$. Then, there exist functions $\varphi_i \in H^s$, $i \in \mathbb{N}$, and measurable sets $(A_d)_{d \in \mathbb{N}_0}$ such that $A_i \cap A_j = \emptyset$, $i \neq j$, $\bigcup_{d \in \mathbb{N}_0} A_d = \mathbb{T}^n$ satisfying:*

(1) $\{T_k \varphi_i : i \in \mathbb{N}, k \in \mathbb{Z}^n\}$ is a Parseval frame for V_s;
(2) for $i > d$, $\mathcal{T}_s \varphi_i(t) = \mathbf{0}$, for a.e. $t \in A_d$;
(3) $\{\mathcal{T}_s \varphi_1(t), \ldots, \mathcal{T}_s \varphi_d(t)\}$ is an orthonormal basis for $J_s(t)$, for a.e. $t \in A_d$;
(4) $\dim J_s(t) = d$ for a.e. $t \in A_d$.

Proof By Theorem 2.3, there exist $\varphi_i \in H^s$, $i \in \mathbb{N}$, such that $\{T_k \varphi_i : i \in \mathbb{N}, k \in \mathbb{Z}^n\}$ is a Parseval frame for V_s. By Theorem 2.2, $\{\mathcal{T}_s \varphi_i(t) : i \in \mathbb{N}\}$ is a Parseval frame for $J_s(t)$ for a.e. $t \in \mathbb{T}^n$.

Since $\sigma_{V_s^{i+1}} \subset \sigma_{V_s^i}$ for every $i \in \mathbb{N}$, we can define a family of disjoint sets $(A_d)_{d \in \mathbb{N}_0}$ as follows: $A_0 := \mathbb{T}^n \setminus \sigma_{V_s}$ and $A_d := \sigma_{V_s^d} \setminus \sigma_{V_s^{d+1}}$ for $d > 0$. Since $\dim J_s(t) < +\infty$, we have $\sum_{i \in \mathbb{N}} \|\mathcal{T}_s \varphi_i(t)\|_{\ell_s^2} < +\infty$, and thus, $\bigcap_{i \in \mathbb{N}} \sigma_{V_s^i} = \emptyset$. Consequently, $\bigcup_{d \in \mathbb{N}_0} A_d = \mathbb{T}^n$.

It is obvious that (2) is true for $d > 0$. Further, by (1), Theorem 2.3 and Lemma 2.1, the assertion (3) follows, and thus $\dim J_s(t) = d$ for a.e. $t \in A_d$. For $d = 0$, we have $J_s(t) = \{\mathbf{0}\}$ for a.e. $t \in A_0$. □

2.3 Shift-preserving Operators and Range Operators

Let V_s be a shift-invariant subspace of H^s with the range function J_s and associated projection P_{J_s} (see [6]). Following [13] we define the shift-preserving operator and the range operator.

We say that a bounded linear operator $L : V_s \to H^s$ is shift-preserving if $LT_k = T_k L$ for all $k \in \mathbb{Z}^n$.

A range operator on J_s (with values in ℓ_s^2) is a mapping

$$R_s : \mathbb{T}^n \to \{\text{set of bounded operators defined on closed subspaces of } \ell_s^2\},$$

so that the domain of $R_s(t)$ equals $J_s(t)$ for a.e. $t \in \mathbb{T}^n$. The operator R_s is measurable if $t \mapsto R_s(t)P_{J_s}(t)$ is weakly operator measurable, i.e., $t \mapsto \langle R_s(t)P_{J_s}(t)a, b\rangle_{\ell_s^2}$ is measurable scalar function for each $a, b \in \ell_s^2$.

Now we transfer results from [13] related to L^2 spaces to H^s spaces.

Theorem 2.4 ([13], Proposition 4.2) *Let $\varphi \in H^s$ be a tight-frame generator of $S_s(\varphi)$ and an operator $L : S_s(\varphi) \to H^s$ be shift-preserving. Then, for any $g \in L_s^2(\mathbb{T}^n)$,*

$$\big(\mathcal{T}_s L \mathcal{T}_s^{-1}\big)(g\Phi)(t) = g(t)\big(\mathcal{T}_s L \mathcal{T}_s^{-1}\big)\Phi(t), \quad \textit{for a.e. } t \in \mathbb{T}^n, \tag{2.5}$$

where $\Phi = \mathcal{T}_s\varphi$.

Proof For $\Phi = \mathcal{T}_s\varphi$, $\varphi \in S_s(\varphi)$, by Lemma 2.1, we have

$$\begin{aligned}\big(\mathcal{T}_s L \mathcal{T}_s^{-1}\big)(\mathrm{e}^{-2\pi\sqrt{-1}\langle\cdot,k\rangle}\Phi) &= \big(\mathcal{T}_s L T_k\big)\varphi = \big(\mathcal{T}_s T_k L\big)(\mathcal{T}_s^{-1}\mathcal{T}_s\varphi)\\ &= \mathrm{e}^{-2\pi\sqrt{-1}\langle\cdot,k\rangle}\big(\mathcal{T}_s L \mathcal{T}_s^{-1}\big)\Phi.\end{aligned}$$

Thus, for all trigonometric polynomials $p_m(t) = \sum_{|k|\le m} a_k \mathrm{e}^{-2\pi\sqrt{-1}\langle t,k\rangle}$, holds

$$\big(\mathcal{T}_s L \mathcal{T}_s^{-1}\big)(p_m\Phi)(t) = p_m(t)\big(\mathcal{T}_s L \mathcal{T}_s^{-1}\big)\Phi(t), \quad \text{for a.e. } t \in \mathbb{T}^n. \tag{2.6}$$

Using Lemma 2.1, the fact that φ is a tight-frame generator of $V_s = S_s(\varphi)$, and boundedness of L, we get

$$\begin{aligned}\int_{\mathbb{T}^n} |p_m(t)|^2 \big\|(\mathcal{T}_s L \mathcal{T}_s^{-1})\Phi(t)\big\|_{\ell_s^2}^2 \,\mathrm{d}t &= \int_{\mathbb{T}^n} \big\|(\mathcal{T}_s L \mathcal{T}_s^{-1})(p_m\Phi)(t)\big\|_{\ell_s^2}^2 \,\mathrm{d}t\\ &= \big\|(\mathcal{T}_s L \mathcal{T}_s^{-1})(p_m\Phi)\big\|_{H(\mathbb{T}^n,\ell_s^2)}^2\\ &\le C^2 \big\|p_m\Phi\big\|_{H(\mathbb{T}^n,\ell_s^2)}^2\\ &= C^2 \int_{\mathbb{T}^n} |p_m(t)|^2 \|\Phi(t)\|_{\ell_s^2}^2 \,\mathrm{d}t\\ &= C^2 \int_{\mathbb{T}^n} |p_m(t)|^2 \mathbf{1}_{\sigma_{S_s(\varphi)}}(t)\,\mathrm{d}t < +\infty.\end{aligned} \tag{2.7}$$

For every $r \in L^\infty(\mathbb{T}^n)$, there exists a sequence of polynomials $(p^i_{m_i})_{i\in\mathbb{N}}$ such that: $\|p^i_{m_i}\|_\infty \le \|r\|_\infty$, $i \in \mathbb{N}$, and $\lim_{i\to+\infty} p^i_{m_i}(t) = r(t)$, for a.e. $t \in \mathbb{T}^n$. Thus, from (2.7), we have

$$\int_{\mathbb{T}^n} |r(t)|^2 \big\|(\mathcal{T}_s L \mathcal{T}_s^{-1})\Phi(t)\big\|^2_{\ell^2_s}\, \mathrm{d}t \le C^2 \int_{\mathbb{T}^n} |r(t)|^2 \|\Phi(t)\|^2_{\ell^2_s}\, \mathrm{d}t,$$

so

$$\big\|(\mathcal{T}_s L \mathcal{T}_s^{-1})\Phi(t)\|_{\ell^2_s} \le C\|\Phi(t)\|_{\ell^2_s} \quad \text{for a.e. } t \in \mathbb{T}^n. \tag{2.8}$$

Therefore, we can choose a sequence $(p^i_{m_i})_{i\in\mathbb{N}}$ such that $p^i_{m_i} \to g$ in $L^2_s(\mathbb{T}^n)$ and additionally

$$\lim_{i\to+\infty} p^i_{m_i}(t) = g(t), \quad \lim_{i\to+\infty} \big(\mathcal{T}_s L \mathcal{T}_s^{-1}\big)(p^i_{m_i}\Phi)(t) = \big(\mathcal{T}_s L \mathcal{T}_s^{-1}\big)(g\Phi)(t), \tag{2.9}$$

for a.e. $t \in \mathbb{T}^n$. Finally, from (2.6), using (2.8) and (2.9), it follows that (2.5) holds for a general functions in $L^2_s(\mathbb{T}^n)$. □

Let us mention a consequence of the theorems 2.4 and 2.1.

Corollary 2.1 *Let $V_s \subset H^s$ be a shift-invariant space and let $L : V_s \to H^s$ be the shift-preserving operator. Then, for any $\Phi \in \mathcal{T}_s V_s$ and measurable function g so that $g\Phi \in H(\mathbb{T}^n, \ell^2_s)$ (and thus, $g\Phi \in \mathcal{T}_s V_s$), holds*

$$\big(\mathcal{T}_s L \mathcal{T}_s^{-1}\big)(g\Phi)(t) = g(t)\big(\mathcal{T}_s L \mathcal{T}_s^{-1}\big)\Phi(t), \quad \textit{for a.e. } t \in \mathbb{T}^n.$$

Theorem 2.5 ([13], Theorem 4.5) *Let $V_s \subset H^s$ be a shift-invariant space with the range function J_s.*

(1) Let $L : V_s \to H^s$ be a shift-preserving operator. Then, there exists a measurable range operator R_s on J_s so that

$$(\mathcal{T}_s L) f(t) = R_s(t)(\mathcal{T}_s f(t)) \quad \textit{for a.e. } t \in \mathbb{T}^n,\ f \in V_s. \tag{2.10}$$

(2) Let R_s be a measurable range operator on J_s such that

$$\operatorname*{ess\,sup}_{t\in\mathbb{T}^n} \|R_s(t)\| < +\infty.$$

Then, there exists a shift-preserving operator $L : V_s \to H^s$ that satisfies (2.10).

The correspondence between L and R_s is one-to-one under the convention that the range operators are identified if they are equal almost everywhere. Furthermore, $\operatorname*{ess\,sup}_{t\in\mathbb{T}^n} \|R_s(t)\| = \|L\|$.

Proof

(1) From Theorem 2.3, we have that $V_s = \oplus_{i=1}^{+\infty} S_s(\varphi_i)$, where φ_i is a tight-frame generator of $S_s(\varphi_i)$.

First, we observe $V_s^k := \oplus_{i=1}^{k} S_s(\varphi_i)$ with a range function J_s^k. Let $\Phi_i = \mathcal{T}_s\varphi_i$. Then $\{\Phi_1(t), \dots, \Phi_k(t)\}\backslash\{\mathbf{0}\}$ (if $t \notin \sigma_{V_s^k}$, then $\Phi_i(t) = \mathbf{0}$, $i = 1, \dots, k$) is an orthonormal basis of $J_s^k(t)$ for a.e. $t \in \mathbb{T}^n$. We define $R_s^k(t) : J_s^k(t) \to \ell_s^2$ with

$$R_s^k(t)\Big(\sum_{i=1}^{k} \alpha_i \Phi_i(t)\Big) = \sum_{i=1}^{k} \alpha_i (\mathcal{T}_s L \mathcal{T}_s^{-1})\Phi_i(t), \quad \alpha_i \in \mathbb{C}, i = 1, \dots, k.$$

Note, by (2.8), it follows that $R_s^k(t)$ is a well-defined operator. Now, let $f \in V_s^k$ be arbitrary. Then there exist $f_i \in S_s(\varphi_i)$, $i = 1, \dots, k$, such that $f = f_1 + \dots + f_k$, and $\mathcal{T}_s f = \mathcal{T}_s f_1 + \dots + \mathcal{T}_s f_k = g_1\Phi_1 + \dots + g_k\Phi_k$, for some $g_i \in L_s^2(\mathbb{T}^n)$. Therefore, by Theorem 2.4, we have

$$\begin{aligned}(\mathcal{T}_s L) f(t) = (\mathcal{T}_s L \mathcal{T}_s^{-1})\Big(\sum_{i=1}^{k} g_i\Phi_i\Big)(t) &= \sum_{i=1}^{k} g_i(t)(\mathcal{T}_s L \mathcal{T}_s^{-1})\Phi_i(t)\\ &= \sum_{i=1}^{k} g_i(t) R_s^k(t)\big(\Phi_i(t)\big)\\ &= \sum_{i=1}^{k} R_s^k(t)\big(g_i(t)\Phi_i(t)\big)\\ &= R_s^k(t)\big(\mathcal{T}_s f(t)\big). \end{aligned} \tag{2.11}$$

Since $\mathcal{T}_s$ is the isometry, we conclude that R_s^k is measurable.

Let $\|L\| \le C$, where C is some positive constant. We will prove that $\|R_s^k(t)\| \le C$ for a.e. $t \in \mathbb{T}^n$. First, we will prove that

$$\operatorname*{ess\,sup}_{t\in\mathbb{T}^n} \|R_s^k(t)(\Psi_\alpha(t))\|_{\ell_s^2} \le C, \tag{2.12}$$

where $\Psi_\alpha \in H(\mathbb{T}^n, \ell_s^2)$ is given by $\Psi_\alpha(t) := \sum_{i=1}^{k} \alpha_i\Phi_i(t)$ for $\alpha = (\alpha_1, \dots, \alpha_k) \in \mathbb{S} := \{\alpha \in \mathbb{C}^k : |\alpha_1|^2 + \dots + |\alpha_k|^2 = 1\}$. Obviously, $\|\Psi_\alpha(t)\|_{\ell_s^2} = 1$.

To obtain a contradiction, suppose that (2.12) is false. Then, there exist $\varepsilon > 0$ and a measurable set $A \subset \mathbb{T}^n$ with $|A| > 0$, such that for $t \in A$,

$\|R_s^k(t)(\Psi_\alpha(t))\|_{\ell_s^2} > C + \varepsilon$. Let $\Psi := \Psi_\alpha \mathbf{1}_A$ and $\psi := \mathcal{T}_s^{-1}\Psi \in V_s^k$. Since $\mathcal{T}_s$ is isometry and $\|L\| \le C$, we get

$$\|(\mathcal{T}_s L)\psi\|_{H(\mathbb{T}^n,\ell_s^2)} = \|L\psi\|_{H^s} \le C\|\psi\|_{H^s} = C\|\Psi\|_{H(\mathbb{T}^n,\ell_s^2)}.$$

But, by (2.11),

$$\begin{aligned}\|(\mathcal{T}_s L)\psi\|^2_{H(\mathbb{T}^n,\ell_s^2)} &= \int_{\mathbb{T}^n} \|R_s^k(t)(\Psi(t))\|^2_{\ell_s^2}\, \mathrm{d}t = \int_A \|R_s^k(t)(\Psi_\alpha(t))\|^2_{\ell_s^2}\, \mathrm{d}t \\ &\ge (C+\varepsilon)^2 \int_A \mathrm{d}t = (C+\varepsilon)^2 \int_A \|\Psi_\alpha(t)\|^2_{\ell_s^2}\, \mathrm{d}t \\ &= (C+\varepsilon)^2 \|\Psi\|^2_{H(\mathbb{T}^n,\ell_s^2)},\end{aligned}$$

which is a contradiction. Thus, (2.12) is true. Now, by (2.12), for a dense subset $(\alpha_n)_{n\in\mathbb{N}}$ of $\mathbb{S}$, we have

$$\begin{aligned}\operatorname*{ess\,sup}_{t\in\mathbb{T}^n} \|R_s^k(t)\| &= \operatorname*{ess\,sup}_{t\in\mathbb{T}^n} \sup_{\alpha\in\mathbb{S}} \|R_s^k(t)(\Psi_\alpha(t))\|_{\ell_s^2} \\ &= \operatorname*{ess\,sup}_{t\in\mathbb{T}^n} \sup_{n\in\mathbb{N}} \|R_s^k(t)(\Psi_{\alpha_n}(t))\|_{\ell_s^2} \le C.\end{aligned}$$

Hence, $\|R_s^k(t)\| \le C$ for a.e. $t \in \mathbb{T}^n$.

Let $\ell \le k$. Since $R_s^\ell(t) = R_s^k(t)|_{J_s^\ell(t)}$, we define $R_s(t) : \bigcup_{\ell\in\mathbb{N}} J_s^\ell(t) \to \ell_s^2$ with $R_s(t)(a) = R_s^\ell(t)(a)$, $a \in J_s^\ell(t)$ for some $\ell \in \mathbb{N}$. Then,

$$\|R_s(t)(a)\|_{\ell_s^2} \le C\|a\|_{\ell_s^2}, \quad a \in \bigcup_{\ell\in\mathbb{N}} J_s^\ell(t).$$

Therefore, we can uniquely extend $R_s(t)$ to $R_s(t) : J_s(t) \to \ell_s^2$ with $\|R_s(t)\| \le C$, using the fact that

$$\overline{\bigcup_{\ell\in\mathbb{N}} J_s^\ell(t)} = J_s(t).$$

Finally, let us show that (2.10) holds. Choose any $f \in V_s$ and $(f_k)_{k\in\mathbb{N}}$, $f_k \in V_s^k$, such that $\lim_{k\to+\infty} f_k = f$ in H^s, $\lim_{k\to+\infty} \mathcal{T}_s f_k(t) = \lim_{k\to+\infty} \mathcal{T}_s f(t)$ and $\lim_{k\to+\infty}(\mathcal{T}_s L) f_k(t) = \lim_{k\to+\infty}(\mathcal{T}_s L) f(t)$ for a.e. $t \in \mathbb{T}^n$. Thus, for this sequence, by the previous construction and (2.11), we have

$$(\mathcal{T}_s L) f_k(t) = R_s(t)(\mathcal{T}_s f_k(t)).$$

Let $k \to +\infty$. It follows that (2.10) holds.

(2) Let R_s be a measurable range operator on J_s such that

$$\operatorname*{ess\,sup}_{t\in\mathbb{T}^n} \|R_s(t)\| = C < +\infty.$$

Then, $R_s(t)(\mathcal{T}_s f(t))$ is also measurable operator and

$$\begin{aligned}
\|R_s(\mathcal{T}_s f)\|^2_{H(\mathbb{T}^n,\ell^2_s)} &= \int_{\mathbb{T}^n} \|R_s(t)(\mathcal{T}_s f(t))\|^2_{\ell^2_s}\,\mathrm{d}t \\
&\leq \operatorname*{ess\,sup}_{t\in\mathbb{T}^n} \|R_s(t)\|^2 \int_{\mathbb{T}^n} \|\mathcal{T}_s f(t)\|^2_{\ell^2_s}\,\mathrm{d}t \\
&= C^2\|\mathcal{T}_s f\|^2_{H(\mathbb{T}^n,\ell^2_s)} = C^2\|f\|^2_{H^s}. \qquad (2.13)
\end{aligned}$$

For $L : V_s \to H^s$ given by $Lf = \mathcal{T}_s^{-1}R_s(\mathcal{T}_s f)$, we have: L is linear; $\|Lf\|_{H^s} \leq C\|f\|_{H^s}$ (by (2.13)); L satisfies (2.10) and since

$$\begin{aligned}
(\mathcal{T}_s L T_k)f(t) = R_s(t)(\mathcal{T}_s T_k f(t)) &= R_s(t)(\mathrm{e}^{-2\pi\sqrt{-1}\langle t,k\rangle}\mathcal{T}_s f(t)) \\
&= \mathrm{e}^{-2\pi\sqrt{-1}\langle t,k\rangle} R_s(t)(\mathcal{T}_s f(t)) \\
&= \mathrm{e}^{-2\pi\sqrt{-1}\langle t,k\rangle}\mathcal{T}_s(Lf)(t) \\
&= (\mathcal{T}_s T_k L)f(t),
\end{aligned}$$

it follows that L is a shift-preserving operator.

Finally, from (2.10), it follows immediately that there is a unique correspondence between the operators L and R_s.

□

Theorem 2.6 ([13], Theorem 4.6) *Let $V_s \subset H^s$ be a shift-invariant space with the range function J_s, and let $L : V_s \to H^s$ be a shift-preserving operator with the corresponding range operator R_s on J_s. Then, for some $C > 0$,*

$$\|Lf\|_{H^s} \geq C\|f\|_{H^s}, \quad f \in V_s, \qquad (2.14)$$

if and only if

$$\|R_s(t)(a)\|_{\ell^2_s} \geq C\|a\|_{\ell^2_s}, \quad a \in J_s(t),\ \textit{for a.e. } t \in \mathbb{T}^n, \qquad (2.15)$$

for some $C > 0$.

Proof If (2.15) holds, then using (2.10) we have

$$\begin{aligned}\|Lf\|_{H^s}^2 &= \|(\mathcal{T}_s L)f\|_{H(\mathbb{T}^n,\ell_s^2)}^2 = \int_{\mathbb{T}^n} \|R_s(t)(\mathcal{T}_s f(t))\|_{\ell_s^2}^2 \,\mathrm{d}t \\ &\geq \int_{\mathbb{T}^n} C^2 \|\mathcal{T}_s f(t)\|_{\ell_s^2}^2 \,\mathrm{d}t = C^2 \|\mathcal{T}_s f\|_{H(\mathbb{T}^n,\ell_s^2)}^2 = C^2 \|f\|_{H^s}^2, \quad f \in V_s.\end{aligned}$$

Assume, now, that (2.14) holds. Let $\{a_1, a_2, \ldots\}$ be a dense subset of ℓ_s^2. We will prove that for a.e. $t \in \mathbb{T}^n$,

$$\|R_s(t)(P_{J_s}(t)a_i)\|_{\ell_s^2} \geq C\|P_{J_s}(t)a_i\|_{\ell_s^2}, \quad i \in \mathbb{N}. \tag{2.16}$$

Suppose that (2.16) is not valid. Then there exists a measurable set $A \subset \mathbb{T}^n$ with $|A| > 0$, $i_0 \in \mathbb{N}$, and $\varepsilon > 0$ such that, for $t \in A$, we have

$$\|R_s(t)(P_{J_s}(t)a_{i_0})\|_{\ell_s^2} \leq (C - \varepsilon)\|P_{J_s}(t)a_{i_0}\|_{\ell_s^2}.$$

Let $f \in V_s$ be a function such that $\mathcal{T}_s f(t) = \mathbf{1}_A(t) P_{J_s}(t)a_{i_0}$. Then, using (2.10) again, we have

$$\begin{aligned}\|Lf\|_{H^s}^2 = \|(\mathcal{T}_s L)f\|_{H(\mathbb{T}^n,\ell_s^2)}^2 &= \int_{\mathbb{T}^n} \|R_s(t)(\mathcal{T}_s f(t))\|_{\ell_s^2}^2 \,\mathrm{d}t \\ &\leq (C-\varepsilon)^2 \int_A \|P_{J_s}(t)a_{i_0}\|_{\ell_s^2}^2 \,\mathrm{d}t \\ &= (C-\varepsilon)^2 \int_{\mathbb{T}^n} \|\mathcal{T}_s f(t)\|_{\ell_s^2}^2 \,\mathrm{d}t \\ &= (C-\varepsilon)^2 \|f\|_{H^s}^2,\end{aligned}$$

contrary to (2.14). □

Corollary 2.2 *Let L be a shift-preserving operator, and let R_s be its corresponding range operator. Then, L is an isometry if and only if $R_s(t)$ is an isometry, for a.e. $t \in \mathbb{T}^n$.*

Proof Using (2.10) we get

$$\begin{aligned}\|(\mathcal{T}_s L)f\|_{H(\mathbb{T}^n,\ell_s^2)}^2 = \int_{\mathbb{T}^n} \|(\mathcal{T}_s L)f(t)\|_{\ell_s^2}^2 \,\mathrm{d}t &= \int_{\mathbb{T}^n} \|R_s(t)(\mathcal{T}_s f(t))\|_{\ell_s^2}^2 \,\mathrm{d}t \\ &= \|R_s(\mathcal{T}_s f)\|_{H(\mathbb{T}^n,\ell_s^2)}^2.\end{aligned}$$

Thus, since

$$\|R_s(\mathcal{T}_s f)\|_{H(\mathbb{T}^n,\ell_s^2)} = \|(\mathcal{T}_s L) f\|_{H(\mathbb{T}^n,\ell_s^2)} = \|Lf\|_{H^s} \leq C\|f\|_{H^s}$$
$$= C\|\mathcal{T}_s f\|_{H(\mathbb{T}^n,\ell_s^2)},$$

$C > 0$, the assertion follows. □

Theorem 2.7 ([13], Theorem 4.8) *Let $V_s \subset H^s$ be a shift-invariant space with the associated range function J_s and $L : V_s \to V_s$ be a shift-preserving operator with corresponding range operator R_s.*

(1) The adjoint operator $L^* : V_s \to V_s$ is also shift-preserving and with $R_s^*(t) = (R_s(t))^*$, for a.e. $t \in \mathbb{T}^n$, is given the corresponding range operator R_s^*.
(2) The operator L is self-adjoint and $\sigma(L) \subseteq [A, B]$ if and only if $R_s(t)$ is self-adjoint and $\sigma(R_s(t)) \subseteq [A, B]$ for a.e. $t \in \mathbb{T}^n$, where $A, B \in \mathbb{R}$ such that $A \leq B$.
(3) The operator L is unitary if and only if $R_s(t)$ is a unitary operator for a.e. $t \in \mathbb{T}^n$.

Proof

(1) It is obvious that R_s^* is measurable and uniformly bounded on J_s. From Theorem 2.5, we have that there exists a shift-preserving operator $L^\bullet : V_s \to H^s$ such that $(\mathcal{T}_s L^\bullet) f(t) = R_s^*(t)(\mathcal{T}_s f(t))$, $f \in V_s$ and a.e. $t \in \mathbb{T}^n$. Thus, for $f, g \in V_s$, we get

$$\langle Lf, g\rangle_{H^s} = \langle (\mathcal{T}_s L) f, \mathcal{T}_s g\rangle_{H(\mathbb{T}^n,\ell_s^2)} = \int_{\mathbb{T}^n} \langle R_s(t)(\mathcal{T}_s f(t)), \mathcal{T}_s g(t)\rangle_{\ell_s^2}\, dt$$
$$= \int_{\mathbb{T}^n} \langle \mathcal{T}_s f(t), R_s^*(t)(\mathcal{T}_s g(t))\rangle_{\ell_s^2}\, dt$$
$$= \langle \mathcal{T}_s f, (\mathcal{T}_s L^\bullet) g\rangle_{H(\mathbb{T}^n,\ell_s^2)} = \langle f, L^\bullet g\rangle_{H^s}.$$

(2) From the part (1), we have $L^* = L$ if and only if $R_s^*(t) = R_s(t)$, for a.e. $t \in \mathbb{T}^n$. Let $\sigma(L) \subseteq [A, B]$, i.e., for all $f \in V_s$ we have

$$A\|f\|_{H^s}^2 \leq \langle Lf, f\rangle_{H^s} = \langle \mathcal{T}_s Lf, \mathcal{T}_s f\rangle_{H(\mathbb{T}^n,\ell_s^2)}$$
$$= \int_{\mathbb{T}^n} \langle R_s(t)(\mathcal{T}_s f(t)), \mathcal{T}_s f(t)\rangle_{\ell_s^2}\, dt \leq B\|f\|_{H^s}^2. \tag{2.17}$$

Analysis similar to that in the proof of Theorem 2.6 shows that the assertion holds.

In the opposite direction, suppose that $\sigma(R_s(t)) \subseteq [A, B]$ for a.e. $t \in \mathbb{T}^n$, i.e., for $f \in V_s$,

$$A\|\mathcal{T}_s f(t)\|_{\ell_s^2}^2 \leq \langle R_s(t)(\mathcal{T}_s f(t)), \mathcal{T}_s f(t)\rangle_{\ell_s^2} \leq B\|\mathcal{T}_s f(t)\|_{\ell_s^2}^2, \quad \text{for a.e. } t \in \mathbb{T}^n.$$

We obtain (2.17) by the integration of aforementioned inequalities over $\mathbb{T}^n$.

(3) The statement follows from (2). L is unitary, $\sigma(LL^*) = \sigma(L^*L) = \{1\}$ if and only if $\sigma(R_s(t)R_s^*(t)) = \sigma(R_s^*(t)R_s(t)) = \{1\}$ for a.e. $t \in \mathbb{T}^n$, i.e. $R_s(t)$ is unitary for a.e. $t \in \mathbb{T}^n$.

□

Similarly as in [13], we have the next propositions.

Proposition 2.3 *Let $V_s \subset H^s$ be a shift-invariant space, and let L be its shift-preserving operator. For $\widetilde{V}_s = \overline{L(V_s)}$, we have $\dim_{\widetilde{V}_s}(t) \leq \dim_{V_s}(t)$ for a.e. $t \in \mathbb{T}^n$.*

Proof Let $V_s = S_s(\mathcal{A}_{I,s})$. Using the theorems 2.5 and 2.1, the range function $\widetilde{J}_s$ of $\widetilde{V}_s = S_s(\{L\varphi : \varphi \in \mathcal{A}_{I,s}\})$ satisfies

$$\begin{aligned}\widetilde{J}_s(t) &= \overline{\operatorname{span}}\{\mathcal{T}_s\phi(t) : \phi \in \{L\varphi : \varphi \in \mathcal{A}_{I,s}\}\} = \overline{\operatorname{span}}\{(\mathcal{T}_s L)\varphi(t) : \varphi \in \mathcal{A}_{I,s}\} \\ &= \overline{\operatorname{span}}\{R_s(t)(\mathcal{T}_s\varphi(t)) : \varphi \in \mathcal{A}_{I,s}\} = \overline{R_s(t)(J_s(t))}, \quad \text{for a.e. } t \in \mathbb{T}^n.\end{aligned}$$

Thus, we have that $\dim \widetilde{J}_s(t) \leq \dim J_s(t)$, for a.e. $t \in \mathbb{T}^n$.

On the other hand, by Theorem 2.3, $\dim_{V_s}(t) = \sum_{i\in\mathbb{N}} \|\mathcal{T}_s\varphi_i(t)\|_{\ell_s^2}$, for a.e. $t \in \mathbb{T}^n$, which completes the proof. □

Proposition 2.4 *Let $V_s, \widetilde{V}_s \subset H^s$ be the shift-invariant spaces. Then,*

$$\dim_{V_s}(t) = \dim_{\widetilde{V}_s}(t) \quad \textit{for a.e. } t \in \mathbb{T}^n$$

if and only if there exists an operator $L : V_s \to \widetilde{V}_s$, which is a shift-preserving isomorphism (or isometry).

Proof Suppose that V_s and $\widetilde{V}_s$ are shift-invariant spaces such that $\dim_{V_s}(t) = \dim_{\widetilde{V}_s}(t)$ for a.e. $t \in \mathbb{T}^n$. Using Theorem 2.3, the spaces V_s and $\widetilde{V}_s$ can be decomposed as

$$V_s = \bigoplus_{i\in\mathbb{N}} S_s(\varphi_i), \quad \widetilde{V}_s = \bigoplus_{i\in\mathbb{N}} S_s(\widetilde{\varphi}_i),$$

where φ_i and $\widetilde{\varphi}_i$ are tight-frame generators, and $\sigma_{S_s(\varphi_i)} = \sigma_{S_s(\widetilde{\varphi}_i)}$, $i \in \mathbb{N}$. Let $L_i : S_s(\varphi_i) \to S_s(\widetilde{\varphi}_i)$ be defined by $L_i(T_k\varphi_i) = T_k\widetilde{\varphi}_i$, $i \in \mathbb{N}$. Now, for arbitrary

$(c_k)_{k\in\mathbb{Z}^n} \in \ell_s^2$ with finite number of elements, which are different from zero, we have

$$\begin{aligned}
\left\| \sum_{k\in\mathbb{Z}^n} c_k T_k \varphi_i \right\|_{H^s}^2 &= \left\| \sum_{k\in\mathbb{Z}^n} c_k \mathrm{e}^{-2\pi\sqrt{-1}\langle t,k\rangle} \mathcal{T}_s \varphi_i(t) \right\|_{H(\mathbb{T}^n,\ell_s^2)}^2 \\
&= \int_{\mathbb{T}^n} \left| \sum_{k\in\mathbb{Z}^n} c_k \mathrm{e}^{-2\pi\sqrt{-1}\langle t,k\rangle} \right|^2 \|\mathcal{T}_s \varphi_i(t)\|_{\ell_s^2}^2 \,\mathrm{d}t \\
&= \int_{\mathbb{T}^n} \left| \sum_{k\in\mathbb{Z}^n} c_k \mathrm{e}^{-2\pi\sqrt{-1}\langle t,k\rangle} \right|^2 \|\mathcal{T}_s \widetilde{\varphi}_i(t)\|_{\ell_s^2}^2 \,\mathrm{d}t \\
&= \left\| \sum_{k\in\mathbb{Z}^n} c_k T_k \widetilde{\varphi}_i \right\|_{H^s}^2 = \left\| L_i\Big(\sum_{k\in\mathbb{Z}^n} c_k T_k \varphi_i \Big) \right\|_{H^s}^2 .
\end{aligned}$$

Therefore, L_i is a shift-preserving isometry. Then, the operator $L := \bigoplus_{i\in\mathbb{N}} L_i$ has the desired properties.

In the opposite direction, assume that $L : V_s \to \widetilde{V}_s$ is a shift-preserving isomorphism. Then, applying Proposition 2.3 to $L : V_s \to \widetilde{V}_s$ and $L^{-1} : \widetilde{V}_s \to V_s$, we get $\dim_{V_s}(t) = \dim_{\widetilde{V}_s}(t)$, for a.e. $t \in \mathbb{T}^n$. □

2.4 The Frame Operator

Let $E_s(\mathcal{A}_{I,s})$ be a Bessel family for $S_s(\mathcal{A}_{I,s})$. The operator

$$K : S_s(\mathcal{A}_{I,s}) \to \ell_s^2(\mathbb{Z}^n \times I)$$

given by $Kf = \left(\frac{\langle f, T_k\varphi_i\rangle_{H^s}}{\mu_s(k+i)} \right)_{(k,i)\in\mathbb{Z}^n\times I}$, has the dual $K^* : \ell_s^2(\mathbb{Z}^n \times I) \to S_s(\mathcal{A}_{I,s})$ defined by

$$K^* c = \sum_{(k,i)\in\mathbb{Z}^n\times I} c_{k,i} T_k \varphi_i \mu_s(k+i).$$

In order to distinguish the shift-invariant operator defined by K^*K and a general shift-invariant operator L defined in the previous section, we use the notation $L_0 = K^*K$. The frame operator $L_0 : S_s(\mathcal{A}_{I,s}) \to S_s(\mathcal{A}_{I,s})$ given by $L_0 = K^*K$ is self-adjoint and

$$L_0 f = \sum_{(k,i)\in\mathbb{Z}^n\times I} \langle f, T_k\varphi_i\rangle_{H^s} T_k \varphi_i,$$

where the convergence is unconditional in H^s.

Let $t \in \mathbb{T}^n$ be fixed and consider $(\mathcal{T}_s\varphi_i(t))_{i\in I}$. The corresponding Gramian $G_s : \ell^2(I) \to \ell^2(I)$ and the dual Gramian $\widetilde{G}_s : \ell_s^2(\mathbb{Z}^n) \to \ell_s^2(\mathbb{Z}^n)$ are defined by $G_s = N_s^* N_s$ and $\widetilde{G}_s = N_s N_s^*$, respectively, where $N_s : \ell^2(I) \to \ell_s^2(\mathbb{Z}^n)$ is given by $N_s(c) = \sum_{i\in I} c_i \mathcal{T}_s\varphi_i(t)$, and $N_s^* : \ell_s^2(\mathbb{Z}^n) \to \ell^2(I)$ is adjoint of N_s given by $N_s^*(a) = \left(\langle a, \mathcal{T}_s\varphi_i(t)\rangle_{\ell_s^2}\right)_{i\in I}$ (see [6] for more details).

Remark 2.1 Obviously, if $(\mathcal{T}_s\varphi_i)_{i\in I}$ is a Bessel family, then the operator N_s (i.e. N_s^*) is bounded, and vice versa.

The following two assertions are necessary for the proof of Theorem 2.9.

Lemma 2.2 ([6]) *Let $E_s(\mathcal{A}_{I,s})$ be a Bessel family. Then, for all $f \in \mathcal{A}_{I,s}$, we have*

$$\sum_{\varphi\in\mathcal{A}_{I,s}} \sum_{k\in\mathbb{Z}^n} \left|\langle T_k\varphi, f\rangle_{H^s}\right|^2 = \sum_{\varphi\in\mathcal{A}_{I,s}} \int_{\mathbb{T}^n} \left|\langle \mathcal{T}_s\varphi(t), \mathcal{T}_s f(t)\rangle_{\ell_s^2}\right|^2 dt.$$

Theorem 2.8 ([13], Theorem 5.1) *Let $E_s(\mathcal{A}_{I,s})$ be a Bessel family for $S_s(\mathcal{A}_{I,s})$, and let J_s be the range function of $S_s(\mathcal{A}_{I,s})$. Then, $L_0 = K^*K$ is self-adjoint, shift-preserving operator with the range operator $R_s(t) = \widetilde{G}_s(t)|_{J_s(t)}$, where $\widetilde{G}_s(t)$ is the dual Gramian of $\{\mathcal{T}_s\varphi_i(t) : i \in I\}$ for a.e. $t \in \mathbb{T}^n$.*

Proof Since

$$\begin{aligned} L_0 T_\ell f &= \sum_{(k,i)\in\mathbb{Z}^n\times I} \langle T_\ell f, T_k\varphi_i\rangle_{H^s} T_k\varphi_i = \sum_{(k,i)\in\mathbb{Z}^n\times I} \langle f, T_{k-\ell}\varphi_i\rangle_{H^s} T_k\varphi_i \\ &= \sum_{(k,i)\in\mathbb{Z}^n\times I} \langle f, T_k\varphi_i\rangle_{H^s} T_{k+\ell}\varphi_i, \end{aligned}$$

for all $\ell \in \mathbb{Z}^n$, we have $L_0 T_\ell = T_\ell L_0$, i.e. L_0 is shift-preserving. Further,

$$\begin{aligned} \|Kf\|_{\ell_s^2}^2 = \langle Kf, Kf\rangle_{\ell_s^2} &= \langle L_0 f, f\rangle_{H^s} \\ &= \langle (\mathcal{T}_s L_0) f, \mathcal{T}_s f\rangle_{H(\mathbb{T}^n, \ell_s^2)} \\ &= \int_{\mathbb{T}^n} \langle \widetilde{R}_s(t)\big(\mathcal{T}_s f(t)\big), \mathcal{T}_s f(t)\rangle_{\ell_s^2} dt, \end{aligned} \tag{2.18}$$

for $f \in S_s(\mathcal{A}_{I,s})$, where $\widetilde{R}_s$ is the range operator of L_0. By Lemma 2.2, we get

$$\begin{aligned} \|Kf\|_{\ell_s^2}^2 &= \left\| \left(\frac{\langle f, T_k\varphi_i\rangle_{H^s}}{\mu_s(k+i)} \right)_{(k,i)\in\mathbb{Z}^n\times I} \right\|_{\ell_s^2}^2 \\ &= \sum_{(k,i)\in\mathbb{Z}^n\times I} \left|\langle f, T_k\varphi_i\rangle_{H^s}\right|^2 \end{aligned}$$

$$= \sum_{i\in I} \int_{\mathbb{T}^n} \left| \langle \mathcal{T}_s f(t), \mathcal{T}_s \varphi_i(t) \rangle_{\ell_s^2} \right|^2 \mathrm{d}t$$

$$= \int_{\mathbb{T}^n} \sum_{i\in I} \langle \mathcal{T}_s f(t), \mathcal{T}_s \varphi_i(t) \rangle_{\ell_s^2} \overline{\langle \mathcal{T}_s f(t), \mathcal{T}_s \varphi_i(t) \rangle_{\ell_s^2}} \mathrm{d}t$$

$$= \int_{\mathbb{T}^n} \left\langle \left(\langle \mathcal{T}_s f(t), \mathcal{T}_s \varphi_i(t) \rangle_{\ell_s^2} \right)_{i\in I}, \left(\langle \mathcal{T}_s f(t), \mathcal{T}_s \varphi_i(t) \rangle_{\ell_s^2} \right)_{i\in I} \right\rangle_{\ell^2} \mathrm{d}t$$

$$= \int_{\mathbb{T}^n} \langle N_s^*(\mathcal{T}_s f(t)), N_s^*(\mathcal{T}_s f(t)) \rangle_{\ell^2} \mathrm{d}t$$

$$= \int_{\mathbb{T}^n} \langle N_s N_s^*(\mathcal{T}_s f(t)), \mathcal{T}_s f(t) \rangle_{\ell_s^2} \mathrm{d}t$$

$$= \int_{\mathbb{T}^n} \langle \widetilde{G}_s(t) \mathcal{T}_s f(t), \mathcal{T}_s f(t) \rangle_{\ell_s^2} \mathrm{d}t. \tag{2.19}$$

Combining (2.18) with (2.19), we obtain

$$\int_{\mathbb{T}^n} \langle \big(\widetilde{R}_s(t) - \widetilde{G}_s(t)|_{J_s(t)}\big)\big(\mathcal{T}_s f(t)\big), \mathcal{T}_s f(t) \rangle_{\ell_s^2} \mathrm{d}t = 0, \quad f \in S_s(\mathcal{A}_{I,s}).$$

Thus, $\widetilde{R}_s(t) = \widetilde{G}_s(t)|_{J_s(t)} := R_s(t)$ for a.e. $t \in \mathbb{T}^n$. □

Theorem 2.9 ([13], Theorem 5.2) *Let $E_s(\mathcal{A}_{I,s})$ be a frame with constants A, B. Then, $E_s(\mathcal{B}_{I,s})$ is its dual frame with constants B^{-1}, A^{-1}, where $\mathcal{B}_{I,s} = \{\theta_i : \theta_i = L_0^{-1}\varphi_i, i \in I\}$. Furthermore,*

$$\mathcal{T}_s \theta_i(t) = R_s^{-1}(t)\big(\mathcal{T}_s \varphi_i(t)\big), \quad \textit{for a.e. } t \in \mathbb{T}^n, i \in I. \tag{2.20}$$

Proof Since

$$\sum_{(k,i)\in\mathbb{Z}^n\times I} |\langle f, L_0^{-1} T_k \varphi_i \rangle_{H^s}|^2 = \sum_{(k,i)\in\mathbb{Z}^n\times I} |\langle L_0^{-1} f, T_k \varphi_i \rangle_{H^s}|^2$$

$$= \sum_{(k,i)\in\mathbb{Z}^n\times I} \langle L_0^{-1} f, T_k \varphi_i \rangle_{H^s} \overline{\langle L_0^{-1} f, T_k \varphi_i \rangle}_{H^s}$$

$$= \sum_{(k,i)\in\mathbb{Z}^n\times I} \langle \langle L_0^{-1} f, T_k \varphi_i \rangle_{H^s} T_k \varphi_i, L_0^{-1} f \rangle_{H^s}$$

$$= \left\langle \sum_{(k,i)\in\mathbb{Z}^n\times I} \langle L_0^{-1} f, T_k \varphi_i \rangle_{H^s} T_k \varphi_i, L_0^{-1} f \right\rangle_{H^s}$$

$$= \langle L_0(L_0^{-1} f), L_0^{-1} f \rangle_{H^s} = \langle L_0^{-1} f, f \rangle_{H^s},$$

we get

$$B^{-1}\|f\|_{H^s} \leq \sum_{(k,i)\in\mathbb{Z}^n\times I} \left|\langle f, L_0^{-1}T_k\varphi_i\rangle_{H^s}\right|^2 = \langle L_0^{-1}f, f\rangle_{H^s} \leq A^{-1}\|f\|_{H^s}. \tag{2.21}$$

Consequently, $\{L_0^{-1}T_k\varphi_i : k \in \mathbb{Z}^n, i \in I\}$ is a dual frame of $E_s(\mathcal{A}_{I,s})$ with constants B^{-1}, A^{-1}. Furthermore, by Theorem 2.8, L_0 is a shift-preserving operator. Thus, for any $F = L_0 f \in H^s$, we have

$$L_0^{-1}T_kF = L_0^{-1}T_kL_0f = L_0^{-1}L_0T_kf = T_kf = T_kL_0^{-1}F, \quad k \in \mathbb{Z}^n.$$

Hence, L_0^{-1} is shift-preserving. Now, by (2.21), it follows that $E_s(\mathcal{B}_{I,s})$ is a dual frame of $E_s(\mathcal{A}_{I,s})$ with constants B^{-1}, A^{-1}. Moreover,

$$\sum_{(k,i)\in\mathbb{Z}^n\times I} \langle f, T_k\theta_i\rangle_{H^s} T_k\varphi_i = \sum_{(k,i)\in\mathbb{Z}^n\times I} \langle f, T_k\varphi_i\rangle_{H^s} T_k\theta_i = f \in S_s(\mathcal{A}_{I,s})$$

holds with the unconditional convergence in H^s.

Finally, (2.20) follows from (2.10) and Theorem 2.8. □

Remark 2.2 Let $E_s(\mathcal{A}_{I,s})$ be a Riesz family with constants A, B. Then, the dual $E_s(\mathcal{B}_{I,s})$ is a Riesz family with constants B^{-1}, A^{-1}. Moreover,

$$\langle T_k\varphi_i, T_\ell\theta_j\rangle_{H^s} = \delta_{k,\ell}\delta_{i,j}, \quad k, \ell \in \mathbb{Z}^n,\ i, j \in I,$$

where $\delta_{i,j} = \begin{cases} 1, & i = j, \\ 0, & i \neq j. \end{cases}$

2.5 Structural Theorems

This section will be devoted to the main results of [6].

Recall that $\mathcal{D}_{L^2}(\mathbb{R}^n) = \bigcap_{s\geq 0} H^s$ and $\mathcal{D}'_{L^2}(\mathbb{R}^n) = \bigcup_{s\geq 0} H^{-s}$ (see for instance, [26, 28]). Note that V_s is a closed subspace of H^s, $s \in \mathbb{R}$, so it is also a separable Hilbert space. The space of periodic L^2-functions, $L^2_{per}(\mathbb{R}^n)$, is defined by

$$L^2_{per}(\mathbb{R}^n) = \left\{ g : g(\cdot) = \sum_{k\in\mathbb{Z}^n} a_k e^{-2\pi\sqrt{-1}\langle\cdot,k\rangle},\ (a_k)_{k\in\mathbb{Z}^n} \in \ell^2 \right\}.$$

Theorem 2.10 ([6]) *Assume that $E_s(\mathcal{A}_{I,s})$ is a frame for V_s and that $E_s(\mathcal{B}_{I,s})$ is its dual frame, where $\mathcal{B}_{I,s} = \{\theta_i : \theta_i = L_0^{-1}\varphi_i, i \in I\}$. Then, $\mathcal{F}(V_s)$ is the set of*

Fourier transforms of elements $f \in \mathcal{D}'_{L^2}(\mathbb{R}^n)$ *so that*

$$\widehat{f}(\cdot) = \sum_{i \in I} \widehat{\varphi}_i(\cdot) \sum_{k \in \mathbb{Z}^n} a_k^i e^{-2\pi\sqrt{-1}\langle \cdot, k \rangle},$$

where $(a_k^i)_{k \in \mathbb{Z}^n} \in \ell^2$ *is given by*

$$a_k^i = \int_{\mathbb{R}^n} \widehat{f}(x) e^{2\pi\sqrt{-1}\langle k, x \rangle} \overline{\widehat{\theta}_i}(x) \mu_s^2(x)\, dx, \quad k \in \mathbb{Z}^n, i \in I. \tag{2.22}$$

Equivalently, it is equal to the space of elements $f \in \mathcal{D}'_{L^2}(\mathbb{R}^n)$*, which Fourier transforms have the form*

$$\widehat{f} = \sum_{i \in I} f_i g_i, \quad g_i \in L^2_{per}(\mathbb{R}^n),$$

where $f_i = \widehat{\varphi}_i \in L^2_s(\mathbb{R}^n)$*,* $i \in I$*, and* g_i*,* $i \in I$*, have the expansions*

$$g_i(\cdot) = \sum_{k \in \mathbb{Z}^n} a_k^i e^{-2\pi\sqrt{-1}\langle \cdot, k \rangle},$$

with a_k^i *determined by (2.22).*

Instead of notation V_s^p in [24] (and V_p in [12]), we use $\mathcal{V}_s^2$, for $p = 2$. The main properties of the space $\mathcal{V}_s^2$ were announced in [24], where the weighted spaces V_s^p, $p \in [1, +\infty)$, were considered.

So, assume that $p = 2$. In the case $s = 0$, we assume that $\psi^i \in \mathcal{L}^\infty$, $i = 1, \ldots, r$, where

$$\mathcal{L}^\infty = \Big\{\psi : \|\psi\|_{\mathcal{L}^\infty} = \sup_{t \in \mathbb{T}^n} \sum_{j \in \mathbb{Z}^n} |\psi(t + j)| < +\infty\Big\}.$$

By [12], $\mathcal{V}^2 = \big\{f : f = \sum_{i=1}^r \sum_{k \in \mathbb{Z}^n} c_k^i T_k \psi^i,\ (c_k^i)_{k \in \mathbb{Z}^n} \in \ell^2,\ i = 1, \ldots, r\big\}$.

Theorem 2.11 ([6]) *Assume that* $\mathcal{A}_r = \{\psi^i : i = 1, \ldots, r\} \subset L^2(\mathbb{R}^n) \cap \mathcal{L}^\infty$*. Then,*

$$\mathcal{V}^2 = V_0 = S_0(\mathcal{A}_r),$$

if $\mathcal{V}^2$ *is closed in* $L^2(\mathbb{R}^n)$*.*

Now, we consider weighted versions. Let $s > 0$ be fixed. We will introduce several assumptions on generators ψ^i, $i = 1, \dots, r$, in order to have that their linear combinations determine subspaces of H^s and of L^2_s:

$$\psi^i \in H^s \cap L^2_s, \quad i = 1, \dots, r. \tag{2.23}$$

Moreover, in order to have the same assumptions as in [12] (and [24]), we assume, as in the previous assumption, that

$$\psi^i \in \mathcal{L}^\infty, \quad i = 1, \dots, r. \tag{2.24}$$

Let

$$\mathcal{V}^2_s = \left\{ f : f = \sum_{i=1}^{r} \sum_{k \in \mathbb{Z}^n} c^i_k T_k \psi^i, \ (c^i_k)_{k \in \mathbb{Z}^n} \in \ell^2_s, \ i = 1, \dots, r \right\}, \tag{2.25}$$

by recalling the notation in [24].

Theorem 2.12 ([6]) *Assume that $s > 0$, (2.23) and (2.24) hold.*
(1) *Assume that*

$$\mathcal{V}^2_s \text{ and } \mathcal{F}(\mathcal{V}^2_s) \text{ are closed in } L^2_s.$$

Then,

$$\mathcal{V}^2_s \subset H^s \text{ and } \mathcal{V}^2_s = V_s = S_s(\mathcal{A}_{r,s}).$$

In particular, any element $f \in V_s$ has the frame expansion as in (2.25).
(2) *Assume that $s > 1/2$ and that $\mathcal{V}^2_s$ is closed in L^2_s. Then, $\mathcal{F}(\mathcal{V}^2_s)$ is closed in L^2_s, and both assertions in (1) hold true.*

Concerning the duality, we have the following assertion.

Theorem 2.13 ([6]) *Assume that $s > 0$, (2.23), and (2.24) hold. Moreover, assume that the conditions of assertion (1) or conditions of assertion (2) of Theorem 2.12 hold. Then in (both cases),*
(1) $(\mathcal{V}^2_s)' = \mathcal{V}^2_{-s}$, *where $\mathcal{V}^2_{-s}$ is the space of formal series of the form*

$$F(\cdot) = \sum_{i=1}^{r} \sum_{k \in \mathbb{Z}^n} b^i_k \psi^i(\cdot - k), \quad \sum_{i=1}^{r} \sum_{k \in \mathbb{Z}^n} |b^i_k|^2 (1 + |k|^2)^{-s} < +\infty,$$

with the dual pairing

$$\langle F, f \rangle = \sum_{i=1}^{r} \sum_{k \in \mathbb{Z}^n} b^i_k c^i_k \quad (f \text{ is of the form given in (2.25)}).$$

(2) $\mathcal{V}^2_{-s} = V_{-s}$.

In order to consider the intersections of V_s, $s \geq 0$, instead of conditions (2.23) and (2.24), we assume

$$\psi^i \in \mathcal{S}(\mathbb{R}^n), \quad i = 1, \dots, r. \tag{2.26}$$

Theorem 2.14 ([6]) *Assume that (2.26) holds. Then,*

$$\bigcap_{s \geq 0} \mathcal{V}^2_s = \bigcap_{s \geq 0} V_s,$$

and the expansion for their elements has the form as in (2.25) with

$$\sup_{k \in \mathbb{Z}^n} |c^i_k||k|^s < +\infty, \quad i = 1, \dots, r, \textit{ for every } s \geq 0.$$

Recall that the space $\mathcal{P}(\mathbb{R}^n) = \mathcal{P}$ of periodic smooth test functions (with period one in any variable) is given by

$$\mathcal{P} = \left\{ \phi : \phi(\cdot) = \sum_{k \in \mathbb{Z}^n} a_k e^{-2\pi\sqrt{-1}\langle \cdot, k\rangle}, \ (a_k)_{k \in \mathbb{Z}^n} \in \ell^2_s \text{ for every } s \in \mathbb{Z} \right\},$$

while its dual space $\mathcal{P}'(\mathbb{R}^n) = \mathcal{P}'$ is given by

$$\mathcal{P}' = \left\{ \phi : \phi(\cdot) = \sum_{k \in \mathbb{Z}^n} a_k e^{-2\pi\sqrt{-1}\langle \cdot, k\rangle}, \ (a_k)_{k \in \mathbb{Z}^n} \in \ell^2_{-s} \text{ for some } s > 0 \right\}.$$

A direct consequence of Theorem 2.12 is the following assertion.

Corollary 2.3 ([6]) *Assume that (2.26) holds. Then*

$$\mathcal{F}\Big(\bigcap_{s \geq 0} \mathcal{V}^2_s\Big)$$
$$= \left\{ \sum_{i=1}^{r} \widehat{\psi}^i(\cdot) \sum_{k \in \mathbb{Z}^n} c^i_k e^{-2\pi\sqrt{-1}\langle \cdot, k\rangle} : (c^i_k)_{k \in \mathbb{Z}^n} \in \ell^2_s, i = 1, \dots, r, \textit{ for every } s \geq 0 \right\},$$

where $\sum_{k \in \mathbb{Z}^n} c^i_k e^{-2\pi\sqrt{-1}\langle \cdot, k\rangle} \in \mathcal{P}$, $i = 1, \dots, r$.

Concerning the duality, by Theorem 2.13, we have the next corollary.

Corollary 2.4 ([6]) *Assume that (2.26) holds. Then* $V_s' = \mathcal{V}_{-s}^2$, $\bigcup_{s\geq 0} V_s' = \bigcup_{s\geq 0} \mathcal{V}_{-s}^2$ *and*

$$\mathcal{F}\Big(\bigcup_{s\leq 0} \mathcal{V}_s^2\Big)$$

$$=\Big\{\sum_{i=1}^{r} \widehat{\psi}^i(\cdot) \sum_{k\in\mathbb{Z}^n} c_k^i \mathrm{e}^{-2\pi\sqrt{-1}\langle\cdot,k\rangle} : (c_k^i)_{k\in\mathbb{Z}^n} \in \ell_s^2, i = 1,\ldots,r, \textit{for some } s \leq 0\Big\},$$

where $\sum_{k\in\mathbb{Z}^n} c_k^i \mathrm{e}^{-2\pi\sqrt{-1}\langle\cdot,k\rangle} \in \mathcal{P}'$, $i = 1,\ldots,r$.

Note that the assumption $\psi^i \in \mathcal{S}(\mathbb{R}^n)$ implies a well-defined product of a smooth function and a (periodic) Schwartz distribution.

2.6 Spectral Analysis of the Range Operator

This section contains relevant concepts and auxiliary results needed to prove theorems in the last section. These specialize to the results of [4] if $s = 0$. Ideas used in proofs are in much the same way as in [4].

Theorem 2.15 ([4]) *Let* $A = A(t)$ *be an* $n \times n$ *matrix of measurable functions defined on a measurable set* B. *Then, there exist* n *measurable functions* $\lambda_i : B \to \mathbb{C}$, $i = 1,\ldots,n$, *such that* $\lambda_1(t),\ldots,\lambda_n(t)$ *are eigenvalues for matrix* $A(t)$, *for a.e.* $t \in B$.

Theorem 2.16 ([4], Proposition 3.4, Theorem 5.4) *Let* J_s *be a range function for the shift-invariant space* $V_s \subset H^s$ *such that* $\dim J_s(t) = d < +\infty$ *for a.e.* $t \in \Omega$, *where* $\Omega \subseteq \mathbb{T}^n$ *is measurable set, and let* $L : V_s \to V_s$ *be a shift-preserving operator with the corresponding range operator* R_s. *Then, there exist* d^2 *measurable bounded functions* $(R_s^{ij})_{i,j=1}^d$ *defined on* Ω, *such that*

$$[R_s(t)] = \begin{bmatrix} R_s^{11}(t) & R_s^{12}(t) & \cdots & R_s^{1d}(t) \\ R_s^{21}(t) & R_s^{22}(t) & \cdots & R_s^{2d}(t) \\ \vdots & \vdots & \ddots & \vdots \\ R_s^{d1}(t) & R_s^{d2}(t) & \cdots & R_s^{dd}(t) \end{bmatrix}_{d\times d},$$

for a.e. $t \in \Omega$. *Moreover, there exist* d *measurable functions* $\lambda_s^i : \Omega \to \mathbb{C}$, $i = 1,\ldots,d$, *so that* $\lambda_s^1(t),\ldots,\lambda_s^d(t)$ *are eigenvalues for* $R_s(t)$ *for a.e.* $t \in \Omega$.

Proof Let us consider $(A_d)_{d\in\mathbb{N}_0}$ and $\varphi_i \in H^s$, $i \in \mathbb{N}$, from Proposition 2.2. Since $\{\mathcal{T}_s\varphi_1(t), \ldots, \mathcal{T}_s\varphi_d(t)\}$ is the orthonormal basis for $J_s(t)$ for a.e. $t \in A_d$, we can conclude that $R_s(t)$ has a matrix representation with elements

$$R_s^{ij}(t) := \big\langle R_s(t)\mathcal{T}_s\varphi_j(t), \mathcal{T}_s\varphi_i(t)\big\rangle_{\ell_s^2},$$

for a.e. $t \in A_d$. Obviously, $R_s^{ij}(t)$ is measurable for a.e. $t \in A_d$. Since L is bounded, by Theorem 2.5, we have $|R_s^{ij}(t)| \leq \|L\|$, $i, j = 1, \ldots, d$, for a.e. $t \in A_d$. Clearly, the set Ω is included in A_d.

Let $\nu_s(t) : J_s(t) \to \mathbb{C}^d$ be given by $\nu_s(t)\big(\mathcal{T}_s\varphi_i(t)\big) = e_i$, for a.e. $t \in A_d$, $i = 1, \ldots, d$, where $\{e_1, \ldots, e_d\}$ is the canonical basis for $\mathbb{C}^d$. In this way, the unique connection between bases is established, for a.e. $t \in A_d$. Then, $R_s(t) = \nu_s(t)^{-1}[R_s(t)]\nu_s(t)$, for a.e. $t \in \Omega$. For a measurable function $\lambda_s : \Omega \to \mathbb{C}$ holds

$$\ker\big(R_s(t) - \lambda_s(t)1(t)\big) = \ker\big(\nu_s(t)^{-1}([R_s(t)] - \lambda_s(t)1(t))\nu_s(t)\big), \quad \text{a.e. } t \in \Omega.$$

Therefore, by Theorem 2.15, the assertion follows. □

For a bounded measurable range operator R_s, we will denote by $\mathfrak{R}_\lambda(t)$ the set of all eigenvalues of $R_s(t)$. The length of FSI space $V_s \subset H^s$, denoted by $\Gamma(V_s)$, is defined to be the smallest $r \in \mathbb{N}$ such that $V_s = S_s(\varphi_1, \ldots, \varphi_r)$, where $\varphi_1, \ldots, \varphi_r \in H^s$. In the notation of [4], the equivalent definition is $\Gamma(V_s) = \operatorname{ess\,sup}_{t\in\mathbb{T}^n} \dim J_s(t)$, where J_s corresponds to V_s.

Theorem 2.17 ([4], Theorem 5.5) *Let J_s be a range function so that $\dim J_s(t) < +\infty$ for a.e. $t \in \mathbb{T}^n$, and let $R_s(t) : J_s(t) \to J_s(t)$ be a bounded measurable range operator. Then, there exist $\lambda_s^i \in L^\infty(\mathbb{T}^n)$, $i \in \mathbb{N}$, such that $\lambda_s^i(t) \neq \lambda_s^j(t)$, $i \neq j$, for a.e. $t \in \mathbb{T}^n$, and if*

$$A_{d,q} := \{t \in A_d : \operatorname{card}(\mathfrak{R}_\lambda(t)) = q\},$$

where the sequence $(A_d)_{d\in\mathbb{N}_0}$ is given by Proposition 2.2, then

$$\mathfrak{R}_\lambda(t) = \{\lambda_s^1(t), \ldots, \lambda_s^q(t)\}$$

for a.e. $t \in A_{d,q}$, for all $q \leq d$ and $d, q \in \mathbb{N}$.

Proof By Proposition 2.2, $A_i \cap A_j = \emptyset$, $i \neq j$, and $\sigma_{V_s} = \bigcup_{d\in\mathbb{N}} A_d$. Moreover, by Theorem 2.16, for every $d \in \mathbb{N}$ there exist d measurable functions $\lambda_s^{d,1}, \ldots, \lambda_s^{d,d} : A_d \to \mathbb{C}$, which are the eigenvalues of $R_s(t)$ for a.e. $t \in A_d$. Let us regard $d \in \mathbb{N}$ as fixed and, for all $q \leq d$, define

$$A_{d,q} := \big\{t \in A_d : \operatorname{card}\{\lambda_s^{d,1}(t), \ldots, \lambda_s^{d,d}(t)\} = q\big\}.$$

These sets are measurable and disjoint and $A_d = \bigcup_{q=1}^{d} A_{d,q}$. Now, there exist measurable functions $\lambda_s^{d,q;1}, \ldots, \lambda_s^{d,q;q} : A_{d,q} \to \mathbb{C}$, which are eigenvalues of $R_s(t)$ such that $\lambda_s^{d,q;i}(t) \neq \lambda_s^{d,q;j}(t)$, $i \neq j$, for a.e. $t \in A_{d,q}$. Further, since R_s is bounded, there exists a constant $C > 0$ such that $|\lambda_s^{d,q;i}(t)| \leq C$ for a.e. $t \in A_{d,q}$ and $i \leq q \leq d$. Let us define the functions $\lambda_s^i : \mathbb{T}^n \to \mathbb{C}$ by

$$\lambda_s^i(t) := \begin{cases} \lambda_s^{d,q;i}(t), & t \in A_{d,q},\ i \leq q \leq d, \\ C + i, & \text{otherwise.} \end{cases}$$

Then, $\lambda_s^i(t) \neq \lambda_s^j(t)$, $i \neq j$, for a.e. $t \in \mathbb{T}^n$ and $\lambda_s^i \in L^\infty(\mathbb{T}^n)$, $i \in \mathbb{N}$. Moreover, $\lambda_s^i(t)$ is the eigenvalue of $R_s(t)$ for a.e. $t \in A_{d,q}$, since

$$\ker\big(R_s(t) - \lambda_s^i(t)1(t)\big) = \ker\big(R_s(t) - \lambda_s^{d,q;i}(t)1(t)\big),$$

for a.e. $t \in A_{d,q}$ and all $i \leq q \leq d$. Otherwise, since $\lambda_s^i(t) = C + i$ is not the eigenvalue of $R_s(t)$, we have $\ker\big(R_s(t) - \lambda_s^i(t)1(t)\big) = \{\mathbf{0}\}$. □

Remark 2.3 (1) If J_s is the range function for FSI space $V_s \subset H^s$, then $|A_d| = 0$, for every $d > \Gamma(V_s)$. Thus, for $q \in \mathbb{N}$, we define

$$B_q := \bigcup_{d=q}^{+\infty} A_{d,q}, \quad \text{and} \quad \mathbf{q} := \max\{q \in \mathbb{N} : |B_q| \neq 0\}. \tag{2.27}$$

(2) Let us define $C_i := \bigcup_{q=i}^{+\infty} B_q$, $i \in \mathbb{N}$. The set C_i contains $t \in \sigma_{V_s}$ such that the range operator $R_s(t)$ has at least i different eigenvalues. By the proof of Theorem 2.17, it is clear that $C_i = \big\{t \in \mathbb{T}^n : \ker(R_s(t) - \lambda_s^i(t)1(t)) \neq \{\mathbf{0}\}\big\}$, $i \in \mathbb{N}$, and $|C_i| = 0$ for $i > \mathbf{q}$.

2.7 s-Diagonalization and Dynamical Sampling for Shift-preserving Operators

The definition of s-diagonalization was first introduced by Aguilera et al. in [4]. In this section, we follow the notation used in [4] and [5] and adapt the definition for the space H^s and give conditions under which the shift-preserving operator will be s-diagonalizable. In the last part of this section, we show that iterations of the shift-preserving operator on a finite set of functions produce a frame for V_s.

Let $\psi : \mathbb{T}^n \to \mathbb{C}$ be measurable and M^s_ψ be a multiplication operator $M^s_\psi : H(\mathbb{T}^n, \ell^2_s) \to H(\mathbb{T}^n, \ell^2_s)$,

$$M^s_\psi \mathcal{T}_s f(t) = \left(\frac{\psi(t)\widehat{g}(t+k)}{\mu_s(k)}\right)_{k\in\mathbb{Z}^n}, \qquad t \in \mathbb{T}^n,$$

where $g = (1 - \frac{\Delta}{4\pi^2})^{s/2} f$. It is continuous if and only if $\psi \in L^\infty(\mathbb{T}^n)$.

For a sequence $a = (a_k)_{k\in\mathbb{Z}^n} \in \ell^2_s$, we define

$$\widehat{a}(\cdot) = \sum_{k\in\mathbb{Z}^n} a_k \mathrm{e}^{-2\pi\sqrt{-1}\langle k,\cdot\rangle}$$

and an operator $\Lambda^s_a : H^s \to H^s$ by $\Lambda^s_a = \sum_{k\in\mathbb{Z}^n} a_k T_k$. If $\widehat{a}(\cdot) \in L^\infty(\mathbb{T}^n)$, then the sequence $a = (a_k)_{k\in\mathbb{Z}^n} \in \ell^2_s$ is called a sequence with bounded spectrum. We say that Λ^s_a is an s-eigenvalue of L if $V^a_s := \{f \in H^s : Lf = \Lambda^s_a f\} \neq \{\mathbf{0}\}$, where $a = (a_k)_{k\in\mathbb{Z}^n} \in \ell^2_s$ is a sequence with bounded spectrum. The space V^a_s is called s-eigenspace associated to Λ^s_a. Note that "s" in "s-eigenvalue" and "s-eigenspace" has nothing to do with "s" in the space H^s, even though we use the same letter.

The following lemma will be useful in the sequel.

Lemma 2.3 *Let $a = (a_k)_{k\in\mathbb{Z}^n} \in \ell^2_s$. Then, $\Lambda^s_a = \mathcal{T}_s^{-1} M^s_{\widehat{a}} \mathcal{T}_s : H^s \to H^s$ is a linear bounded operator if and only if a has a bounded spectrum.*

Proof Let $f \in H^s$, then $(1 - \Delta/(4\pi^2))^{s/2} f =: g \in L^2$ and

$$\begin{aligned}
\mathcal{T}_s^{-1} M^s_{\widehat{a}} \mathcal{T}_s f(t) &= \mathcal{T}_s^{-1}\left(\frac{\widehat{a}(t)\widehat{g}(t+j)}{\mu_s(j)}\right)_{j\in\mathbb{Z}^n} \\
&= \sum_{k\in\mathbb{Z}^n} a_k \mathcal{T}_s^{-1} \mathrm{e}^{-2\pi\sqrt{-1}\langle k,t\rangle}\left(\frac{\widehat{g}(t+j)}{\mu_s(j)}\right)_{j\in\mathbb{Z}^n} \\
&= \sum_{k\in\mathbb{Z}^n} a_k \mathcal{T}_s^{-1} \mathrm{e}^{-2\pi\sqrt{-1}\langle k,t\rangle} \mathcal{T}_s f(t) = \sum_{k\in\mathbb{Z}^n} a_k T_k f(t) = \Lambda^s_a f(t),
\end{aligned}$$

where we used Lemma 2.1. Hence, $\Lambda^s_a = \mathcal{T}_s^{-1} M^s_{\widehat{a}} \mathcal{T}_s$. Linearity simply follows from the definition of operator Λ^s_a. Since $\mathcal{T}_s$ is an isometric isomorphism (see Lemma 2.1), the statement follows. □

Next tree lemmas are necessary for Theorem 2.18.

Lemma 2.4 *For $f \in V^a_s$, we have $R_s(t)\big(\mathcal{T}_s f(t)\big) = \widehat{a}(t)\mathcal{T}_s f(t)$, for a.e. $t \in \mathbb{T}^n$.*

Proof Using (2.10), Lemma 2.3 and $Lf = \Lambda^s_a f$, we have

$$R_s(t)\big(\mathcal{T}_s f(t)\big) = \mathcal{T}_s\big(Lf\big)(t) = \mathcal{T}_s\big(\Lambda^s_a f\big)(t) = \mathcal{T}_s\big(\mathcal{T}_s^{-1} M^s_{\widehat{a}} \mathcal{T}_s f\big)(t) = \widehat{a}(t)\mathcal{T}_s f(t),$$

for a.e. $t \in \mathbb{T}^n$. □

Lemma 2.5 *Let $V_s \subset H^s$ be a shift-invariant space. Then, there exists $\varphi \in V_s$ such that* $\operatorname{supp} \|\mathcal{T}_s\varphi(\cdot)\|_{\ell^2_s} = \sigma_{V_s}$.

Proof The proof is a direct consequence of [19, Proposition 2.9] and therefore is omitted. □

The next lemma may be proved in much the same way as [4, Proposition 3.5], so the proof is omitted.

Lemma 2.6 *Let J_s be a range function such that* $\dim J_s(t) < +\infty$ *and let $R_s(t) : J_s(t) \to J_s(t)$ be a measurable range operator for a.e. $t \in \mathbb{T}^n$. Then, the mapping $t \mapsto \ker(R_s(t))$, $t \in \mathbb{T}^n$, is a measurable range function.*

Using the lemmas 2.4, 2.5, and 2.6, we get the following statement.

Theorem 2.18 ([4], Proposition 4.5) *Let J_s be a range function for shift-invariant space $V_s \subset H^s$ such that* $\dim J_s(t) < +\infty$ *for a.e. $t \in \mathbb{T}^n$. Let $a \in \ell^2_s$ be a sequence of bounded spectrum and let $L : V_s \to V_s$ be a bounded shift-preserving operator. If Λ^s_a is an s-eigenvalue of L, then $\Lambda^s_a(t) := \widehat{a}(t)$ is an eigenvalue of $R_s(t)$ for a.e. $t \in \sigma_{V^a_s}$. Moreover,*

$$J^a_s(t) := \ker\big(R_s(t) - \Lambda^s_a(t)1(t)\big), \quad \textit{for a.e. } t \in \mathbb{T}^n,$$

is the measurable range function for V^a_s.

Remark 2.4 From the previous results, it follows that:

(1) V^a_s is a shift-invariant subspace of V_s;
(2) $LV^a_s \subseteq V^a_s$;
(3) if $V^a_s \cap V^b_s = \{\mathbf{0}\}$, then $\widehat{a}(t) \neq \widehat{b}(t)$ a.e. in $\sigma_{V^a_s} \cap \sigma_{V^b_s}$.

In the remainder of this section, we assume $V_s \subset H^s$ to be an FSI space with the range function J_s and $L : V_s \to V_s$ to be a shift-preserving operator with the corresponding range operator R_s.

We will say that $(V_s, L, a^1, \ldots, a^p)$ is an s-diagonalization of L, i.e., the operator L is s-diagonalizable, if there exists natural number p so that $\Lambda^s_{a^i}$, $i = 1, \ldots, p$, are s-eigenvalues of L and $V_s = V^{a^1}_s \oplus V^{a^2}_s \oplus \cdots \oplus V^{a^p}_s$, where $\oplus$ denote direct sum and a^i, $i = 1, \ldots, p$, are sequences of bounded spectrum.

Theorem 2.19 ([4], Theorem 6.4) *If L is s-diagonalizable, then $R_s(t)$ is diagonalizable for a.e. $t \in \sigma_{V_s}$.*

Proof Assume that $(V_s, L, a^1, \ldots, a^p)$ is an s-diagonalization of L. Then, by Theorem 2.18, $\lambda^s_{a^\ell}(t) := \widehat{a^\ell}(t)$ is an eigenvalue of $R_s(t)$ for a.e. $t \in \sigma_{V^{a^\ell}_s}$ and $J^{a^\ell}_s(t) := \ker\big(R_s(t) - \widehat{a^\ell}(t)1(t)\big)$ is eigenspace for $R_s(t)$, for a.e. $t \in \mathbb{T}^n$, for $\ell = 1, \ldots, p$. We only need to show that $J_s(t) = J^{a^1}_s(t) \oplus \cdots \oplus J^{a^p}_s(t)$, for a.e. $t \in \mathbb{T}^n$.

We check at once that $J_s^{a^1}(t) + \cdots + J_s^{a^p}(t) \subseteq J_s(t)$, which is clear from

$$V_s^{a^\ell} \subseteq V_s \Rightarrow J_s^{a^\ell}(t) \subseteq J_s(t), \quad \ell = 1, \ldots, p.$$

For $\phi \in V_s$, we have $\phi = \phi_1 + \cdots + \phi_p$, where $\phi_\ell \in V_s^{a^\ell}$, $\ell = 1, \ldots, p$. Thus, $\mathcal{T}_s\phi(t) = \mathcal{T}_s\phi_1(t) + \cdots + \mathcal{T}_s\phi_p(t)$ for a.e. $t \in \mathbb{T}^n$, and therefore,

$$J_s(t) = \overline{\text{span}}\{\mathcal{T}_s\varphi(t) : \varphi \in \mathcal{A}_{r,s}\} \subseteq \overline{J_s^{a^1}(t) + \cdots + J_s^{a^p}(t)} = J_s^{a^1}(t) + \cdots + J_s^{a^p}(t),$$

where $V_s = S_s(\mathcal{A}_{r,s})$. The second equality follows from the fact that $\dim J_s(t) < +\infty$ for a.e. $t \in \mathbb{T}^n$.

From the aforementioned, we conclude that $J_s(t) = J_s^{a^1}(t) + \cdots + J_s^{a^p}(t)$, for a.e. $t \in \mathbb{T}^n$. Using Proposition 2.1, we get $J_s(t) = J_s^{a^1}(t) \oplus \cdots \oplus J_s^{a^p}(t)$, for a.e. $t \in \mathbb{T}^n$, i.e. the sum is direct. □

Theorem 2.20 ([4], Theorem 6.8) *If the operator $R_s(t)$ is diagonalizable for a.e. $t \in \sigma_{V_s}$, then there exist sequences $(a^i)_{i=1}^{\mathbf{q}}$ of bounded spectrum, where $\mathbf{q}$ is given by (2.27), such that $J_s^{a^i}(t) := \ker\big(R_s(t) - \widehat{a}^i(t)1(t)\big)$, $i = 1, \ldots, \mathbf{q}$, are measurable range functions and*

(1) $J_s(t) = J_s^{a^1}(t) \oplus \cdots \oplus J_s^{a^{\mathbf{q}}}(t)$, for a.e. $t \in \mathbb{T}^n$, where $\oplus$ denote direct sum;
(2) the sets $C_i = \big\{t \in \sigma_{V_s} : J_s^{a^i}(t) \neq \{\mathbf{0}\}\big\}$, $i = 1, \ldots, \mathbf{q}$, satisfy $|C_i| > 0$ and $C_{i+1} \subset C_i$, $i = 1, \ldots, \mathbf{q} - 1$.

Proof (1) Let $\Gamma(V_s) = r$. By Theorem 2.17 and Remark 2.3, there exist measurable functions $\lambda_s^1, \ldots, \lambda_s^{\mathbf{q}} \in L^\infty(\mathbb{T}^n)$ such that

$$J_s(t) = \bigoplus_{i=1}^{\mathbf{q}} \ker\big(R_s(t) - \lambda_s^i(t)1(t)\big), \quad \text{for a.e. } t \in \mathbb{T}^n.$$

Indeed, if $t \in A_{d,q}$, then (see the proof of Theorem 2.17)

$$\bigoplus_{i=1}^{\mathbf{q}} \ker\big(R_s(t) - \lambda_s^i(t)1(t)\big) = \bigoplus_{i=1}^{q} \ker\big(R_s(t) - \lambda_s^{d,q;i}(t)1(t)\big) \oplus \bigoplus_{i=q+1}^{\mathbf{q}} \{\mathbf{0}\} = J_s(t),$$

since $R_s(t)$ is diagonalizable and $\{\lambda_s^{d,q;1}(t), \ldots, \lambda_s^{d,q;q}(t)\}$ is the set of eigenvalues of $R_s(t)$ on $A_{d,q}$, for a.e. $t \in \mathbb{T}^n$. If $t \notin \sigma_{V_s}$, then $\ker\big(R_s(t) - \lambda_s^i(t)1(t)\big) = \{\mathbf{0}\}$, $i = 1, \ldots, \mathbf{q}$, and $J_s(t) = \{\mathbf{0}\}$.

Moreover, since $\lambda_s^i \in L^\infty(\mathbb{T}^n)$, there exists a sequence $a^i = (a_k^i)_{k\in\mathbb{Z}^n} \in \ell_s^2$ of bounded spectrum so that $\lambda_s^i(t) = \widehat{a}^i(t)$ for a.e. $t \in \mathbb{T}^n$ and the function $J_s^{a^i}(t) = \ker\big(R_s(t) - \widehat{a}^i(t)1(t)\big)$ is measurable. Hence, (1) holds.

(2) It follows from Remark 2.3 (2).

□

Now we can prove a theorem that gives necessary conditions for s-diagonalization of shift-preserving operator L. The main idea of the proof is presented in [4].

Theorem 2.21 ([4], Theorem 6.16) *If $L : V_s \to V_s$ is a normal shift-preserving operator, then L is s-diagonalizable and*

$$L = \sum_{i=1}^{p} \Lambda^s_{a^i} P_{V_s^{a^i}},$$

where $(V_s, L, a^1, \ldots, a^p)$ is an s-diagonalization of L and $P_{V_s^{a^i}} : V_s \to V_s^{a^i}$, $i = 1, \ldots, p$, are the orthogonal projections.

Proof Since L is normal and V_s is an FSI space, by Theorem 2.7, the range operator $R_s(t)$ is normal for a.e. $t \in \mathbb{T}^n$, and therefore, the range operator $R_s(t)$ is diagonalizable for a.e. $t \in \mathbb{T}^n$, and moreover, its eigenspaces are orthogonal.

Let $\mathbf{q}$ be given by (2.27). By Theorem 2.20, we have $J_s(t) = J_s^{a^1}(t) \oplus \cdots \oplus J_s^{a^{\mathbf{q}}}(t)$ for a.e. $t \in \mathbb{T}^n$, where $J_s^{a^i}(t) = \ker\left(R_s(t) - \widehat{a}^i(t)1(t)\right)$ is the measurable range function, and a^i is the sequence of bounded spectrum for all $i = 1, \ldots, \mathbf{q}$.

Using Theorem 2.18, we get that $V_s^{a^i} = \{f \in H^s : Lf = \Lambda^s_{a^i} f\} \neq \{\mathbf{0}\}$ is the shift-invariant space with the measurable range function $J_s^{a^i}$. The orthogonality between the eigenspaces $J_s^{a^i}(t)$, $i = 1, \ldots, \mathbf{q}$, for a.e. $t \in \mathbb{T}^n$, implies the orthogonality between the s-eigenspaces $V_s^{a^i}$, $i = 1, \ldots, \mathbf{q}$. Finally, by Proposition 2.1 (2), we get $V_s = V_s^{a^1} \oplus \cdots \oplus V_s^{a^{\mathbf{q}}}$, and consequently, the shift-preserving operator L is s-diagonalizable.

If $(V_s, L, a^1, \ldots, a^p)$ is any s-diagonalization for L, since the eigenspaces of $R_s(t)$ are orthogonal for a.e. $t \in \mathbb{T}^n$, we get that the s-eigenspaces $V_s^{a^i}$ are orthogonal and $V_s = V_s^{a^1} \oplus \cdots \oplus V_s^{a^p}$. Thus, the assertion holds. □

Theorem 2.22 ([5], Proposition 2.16) *If the operator $L : V_s \to V_s$ is normal, then the following assertions hold.*

(1) The operators L and its adjoint L^* are s-diagonalizable.
(2) If the operator Λ^s_a is an s-eigenvalue of operator L, then its adjoint, denoted by Λ^{s*}_a, is an s-eigenvalue of L^* and $V_s^a = \{f \in H^s : L^* f = \Lambda^{s*}_a f\}$. Moreover, $\Lambda^{s*}_a = \Lambda^s_{\widetilde{a}}$, where $\widetilde{a} \in \ell^2_s$ and $\widetilde{a}_k := \overline{a}_{-k}$, $k \in \mathbb{Z}^n$.
(3) If $(V_s, L, a^1, \ldots, a^p)$ is an s-diagonalization for operator L, then $(V_s, L^*, \widetilde{a}^1, \ldots, \widetilde{a}^p)$ is an s-diagonalization for L^*.

Proof

(1) The adjoint L^* of normal operator L is also normal, and thus, by Theorem 2.21, L and L^* are s-diagonalizable.

(2) Since

$$\overline{\overline{a}}(t) = \sum_{k\in\mathbb{Z}^n} \overline{a}_k e^{2\pi\sqrt{-1}\langle k,t\rangle} = \sum_{k\in\mathbb{Z}^n} \overline{a}_{-k} e^{-2\pi\sqrt{-1}\langle k,t\rangle},$$

write $\widetilde{a}_k := \overline{a}_{-k}$, $k \in \mathbb{Z}^n$. It is clear that $\widetilde{a} = (\widetilde{a}_k)_{k\in\mathbb{Z}^n} \in \ell_s^2$. For Λ_a^s which is an s-eigenvalue of L and $\varphi_1, \varphi_2 \in V_s$, we get

$$\langle \Lambda_a^s \varphi_1, \varphi_2 \rangle_{H^s} = \langle \varphi_1, \Lambda_{\widetilde{a}}^s \varphi_2 \rangle_{H^s},$$

i.e., $\Lambda_a^{s*} = \Lambda_{\widetilde{a}}^s$. Moreover, since $L - \Lambda_a^s$ is a normal operator, we get $\ker(L^* - \Lambda_{\widetilde{a}}^s) = \ker((L - \Lambda_a^s)^*) = \ker(L - \Lambda_a^s) = V_s^a \neq \{\mathbf{0}\}$, and the assertion holds.

(3) Let $(V_s, L, a^1, \dots, a^p)$ be an s-diagonalization for L. By (2), we have $\ker(L^* - \Lambda_{\widetilde{a}^j}^s) = V_s^{a^j}$, $j = 1, \dots, p$, and therefore, $V_s = V_s^{a^1} \oplus \cdots \oplus V_s^{a^p}$ is decomposition on s-eigenspaces for L^*.

□

We introduce the notion of $M = \{1, 2, \dots, m\}$, $D = \{0, 1, \dots, d-1\}$ and $V_s^b := \{f \in H^s : L^* f = \Lambda_b^s f\}$, where L^* is the adjoint operator of shift-preserving operator $L : V_s \to V_s$ and $b \in \ell_s^2$ is a sequence of bounded spectrum such that Λ_b^s is an s-eigenvalue of L^*. By Theorem 2.22, if L is normal, then $\Lambda_b^s = \Lambda_{\widetilde{a}}^s$, where Λ_a^s is an s-eigenvalue of L.

The following assertion is a restatement of [5, Theorem 3.2].

Theorem 2.23 *Let $V_s = S_s(\varphi_1, \dots, \varphi_d) \subset H^s$ be a shift-invariant space, and let $J_s^b(t) := \ker\left(R_s^*(t) - \widehat{b}(t)1(t)\right)$, where $\widehat{b}(t)$ is eigenvalue of $R_s^*(t)$ for a.e. $t \in \sigma_{V_s^b}$. If $\{R_s^j(t)(\mathcal{T}_s\phi_i(t)) : \phi_i \in V_s, i \in M, j \in D\}$ is a frame for $J_s(t)$ with a frame bounds $A, B > 0$, for a.e. $t \in \sigma_{V_s^b}$, then $\{P_{J_s^b}(t)(\mathcal{T}_s\phi_i(t)) : \phi_i \in V_s, i \in M\}$ is a frame for $J_s^b(t)$ with bounds $A/C(t)$, $B/C(t)$, for a.e. $t \in \sigma_{V_s^b}$, where $C(t) = \sum_{j\in D} |\widehat{b}(t)|^{2j}$.*

Proof The assertion follows from the next equalities.

$$\begin{aligned}
&\sum_{j\in D}\sum_{i\in M} \left|\left\langle \mathcal{T}_s\phi(t), R_s^j(t)(\mathcal{T}_s\phi_i(t))\right\rangle_{\ell_s^2}\right|^2 \\
&\quad= \sum_{j\in D}\sum_{i\in M} \left|\left\langle R_s^{*^j}(t)(\mathcal{T}_s\phi(t)), \mathcal{T}_s\phi_i(t)\right\rangle_{\ell_s^2}\right|^2 \\
&\quad= \sum_{j\in D}\sum_{i\in M} \left|\left\langle \widehat{b}(t)^j(\mathcal{T}_s\phi(t)), \mathcal{T}_s\phi_i(t)\right\rangle_{\ell_s^2}\right|^2 \\
&\quad= \sum_{j\in D} |\widehat{b}(t)|^{2j} \sum_{i\in M} \left|\left\langle \mathcal{T}_s\phi(t), P_{J_s^b}(t)(\mathcal{T}_s\phi_i(t))\right\rangle_{\ell_s^2}\right|^2,
\end{aligned}$$

where $\phi \in V_s^b$ and $\widehat{b}(t)$ is eigenvalue of $R_s^*(t)$, for a.e. $t \in \sigma_{V_s^b}$. □

Now, we can prove our main result.

Theorem 2.24 ([5], Theorem 3.6) *Let $V_s = S_s(\varphi_1, \dots, \varphi_d) \subset H^s$ be a shift-invariant space, and let $L : V_s \to V_s$ be a shift-preserving operator with the corresponding range operator R_s. If*

$$\{L^j \phi_i : \phi_i \in V_s, i \in M, j \in D\}$$

is a frame generator set for V_s with bounds $A, B > 0$, then

$$\{P_{V_s^b} \phi_i : \phi_i \in V_s, i \in M\}$$

is a frame generator set for V_s^b with bounds $A\big(\sum_{j=0}^{d-1} \|L\|^{2j}\big)^{-1}$ and B.

Proof Assume that $\{L^j \phi_i : \phi_i \in V_s, i \in M, j \in D\}$ is a frame generator set for V_s with bounds $A, B > 0$. Using Theorem 2.2 and the equality (2.10), we get

$$\{\mathcal{T}_s\big(L^j \phi_i\big)(t) : i \in M, j \in D\} = \{R_s^j(t)\big(\mathcal{T}_s \phi_i(t)\big) : i \in M, j \in D\}$$

is a frame for $J_s(t)$ with the same frame bounds, for a.e. $t \in \mathbb{T}^n$. If Λ_b^s is an s-eigenvalue of L^*, then, using Theorems 2.7 and 2.18, it follows that $\widehat{b}(t)$ is an eigenvalue of $R_s^*(t)$ for a.e. $t \in \sigma_{V_s^b}$. By Theorem 2.23, we conclude that

$$\Big\{P_{J_{V_s^b}}(t)\big(\mathcal{T}_s \phi_i(t)\big) : i \in M, j \in D\Big\}$$

is a frame for $J_s^b(t)$ with bounds $\frac{A}{C(t)}$ and $\frac{B}{C(t)}$ for a.e. $t \in \sigma_{V_s^b}$, where $C(t) = \sum_{j=0}^{d-1} |\widehat{b}(t)|^{2j}$. Using Theorem 2.5, we get

$$1 \le C(t) = \sum_{j=0}^{d-1} |\widehat{b}(t)|^{2j} \le \sum_{j=0}^{d-1} \|R_s(t)\|^{2j} \le \sum_{j=0}^{d-1} \|L\|^{2j}.$$

Therefore, using [6, Theorem 2.1(1)] and Theorem 2.2, the assertion follows. □

Finally, we will prove that the equivalence in Theorem 2.24 holds under additional assumptions. We will say that shift-preserving operator $L : V_s \to V_s$ has the spectral property if there exists $C > 0$ so that $|\lambda_s' - \lambda_s| \ge C$, for all $\lambda_s' \ne \lambda_s$, $\lambda_s, \lambda_s' \in \mathfrak{R}_\lambda(t)$, for a.e. $t \in \sigma_{V_s}$.

Theorem 2.25 ([5], Theorem 3.8) *Let $V_s = S_s(\varphi_1, \dots, \varphi_d) \subset H^s$ be a shift-invariant space, and let $L : V_s \to V_s$ be a normal shift-preserving operator, which satisfies the spectral property, with the corresponding range operator R_s. The set*

$$\{L^j \phi_i : \phi_i \in V_s, i \in M, j \in D\}$$

is a frame generator set for V_s *if and only if*

$$\{P_{V_s^b}\phi_i : \phi_i \in V_s, i \in M\}$$

is a frame generator set for V_s^b *with the same frame bounds for every* s*-eigenvalue* Λ_b^s *of* L^*.

Proof On account of Theorem 2.24, it remains to prove that the opposite implication holds.

Since L is normal operator, by Theorem 2.22, L^* is s-diagonalizable, and we can construct an s-diagonalization $(V_s, L^*, b^1, \dots, b^{\mathbf{q}})$ for L^* so that $\sigma_{V_s^{b^{\ell+1}}} \subseteq \sigma_{V_s^{b^\ell}}$, $\ell = 1, \dots, \mathbf{q}-1$, and

$$|\widehat{b^i}(t) - \widehat{b^j}(t)| \geq C, \quad i \neq j, \tag{2.28}$$

for some constant $C > 0$, for a.e. $t \in \mathbb{T}^n$, $i, j = 1, \dots, \mathbf{q}$ (see Theorem 2.20 and Remark 2.4 (3)).

Let $\{P_{V_s^{b^\ell}}\phi_i : i \in M\}$ be a frame generator set for $V_s^{b^\ell}$, $\ell = 1, \dots, \mathbf{q}$, with frame bounds $A, B > 0$. By Theorem 2.2, $\{\mathcal{T}_s(P_{V_s^{b^\ell}}\phi_i)(t) : i \in M\}$ is a frame for $J_s^{b^\ell}(t)$ with the same frame bounds for a.e. $t \in \mathbb{T}^n$, $\ell = 1, \dots, \mathbf{q}$. Using [6, Theorem 2.1 (1)], we get that $\{P_{J_s^{b^\ell}}(t)(\mathcal{T}_s\phi_i(t)) : i \in M\}$ is a frame for $J_s^{b^\ell}(t)$ with the same frame bounds for a.e. $t \in \mathbb{T}^n$, $\ell = 1, \dots, \mathbf{q}$.

Let $A_\ell := \sigma_{V_s^{b^\ell}} \backslash \sigma_{V_s^{b^{\ell+1}}}$ for $\ell = 1, \dots, \mathbf{q}-1$, and $A_{\mathbf{q}} := \sigma_{V_s^{b^{\mathbf{q}}}}$. Then, $\mathbb{T}^n = \bigcup_{\ell=1}^{\mathbf{q}} A_\ell \cup \left(\mathbb{T}^n \backslash \sigma_{V_s}\right)$.

Fix $\widetilde{\ell} \in \{1, \dots, \mathbf{q}\}$. Then, $\{P_{J_s^{b^\ell}}(t)(\mathcal{T}_s\phi_i(t)) : i \in M\}$ is a frame for $J_s^{b^\ell}$ for a.e. $t \in A_\ell$, $\ell = 1, \dots, \widetilde{\ell}$. By [5, Theorem 3.5], $\{R_s^j(t)(\mathcal{T}_s\phi_i(t)) : i \in M, j \in D\}$ is a frame for $J_s(t)$ for a.e. $t \in A_{\widetilde{\ell}}$ with bounds

$$A\left(\frac{\mathbf{q}}{\beta(t)}\sum_{i=0}^{\widetilde{\ell}-1}\binom{\widetilde{\ell}-1}{i}^2 \|R_s(t)\|^{2i}\right)^{-1} \quad \text{and} \quad B\left(\widetilde{\ell}\sum_{j=0}^{d-1}\|R_s(t)\|^{2j}\right),$$

where $\beta(t) = \min_{1\leq \ell \leq \widetilde{\ell}} \prod_{i=1, i\neq \ell}^{\widetilde{\ell}} |\widehat{b^\ell}(t) - \widehat{b^i}(t)|^2$.

There is no loss of generality in assuming $C < 1$ in (2.28). Then, $C^{2\mathbf{q}} \leq C^{2\widetilde{\ell}} \leq \beta(t)$ for a.e. $t \in A_{\widetilde{\ell}}$, and we get

$$A\left(\frac{\mathbf{q}}{C^{2\mathbf{q}}}\sum_{i=0}^{\mathbf{q}-1}\binom{\mathbf{q}-1}{i}^2 \|L\|^{2i}\right)^{-1} \leq A\left(\frac{\mathbf{q}}{\beta(t)}\sum_{i=0}^{\widetilde{\ell}-1}\binom{\widetilde{\ell}-1}{i}^2 \|R_s(t)\|^{2i}\right)^{-1},$$

$$B\left(\widetilde{\ell}\sum_{j=0}^{d-1}\|R_s(t)\|^{2j}\right) \leq B\left(\mathbf{q}\sum_{j=0}^{d-1}\|L\|^{2j}\right),$$

for a.e. $t \in A_{\widetilde{\ell}}$. For every set $A_{\widetilde{\ell}}$, $\widetilde{\ell} = 1, \dots, \mathbf{q}$, we have the same frame bounds, and therefore, $\{R_s^j(t)(\mathcal{T}_s\phi_i(t)) : i \in M, j \in D\}$ is a frame for $J_s(t)$ for a.e. $t \in \mathbb{T}^n$. Finally, using (2.10) and Theorem 2.2, we conclude that the family $\{L^j\phi_i : i \in M, j \in D\}$ is a frame generator for V_s with the frame bounds $A\big(\frac{\mathbf{q}}{C^{2\mathbf{q}}}\sum_{i=0}^{\mathbf{q}-1}\binom{\mathbf{q}-1}{i}^2\|L\|^{2i}\big)^{-1}$, $B\big(\mathbf{q}\sum_{j=0}^{d-1}\|L\|^{2j}\big)$. □

Acknowledgments This research was supported by the Science Fund of the Republic of Serbia, #GRANT No 2727, *Global and local analysis of operators and distributions*—GOALS. The authors are also supported by Serbian Ministry of Science and Technology (grant 451-03-66/2024-03/200122 and 451-03-65/2024-03/200122), and project F10 of the Serbian Academy of Sciences and Arts.

References

1. Aceska, R., Aldroubi, A., Davis, J., Petrosyan, A.: Dynamical sampling in shift-invariant spaces. Contemp. Math. AMS **603**, 139–148 (2012)
2. Aceska, R., Petrosyan, A., Tang, S.: Multidimensional signal recovery in discrete evolution systems via spatiotemporal trade off. Sampl. Theory Signal Image Process. **14**(2), 153–169 (2015)
3. Aceska, R., Tang, S.: Dynamical sampling in hybrid shift invariant spaces. In: Furst, V., Kornelson, K.A., Weber, E.S. (eds.) Operator Methods in Wavelets, Tilings, and Frames, vol. 626 of Contemporary Mathematics. American Mathematical Society, Providence (2014)
4. Aguilera, A., Cabrelli, C., Carbajal, D., Paternostro, V.: Diagonalization of shift-preserving operators. Advan. in Math. (2021). https://doi.org/10.1016/j.aim.2021.107892
5. Aguilera, A., Cabrelli, C., Carbajal, D., Paternostro, V.: Dynamical sampling for shift-preserving operators. Appl. Comput. Harmon. Anal. **51**, 258–274 (2021)
6. Aksentijević, A., Aleksić, S., Pilipović, S.: The structure of shift-invariant subspaces of Sobolev spaces. Theor. Math. Phys. **218**, 177–191 (2024)
7. Aldroubi, A., Cabrelli, C., Molter, U., Tang, S.: Dynamical sampling. Appl. Comput. Harmon. Anal. **42**(3), 378–401 (2017)
8. Aldroubi, A., Davis, J., Krishtal, I.: Dynamical sampling: time-space trade-off. Appl. Comput. Harmon. Anal. **34**(3), 495–503 (2013)
9. Aldroubi, A., Davis, J., Krishtal, I.: Exact reconstruction of signals in evolutionary systems via spatiotemporal trade-off. J. Fourier Anal. Appl. **21**(1), 11–31 (2015)
10. Aldroubi, A., Petrosyan, A.: Dynamical sampling and systems from iterative actions of operators. In: Frames and Other Bases in Abstract and Function Spaces, vol. 1, pp. 15–26. Springer, Cham (2017)
11. Aldroubi, A., Cabrelli, C., Molter, U., Petrosyan, A.: Local-to-global frames and applications to dynamical sampling problem. Excursions in Harmonic Analysis, Volume 6: In Honor of John Benedetto's 80th Birthday (Applied and Numerical Harmonic Analysis), 1st edn., pp. 211–220 (2021)
12. Aldroubi, A., Sun, Q., Tang, W.: p-frames and shift-invariant subspaces of L^p. J. Fourier Anal. Appl. **7**, 1–21 (2001)
13. Bownik, M.: The structure of shift-invariant subspaces of $L^2(\mathbb{R}^n)$. J. Funct. Anal. **177**, 282–309 (2000)
14. Cabrelli, C., Molter, U., Paternostro, V., Philipp, F.: Dynamical sampling on finite index sets. J. Anal. Math. **140**, 637–667 (2020)

15. Christensen, O.: An Introduction to Frames and Riesz Bases. Applied and Numerical Harmonic Analysis. Birkhäuser, Basel, Second expanded edition (2016)
16. Christensen, O., Hasannasab, M., Philipp, F.: Frame properties of operator orbits. Math. Nachr. **293**(1), 52–66 (2020)
17. Christensen, O., Hasannasab, M., Rashidi, E.: Dynamical sampling and frame representations with bounded operators. J. Math. Anal. Appl. **463**(2), 634–644 (2018)
18. Christensen, O., Hasannasab, M., Stoeva, D.T.: Operator representations of sequences and dynamical sampling. Sampl. Theory Signal Image Process. **17**, 29–42 (2018)
19. de Boor, C., DeVore, R.A., Ron, A.: The structure of finitely generated shift-invariant spaces in $L^2(\mathbb{R}^d)$. J. Funct. Anal. **119**, 37–78 (1994)
20. de Boor, C., DeVore, R.A., Ron, A.: Approximation from shift-invariant subspaces of $L^2(\mathbb{R}^d)$. Trans. Am. Math. Soc. **341**, 787–806 (1994)
21. Gröchening, K.: Foundations of Time-Frequency Analysis. Birkhauser, Boston (2001)
22. Helson, H.: Lectures on Invariant Subspaces. Academic Press, New York, London (1964)
23. Philipp, F.: Bessel orbits of normal operators. J. Math. Anal. Appl. **44**(2), 767–785 (2017)
24. Pilipović, S., Simić, S.: Frames for weighted shift-invariant spaces. Mediterr. J. Math. **9**, 897–912 (2012)
25. Ron, A., Shen, Z.: Frames and stable bases for shift-invariant subspaces of $L^2(\mathbb{R}^d)$. Can. J. Math. **47**, 1051–1094 (1995)
26. Schwartz, L.: Théorie des Distributions. Hermann, Paris (1966)
27. Tang, S.: System identification in dynamical sampling. Adv. Comput. Math. **43**(3), 555–580 (2017)
28. Vladimirov, V.S.: Generalized Functions in Mathematical Physics, Transl. by Yankovskii, G. "Mir", Moscow, xii+362 pp. (1979)

Chapter 3
Multicriteria Model with a Dominant Criterion in Satisfaction and Non-Satisfaction

Miloljub Albijanić, Miodrag M. Žižović, Dragan Pamučar, and Mališa R. Žižović

Dedicated to Academician Gradimir V. Milovanović on the occasion of his 75th birthday

3.1 Introduction

Optimization problems, or problems of finding optimal solutions, are issues that are analyzed and solved on a daily basis. Decision-making problems and the selection of the optimal alternative are often based on multiple criteria and are encountered in various fields such as technical systems, economic trends, and logistics systems. Common to all multicriteria optimization problems is the attempt to find a solution that satisfies all specified criteria to the greatest extent possible, while respecting predefined constraints. Most such problems do not have a unique and global solution; in other words, there is no optimal solution that satisfies all criteria simultaneously.

In most multicriteria problems, some criteria are partially or completely conflicting. Also, in certain situations, the criteria can be diverse, subjectively defined, and represented in different units of measurement. Therefore, a unique solution is often defined based on subjective assessments of the decision-maker regarding the

M. Albijanić
FEFA Faculty, University Metropolitan, Belgrade, Serbia
e-mail: malbijanic@fefa.edu.rs

M. M. Žižović
Economics Institute, Belgrade, Serbia

D. Pamučar (✉)
Faculty of Organizational Sciences, University of Belgrade, Belgrade, Serbia
e-mail: dragan.pamucar@fon.bg.ac.rs

M. R. Žižović
Faculty of Technical Sciences, University of Kragujevac, Čačak, Serbia

M. P. Stanić et al. (eds.), *Analysis, Approximation, Optimization: Computation and Applications*, Springer Optimization and Its Applications 224,
https://doi.org/10.1007/978-3-031-85743-0_3

acceptability of a particular solution [4]. The solution accepted by the decision-maker is called the best or preferred solution—the solution that is the best under given conditions.

In the last few decades, numerous researchers have been attempting to develop multicriteria algorithms with the aim of making rational decisions while minimizing the subjective assessments of decision-makers [2, 7, 16]. In the second half of the last century, several basic multicriteria models were presented, which today represent traditional models, among which are Elimination Et Choix Traduisant la Réalité (ELECTRE) [14], Analytical Hierarchy Process (AHP) [15], Technique for Order Preference by Similarity to Ideal Solution (TOPSIS) [7], and Complex Proportional Assessment (COPRAS) [18]. Traditional multicriteria techniques differ in their methodological frameworks. For example, the AHP model is based on subjective pairwise comparisons and checks the consistency of these comparisons, based on which the quality of the obtained solution is defined [3, 9, 17]. On the other hand, the TOPSIS model is based on the idea of defining ideal Euclidean distances of the considered set of alternatives from ideal and anti-ideal values [8]. The COPRAS model involves grouping the criteria into benefit and cost groups and suggests the application of an aggregation function in a way that defines the percentage suitability of an alternative and its final rank.

In the last two decades, several interesting newer approaches have been developed to solve multicriteria analysis problems, among which we will highlight methods with pre-expressed preferences. The most well-known representatives of this group of methods are: Multi-Criteria Optimization and Compromise Solution (VIKOR) [10], COmbinative Distance-based ASsessment (CODAS) [5], Multi-Attributive Border Approximation area Comparison (MABAC) [11], MultiAtributive Ideal-Real Comparative Analysis (MAIRCA) [12], and Ranking of Alternatives through Functional mapping of criterion subintervals into a Single Interval (RAFSI) [20]. All the presented multicriteria techniques have demonstrated their effectiveness in solving classical multicriteria problems [1, 6, 13, 19]. However, one limitation of the presented models is the inability to solve multicriteria problems where there is a dominant criterion.

Žižović et al. [22, 23] and Žižović [21] in their works have considered multicriteria problems where one of the criteria is dominant in the following sense: It is expected that this criterion in a given multicriteria problem is either fully satisfied or satisfactorily satisfied. If this criterion is not met, then the problem itself does not make sense. In other words, alternatives that do not meet this criterion are not significant or usable for solving this problem, regardless of their values in other criteria.

The methodology used to solve this multicriteria problem involved defining constant weight coefficients for all criteria for complete fulfillment of the dominant criterion or for some interval with sufficiently good fulfillment of this criterion. For alternatives with lower fulfillment according to this criterion, the calculation of new weight coefficients was done in a way that the value of the weight coefficient of the dominant criterion was increased based on the reduction in fulfillment of the alternative's value according to this dominant criterion, while the other weight coefficients were proportionally reduced (compared to their initial values). Certainly, if the value of the alternative according to this dominant criterion were to fall to zero, then the weight coefficient of the dominant criterion would take the value of 1 (while the other criteria would have weight coefficients equal to zero). Therefore, according to this methodology, an alternative that is not fulfilled at all has a value of zero in the previously mentioned works.

In this chapter, a new problem is considered in which non-fulfillment according to, again, we can say, a dominant criterion is not tolerated similarly as in the previous case, but now the complete fulfillment of this criterion is considered super-desirable and even more important than all other criteria in this multicriteria problem. Therefore, these problems face a challenge with fulfillment, which can be observed as being very rare. An example of such a problem could be the choice of therapy for treating certain diseases, where the main or dominant criterion is the effectiveness of the drugs in curing the patient (where 0 means that the drug has no effect, and 1 represents complete cure), while other criteria may include the availability of the drug, the cost of the drug, the tolerance of the drug, etc. Here, it is evident that constant values for criteria must be defined within a reasonable interval, which is clearly a closed subinterval of the interval (0, 1)! Therefore, variable values of the dominant criterion and other criteria occur for decreased values, i.e., values to the left of the minimum point in the subinterval of constant values, and also for increased values of the dominant criterion for alternatives that are greater than the maximum point in the subinterval with constant weight coefficients. To satisfy this, we need a new methodology.

3.2 New Multicriteria Approach with a Dominant Criterion

For the presentation of this multicriteria method, it is necessary to make some assumptions related to the decision matrix that contains alternatives $A_1, \dots, A_m$ and criteria $C_1, \dots, C_n$, $m, n \in \mathbb{N}$, $m, n \geq 2$ (Table 3.1).

Understood, the values a_{ij}, $i = 1, \dots, m$, $j = a, \dots, n$ represent the characteristics of alternative A_i according to criterion C_j, and here these are numbers between 0 and 1, that is, our decision matrix is normalized. Additionally, all criteria are transformed into maximization type. More details on the normalization of the decision matrix can be found in the monograph.

Table 3.1 Decision matrix

	C_1	C_2	$\dots$	C_n
A_1	a_{11}	a_{12}	$\dots$	a_{1n}
A_2	a_{21}	a_{22}	$\dots$	a_{2n}
$\vdots$	$\vdots$	$\vdots$	$\ddots$	$\vdots$
A_m	a_{m1}	a_{m2}	$\dots$	a_{mn}

We will assume that criterion C_1 is the dominant criterion in terms of fulfillment and non-fulfillment, as mentioned in the previous section. We will assume that the other criteria are ranked in strength from C_2 to C_n.

Now we present the new method step by step.

Step 1 We will assume that $[P, Q]$ is a subinterval of the interval [0, 1] where we have expected values of alternatives for criterion C_1. In this case, the interval is called the reduced fulfillment interval, and the interval $(Q, 1]$ is called the increased fulfillment interval of criterion C_1.

Step 2 We determine the weight coefficients of criteria $C_1, C_2, \ldots, C_n$ on the interval $[P, Q]$. These are positive numbers $w_1, w_2, \ldots, w_n$ such that their sum is 1. According to the assumption about the strength of the criteria, they are ordered by size (although in general, in the interval of expected fulfillment, w_1 is not necessarily the largest).

Determining the weight coefficients in this step is a matter for the decision-maker!

To define the weight coefficients in the intervals of reduced and increased fulfillment, it is necessary to define the following two functions (which we call by a common name—the linkage functions).

Step 3 Let f be a function that maps the interval [0, 1] onto itself with the following properties:

- $f(0) = 0$;
- $f(x)$ is a non-decreasing function on the interval $[0, P)$;
- $f(x) = 1$ on the interval $[P, Q]$;
- $f(x)$ is a non-increasing function on the interval $(Q, 1]$;
- $f(1) = 0$.

Step 4 Let F be a function that maps the interval [0, 1] to the interval $\left[1, \frac{1}{w_1}\right]$ with the following properties:

- $F(0) = \frac{1}{w_1}$;
- $F(x)$ is a non-increasing function on the interval $[0, P)$;
- $F(x) = 1$ on the interval $[P, Q]$;
- $F(x)$ is a non-decreasing function on the interval $(Q, 1]$;
- $F(1) = \frac{1}{w_1}$ (Figs. 3.1 and 3.2).

Note 1 Between these two functions, the following relationships exist:

$$F(x) = \frac{1 + f(x)(w_1 - 1)}{w_1}. \tag{3.1}$$

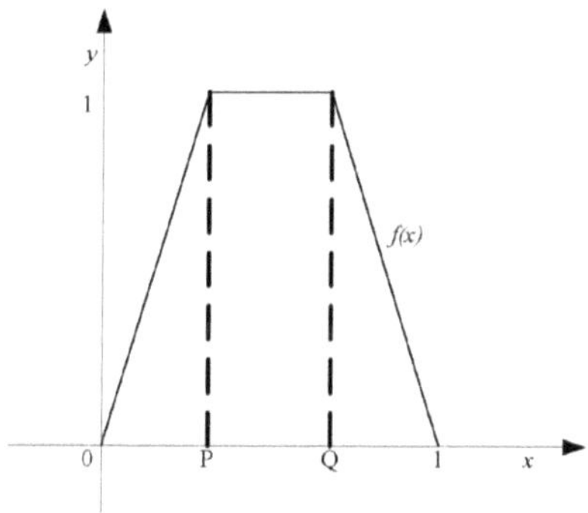

Fig. 3.1 Example of the function f

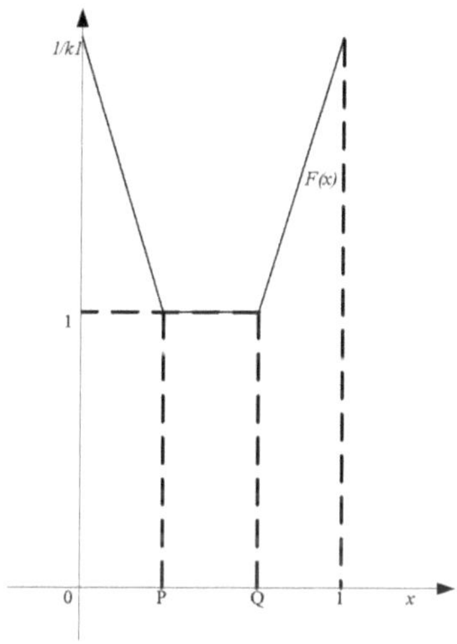

Fig. 3.2 Example of the function F

That is

$$f(x) = \frac{1 - w_1 F(x)}{1 - w_1}. \tag{3.2}$$

Note 2 It is obvious that by specifying one of these functions, we alson obtain the other using relations (3.1) and (3.2).

Step 5 If A_i is an alternative with values according to criteria $(a_{i1}, a_{i2}, \ldots, a_{in})$ then we use the weight coefficients for it

$$w_1, w_2, \ldots, w_n$$

if a_{i1} is from the interval $[P, Q]$, i.e.,

$$W_1(a_{i1}) = F(a_{i1}) w_1, \tag{3.3}$$

$$W_k(a_{i1}) = f(a_{i1})w_k, \quad k = 2, 3, \ldots, n, \quad \text{if } a_{i1} \text{ is outside of this interval} \tag{3.4}$$

Note 3 Alternatives with different values for the first criterion have different weight coefficients (except for alternatives that have values within the expected interval for the first criterion).

For alternatives for which the weight coefficients are calculated according to formulas (3.3) and (3.4), the following holds true.

Sentence 1 If the weight coefficients are calculated according to formulas (3) and (4), then

$$W_1(a_{i1}) + W_2(a_{i1}) + \cdots + W_n(a_{i1}) = 1.$$

Proof: Starting from (3.1)-(3.3), and (3.4), we have

$$\begin{aligned} W_1(a_{i1}) + W_2(a_{i1}) + \cdots + W_n(a_{i1}) &= F(a_{i1})w_1 + f(a_{i1})w_2 + \cdots + f(a_{i1})w_n \\ &= F(a_{i1})w_1 + f(a_{i1})(w_2 + \cdots + w_n) \\ &= F(a_{i1})w_1 + f(a_{i1})(1 - w_1) \\ &= F(a_{i1})w_1 + \frac{1 - w_1 F(x)}{1 - w_1}(1 - w_1) \\ &= 1. \end{aligned}$$

Step 6 For any alternative A_i, the weight coefficients are calculated (we will denote them as: $W_1, W_2, \ldots, W_n$) and then calculate

$$V(A_i) = W_1 a_{i1} + W_2 a_{i2} + \cdots + W_n a_{in}. \tag{3.5}$$

Clearly, this formula can also be expressed in the following way:

$$V(A_i) = F(a_{i1})w_1 a_{i1} + f(a_{i1})(a_{i2}w_2 + \cdots + a_{in}w_n). \tag{3.6}$$

It holds true.

Statement 2 For every alternative Ai with a positive a_{i1} less than 1, $V(A_i)$ is positive and less than 1.

Proof It is obvious that $V(A_i)$ is less than $W_1 + \cdots + W_n = 1$ since a_{i1} is at least less than 1 making $W_1 a_{i1}$ less than W_1.

From the definition of the function $F(x)$, i.e., from $F(1) = \frac{1}{w_1}$, it directly follows that $V(A_i) = 1$ if and only if $a_{i1} = 1$. □

Step 7 If the values of alternatives are calculated (cases where the value according to the first criterion is less than 1), then alternative A_i is ranked ahead of alternative

A_j if $V(A_i)$ is greater than $V(A_j)$. If $V(A_i) = V(A_j)$ then A_i is ranked ahead of A_j if a_{i1} is greater than a_{j1}. If $a_{i1} = a_{j1}$, then we consider a_{i2} and a_{j2}, and the better alternative is the one with a higher value on the second criterion and so on until the end. It follows that it is not possible to determine a better alternative except in the case where all their values are the same. In the case of two alternatives with a value of 1 according to the first criterion, then their alternative values are 1. The previous procedure is then observed from the second criterion onward, just as in the previous case.

3.3 Example: Selection of a Spa for Recovery

An illustration of the application of this method is provided in a hypothetical example created based on some real experiences of users of spa resorts for vacation, rehabilitation, recovery, leisure, etc., which have a tradition in Serbian spas spanning over millennia. The assumption is that we have 10 visitors to spa resorts who share a common doctor. The doctor provides each of them with a recommendation (in the form of an assessment of how each of the four spas they can use will impact their health).

The assessment ranges from 0.0 to 1, where 0.0 means that the spa will not contribute to an improvement in the condition, and 1 indicates that the spa will provide the patient with complete recovery (for example, 0.60 would mean that the patient has a 60% recovery in that spa).

This assessment denoted as C_1 is the first criterion for deciding which spa each of the 10 visitors will choose. For this criterion, we assume that it is dominant, with an expected effect of the spa for the patient in the interval [0.6, 0.8]. The interval [0, 0.6) represents a reduced effect interval, while the interval (0.8, 1] represents an increased effect. We assume that the weight coefficient for the first criterion in the interval of expected effect is a constant $w_1 = 0.3$.

In addition to this dominant criterion for evaluation, regular criteria are used.

C_2 The cost of staying at the spa is also given as a number between 0 and 1, where 0 means it is not acceptable, and 1 means that the cost is popular and generally acceptable. The weight coefficient for the cost is $w_2 = 0.25$.

C_3 The quality of spa services is assessed on an interval (also given as a number between 0 and 1, where 0 means it is not acceptable, and 1 means the quality is outstanding for the offered price). The weight coefficient for quality is $w_3 = 0.20$.

C_4 The rating of entertainment and non-medical facilities is also given as a number between 0 and 1, where 0 means there are no such facilities, and 1 means the entertainment is outstanding, and there is enough of it for the popularly offered price. The weight coefficient for the rating of entertainment and non-medical facilities is $w_4 = 0.15$.

C_5 The rating of the cost and quality of the journey to the spa is also given as a number between 0 and 1, where 0 means it is not acceptable due to the journey being too strenuous, expensive, and unacceptable for almost all potential guests, and

Table 3.2 Ratings of spa places according to non-dominant criteria

	C_1	C_2	C_3	C_4	C_5
A_1		0.6	0.8	0.5	0.6
A_2		0.7	0.7	0.55	0.55
A_3		0.8	0.6	0.6	0.7
A_4		0.9	0.5	0.7	0.9

1 means the journey is comfortable, of outstanding quality for the popularly offered price. The weight coefficient for the quality of the journey to the spa is $w_5 = 0.10$.

So, all criteria are formulated as maximization type. The weight coefficients are collectively evaluated by the entire group of 10 prospective spa visitors.

The entire group collectively assessed four spas, A_1, A_2, A_3 and A_4, according to criteria C_2, C_3, C_4 and C_5, and these ratings are provided in the Table 3.2.

The ratings of the spa's effect for each individual were provided by the doctor and will be used to calculate the effectiveness values for each of the spas for each visitor.

The assessment of spas is carried out according to the formula (3.6). The linkage functions are given by the following formulas:

$$F(x) = \begin{cases} \frac{10}{3} - \frac{35}{9}x, & \text{for } 0 \leq x \leq 0.6, \\ 1, & \text{for } 0.6 \leq x \leq 0.8, \\ \frac{25}{3} - \frac{35}{3}x, & \text{for } 0.8 \leq x \leq 1. \end{cases}$$

Now, using the doctor's assessments for the effectiveness of each spa for users, we can perform the desired evaluation.

User U_1 has their first column in Table 3.2 with values:

- Spa A_1 rating 0.7;
- Spa A_2 rating 0.6;
- Spa A_3 rating 0.45;
- Spa A_4 rating 0.5.

Now, we simply apply formula (3.6) and obtain

$$V(A_1) = 0.655, \quad V(A_2) = 0.635, \quad V(A_3) = 0.5737, \quad V(A_4) = 0.6417.$$

We conclude that for user U_1, the best spa is $A1$.

User U_2 has their first column in Table 3.2 with values:

- Spa A_1 rating 0.9;
- Spa A_2 rating 0.7;
- Spa A_3 rating 0.7;
- Spa A_4 rating 0.6.

Now, we simply apply formula (3.6) and obtain

$$V(A_1) = 0.8975, \quad V(A_2) = 0.6625, \quad V(A_3) = 0.69, \quad V(A_4) = 0.7.$$

We conclude that for user U_2, the best spa is A_1.

User U_3 has their first column in Table 3.2 with values:

- Spa A_1 rating 0.8;
- Spa A_2 rating 0.8;
- Spa A_3 rating 0.8;
- Spa A_4 rating 0.8.

Now we apply formula (3.6) and obtain

$$V(A_1) = 0.685, \quad V(A_2) = 0.6925, \quad V(A_3) = 0.72, \quad V(A_4) = 0.76.$$

We conclude that for user U_3, the best spa is A_4.

User U_4 has their first column in Table 3.2 with values:

- Spa A_1 rating 0.5;
- Spa A_2 rating 0.4;
- Spa A_3 rating 0.3;
- Spa A_4 rating 0.2.

Now we apply formula (3.6) and obtain

$$V(A_1) = 0.5792, \quad V(A_2) = 0.5150, \quad V(A_3) = 0.4350, \quad V(A_4) = 0.2130.$$

We conclude that for user U_4, the best spa is A_1.

User U_5 has their first column in Table 3.2 with values:

- Spa A_1 rating 0.7;
- Spa A_2 rating 0.68;
- Spa A_3 rating 0.4;
- Spa A_4 rating 0.5.

Now we apply formula (3.6) and obtain

$$V(A_1) = 0.655, \quad V(A_2) = 0.6565, \quad V(A_3) = 0.5334, \quad V(A_4) = 0.4550.$$

We conclude that for user U_5, the best spa is A_2.

User U_6 has their first column in Table 3.2 with values:

- Spa A_1 rating 0.7;
- Spa A_2 rating 0.6;
- Spa A_3 rating 0.8;
- Spa A_4 rating 0.5.

Now we apply formula (3.6) and obtain

$$V(A_1) = 0.655, \quad V(A_2) = 0.6325, \quad V(A_3) = 0.7200, \quad V(A_4) = 0.6417.$$

We conclude that for user U_6, the best spa is A_3.

User U_7 has their first column in Table 3.2 with values:

- Spa A_1 rating 0.6;
- Spa A_2 rating 0.75;
- Spa A_3 rating 0.5;
- Spa A_4 rating 0.7.

Now we apply formula (3.6) and obtain

$$V(A_1) = 0.625, \quad V(A_2) = 0.6775, \quad V(A_3) = 0.6083, \quad V(A_4) = 0.7300.$$

We conclude that for user U_7, the best spa is A_4.

User U_8 has their first column in Table 3.2 with values:

- Spa A_1 rating 0.6;
- Spa A_2 rating 0.9;
- Spa A_3 rating 0.7;
- Spa A_4 rating 0.6.

Now we apply formula (3.6) and obtain

$$V(A_1) = 0.625, \quad V(A_2) = 0.8113, \quad V(A_3) = 0.5737, \quad V(A_4) = 0.6417.$$

We conclude that for user U_8, the best spa is A_2.

User U_9 has their first column in Table 3.2 with values:

- Spa A_1 rating 0.4;
- Spa A_2 rating 0.5;
- Spa A_3 rating 0.9;
- Spa A_4 rating 0.6.

Now we apply formula (3.6) and obtain

$$V(A_1) = 0.5100, \quad V(A_2) = 0.5854, \quad V(A_3) = 0.8250, \quad V(A_4) = 0.7000.$$

We conclude that for user U_9, the best spa is A_3.

User U_{10} has their first column in Table 3.2 with values:

- Spa A_1 rating 0.7;
- Spa A_2 rating 0.65;
- Spa A_3 rating 0.6;
- Spa A_4 rating 0.5.

Now we apply formula (3.6) and obtain

$$V(A_1) = 0.655, \quad V(A_2) = 0.6475, \quad V(A_3) = 0.6600, \quad V(A_4) = 0.6417.$$

We conclude that for user U_{10}, the best spa is A_3.

So,

- U_1, U_2 and U_4 go to spa A_1,
- U_5 and U_8 go to spa A_2,
- U_6, U_9 and U_{10}, go to spa A_3,
- U_3 and U_7 go to spa A_4.

In this example, it can be seen that medical conditions and services play the most significant role in determining the spa resort for individual patients, although it is not always decisive.

3.4 Conclusion

This multicriteria analysis method is extremely important in certain situations where other methods simply do not provide an answer. It can be understood as an enhancement of the method with a dominant criterion given in the Žižović et al. [13]. paper (the case when the interval of expected values is translated into the unit). Conversely, it is not possible; in fact, problems that can be solved here are not solvable in other methods. Clearly, if we were to expand the interval of expected values to the entire closed interval between 0 and 1, we would arrive at the weighted sum method of multicriteria analysis. It remains to test this method in situations where we do not have all the information regarding finding the weight coefficients and in applications in such situations in general. Also, the presented methodology does not consider the presence of uncertainty in the information. Therefore, it is necessary to direct future research toward expanding the mathematical model by applying fuzzy and rough sets.

Appendix

It can be relatively easily proven the following highly important.

Theorem 3.1 *If a multicriteria model with alternatives $A_1, A_2, \ldots, A_m$ and criteria $C_1, C_2, \ldots, C_n$ with a dominant criterion (according to this paper) C_1 is solved using this proposed method, and then expanded by adding new alternatives $B_1, \ldots, B_s$ and solved again, the order of the initial alternatives $A_1, A_2, \ldots, A_m$ remains unchanged.*

This fact has one very important consequence.

Consequence The introduction of new alternatives into the model does not affect the order of existing alternatives in the model.

References

1. Ali, J., Naser Hussain, K., Alnoor, A., Muhsen, Y.R., Atiyah, A.G.: Benchmarking methodology of banks based on financial sustainability using CRITIC and RAFSI techniques. Decis. Mak. Appl. Manag. Eng. **7**(1), 315–341 (2024)
2. Badi, I., Bayane Bouraima, M., Qiu, Y., Stević, Ž.: Optimizing priority sequencing rules in parallel machine scheduling: an evaluation and selection approach using hybrid MCDM techniques. Decis. Mak. Adv. **2**(1), 22–31 (2024)
3. Badi, I., Stević, Ž, Bayane Bouraima, M. Badi, I.: Overcoming obstacles to renewable energy development in Libya: an MCDM Approach towards effective strategy formulation. Decis. Mak. Adv. **1**(1), 17–24 (2023)
4. Dağinstanli, H.A., Kurtay, K.G.: Facility location selection for ammunition depots based on GIS and Pythagorean fuzzy WASPAS J. Oper. Intell. **2**(1), 36–49 (2024)
5. Ghorabaee, M.K., Zavadskas, E.K., Turskis, Z., Antucheviciene, J.: A new combinative distance-based assessment (codas) method for multi-criteria decision-making. Econom. Comput. Econom. Cybernet. Stud. Res. **50**(3), 25–44 (2016)
6. Işik, C., Türkkan, M., Marbou, S., Gül, S.: Istanbul Stock Exchange with MCDM Methods Stock market performance evaluation of listed food and beverage companies in Istanbul stock exchange with MCDM methods. Decis. Mak.: Appl. Manag. Eng. **7**(2), 35–64 (2024)
7. Kahraman, C., Büyüközkan, G., Ateş, N.Y.: A two phase multi-attribute decision-making approach for new product introduction. Inform. Sci. **177**(7), 1567–1582 (2007)
8. Narang, M., Kumar, A., Dhawan, R.: A fuzzy extension of MEREC method using parabolic measure and its applications. J. Decis. Anal. Intell. Comput. **3**(1), 33–46 (2023)
9. Nezhad, M. Z., Nazarian-Jashnabadi , J., Rezazadeh, J., Mehraeen, M., Bagheri, R.: Assessing dimensions influencing iot implementation readiness in industries: a fuzzy DEMATEL and fuzzy AHP analysis. J. Decis. Anal. Intell. Comput. **3**(1), 33–46 (2023) **1**(1), 102–123 (2023)
10. Opricovic, S., Tzeng, G-H.: Compromise solution by mcdm methods: comparative analysis of vikor and topsis. Eur. J. Oper. Res. **156**(2), 445–455 (2004)
11. Pamučar, D., Ćirović, G.: The selection of transport and handling resources in logistics centers using multi-attributive border approximation area comparison (mabac). Expert Syst. Appl. **42**(6), 3016–3028 (2015)
12. Pamučar, D., Pejčić Tarle, S., Parezanović, T.: New hybrid multi-criteria decision-making dematelmairca model: sustainable selection of a location for the development of multimodal logistics centre. Econom. Res.-Ekonomska istraživanja **31**(1), 1641–1665 (2018)
13. Rakić, M., Žižovic, M. M., Miljković, B., Njegus, A., Žižović, M. R., Djordjević, I.: Multi-criteria selection of standards for system analyst activities in organizations. Facta Univ. Ser. Mech. Eng. **21**(3), 433–451 (2023)
14. Roy, B.: Classement et choix en présence de points de vue multiples. Revue française d'informatique et de recherche opérationnelle **2**(8), 57–75 (1968)
15. Saaty, L.T.: Analytic Hierarchy Process. McGraw-Hill, New York (1980)
16. Vahidinia, A., Hasani, A.: A comprehensive evaluation model for smart supply chain based on the hybrid multi-criteria decision-making method J. Soft Comput. Decis. Anal. **1**(1), 219–237 (2023)
17. Younis Al-Z., H.K., Konur, M.: Fuzzy analytic hierarchal process for sustainable public transport system J. Oper. Intell. **1**(1), 1–10 (2023)

18. Zavadskas, E.K., Govindan, K., Antucheviciene, J., Turskis, Z.: Hybrid multiple criteria decision-making methods: A review of applications for sustainability issues. Econom. Res.-Ekonomska istraživanja **29**(1), 857–887 (2016)
19. Zhou, B., Chen, J., Wu, Q., Pamučar, D., Wang, W., Zhou, L.: Risk priority evaluation of power transformer parts based on hybrid FMEA framework under hesitant fuzzy environment. Facta Univ. Ser. Mech. Eng. **20**(2) 399–420 (2022)
20. Žižović, M., Pamučar, D., Albijanić, M.: Prasenjit Chatterjee, Pribićević, I..: Eliminating rank reversal problem using a new multi-attribute model–the rafsi method. Mathematics **8**(6), 1015 (2020)
21. Žižović, M.M.: Methodology for evaluation of spa by visitors dominantly based on the quality of their rehabilitation. Quaestus **12**, 29–35 (2018)
22. Žižović, M.M, Damljanović, N., Žižović, M.R.: Multi-criteria decision making method for models with the dominant criterion. Filomat **31**(10), 2981–2989 (2017)
23. Žižović, M.M, Albijanić, M., Jovanović, V., Žižović, M.R.: A new method of multi-criteria analysis for evaluation and decision making by dominant criterion. Informatica **30**(4), 819–832 (2019)

Chapter 4
Enhanced Averaged Quadrature Rules with Application to Error Estimation

Hanan Almutairi, Dušan Lj. Djukić, Rada M. Mutavdžić Djukić, Miroslav S. Pranić, Lothar Reichel, and Miodrag M. Spalević

Dedicated to Academician Gradimir V. Milovanović on the occasion of his 75th birthday

4.1 Introduction

Orthogonal polynomials and quadrature play an important role in computational mathematics; see, e.g., Gautschi [1] for discussions of many applications. Milovanović has made numerous significant contributions to both the analysis and applications of orthogonal polynomials and quadrature; see, e.g., [2–8] for some recent publications.

Research of D.Lj. Djukić, R.M. Mutavdžić Djukić, and M.M. Spalević is supported in part by the Serbian Ministry of Science, Technological Development, and Innovation, according to Contract 451-03-47/2023-01/200105 dated on 3 February, 2023

H. Almutairi
Department of Mathematical Sciences, Kent State University, Kent, OH, USA
e-mail: halmutai@kent.edu

D. Lj. Djukić · R. M. Mutavdžić Djukić · M. M. Spalević
University of Belgrade, Faculty of Mechanical Engineering, Department of Mathematics, Belgrade, Serbia
e-mail: ddjukic@mas.bg.ac.rs; rmutavdzic@mas.bg.ac.rs; mspalevic@mas.bg.ac.rs

M. S. Pranić
University of Banja Luka, Faculty of Science, Department of Mathematics and Informatics, Banja Luka, Bosnia and Herzegovina
e-mail: pranic77m@yahoo.com

L. Reichel (✉)
Kent State University, Department of Mathematical Sciences, Kent, OH, USA
e-mail: reichel@math.kent.edu

M. P. Stanić et al. (eds.), *Analysis, Approximation, Optimization: Computation and Applications*, Springer Optimization and Its Applications 224,
https://doi.org/10.1007/978-3-031-85743-0_4

This chapter is concerned with error estimation for Gauss quadrature rules. Let $d\mu$ be a nonnegative measure with support on the real axis such that the integral

$$\mathcal{I}(f) := \int_a^b f(x)\,\mathrm{d}\mu(x), \quad -\infty \leq a < b \leq \infty, \tag{4.1}$$

is a *positive definite* linear functional on the space of algebraic polynomials $\mathbb{P}$, i.e., such that the bilinear form

$$(f, g) := \mathcal{I}(fg)$$

is an inner product on $\mathbb{P}$. Without loss of generality, we may assume that $\mathcal{I}(1) = \int d\mu = 1$. Let $p_0, p_1, p_2, \ldots$ denote the sequence of monic orthogonal polynomials with respect to the functional $\mathcal{I}$, i.e.,

$$\mathcal{I}(p_j p_k) = (p_j, p_k) \begin{cases} > 0, & j = k, \\ = 0, & j \neq k. \end{cases}$$

Thus, p_j is of degree j with leading coefficient one. The polynomials p_j satisfy a three-term recursion relation of the form

$$\begin{aligned} p_1(x) &= (x - a_0)p_0(x), \quad p_0(x) = 1, \\ p_{i+1}(x) &= (x - a_i)p_i(x) - b_i p_{i-1}(x), \quad i = 1, 2, \ldots, \end{aligned} \tag{4.2}$$

with $b_i > 0$ and $a_i \in \mathbb{R}$. The recursion relations for $p_0, p_1, \ldots, p_n$ can be expressed as

$$x \begin{bmatrix} p_0(x) \\ p_1(x) \\ \vdots \\ p_{n-1}(x) \end{bmatrix} = J_n \begin{bmatrix} p_0(x) \\ p_1(x) \\ \vdots \\ p_{n-1}(x) \end{bmatrix} + \begin{bmatrix} 0 \\ \vdots \\ 0 \\ p_n(x) \end{bmatrix},$$

where

$$J_n = \begin{bmatrix} a_0 & 1 & & & 0 \\ b_1 & a_1 & 1 & & \\ & & \ddots & & \\ & & \ddots & \ddots & 1 \\ 0 & & & b_{n-1} & a_{n-1} \end{bmatrix} \in \mathbb{R}^{n \times n}. \tag{4.3}$$

The matrix J_n can be symmetrized by a real diagonal similarity transformation. Denote the symmetrized tridiagonal matrix so obtained by T_n.

Let the function f be continuous on the convex hull of the support of the measure $d\mu$. The n-point Gauss quadrature rule for the approximation of the functional $\mathcal{I}$ is of the form

$$\mathcal{G}_n(f) := \sum_{i=1}^{n} f(x_i)w_i \tag{4.4}$$

and is characterized by the property that

$$\mathcal{I}(f) = \mathcal{G}_n(f), \quad \forall f \in \mathbb{P}_{2n-1}, \tag{4.5}$$

where $\mathbb{P}_{2n-1}$ denotes the set of all polynomials of degree at most $2n-1$; see Gautschi [1] or Szegő [9] for discussions on Gauss quadrature.

It is well known that the eigenvalues and the square of the first component of the normalized eigenvectors of T_n are the nodes and weights of the Gauss rule (4.4), respectively. They can be computed quite efficiently with the Golub–Welsch algorithm (see, e.g., [1, 10, 11]) or by a method described by Laurie [12].

The Gauss rule (4.4) can be expressed as

$$\mathcal{G}_n(f) = e_1^T f(T_n) e_1, \tag{4.6}$$

where $e_1 = [1, 0, \ldots, 0]^T \in \mathbb{R}^n$ denotes the first axis vector, and the superscript ^{T} stands for transposition. This representation allows the evaluation of the Gauss rule without computing the spectral factorization of T_n. This may be attractive for some functions f. More generally,

$$\mathcal{I}(f) = e_1^T f(\hat{T}_n) e_1, \quad \forall f \in \mathbb{P}_{2n-1}, \tag{4.7}$$

for any tridiagonal matrix

$$\hat{T}_n = \begin{bmatrix} \hat{a}_0 & \hat{c}_1 & & & 0 \\ \hat{b}_1 & \hat{a}_1 & \hat{c}_2 & & \\ & \ddots & \ddots & \ddots & \\ & & \hat{b}_{n-2} & \hat{a}_{n-2} & \hat{c}_{n-1} \\ 0 & & & \hat{b}_{n-1} & \hat{a}_{n-1} \end{bmatrix} \in \mathbb{R}^{n\times n}, \tag{4.8}$$

such that the polynomials

$$\begin{aligned} \hat{c}_1 \hat{p}_1(x) &= (x - \hat{a}_0)\hat{p}_0(x), \quad \hat{p}_0(x) = 1, \\ \hat{c}_{i+1}\hat{p}_{i+1}(x) &= (x - \hat{a}_i)\hat{p}_i(x) - \hat{b}_i \hat{p}_{i-1}(x), \quad i = 1, 2, \ldots, n-1, \end{aligned} \tag{4.9}$$

are orthogonal with respect to the functional $\mathcal{I}$. The matrices $\hat{T}_n$ and J_n are related in the following way:

$$a_i = \hat{a}_i, \quad b_i = \hat{b}_i \hat{c}_i. \tag{4.10}$$

An analogous discussion holds in the case when the functional $\mathcal{I}$ is not positive definite but *quasi-definite* on $\mathbb{P}_{n-1}$. The functional $\mathcal{I}$ is said to be quasi-definite if all $k \times k$ leading principal submatrices (for $k = 1, 2, \ldots, n$) of the Hankel matrix

$$H_n = \begin{bmatrix} \mu_0 & \mu_1 & \mu_2 & \mu_3 & \cdots & \mu_{n-1} \\ \mu_1 & \mu_2 & \mu_3 & \cdots & & \mu_n \\ \mu_2 & \mu_3 & & & & \vdots \\ \mu_3 & & & & & \\ \vdots & & & & \ddots & \\ \mu_{n-1} & \cdots & & & & \mu_{2n-2} \end{bmatrix}$$

determined by the moments

$$\mu_j := \mathcal{I}(x^j), \quad j = 0, 1, 2, \ldots ,$$

are nonsingular; see, e.g., [13]. The crucial difference with respect to the case of a positive definite functional $\mathcal{I}$ is that the coefficients b_i in the three-term recurrence relation (4.2) are not necessarily positive; they only are known to be nonvanishing. If $\mathcal{I}$ is real-valued functional, then the coefficients b_i and a_i are real for all i. It follows that the symmetrization of the matrix (4.8) may have complex entries, and the eigenvalues of the symmetrized matrix may be complex-valued and have multiplicity larger than 1. However, formula (4.7) remains valid, and we may consider it an n-point Gauss quadrature rule for $\mathcal{I}$. We remark that if the matrix $\hat{T}_n$ has multiple eigenvalues, then expression (4.7) is not of the form (4.4); see, for example, [14, p. 10]. Since the matrix $\hat{T}_n$ is unreduced, i.e., all subdiagonal entries are nonvanishing, this matrix has n distinct eigenvalues if and only if it is diagonalizable, in which case the Gauss rule is of the form (4.4). Non-real nodes and weights in (4.4) (if they exist) come in complex conjugate pairs. Therefore, the value of $\mathcal{G}_n(f)$ is a real number if the integrand f is real-valued.

This chapter is concerned with enhanced averaged rules that are associated with a quasi-definite functional $\mathcal{I}$. Consider the enhanced averaged rule that is the average of the $(n+k)$-point generalized anti-Gauss quadrature rule

$$\widetilde{\mathcal{G}}_{n+k}^{(k)}(f) := \sum_{i=1}^{n+k} f\left(\widetilde{x}_i^{(k)}\right) \widetilde{w}_i^{(k)} \tag{4.11}$$

and the Gauss rule (4.4). The rule (4.11) satisfies

$$\left(\mathcal{I} - \widetilde{\mathcal{G}}_{n+k}^{(k)}\right)(f) = -(\mathcal{I} - \mathcal{G}_n)(f), \quad \forall f \in \mathbb{P}_{2n+2k-1}, \tag{4.12}$$

for some $k \geq 2$; see Sect. 4.2. This rule was introduced in [15] for a positive definite functional $\mathcal{I}$ and generalizes the $(n+1)$-point anti-Gauss rule described by Laurie [16]. The latter is obtained when $k = 1$. This chapter is concerned with the more general case when the functional $\mathcal{I}$ is allowed to be quasi-definite. It follows from (4.5) that

$$\widetilde{\mathcal{G}}_{n+k}^{(k)}(f) = \mathcal{I}(f), \quad \forall f \in \mathbb{P}_{2n-1}. \tag{4.13}$$

Wc can express (4.12) as

$$\widetilde{\mathcal{G}}_{n+k}^{(k)}(f) = (2\mathcal{I} - \mathcal{G}_n)(f), \qquad \forall f \in \mathbb{P}_{2n+2k-1}.$$

This shows that when the quadrature rule $\widetilde{\mathcal{G}}_{n+k}^{(k)}$ exists, we may consider it an $(n+k)$-point Gauss quadrature rule associated with the functional $2\mathcal{I} - \mathcal{G}_n$. Existence of the quadrature rule $\widetilde{\mathcal{G}}_{n+k}^{(k)}$ is guaranteed if the functional $2\mathcal{I} - \mathcal{G}_n$ is quasi-definite on $\mathbb{P}_{n+k-1}$.

Let $\pi_0, \pi_1, \pi_2, \ldots, \pi_{n+k}$ be the unique family of monic orthogonal polynomials associated with the bilinear form

$$\langle f, g \rangle := (2\mathcal{I} - \mathcal{G}_n)(fg).$$

These polynomials satisfy the recursion formulas

$$x \begin{bmatrix} \pi_0(x) \\ \pi_1(x) \\ \vdots \\ \pi_{n+k-1}(x) \end{bmatrix} = \widetilde{J}_{n+k}^{(k)} \begin{bmatrix} \pi_0(x) \\ \pi_1(x) \\ \vdots \\ \pi_{n+k-1}(x) \end{bmatrix} + \begin{bmatrix} 0 \\ \vdots \\ 0 \\ \pi_{n+k}(x) \end{bmatrix},$$

where

$$\widetilde{J}_{n+k}^{(k)} = \begin{bmatrix} \alpha_0 & 1 & & & 0 \\ \beta_1 & \alpha_1 & 1 & & \\ & & \ddots & & \\ & & \ddots & \ddots & 1 \\ 0 & & & \beta_{n+k-1} & \alpha_{n+k-1} \end{bmatrix} \in \mathbb{R}^{(n+k)\times(n+k)} \tag{4.14}$$

for $n \geq 2$ and $k \geq 1$. Then, we have

$$\widetilde{\mathcal{G}}_{n+k}^{(k)}(f) = e_1^T f\left(\widetilde{J}_{n+k}^{(k)}\right) e_1.$$

There is an interesting relation between the matrices J_n and $\widetilde{J}_{n+k}^{(k)}$. Note that the functionals $\mathcal{I}$ and $2\mathcal{I} - \mathcal{G}_n$ coincide on $\mathbb{P}_{2n-1}$. This means that $a_i = \alpha_i$ and $b_i = \beta_i$ for $i = 0, 1, \ldots, n-1$. Only the last k rows of $\widetilde{J}_{n+k}^{(k)}$ have to be computed to determine the quadrature rule $\widetilde{\mathcal{G}}_{n+k}^{(k)}$ if we already know the matrix J_n.

We are in a position to introduce the enhanced averaged quadrature rule

$$\mathcal{A}_{n+k}^{(k)} := \frac{1}{2}\left(\mathcal{G}_n + \widetilde{\mathcal{G}}_{n+k}^{(k)}\right),$$

associated with quasi-definite functionals $\mathcal{I}$. These rules have previously been defined for positive definite functionals in [15]. They generalize the averaged rule $\mathcal{A}_{n+1}^{(1)}$ defined by Laurie [16]. We will show in Sect. 4.2 that the averaged rule $\mathcal{A}_{n+k}^{(k)}$ integrates a larger set of polynomials exactly than $\mathcal{G}_n$ and $\widetilde{\mathcal{G}}_{n+k}^{(k)}$. Therefore, the difference

$$\mathcal{A}_{n+k}^{(k)}(f) - \mathcal{G}_n(f) \tag{4.15}$$

may be a useful estimate of the quadrature error

$$\mathcal{I}(f) - \mathcal{G}_n(f). \tag{4.16}$$

We illustrate in Sect. 4.4 that this is indeed the case.

Having a fairly accurate estimate of the quadrature error (4.16) is important in many applications in which Gauss quadrature rules are used, because this allows us to determine a suitable number of nodes, n, in the Gauss rule (4.4). We note that a too small value of n yields an approximation of $\mathcal{I}(f)$ of insufficient accuracy, while a too large value of n requires that unnecessarily many evaluations of the integrand f are carried out. The latter may increase the computational cost significantly when it is expensive to evaluate f.

The estimation of the quadrature error in an n-point Gauss quadrature rule has received considerable attention in the literature. The classical approach to estimate this error is to evaluate a $(2n+1)$-node Gauss–Kronrod rule associated with the Gauss rule; see Notaris [17] for an insightful discussion on this approach. However, Gauss–Kronrod rules are not guaranteed to exist for all measures $d\mu$ and numbers of nodes n, and they are somewhat complicated to compute when they exist. Therefore, the application of the optimal averaged rule S_{2n+1} associated with the Gauss rule (4.4) has recently attracted some attention. These rules were proposed by Spalević [18] for positive definite functionals $\mathcal{I}$. The rule S_{2n+1} has $2n+1$ real nodes and exists for all positive definite functionals. Moreover, it is easy to evaluate; see [19].

The application of these rules for more general measures is discussed in [20]. The difference

$$\mathcal{S}_{2n+1}(f) - \mathcal{G}_n(f) \tag{4.17}$$

furnishes in many situations a quite accurate estimate of the quadrature error (4.16); see [21] for computed examples. However, when the integrand f is not smooth, the quality of the quadrature error estimate (4.17) may be poor. We therefore are interested in investigating the performance of the error estimate (4.15).

We also derive error estimates that are analogous to (4.15) for approximations of matrix functionals of the form

$$\mathcal{L}(f) = \mathcal{L}^{(A,u,v)}(f) = u^T f(A)v, \tag{4.18}$$

where $A \in \mathbb{R}^{N\times N}$ is a large possibly nonsymmetric matrix, f is a function such that $f(A) \in \mathbb{R}^{N\times N}$ is well defined, and $u, v \in \mathbb{R}^N$ such that $u^T v = 1$. Obviously, $\mathcal{L}$ is a real-valued linear functional.

This chapter is organized as follows. Section 4.2 discusses properties of generalized anti-Gauss rules and describes the entries of the matrix (4.14). Section 4.3 reviews how the non-symmetric Lanczos process determines a nonsymmetric tridiagonal matrix that is analogous to the matrix (4.3). We derive enhanced averaged quadrature rules for the approximation of (4.18). Section 4.4 presents a few computed examples, and Sect. 4.5 contains concluding remarks.

4.2 Enhanced Anti-Gauss Rules $\widetilde{\mathcal{G}}^{(k)}_{n+k}$ for $k < 4$

This section reviews results from [15, 16], where it is shown how to compute the coefficients α_n, α_{n+1}, α_{n+2}, β_n, β_{n+1}, and β_{n+2} in (4.14) when the matrix J_{n+3} is known. These formulas are in [15] derived under the assumption that the functional $\mathcal{I}$ is positive definite. Here we show that they are valid in the more general case when $\mathcal{I}$ is quasi-definite.

Laurie showed in [16] that

$$\alpha_n = a_n, \quad \beta_n = 2b_n,$$

which indicates that the quadrature rule $\widetilde{\mathcal{G}}^{(1)}_{n+1}$ exists whenever $\mathcal{I}$ is quasi-definite. If $\mathcal{I}$ is positive definite, then the nodes of $\widetilde{\mathcal{G}}^{(1)}_{n+1}$ are real and simple.

We turn to the rule $\widetilde{\mathcal{G}}^{(2)}_{n+2}$. This rule is defined by the matrix $\widetilde{J}^{(2)}_{n+2}$. The expressions for the entries

$$\alpha_{n+1} = \frac{a_{n+1}b_{n+1} - a_{n-1}b_n}{\beta_{n+1}},$$
$$\beta_{n+1} = b_{n+1} - b_n,$$

of this matrix are shown in [15] for the situation when $\mathcal{I}$ is positive definite. These expressions also hold when $\mathcal{I}$ is quasi-definite. Note that when β_{n+1} vanishes, the coefficient α_{n+1} is not defined. It is not straightforward to define the generalized anti-Gauss rule in this situation. When $\beta_{n+1} > 0$ and $\mathcal{I}$ is positive definite, the rule $\widetilde{\mathcal{G}}^{(1)}_{n+1}$ exists and has real and simple nodes.

The formulas for the entries α_{n+2} and β_{n+2} of $\widetilde{J}^{(3)}_{n+3}$ are more complicated. They are in [15, Theorem 2.1] shown to be

$$\beta_{n+2} = \frac{\beta_{n+1}(b_{n+2}b_{n+1} - b_n b_{n-1}) - b_{n+1}b_n(a_{n+1} - a_{n-1})^2}{\beta_{n+1}^2}$$

and

$$\alpha_{n+2} = \frac{b_{n+2}b_{n+1}(a_{n+2} + 2a_{n+1} - 2\alpha_{n+1}) + b_n b_{n-1}(2\alpha_{n+1} - 2a_{n-1} - a_{n-2})}{\beta_{n+2}\beta_{n+1}} + \frac{a_{n+1}b_{n+1}(a_{n+1} - \alpha_{n+1})^2 - a_{n-1}b_n(a_{n-1} - \alpha_{n+1})^2}{\beta_{n+2}\beta_{n+1}},$$

when $\mathcal{I}$ is positive definite. These expressions for β_{n+2} and α_{n+2} also are valid when $\mathcal{I}$ is quasi-definite. It is possible, but tedious, to compute the entries of the matrices $\widetilde{J}^{(k)}_{n+k}$ for $k \geq 4$.

It is shown in [15, Section 3] that the enhanced averaged rule $\mathcal{A}^{(k)}_{n+k}$ satisfies

$$\mathcal{A}^{(k)}_{n+k}(f) = \mathcal{I}(f), \quad \forall f \in \mathbb{P}_{2n+2k-1},$$

when $\mathcal{I}$ is positive definite, and this formula remains valid when $\mathcal{I}$ is quasi-definite. It follows that the enhanced averaged rule $\mathcal{A}^{(k)}_{n+k}$ is exact for polynomials of higher degree than the Gauss and generalized anti-Gauss rules that determine it; cf. (4.5) and (4.13).

4.3 Enhanced Averaged Rules for Measures with Support in the Complex Plane

We consider the approximation of the functional (4.18) with $A \in \mathbb{R}^{N\times N}$ and the vectors $u, v \in \mathbb{R}^N$ such that $u^T v = 1$. We assume that N is large so that the evaluation of (4.18) by computing $f(A)$ is too expensive to be attractive.

The functional (4.18) can be associated with a measure with support in the complex plane. Assume for the moment that the matrix A has the spectral factorization

$$A = S\Lambda S^{-1},$$

where the columns of $S \in \mathbb{C}^{N\times N}$ are linearly independent eigenvectors of A, and the diagonal entries of

$$\Lambda = \text{diag}[\lambda_1, \lambda_2, \ldots, \lambda_N]$$

are eigenvalues. Since the matrix A is real, the eigenvalues are real or appear in complex conjugate pairs. Substituting the spectral factorization into (4.18) gives

$$\mathcal{L}(f) = \mathcal{L}^{(A,u,v)}(f) = u^T S f(\Lambda) S^{-1} v = \sum_{j=1}^{N} f(\lambda_j)\mu_j\mu'_j,$$

where $[\mu_1, \mu_2, \ldots, \mu_N] = u^T S$ and $[\mu'_1, \mu'_2, \ldots, \mu'_N]^T = S^{-1}v$. The right-hand side can be expressed as an integral

$$\mathcal{L}(f) = \mathcal{L}^{(A,u,v)}(f) = \int f(t)\,\mathrm{d}\mu_{A,u,v}(t),$$

with the measure

$$d\mu_{A,u,v}(t) = \sum_{j=1}^{N} \delta(t-\lambda_j)\mu_j\mu'_j,$$

where $\delta(\cdot)$ denotes the Dirac delta-function. We can determine an approximation of (4.18) by carrying out ℓ steps of the nonsymmetric Lanczos method applied to A with initial vectors u and v. This method is described by Algorithm 1.

We assume that the linear functional $\mathcal{L}$ is quasi-definite on the set of polynomials $\mathbb{P}_{n+k}$ with k small enough so that all quantities of interest can be computed. This is equivalent to the fact that it is possible to carry out $n+k$ steps of Algorithm 1 without breakdown, i.e., without $\beta_j = 0$ for some $1 \le j \le n+k$; see [14, Section 2]. Properties and implementation details of Algorithm 1 are discussed, e.g., by Saad [22] and Ye [23].

Algorithm 1 The nonsymmetric Lanczos process

Require: $A \in \mathbb{R}^{N\times N}$; $u, v \in \mathbb{R}^N$ such that $u^T v = 1$; number of steps ℓ.

$u_0 := v_0 := 0 \in \mathbb{R}^N$; $u_1 := u$; $v_1 := v$; $\beta_0 := 0$; $\gamma_0 := 0$;
for $j = 1, 2, \ldots, \ell$ **do**
  $\alpha_{j-1} := u_j^T A v_j$;
  $r := Av_j - \alpha_{j-1}v_j - \gamma_{j-1}v_{j-1}$;
  $s := A^T u_j - \alpha_{j-1}u_j - \beta_{j-1}u_{j-1}$;
  $\beta_j := |r^T s|^{1/2}$; $\gamma_j := r^T s/\beta_j$;
  $v_{j+1} := r/\beta_j$; $u_{j+1} := s/\gamma_j$;
end for

Ensure: Tridiagonal matrix $T_\ell \in \mathbb{R}^{\ell\times\ell}$ with diagonal entries $\{\alpha_j\}_{j=0}^{\ell-1}$, subdiagonal entries $\{\beta_j\}_{j=1}^{\ell-1}$, and superdiagonal entries $\{\gamma_j\}_{j=1}^{\ell-1}$ (see (4.19) below) and vectors $u_1, u_2, \ldots, u_{\ell+2}$ and $v_1, v_2, \ldots, v_{\ell+2}$ in $\mathbb{R}^N$.

Introduce the matrices for $\ell = n + k$ for a suitable k,

$$U_\ell = [u_1, \ldots, u_\ell] \in \mathbb{R}^{N\times\ell}, \quad V_\ell = [v_1, \ldots, v_\ell] \in \mathbb{R}^{N\times\ell},$$

where the vectors u_j and v_k are determined by Algorithm 1. The vectors u_j and v_k are biorthonormal, i.e., $U_\ell^T V_\ell = I$. Define the tridiagonal matrix

$$T_\ell = \begin{bmatrix} \alpha_0 & \gamma_1 & & & 0 \\ \beta_1 & \alpha_1 & \gamma_2 & & \\ & \ddots & \ddots & \ddots & \\ & & \beta_{\ell-2} & \alpha_{\ell-2} & \gamma_{\ell-1} \\ 0 & & & \beta_{\ell-1} & \alpha_{\ell-1} \end{bmatrix} \in \mathbb{R}^{\ell\times\ell}. \tag{4.19}$$

Its sub-diagonal and super-diagonal entries are nonvanishing.

The recursions of Algorithm 1 can be written as

$$AV_\ell = V_\ell T_\ell + \beta_\ell v_{\ell+1} e_\ell^T,$$
$$A^T U_\ell = U_\ell T_\ell^T + \gamma_\ell u_{\ell+1} e_\ell^T.$$

Moreover, the recursion formulas show that

$$u_{j+1} = q_j(A^T)u, \quad v_{j+1} = p_j(A)v, \quad 0 \le j \le \ell, \tag{4.20}$$

for some polynomials p_j and q_j of degree j. It follows that

$$\mathcal{L}(q_j p_k) = u^T q_j(A) p_k(A) v = u_{j+1}^T v_{k+1} = \begin{cases} 1, & j = k, \\ 0, & j \neq k. \end{cases}$$

This shows that the Lanczos polynomial p_k is orthogonal to any polynomial from $\text{span}\{q_0, q_1, \dots, q_{k-1}\} = \mathbb{P}_{k-1}$ with respect to the functional $\mathcal{L}$. In other words, $p_0, p_1, \dots, p_\ell$ is a family of orthogonal polynomials with respect to $\mathcal{L}$. It follows from (4.20) that the polynomials p_j satisfy the same recursion relations as the vectors v_j, i.e.,

$$\beta_j p_j(x) = (x - \alpha_{j-1}) p_{j-1}(x) - \gamma_{j-1} p_{j-2}(x), \quad j = 1, 2, \dots, \ell + 1,$$

with $p_{-1}(x) := 0$, $\gamma_0 := 0$, and $\beta_0 := 0$. We conclude that

$$\mathcal{L}(f) = e_1^T f(T_\ell) e_1, \quad \forall f \in \mathbb{P}_{2\ell-1}, \tag{4.21}$$

i.e., the right-hand side of (4.21) is an ℓ-point Gauss quadrature rule for approximating $\mathcal{L}$. This quadrature rule is of the form (4.4) if and only if T_ℓ is diagonalizable; see [14] for details.

We can construct the generalized anti-Gauss quadrature rules $\widetilde{\mathcal{G}}_{n+k}^{(k)}$ for $1 \le k \le 3$ in the following way. Carry out $\ell = n + k$ steps of Algorithm 1 to generate the matrix T_{n+k}. Using (4.10), we compute the matrix J_{n+k}, and then by formulas from Sect. 4.2, we update the last k rows of J_{n+k} to construct the matrix $\widetilde{J}_{n+k}^{(k)}$. Thus, we get

$$\widetilde{\mathcal{G}}_{n+k}^{(k)}(f) = e_1^T f\left(\widetilde{J}_{n+k}^{(k)}\right) e_1.$$

4.4 Numerical Examples

This section presents a few computed examples that illustrate the quality of the approximations and error estimates discussed in this chapter. All computations were carried out in double precision arithmetic (i.e., with about 15 significant decimal digits) using MATLAB R2021a on a 64-bit personal computer.

We compute the exact value (up to the influence of round-off errors) of the functional $\mathcal{L}(f) = u^T f(A) v$ for several different matrices A, vectors u and v, and functions f. We will assume that the functions f can be evaluated not only at A, but also at the reduced matrices T_n and $\widetilde{J}_{n+k}^{(k)}$. We compare $u^T f(T_n) v$ to approximations determined by the standard and generalized anti-Gauss rules and the standard and enhanced averaged rules. Estimates of the approximation errors by the techniques discussed in Sects. 4.2 and 4.3 are tabulated.

Example 4.1 This example shows an application of the techniques of this paper to network analysis. Let $A = [\alpha_{i,j}] \in \mathbb{R}^{500 \times 500}$ be the adjacency matrix for the network Air500, which describes flight connections between the top 500 airports within 1 year from July 1, 2007, to June 30, 2008; see [24, 25]. The airports are modeled by vertices v_i, and the flights are modeled by edges in the directed graph determined by the network. The matrix A has the entry $\alpha_{i,j} = 1$ if there is a flight from airport i to airport j; otherwise, $\alpha_{i,j} = 0$. Generally, but not always, $\alpha_{i,j} = 1$ implies that $\alpha_{j,i} = 1$. This makes A nonsymmetric. The number of edges is much smaller than 500^2. Therefore, the adjacency matrix A is sparse. A walk of length k in a graph is a sequence of vertices $v_{i_1}, v_{i_2}, \ldots, v_{i_{k+1}}$ such that there is an edge from vertex v_{i_j} to vertex $v_{i_{j+1}}$ for $j = 1, 2, \ldots, k$. Vertices in a walk may be repeated. The entry $\alpha_{i,j}^{(\ell)}$ of the matrix $A^{\ell} = [\alpha_{i,j}^{(\ell)}]$ is equal to the number of walks of length ℓ starting at vertex i and ending at vertex j. Short walks are generally considered more important than long walks. This motivates the use of the exponential function in network analysis. The subgraph centrality of node v_j is defined as $e_j^T \exp(A) e_j$; see [26]. We are interested in computing the *total communicability* of the graph, which is defined as $v^T \exp(A) v$, where $v = [1, 1, \ldots, 1]^T / \sqrt{500}$; see, e.g., [27].

Table 4.1 displays the relative error of computed approximations of the total communicability of the network. The approximations of the total communicability determined by the enhanced averaged rules $\mathcal{A}_{n,k}^{(k)}(f)$, $k \in \{1, 2, 3\}$, can be seen to be more accurate than the approximations determined by the standard Gauss rule and related Gauss-type formulas. This suggests that one may use the enhanced averaged quadrature rules to estimate the error in the Gauss rule. Table 4.2 reports the relative difference (4.15) for several enhanced averaged rules. All of the differences provide a quite accurate estimate of the quadrature error (4.16) in the Gauss rules. The error estimates obtained with the rules $\mathcal{A}_{n+3}^{(3)}(f)$ are very accurate.

Table 4.1 Example 4.1: Magnitude of the relative errors of computed approximations of $\mathcal{L}(f) = v^T \exp(A) v$, where $A \in \mathbb{R}^{500 \times 500}$ and $v = [1, 1, \ldots, 1]^T / \sqrt{500}$

	$n = 6$	$n = 8$
$\mathcal{A}_{n,1}^{(1)}(f)$	$4.61 \cdot 10^{-6}$	$1.16 \cdot 10^{-9}$
$\mathcal{A}_{n,2}^{(2)}(f)$	$1.11 \cdot 10^{-6}$	$2.38 \cdot 10^{-9}$
$\mathcal{A}_{n,3}^{(3)}(f)$	$1.59 \cdot 10^{-7}$	$8.28 \cdot 10^{-12}$
$\widetilde{\mathcal{G}}_{n,1}^{(1)}(f)$	$4.66 \cdot 10^{-4}$	$2.46 \cdot 10^{-7}$
$\widetilde{\mathcal{G}}_{n,2}^{(2)}(f)$	$4.59 \cdot 10^{-4}$	$2.39 \cdot 10^{-7}$
$\widetilde{\mathcal{G}}_{n,3}^{(3)}(f)$	$4.57 \cdot 10^{-4}$	$2.43 \cdot 10^{-7}$
$\mathcal{G}_n(f)$	$4.57 \cdot 10^{-4}$	$2.44 \cdot 10^{-7}$

Table 4.2 Example 1: Magnitude of the relative differences of enhanced averaged and Gauss rules, where $\mathcal{L}$, A, v, and f are the same as in Table 4.1

		Error	
N		$n = 6$	$n = 8$
500	$\frac{\mathcal{A}_{n+1}^{(1)}(f) - \mathcal{G}_n(f)}{\mathcal{L}(f)}$	$4.61 \cdot 10^{-4}$	$2.45 \cdot 10^{-7}$
	$\frac{\mathcal{A}_{n+2}^{(2)}(f) - \mathcal{G}_n(f)}{\mathcal{L}(f)}$	$4.58 \cdot 10^{-4}$	$2.41 \cdot 10^{-7}$
	$\frac{\mathcal{A}_{n+3}^{(3)}(f) - \mathcal{G}_n(f)}{\mathcal{L}(f)}$	$4.57 \cdot 10^{-4}$	$2.44 \cdot 10^{-7}$
	$\frac{\mathcal{L}(f) - \mathcal{G}_n(f)}{\mathcal{L}(f)}$	$4.57 \cdot 10^{-4}$	$2.44 \cdot 10^{-7}$

Table 4.3 Example 2: Magnitude of the relative errors of computed approximations of $\mathcal{L} = v^T f(A)v$ for $A \in \mathbb{R}^{N \times N}$ a nonsymmetric Toeplitz matrix, $f(t) = \exp(t)$, and $v = [1, 1, \ldots, 1]^T/\sqrt{N}$

N		$n = 4$	$n = 6$
200	$\mathcal{A}_{n,1}^{(1)}(f)$	$1.16 \cdot 10^{-7}$	$2.72 \cdot 10^{-11}$
	$\mathcal{A}_{n,2}^{(2)}(f)$	$1.12 \cdot 10^{-9}$	$4.78 \cdot 10^{-14}$
	$\mathcal{A}_{n,3}^{(3)}(f)$	$1.70 \cdot 10^{-10}$	0
	$\widetilde{\mathcal{G}}_{n,1}^{(1)}(f)$	$1.81 \cdot 10^{-5}$	$2.65 \cdot 10^{-9}$
	$\widetilde{\mathcal{G}}_{n,2}^{(2)}(f)$	$1.79 \cdot 10^{-5}$	$2.60 \cdot 10^{-9}$
	$\widetilde{\mathcal{G}}_{n,3}^{(3)}(f)$	$1.79 \cdot 10^{-5}$	$2.60 \cdot 10^{-9}$
	$\mathcal{G}_n(f)$	$1.79 \cdot 10^{-5}$	$2.60 \cdot 10^{-9}$

Table 4.4 Example 2: Magnitude of the relative differences of enhanced averaged and Gauss rules, where $\mathcal{L}$, A, v, and f are the same as in Table 4.3

		Error	
N		$n = 4$	$n = 6$
200	$\frac{\mathcal{A}_{n+1}^{(1)}(f)-\mathcal{G}_n(f)}{\mathcal{L}(f)}$	$1.80 \cdot 10^{-5}$	$2.62 \cdot 10^{-9}$
	$\frac{\mathcal{A}_{n+2}^{(2)}(f)-\mathcal{G}_n(f)}{\mathcal{L}(f)}$	$1.79 \cdot 10^{-5}$	$2.60 \cdot 10^{-9}$
	$\frac{\mathcal{A}_{n+3}^{(3)}(f)-\mathcal{G}_n(f)}{\mathcal{L}(f)}$	$1.79 \cdot 10^{-5}$	$2.60 \cdot 10^{-9}$
	$\frac{\mathcal{L}(f)-\mathcal{G}_n(f)}{\mathcal{L}(f)}$	$1.79 \cdot 10^{-5}$	$2.60 \cdot 10^{-9}$

Example 4.2 Let $N = 200$ and let $A \in \mathbb{R}^{N \times N}$ be the nonsymmetric Toeplitz matrix with first row $[1, 1/2, \ldots, 1/N]$ and first column $[1, 1/2^2, \ldots, 1/N^2]$. We seek to approximate $\mathcal{L}(f) = v^T f(A)v$ by Gauss rules, where $v = [1, \ldots, 1]^T/\sqrt{N} \in \mathbb{R}^N$ and $f(t) = \exp(t)$. Table 4.3 shows the magnitude of the relative errors in the computed approximations. We note that the smallest errors are achieved by the approximations furnished by the enhanced averaged rules. Table 4.4 reports the magnitude of the relative differences determined by (4.15) and (4.16). The table shows all the enhanced averaged rules to provide accurate estimates of the error in the Gauss quadrature rules.

The following examples are concerned with functionals $\mathcal{L} = u^T f(A)v$ with a symmetric matrix A. Generalized anti-Gauss rules and enhanced averaged rules for these kinds of functionals with $u = v$ have previously been introduced in [15]. However, the application of enhanced averaged rules for the estimation of the quadrature error in Gauss rules has not been illustrated before.

Example 4.3 Let $A \in \mathbb{R}^{300 \times 300}$ be the symmetric tridiagonal Toeplitz matrix with first row $[0, 1, 0, \ldots, 0]$. The vector $v \in \mathbb{R}^{300}$ has normally distributed random entries with zeros mean and is scaled to have unit Euclidean norm. We let $f(t) = \frac{\exp(t)}{t^2+1}$. This example allows the application of the quadrature rules of Sect. 4.2; the other examples of this section require the quadratures of Sect. 4.3 based on the nonsymmetric Lanczos process. Table 4.5 depicts the magnitude of the relative errors in the computed approximations. We can observe that the smallest errors are achieved with the enhanced averaged rules. Table 4.6 shows the error

Table 4.5 Example 3: Magnitude of the relative errors of computed approximations of $v^T f(A)v$ for $A \in \mathbb{R}^{N\times N}$ a symmetric banded Toeplitz matrix, $f(t) = \frac{\exp(t)}{(t^2+1)}$, and v with normally distributed random entries

N		$n = 4$	$n = 8$
300	$\mathcal{A}_{n,1}^{(1)}(f)$	$1.10 \cdot 10^{-3}$	$3.83 \cdot 10^{-5}$
	$\mathcal{A}_{n,2}^{(2)}(f)$	$7.26 \cdot 10^{-4}$	$6.88 \cdot 10^{-6}$
	$\mathcal{A}_{n,3}^{(3)}(f)$	$2.15 \cdot 10^{-4}$	$4.00 \cdot 10^{-6}$
	$\widetilde{\mathcal{G}}_{n,1}^{(1)}(f)$	$1.26 \cdot 10^{-2}$	$2.71 \cdot 10^{-4}$
	$\widetilde{\mathcal{G}}_{n,2}^{(2)}(f)$	$1.33 \cdot 10^{-2}$	$2.08 \cdot 10^{-4}$
	$\widetilde{\mathcal{G}}_{n,3}^{(3)}(f)$	$1.52 \cdot 10^{-2}$	$2.02 \cdot 10^{-4}$
	$\mathcal{G}_n(f)$	$1.47 \cdot 10^{-2}$	$1.94 \cdot 10^{-4}$

Table 4.6 Example 3: Magnitude of the relative differences of enhanced averaged and Gauss rules, where $\mathcal{L}$, A, v, and f are the same as in Table 4.5

		Error	
N		$n = 4$	$n = 8$
300	$\frac{\mathcal{A}_{n+1}^{(1)}(f)-\mathcal{G}_n(f)}{\mathcal{L}(f)}$	$1.37 \cdot 10^{-2}$	$2.33 \cdot 10^{-4}$
	$\frac{\mathcal{A}_{n+2}^{(2)}(f)-\mathcal{G}_n(f)}{\mathcal{L}(f)}$	$1.40 \cdot 10^{-2}$	$2.01 \cdot 10^{-4}$
	$\frac{\mathcal{A}_{n+3}^{(3)}(f)-\mathcal{G}_n(f)}{\mathcal{L}(f)}$	$1.49 \cdot 10^{-2}$	$1.98 \cdot 10^{-4}$
	$\frac{\mathcal{L}(f)-\mathcal{G}_n(f)}{\mathcal{L}(f)}$	$1.47 \cdot 10^{-2}$	$1.94 \cdot 10^{-4}$

Table 4.7 Example 4: Magnitude of the relative errors of computed approximations of $u^T f(A)v$ for $A \in \mathbb{R}^{N\times N}$ a symmetric banded Toeplitz matrix, $f(t) = \exp(t)$, and u different from v

N		$n = 4$	$n = 6$
300	$\mathcal{A}_{n,1}^{(1)}(f)$	$5.22 \cdot 10^{-8}$	$1.49 \cdot 10^{-12}$
	$\mathcal{A}_{n,2}^{(2)}(f)$	$4.78 \cdot 10^{-11}$	$5.31 \cdot 10^{-15}$
	$\mathcal{A}_{n,3}^{(3)}(f)$	$2.27 \cdot 10^{-13}$	$5.31 \cdot 10^{-16}$
	$\widetilde{\mathcal{G}}_{n,1}^{(1)}(f)$	$4.70 \cdot 10^{-6}$	$2.51 \cdot 10^{-10}$
	$\widetilde{\mathcal{G}}_{n,2}^{(2)}(f)$	$4.81 \cdot 10^{-6}$	$2.54 \cdot 10^{-10}$
	$\widetilde{\mathcal{G}}_{n,3}^{(3)}(f)$	$4.81 \cdot 10^{-6}$	$2.54 \cdot 10^{-10}$
	$\mathcal{G}_n(f)$	$4.81 \cdot 10^{-6}$	$2.54 \cdot 10^{-10}$

estimates furnished by the averaged rules $\mathcal{A}_{n+k}^{(k)}(f)$ for $k \in \{2, 3\}$ to give very accurate estimates of the errors in $\mathcal{G}_n f$. We remark that the errors in Table 4.6 are close to the relative errors, because $\mathcal{L}(f) = v^T f(A)v = 0.8063$.

Example 4.4 We approximate $\mathcal{L}(f) := u^T f(A)v$, where $f(t) = \exp(t)$, $u = e_2$, $v = e_2 + \frac{e_3}{\sqrt{2}}$, and the matrix $A \in \mathbb{R}^{300\times 300}$ is the same as in Example 4.3. Table 4.7 shows the magnitude of the relative errors in the computed approximations. The table shows the smallest errors to be achieved by the enhanced averaged rules $\mathcal{A}_{n,k}^{(k)}(f)$; the errors are particularly small for $k \in \{2, 3\}$. Table 4.8 shows the differences $\mathcal{A}_{n+k}^{(k)}(f) - \mathcal{G}_n(f)$, $k \in \{1, 2, 3\}$, to be accurate estimates of the quadrature error in $\mathcal{G}_n(f)$.

Table 4.8 Example 4: Magnitude of the relative differences of enhanced averaged and Gauss rules, where $\mathcal{L}$, A, v, and f are the same as in Table 4.7

N		Error $n=4$	$n=6$
300	$\frac{\mathcal{A}^{(1)}_{n+1}(f)-\mathcal{G}_n(f)}{\mathcal{L}(f)}$	$4.75 \cdot 10^{-6}$	$2.53 \cdot 10^{-10}$
	$\frac{\mathcal{A}^{(2)}_{n+2}(f)-\mathcal{G}_n(f)}{\mathcal{L}(f)}$	$4.81 \cdot 10^{-6}$	$2.54 \cdot 10^{-10}$
	$\frac{\mathcal{A}^{(3)}_{n+3}(f)-\mathcal{G}_n(f)}{\mathcal{L}(f)}$	$4.81 \cdot 10^{-6}$	$2.54 \cdot 10^{-10}$
	$\frac{\mathcal{L}(f)-\mathcal{G}_n(f)}{\mathcal{L}(f)}$	$4.81 \cdot 10^{-6}$	$2.54 \cdot 10^{-10}$

Table 4.9 Example 5: Magnitude of the relative errors of computed approximations of $u^T\left(A^2+I\right)^{-1}u$ for $A \in \mathbb{R}^{100\times 100}$ a random symmetric matrix, and u is a random vector

N		$n=3$	$n=5$
100	$\mathcal{A}^{(1)}_{n,1}(f)$	$3.63 \cdot 10^{-4}$	$4.33 \cdot 10^{-6}$
	$\mathcal{A}^{(2)}_{n,2}(f)$	$5.01 \cdot 10^{-5}$	$1.34 \cdot 10^{-6}$
	$\mathcal{A}^{(3)}_{n,3}(f)$	$3.33 \cdot 10^{-7}$	$1.28 \cdot 10^{-8}$
	$\widetilde{\mathcal{G}}^{(1)}_{n,1}(f)$	$3.50 \cdot 10^{-3}$	$7.88 \cdot 10^{-5}$
	$\widetilde{\mathcal{G}}^{(2)}_{n,2}(f)$	$2.90 \cdot 10^{-3}$	$6.74 \cdot 10^{-5}$
	$\widetilde{\mathcal{G}}^{(3)}_{n,3}(f)$	$2.80 \cdot 10^{-3}$	$7.02 \cdot 10^{-5}$
	$\mathcal{G}_n(f)$	$2.80 \cdot 10^{-3}$	$7.01 \cdot 10^{-5}$

Table 4.10 Example 5: Magnitude of the relative differences of enhanced averaged and Gauss rules, where $\mathcal{L}$, A, v, and f are the same as in Table 4.9

N		Error $n=3$	$n=5$
100	$\frac{\mathcal{A}^{(1)}_{n+1}(f)-\mathcal{G}_n(f)}{\mathcal{L}(f)}$	$3.20 \cdot 10^{-3}$	$7.45 \cdot 10^{-5}$
	$\frac{\mathcal{A}^{(2)}_{n+2}(f)-\mathcal{G}_n(f)}{\mathcal{L}(f)}$	$2.80 \cdot 10^{-3}$	$6.88 \cdot 10^{-5}$
	$\frac{\mathcal{A}^{(3)}_{n+3}(f)-\mathcal{G}_n(f)}{\mathcal{L}(f)}$	$2.80 \cdot 10^{-3}$	$7.01 \cdot 10^{-5}$
	$\frac{\mathcal{L}(f)-\mathcal{G}_n(f)}{\mathcal{L}(f)}$	$2.80 \cdot 10^{-3}$	$7.01 \cdot 10^{-5}$

Example 4.5 We approximate $\mathcal{L}(f) := u^T\left(A^2+I\right)^{-1}u$, where u is a random vector of unit norm with normally distributed entries with zero mean, and the symmetric matrix $A \in \mathbb{R}^{100\times 100}$ has randomly generated uniformly distributed real eigenvalues in the interval $[-1, 1]$ and a random orthogonal eigenvector matrix, and $f(t) = \frac{1}{t^2+1}$. The exact value of $\mathcal{L}(f)$ is 0.8934. Table 4.9 shows the magnitude of the relative errors in the computed approximations. The table shows the smallest errors to be achieved by the the enhanced averaged rules $\mathcal{A}^{(k)}_{n,k}(f)$; the errors are particularly small for $k = 3$. Table 4.10 shows the differences $\mathcal{A}^{(k)}_{n+k}(f) - \mathcal{G}_n(f)$, $k \in \{1, 2, 3\}$, to be accurate estimates of the quadrature error in $\mathcal{G}_n(f)$.

4.5 Conclusion

This chapter proposes to use enhanced averaged rules to estimate the error in Gauss quadrature rules. The enhanced averaged rules are modifications of averaged rules introduced by Laurie. They are defined with the aid of generalized anti-Gauss rules. The latter rules, as well as the enhanced averaged rules, are determined for quasi-definite functionals $\mathcal{I}$. In particular, we allow functionals that are associated with measures with support in the complex plane. Computed examples illustrate the performance of the enhanced averaged rules.

References

1. Gautschi, W.: Orthogonal Polynomials: Approximation and Computation. Oxford University Press, Oxford (2004)
2. Gautschi, W., Milovanović, G.V.: A Ramanujan integral and its derivatives: computation and analysis. Math. Comp. **93**, 1297–1300 (2024)
3. Golubović, Z.Lj., Milovanović, G.V.: Quadrature processes for efficient calculation of the Clausen functions. BIT Numer. Math. **63** (2023). Art. 10
4. Milovanović, G.V.: Orthogonal polynomials concerning to the Abel and Lindelöf weights and their modifications on the real line. Montes Taurus J. Pure Appl. Math. **4**, 166–174 (2022)
5. Milovanović, G.V.: Orthogonality on the semicircle: old and new results. Electron. Trans. Numer. Anal. **59**, 99–115 (2023)
6. Milovanović, G.V., Spalević, M.M., Pranić, M.S.: Quadrature Formulae with Multiple Nodes. World Scientific Publishing Co. Pte. Ltd., Hackensack, NJ, 2024
7. Milovanović, G.V., Stanić, M.P., Tomović Mladenović, T.V.: Gaussian type quadrature rules related to the oscillatory modification of generalized Laguerre weight functions. J. Comput. Appl. Math. **437** (2024). Art. 115476
8. Milovanović, G.V., Vasović, N.M.: Some modifications of the Chebyshev measures and the corresponding orthogonal polynomials. Bull. Cl. Sci. Math. Nat. Sci. Math. **47**, 67–83 (2022)
9. Szegő, G.: Orthogonal Polynomials, 4th edn., American Mathematical Society, Providence (1975)
10. Gautschi, W.: The interplay between classical analysis and (numerical) linear algebra – a tribute to Gene H. Golub. Electron. Trans. Numer. Anal. **13**, 119–147 (2002)
11. Golub, G.H., Meurant, G.: Matrices, Moments and Quadrature with Applications. Princeton University Press, Princeton (2010)
12. Laurie, D.P.: Computation of Gauss-type quadrature formulas. J. Comput. Appl. Math. **127**, 201–217 (2001)
13. Chihara, T.S.: An Introduction to Orthogonal Polynomials. Gordon and Breach, New York (1978)
14. Pozza, S., Pranić, M.S., Strakoš, Z.: The Lanczos algorithm and complex Gauss quadrature. Electron. Trans. Numer. Anal. **50**, 1–19 (2018)
15. Pranić, M.S., Reichel, L.: Generalized anti-Gauss quadrature rules. J. Comput. Appl. Math. **284**, 235–243 (2015)
16. Laurie, D.P.: Anti-Gauss quadrature formulas. Math. Comp. **65**, 739–747 (1996)
17. Notaris, S.: Gauss-Kronrod quadrature formulae – a survey of fifty years of research. Electron. Trans. Numer. Anal. **45**, 371–404 (2016)
18. Spalević, M.M.: On generalized averaged Gaussian formulas. Math. Comp. **76**, 1483–1492 (2007)

19. Reichel, L., Spalević, M.M.: A new representation of generalized averaged Gauss quadrature rules. Appl. Numer. Math. **165**, 614–619 (2021)
20. Reichel, L., Spalević, M.M., Tang, T.: Generalized averaged Gauss quadrature rules for the approximation of matrix functionals. BIT Numer. Math. **56**, 1045–1067 (2016)
21. Reichel, L., Spalević, M.M.: Averaged Gauss quadrature formulas: properties and applications. J. Comput. Appl. Math. **410** (2022). Art. 114232
22. Saad, Y.: Iterative Methods for Sparse Linear Systems, 2nd edn. SIAM, Philadelphia (2003)
23. Ye, Q.: A breakdown-free variation of the nonsymmetric Lanczos algorithms. Math. Comp. **62**, 179–207 (1994)
24. Biological Networks Data Sets of Newcastle University. Available at http://www.biological-networks.org/
25. Marcelino, J., Kaiser, M.: Critical paths in a metapopulation model of H1N1: efficiently delaying influenza spreading through flight cancellation. PLoS Curr. **4** (2012). e4f8c9a2e1fca8
26. Estrada, E., Higham, D.J.: Network properties revealed through matrix functions. SIAM Rev. **52**, 696–714 (2010)
27. Benzi, M., Klymko, C.: Total communicability as a centrality measure. J. Complex Networks **1**, 1–26 (2013)

Chapter 5
A Curious Hypergeometric Identity and Perfectness of Meixner–Sorokin System of Weights

Alexander I. Aptekarev, Alexander V. Dyachenko, and Vladimir G. Lysov

Dedicated to Academician Gradimir V. Milovanović on the occasion of his 75th birthday

5.1 Introduction an Statement of the Results

5.1.1 Multiple Orthogonal Polynomials for Perfect Systems

Let $(\mu_1, \dots, \mu_r)$ be a vector of Borel non-negative measures on the real line $\mathbb{R}$. We assume that all the moments are finite:

$$c_{j,k} := \int x^{k-1}\,\mathrm{d}\mu_j(x) \in \mathbb{R}, \quad k \in \mathbb{Z}_+ \setminus \{0\}, \quad j = 1, \dots, r.$$

Given the multi-index $\boldsymbol{n} = (n_1, \dots, n_r) \in \mathbb{Z}_+^r$, a nontrivial polynomial $P_{\boldsymbol{n}}$ of degree at most

$$|\boldsymbol{n}| := n_1 + \cdots + n_r$$

is called a (type II) *multiple orthogonal polynomial* if it satisfies the following orthogonality conditions:

$$\int P_{\boldsymbol{n}}(x) x^k\,\mathrm{d}\mu_j(x) = 0, \quad k = 0, \dots, n_j - 1, \quad j = 1, \dots, r. \tag{5.1}$$

The orthogonality conditions (5.1) reduce to a homogeneous system of $|\boldsymbol{n}|$ linear algebraic equations for $|\boldsymbol{n}| + 1$ coefficients of $P_{\boldsymbol{n}}$. So a multiple orthogonal

A. I. Aptekarev · A. V. Dyachenko (✉) · V. G. Lysov
Keldysh Institute of Applied Mathematics (RAS), Moscow, Russian Federation
e-mail: aptekaa@keldysh.ru; diachenko@sfedu.ru; vlysov@mail.ru

M. P. Stanić et al. (eds.), *Analysis, Approximation, Optimization: Computation and Applications*, Springer Optimization and Its Applications 224,
https://doi.org/10.1007/978-3-031-85743-0_5

polynomial always exists. If for any nontrivial solution of (5.1) we have $\deg P_{\boldsymbol{n}} = |\boldsymbol{n}|$, then the multi-index $\boldsymbol{n}$ is called *normal*. Clearly the multi-index $\boldsymbol{n}$ is normal if and only if $\det H_{\boldsymbol{n}} \neq 0$, where

$$H_{\boldsymbol{n}} := \begin{bmatrix} c_{1,1} & c_{1,2} & \cdots & c_{1,|\boldsymbol{n}|} \\ c_{1,2} & c_{1,3} & \cdots & c_{1,|\boldsymbol{n}|+1} \\ \vdots & \vdots & \ddots & \vdots \\ c_{1,n_1} & c_{1,n_1+1} & \cdots & c_{1,|\boldsymbol{n}|+n_1-1} \\ c_{2,1} & c_{2,2} & \cdots & c_{2,|\boldsymbol{n}|} \\ \vdots & \vdots & \ddots & \vdots \\ c_{r,n_r} & c_{r,n_r+1} & \cdots & c_{r,|\boldsymbol{n}|+n_r-1} \end{bmatrix}. \tag{5.2}$$

For a normal multi-index $\boldsymbol{n}$ the polynomial $P_{\boldsymbol{n}}$ is defined uniquely up to multiplication by a constant:

$$P_{\boldsymbol{n}}(x) = \text{const} \cdot \begin{vmatrix} & & & c_{1,|\boldsymbol{n}|+1} \\ & & & c_{1,|\boldsymbol{n}|+2} \\ & H_{\boldsymbol{n}} & & \vdots \\ & & & c_{r,|\boldsymbol{n}|+n_r} \\ 1 & x & \ldots & x^{|\boldsymbol{n}|} \end{vmatrix}. \tag{5.3}$$

If all the multi-indices $\boldsymbol{n} \in \mathbb{Z}_+^r$ are normal, then $(\mu_1, \ldots, \mu_r)$ is called a *perfect* system of measures. The case $r = 1$ corresponds to the orthogonal polynomials, and any measure with infinite support is perfect (quasi-definite, [6]). The case $r > 1$ is more complicated. For example, if $\mathrm{d}\mu_2/\mathrm{d}\mu_1$ is a rational function, then clearly (μ_1, μ_2) is not perfect. Originally perfect systems appeared in number theory, see [10, 14, 20]. Multiple orthogonal polynomials for perfect systems have applications in the simultaneous Gaussian quadrature rules, see [12, 15, 19]. They satisfy [11, 21] the nearest neighbour recurrence relations, which can be applied to the spectral theory on graphs [3]. They also have numerous applications in random matrices [13].

There are two general classes of perfect systems. The first one is called an Angelesco [2] system and corresponds to the case when the convex hulls of the supports of μ_j are pairwise disjoint:

$$\operatorname{conv}(\operatorname{supp}\mu_j) \cap \operatorname{conv}(\operatorname{supp}\mu_k) = \emptyset, \quad j \neq k.$$

The second one is called an AT-system. The definition [16] implies that all μ_j have common support. For each of these systems, all the zeros of $P_{\boldsymbol{n}}$ are known to be simple and real.

We consider a system of $r = 2$ discrete measures with interlacing supports, which is neither an Angelesco system nor an AT-system. The first measure is

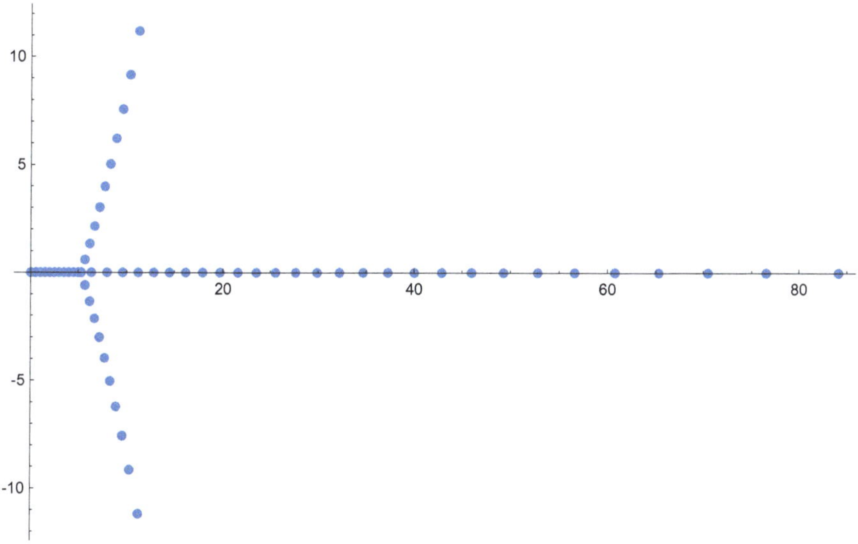

Fig. 5.1 Zeros of the polynomial $P_{30,30}$ with the parameters $\beta_1 = 1$, $\beta_2 = \frac{3}{2}$, $\gamma = \frac{1}{2}$, and $b = \frac{1}{7}$

supported on $\mathbb{Z}_+$, the second one—on $\gamma + \mathbb{Z}_+$ for some $\gamma \in (0, 1)$. For both of the measures, the weight at point x is given by

$$b^x \frac{\Gamma(x+\beta_1)\Gamma(x+\beta_2)}{\Gamma(x+1)\Gamma(x-\gamma+1)}, \quad b \in (0,1), \quad \beta_j > 0.$$

For diagonal indices (n, n), the Rodrigues formula and large n asymptotics for the corresponding multiple orthogonal polynomials were introduced in [17]. An interesting phenomenon is that a part of the zeros is not real, see Fig. 5.1.

An index $\boldsymbol{n} = (n_1, n_2)$ is said to be *marginal* if $n_1 n_2 = 0$. For marginal indices, the corresponding orthogonal polynomials were considered in [8], where it was shown that their recurrence coefficients satisfy Penlevé VI equation.

The main aim of this work is to prove the perfectness of such a system of measures. We apply an approach that is similar to that from our recent work [4]. Our results were partially announced in [5].

5.1.2 *Meixner–Sorokin Polynomials*

Let

$$0 < |b| < 1 \text{ and } \alpha_1,\ \alpha_2,\ \gamma_1-\gamma_2+\alpha_1,\ \gamma_2-\gamma_1+\alpha_2 \in \mathbb{C}\backslash\mathbb{Z}_-, \text{ and } \gamma_1 \neq \gamma_2. \tag{5.4}$$

Choose m to be some non-negative integer and denote

$$R_m(x) := b^x \frac{\Gamma(x-\gamma_1+\alpha_1+m)\Gamma(x-\gamma_2+\alpha_2+m)}{\Gamma(x-\gamma_1+1)\Gamma(x-\gamma_2+1)}. \tag{5.5}$$

The function $R(x) = R_0(x)$ here is a product of two classical Meixner weights on two distinct lattices. In its turn, $R_m(x)$ stands for the product of the same Meixner weights, in which the Meixner parameters α_1, α_2 are increased by m. On each of the lattices $\gamma_j + \mathbb{Z}_+$, $j = 1, 2$, consider the measure

$$\mu_j(y) := \sum_{x \in \gamma_j + \mathbb{Z}_+} R(x)\delta(y-x). \tag{5.6}$$

Then, for $n_1, n_2 \in \mathbb{Z}_+$, the polynomial P_{n_1,n_2} of degree at most $n_1 + n_2$, which is not identically zero, and which satisfies the orthogonality conditions

$$\int P_{n_1,n_2}(x)x^k \, \mathrm{d}\mu_j(x) = 0, \quad k = 0, \ldots, n_j - 1, \quad j = 1, 2, \tag{5.7}$$

we call a *Meixner–Sorokin polynomial*. Such polynomials first appeared in [17] as a generalization of the (classical discrete) Meixner polynomials, see also [7].

These polynomials retain certain important properties of the Meixner polynomials. To state them in a convenient form, we employ the additional parameter $m \in \mathbb{Z}_+$: a polynomial $P_{n_1,n_2}(x; m)$ of degree $\leq n_1 + n_2$ is defined as the orthogonal polynomial in the sense of (5.7) for the weight function R_m, i.e., satisfying

$$\sum\nolimits_{x \in \gamma_j + \mathbb{Z}_+} P_{n_1,n_2}(x; m)x^k R_m(x) = 0, \quad k = 0, \ldots, n_j - 1, \quad j = 1, 2. \tag{5.8}$$

Our main result is stated in Theorems 5.1 and 5.4, and the latter relies on the hypergeometric identity formulated as Theorem 5.5.

5.1.2.1 Rodrigues Formulae and Related Questions

Let $\nabla f(x) = f(x) - f(x-1)$ and $\Delta f(x) = f(x+1) - f(x)$ stand, respectively, for the left and right shift operators.

Theorem 5.1 *Let $m, n \in \mathbb{Z}_+$, and let* (5.4) *hold. If $P_{0,n}(x; m)$ or $P_{n,0}(x; m)$ is the orthogonal polynomial of degree n satisfying* (5.8), *then*

$$P_{m,n+m}(x) = \frac{\nabla^m (R_m(x) P_{0,n}(x; m))}{R(x)}, \tag{5.9a}$$

or, respectively,

$$P_{n+m,m}(x) = \frac{\nabla^m (R_m(x) P_{n,0}(x; m))}{R(x)} \tag{5.9b}$$

is the orthogonal polynomial of degree $n + 2m$ *satisfying* (5.7).

The conditions

$$\alpha_1, \alpha_2, \gamma_1 - \gamma_2 + \alpha_1, \gamma_2 - \gamma_1 + \alpha_2 \in \mathbb{C} \setminus \{0, -1, \ldots, -m\} \tag{5.10}$$

are only needed to make R_m finite on both lattices, while we pose them for every $m \in \mathbb{Z}_+$ in (5.4) to treat all multi-indices. On introducing the polynomials

$$U_m(x) := \frac{R_{m+1}(x)}{R_m(x)} = \prod_{j=1}^{2} \left(x - \gamma_j + \alpha_j + m\right) \quad \text{and} \tag{5.11}$$

$$V(x) := \frac{R_{m+1}(x-1)}{R_m(x)} = \frac{1}{b} \prod_{j=1}^{2} \left(x - \gamma_j\right),$$

the last theorem may be rewritten in a following form more convenient for us.

Theorem 5.2 *Let (5.4) hold. Suppose that (5.8) is satisfied by the polynomials* $P_{n,0}(x; m)$ *and* $P_{0,n}(x; m)$ *of degree* n *for all* $n, m \in \mathbb{Z}_+$. *Then for each* $(n_1, n_2) \in \mathbb{Z}_+^2$ *iterations of the formula*

$$P_{n_1,n_2}(x; m) = U_m(x) P_{n_1-1,n_2-1}(x; m+1) - V(x) P_{n_1-1,n_2-1}(x-1; m+1) \tag{5.12}$$

allow to construct a polynomial $P_{n_1,n_2}(x; m)$ *of degree* $n_1 + n_2$ *satisfying the orthogonality conditions (5.8). More specifically, given* $\nu = 1, 2, \ldots, \min\{n_1, n_2\}$ *the formula (5.12) implies*

$$P_{n_1,n_2}(x; m) = \sum_{k=0}^{\nu} \binom{\nu}{k} P_{n_1-\nu,n_2-\nu}\big(x - \nu + k; m + \nu\big) \cdot \prod_{j=0}^{k-1} U_{m+j}(x) \cdot \prod_{j=0}^{\nu-k-1} \big(-V(x-j)\big), \tag{5.13}$$

where for $k - 1 < 0$ *or* $\nu - k - 1 < 0$, *the empty product is equal to* 1.

In particular, polynomials with diagonal indices $n_1 = n_2$ may be obtained through a discrete analog of the Rodrigues formula. Indeed, both (5.9a) and (5.9b) reduce to $R(x) P_{n,n}(x) = \nabla^n R_n(x)$, while the formula (5.13) for diagonal indices yields

$$P_{n,n}(x) = \sum_{k=0}^{n} \binom{n}{k} \prod_{j=0}^{k-1} U_j(x) \cdot \prod_{j=0}^{n-k-1} \big(-V(x-j)\big). \tag{5.14}$$

The last expression is a (terminating) hypergeometric series: with the notation

$$(a)_n := a \cdot (a+1) \cdots (a+n-1),$$

the expression (5.14) can be written (as is seen by straightforward substitution of the definition of ${}_3F_2$) as

$$\begin{aligned} P_{n,n}(x) &= (-b)^{-n} {}_3F_2\left({-n,\ x-\gamma_1+\alpha_1,\ x-\gamma_2+\alpha_2 \atop x-\gamma_1-n+1,\ x-\gamma_2-n+1} \Big| b \right) \prod_{j=1}^{2} (x-\gamma_j-n+1)_n \\ &= {}_3F_2\left({-n,\quad \gamma_1-x,\quad \gamma_2-x \atop \gamma_1-x-\alpha_1-n+1,\ \gamma_2-x-\alpha_2-n+1} \Big| \tfrac{1}{b} \right) \prod_{k=1}^{2} (x-\gamma_k+\alpha_k)_n . \end{aligned}$$

In its turn, the corresponding weight function satisfies a difference analog of the Pearson equation.

Theorem 5.3 *The function R_m defined earlier satisfies the difference Pearson equation*

$$\nabla\big(U_m(x)R_m(x)\big) = \Delta\big(V(x)R_m(x)\big) = \big(U_m(x) - V(x)\big)R_m(x). \tag{5.15}$$

The formula (5.14) *determines the sequence of Meixner–Sorokin polynomials $P_{n,n}$ satisfying* (5.7). *Moreover, $P_{n,n}$ solves the following third-order difference equation:*

$$U_0(x-3)\nabla^3\big(R(x)P_{n,n}(x)\big) + \sum_{j=0}^{2} A_j(x)\nabla^j\big(R(x)P_{n,n}(x)\big) = 0, \tag{5.16}$$

with the coefficients

$$A_j(x) = \binom{n+3}{3-j}\nabla^{3-j}U_0(x-j) - \binom{n+2}{2-j}\nabla^{2-j}\big(U_0(x-j) - V(x-n-j)\big). \tag{5.17}$$

5.1.2.2 Perfectness of the System of Positive Measures

To make the aforementioned theorems more useful for applications, one needs the corresponding uniqueness result. Here we confine ourselves to the case of positive weights.

Theorem 5.4 *Given $b \in (0,1)$ and $\alpha_1, \alpha_2 > 0$, let $\gamma_1, \gamma_2 \in \mathbb{R}$ satisfy the inequalities $0 < |\gamma_1 - \gamma_2| < 1$ and $-\alpha_2 < \gamma_1 - \gamma_2 < \alpha_1$. Then the orthogonality measures μ_1, μ_2 defined in* (5.6) *are positive and form a perfect system: the corresponding polynomials P_{n_1,n_2} determined by* (5.13) *are unique (up to multiplication by a nonzero constant) polynomials orthogonal in the sense of* (5.7) *with respect to the measures μ_1, μ_2.*

For this theorem, one of the ingredients of the proof is positivity of the measures when the parameters stay within the given ranges. Another ingredient is the Rodrigues-type formulae from Theorem 5.1. The last ingredient is to prove that the polynomials $P_{0,1}(x; m)$ and $P_{1,0}(x; m)$ actually differ (which is the necessary condition for normality of the index $(1, 1)$). The latter then follows from the next theorem.

Theorem 5.5 *If* $b \in \mathbb{C} \setminus [1, +\infty)$ *and* $\gamma \in \mathbb{C} \setminus \mathbb{Z}$, *then*

$$\begin{aligned}(1-b)^{-\alpha_1-\alpha_2} &= {}_1F_0\left(\begin{matrix}\alpha_1+\alpha_2\\ -\end{matrix}\middle| b\right)\\ &= {}_2F_1\left(\begin{matrix}\alpha_1,\ \alpha_2+\gamma\\ 1+\gamma\end{matrix}\middle| b\right){}_2F_1\left(\begin{matrix}\alpha_1-\gamma,\ \alpha_2\\ 1-\gamma\end{matrix}\middle| b\right)\\ &\quad+\frac{b\alpha_1(\alpha_2+\gamma)}{\gamma(1+\gamma)}{}_2F_1\left(\begin{matrix}1+\alpha_1,\ 1+\alpha_2+\gamma\\ 2+\gamma\end{matrix}\middle| b\right){}_2F_1\left(\begin{matrix}\alpha_1-\gamma,\ \alpha_2\\ 1-\gamma\end{matrix}\middle| b\right)\\ &\quad-\frac{b\alpha_2(\alpha_1-\gamma)}{\gamma(1-\gamma)}{}_2F_1\left(\begin{matrix}1+\alpha_1-\gamma,\ 1+\alpha_2\\ 2-\gamma\end{matrix}\middle| b\right){}_2F_1\left(\begin{matrix}\alpha_1,\ \alpha_2+\gamma\\ 1+\gamma\end{matrix}\middle| b\right)\end{aligned} \tag{5.18}$$

In the course of the proof, this identity will be shown to be equivalent to the following elegant relation:

$$\begin{aligned}1 &= {}_2F_1\left(\begin{matrix}\alpha_1,\ -\alpha_2\\ \gamma\end{matrix}\middle| z\right){}_2F_1\left(\begin{matrix}1-\alpha_1,\ \alpha_2\\ 1-\gamma\end{matrix}\middle| z\right)+{}_2F_1\left(\begin{matrix}\alpha_1,\ 1-\alpha_2\\ 1+\gamma\end{matrix}\middle| z\right){}_2F_1\left(\begin{matrix}-\alpha_1,\ \alpha_2\\ -\gamma\end{matrix}\middle| z\right)\\ &\quad-{}_2F_1\left(\begin{matrix}\alpha_1,\ 1-\alpha_2\\ 1+\gamma\end{matrix}\middle| z\right){}_2F_1\left(\begin{matrix}1-\alpha_1,\ \alpha_2\\ 1-\gamma\end{matrix}\middle| z\right).\end{aligned} \tag{5.19}$$

We are not sure that the identities (5.18) and (5.19) are new—nevertheless, we have not seen them in the literature.

5.1.3 Kravchuk–Kravchuk Polynomials as a Special Case of Meixner–Sorokin Polynomials

Take the parameters $b \in (0, 1) \cup (1, +\infty)$ and $N_1, N_2 \in \mathbb{Z}_+$ and $\gamma_1, \gamma_2 \in \mathbb{R}$. The lattices now will be $\gamma_j + \{0, 1, \ldots, N_j\}$. Suppose that the parameters satisfy at least one of the following two conditions:

(a) $N_1 \in \{N_2 - 1, N_2\},\ 0 < \gamma_1 - \gamma_2 < 1$;
(b) $N_2 \in \{N_1 - 1, N_1\},\ 0 < \gamma_2 - \gamma_1 < 1$.

Note that, as in the aforementioned Meixner case, these conditions mean that the nodes of the lattices are interlacing: between each two consecutive nodes of the first lattice, there is precisely one node of the other lattice and vice versa. Consider the product of two Kravchuk's weights:

$$\widetilde{R}(x) := \frac{b^x}{\Gamma(x-\gamma_1+1)\Gamma(N_1-x+\gamma_1+1)} \cdot \frac{1}{\Gamma(x-\gamma_2+1)\Gamma(N_2-x+\gamma_2+1)}, \tag{5.20}$$

and the corresponding measures:

$$\widetilde{\mu}_j(y) := \sum_{x-\gamma_j=0}^{N_j} \widetilde{R}(x)\delta(y-x), \quad j=1,2. \tag{5.21}$$

These measures are positive and supported on finite sets. Accordingly, these systems of measures may have at most $(N_1+1)(N_2+1)$ normal multi-indices (n_1, n_2) varying in $\{0, \dots, N_1\} \times \{0, \dots, N_2\}$, and such a situation may be seen as a "finite analog" to the perfectness. Similarly to the Meixner–Sorokin case, the Kravchuk–Kravchuk polynomial K_{n_1,n_2} is a nontrivial polynomial of degree at most n_1+n_2 satisfying the orthogonality conditions

$$\int K_{n_1,n_2}(x)x^k \, \mathrm{d}\widetilde{\mu}_j(x) = 0, \quad k=0,\dots,n_j-1, \quad j=1,2, \tag{5.22}$$

see [7]. Now, due to

$$\frac{1}{\Gamma(x-\gamma+1)\Gamma(N-x+\gamma+1)} = \frac{\Gamma(x-\gamma-N)}{\pi\Gamma(x-\gamma+1)} \sin\big(\pi(N-x+\gamma+1)\big),$$

the function $\widetilde{R}$ differs from the function $R=R_0$ introduced in (5.5) with $\alpha_1=-N_1$ and $\alpha_2=-N_2$ only by a 1-periodic factor

$$\prod_{j=1}^{2} \frac{\sin\big(\pi(N_j-x+\gamma_j+1)\big)}{\pi}$$
$$= \frac{\cos\big(\pi(N_1-N_2+\gamma_1-\gamma_2)\big) - \cos\big(\pi(N_1+N_2-2x+\gamma_1+\gamma_2)\big)}{2\pi^2}.$$

So, the measures (5.21) are regularisations[1] of the measures (5.6) with $\alpha_1, \alpha_2 \in \mathbb{Z}_-$, while the polynomials K_{n_1,n_2} satisfy the same Rodrigues-type formulae as those in Theorem 5.1 for the Meixner–Sorokin polynomials: we only have to restrain ourselves to $n_j \le N_j$, $j=1,2$, as otherwise, the jth orthogonality measure with shifted parameters would be supported on an empty set (and hence Lemma 5.1 would be inapplicable). The additional allowed range $b>1$ has its counterpart in

[1] This is seen on recalling the conditions (5.10) needed for finiteness of R_m.

the Meixner case as well, cf. [6, p. 177, Eq. (3.6)]. On account of Theorem 5.5, we therefore obtain the following version of Theorem 5.4.

Corollary 5.1 *Given $b \in (0,1) \cup (1,+\infty)$, let $N_1, N_2 \in \mathbb{Z}_+$ and $\gamma_1, \gamma_2 \in \mathbb{R}$ satisfy one of the above conditions (a) and (b). Then all indices $(n_1, n_2) \in \{0, \dots, N_1\} \times \{0, \dots, N_2\}$ are normal for the system of measures $\widetilde{\mu}_1, \widetilde{\mu}_2$ defined in* (5.21).

5.2 Proofs

Lemma 5.1 *Let $m \in \mathbb{Z}_+$ and $0 < |b| < 1$ and $n_1, n_2 \in \mathbb{Z}_+ \setminus \{0\}$. Let also* (5.4) *hold. Then, if Q is a polynomial of degree l, then*

$$\widetilde{Q}(x) := U_m(x)Q(x) - V(x)Q(x-1)$$

is a polynomial of degree $l+2$. Moreover, the polynomial Q satisfies the ortogonality conditions

$$\sum\nolimits_{x \in \gamma_j + \mathbb{Z}_+} Q(x) x^k R_{m+1}(x) = 0, \tag{5.23}$$

for all $k = 0, 1, \dots, n_j - 2$ and $j = 1, 2$ if and only if $\widetilde{Q}$ satisfies

$$\sum\nolimits_{x \in \gamma_j + \mathbb{Z}_+} \widetilde{Q}(x) x^k R_m(x) = 0, \tag{5.24}$$

for all $k = 0, \dots, n_j - 1$ and $j = 1, 2$. (In particular, $\widetilde{Q}$ is always orthogonal to constants (degree-zero polynomials) with respect to the measures μ_1, μ_2 due to its specific relation to U_m and V.)

Proof The degree of $\widetilde{Q}$ cannot exceed $2 + \deg Q$, since $\deg U_m = \deg V = 2$. The leading coefficient $\widetilde{Q}(x)$ equals the leading coefficient of Q multiplied by $1 - \frac{1}{b} \neq 0$, and therefore $\deg \widetilde{Q} = 2 + \deg Q$. Take an integer $j \in \{1, 2\}$ and note that the sums in (5.23) and (5.24) are always absolutely convergent due to the condition $|b| < 1$. For $k = 0, \dots, n_j - 1$,

$$\begin{aligned}
&\sum\nolimits_{x \in \gamma_j + \mathbb{Z}_+} \widetilde{Q}(x) x^k R_m(x) \\
&= \sum\nolimits_{x \in \gamma_j + \mathbb{Z}_+} x^k \cdot \Big(\big(U_m(x) R_m(x)\big) Q(x) - \big(V(x) R_m(x)\big) Q(x-1) \Big) \\
&= \sum\nolimits_{x \in \gamma_j + \mathbb{Z}_+} x^k \cdot \nabla\big(R_{m+1}(x) Q(x)\big) \\
&= -\sum\nolimits_{x \in \gamma_j + \mathbb{Z}_+} R_{m+1}(x) Q(x) \cdot \Big((x+1)^k - x^k \Big),
\end{aligned}$$

where the last equality relies on $R_{m+1}(\gamma_j - 1) = \lim_{x\to+\infty} R_{m+1}(x+\gamma_j) = 0$. Thus,

$$\sum_{x\in\gamma_j+\mathbb{Z}_+} \widetilde{Q}(x)x^k R_m(x) = -\sum_{l=0}^{k-1}\binom{k}{l}\sum_{x\in\gamma_j+\mathbb{Z}_+} Q(x)x^l R_{m+1}(x),$$

where the right-hand side is zero for $k = 0$. The last identity means that the orthogonality condition (5.24) for all $k = 0,\dots,n_j - 1$ is equivalent to the orthogonality condition (5.23) for all $k = 0,\dots,n_j - 2$. □

Proof of Theorem 5.2 For each index (n_1, n_2), the polynomial $P_{n_1,n_2}(x;m)$ is produced by iterations of the formula (5.12) from the polynomial $P_{n_1-\nu,n_2-\nu}(x;m+\nu)$ of degree $n_1+n_2-2\nu$ with $\nu := \min\{n_1,n_2\}$, which corresponds to the marginal index $(n_1-\nu, n_2-\nu)$. By Lemma 5.1, the polynomial $P_{n_1,n_2}(x;m)$ will satisfy the required orthogonality conditions (5.8), and its degree will be equal to n_1+n_2. So, (5.12) is proved.

Now let us prove the formula (5.13). It is trivial when $\nu = 0$, and it reduces to (5.12) when $\nu = 1$. We are in a position to use induction in ν, for which we assume that for some $\nu > 1$

$$\begin{aligned} P_{n_1-1,n_2-1}(x;m+1) = &\sum_{k=0}^{\nu-1}\binom{\nu-1}{k} P_{n_1-\nu,n_2-\nu}\big(x-\nu+k+1;m+\nu\big) \\ &\times \prod_{j=0}^{k-1} U_{m+1+j}(x)\cdot \prod_{j=0}^{\nu-k-2}\big(-V(x-j)\big). \end{aligned}$$

Using this expression and the standard convention $\binom{n-1}{-1} = \binom{n-1}{n} = 0$, the right-hand side of (5.12) may be rewritten as

$$\begin{aligned} &U_m(x)P_{n_1-1,n_2-1}(x;m+1) - V(x)P_{n_1-1,n_2-1}(x-1;m+1) \\ = &\sum_{k=-1}^{\nu-1}\binom{\nu-1}{k} P_{n_1-\nu,n_2-\nu}\big(x-\nu+k+1;m+\nu\big) \\ &\times U_m(x)\prod_{j=1}^{k} U_{m+j}(x)\cdot\prod_{j=0}^{\nu-k-2}\big(-V(x-j)\big) \\ &-\sum_{k=0}^{\nu}\binom{\nu-1}{k} P_{n_1-\nu,n_2-\nu}\big(x-\nu+k;m+\nu\big) \\ &\times V(x)\prod_{j=0}^{k-1} U_{m+1+j}(x-1)\cdot\prod_{j=1}^{\nu-k-1}\big(-V(x-j)\big). \end{aligned}$$

On shifting the summation index $k-1 \mapsto k$ in the first sum, and applying the equality $U_{m+1}(x-1) = U_m(x)$ in the second, we obtain

$$\begin{aligned}
&U_m(x)P_{n_1-1,n_2-1}(x;m+1) - V(x)P_{n_1-1,n_2-1}(x-1;m+1)\\
&= \sum\nolimits_{k=0}^{\nu}\binom{\nu-1}{k-1}P_{n_1-\nu,n_2-\nu}\big(x-\nu+k;m+\nu\big)\prod\nolimits_{j=0}^{k-1}U_{m+j}(x)\\
&\times\prod\nolimits_{j=0}^{\nu-k-1}\big(-V(x-j)\big)\\
&+\sum\nolimits_{k=0}^{\nu}\binom{\nu-1}{k}P_{n_1-\nu,n_2-\nu}\big(x-\nu+k;m+\nu\big)\prod\nolimits_{j=0}^{k-1}U_{m+j}(x)\\
&\times\prod\nolimits_{j=0}^{\nu-k-1}\big(-V(x-j)\big).
\end{aligned}$$

So, it is enough now to gather both sums together and to apply Pascal's formula $\binom{\nu-1}{k-1}+\binom{\nu-1}{k}=\binom{\nu}{k}$ to derive (5.13) from (5.12). □

Proof of Theorem 5.1 Given some $n,k,m \in \mathbb{Z}_+$ with $m \ge 1$, observe that the formula (5.12) may be written as

$$\begin{aligned}
P_{m,n+m}(x;k) &= \frac{R_{k+1}(x)}{R_k(x)}P_{m-1,n+m-1}(x;k+1)\\
&\quad-\frac{R_{k+1}(x-1)}{R_k(x)}P_{m-1,n+m-1}(x-1;k+1)\\
&= \frac{\nabla(R_{k+1}(x)P_{m-1,n+m-1}(x;k+1))}{R_k(x)}.
\end{aligned}$$

After multiplying this formula by $R_k(x)$, its iterations give

$$\begin{aligned}
R_k(x)P_{m,n+m}(x;k) &= \nabla(R_{k+1}(x)P_{m-1,n+m-1}(x;k+1))\\
&= \nabla\nabla(R_{k+2}(x)P_{m-2,n+m-2}(x;k+2))\\
&= \cdots = \nabla^m(R_{k+m}(x)P_{0,n}(x;k+m)),
\end{aligned}$$

which for $k=0$ is equivalent to (5.9a). The reasoning for (5.9b) is essentially the same. So, the theorem's proof now reduces to invoking Lemma 5.1. □

Proof of Theorem 5.3 The double identity

$$U_n(x-1)R_n(x-1) = R_{n+1}(x-1) = V(x)R_n(x)$$

immediately follows from the definition (5.11) of U_{n-1} and V and implies (5.15). In particular, $\nabla^{n+3}\big(U_n(x)R_n(x)\big) = \nabla^{n+2}\big((U_n(x)-V(x))R_n(x)\big)$, whence on applying $U_n(x-n) = U_0(x)$ and using the Leibniz formula $\nabla^n(f(x)g(x)) =$

$\sum_{j=0}^{n}\binom{n}{j}\nabla^j f(x-n+j)\nabla^{n-j}g(x)$ we have

$$\begin{aligned}&\sum\nolimits_{j=0}^{n+3}\binom{n+3}{j}\nabla^j U_0(x-3+j)\cdot\nabla^{n+3-j}R_n(x)\\ =\;&\sum\nolimits_{j=0}^{n+2}\binom{n+2}{j}\nabla^j\big(U_0(x-2+j)-V(x-n-2+j)\big)\cdot\nabla^{n+2-j}R_n(x).\end{aligned}$$

The terms with $j > 2$ on the both sides of the last equality disappear, since $\deg U_0 = \deg V = 2$. Thus,

$$U_0(x-3)\cdot\nabla^{n+3}R_n(x)+\sum\nolimits_{j=0}^{2}A_{2-j}(x)\nabla^{n+2-j}R_n(x)=0, \tag{5.25}$$

where $A_j(x)$ a defined in (5.17). Finally, to obtain (5.16) one can plug (5.9a) with $n = 0$ into (5.25). □

Proof of Theorem 5.5 First, we prove equivalence of the identities (5.18) and (5.19). Application of the Pfaff identity (see [9, p. 1057, **S** 9.131] or [1, p.68, eq.(2.2.6)]) to each ${}_2F_1$ on right-hand side of (5.18) yields

$$\begin{aligned}&(1-b)^{-\alpha_1-\alpha_2}\\ =\;&\Big[(1-b)^{-\alpha_1}{}_2F_1\Big({\alpha_1,\,1-\alpha_2 \atop 1+\gamma}\Big|\tfrac{b}{b-1}\Big)\Big]\cdot\Big[(1-b)^{-\alpha_2}{}_2F_1\Big({1-\alpha_1,\,\alpha_2 \atop 1-\gamma}\Big|\tfrac{b}{b-1}\Big)\Big]\\ &+\frac{b\alpha_1(\alpha_2+\gamma)}{\gamma(1+\gamma)}\Big[(1-b)^{-1-\alpha_1}{}_2F_1\Big({1+\alpha_1,\,1-\alpha_2 \atop 2+\gamma}\Big|\tfrac{b}{b-1}\Big)\Big]\\ &\times\Big[(1-b)^{-\alpha_2}{}_2F_1\Big({1-\alpha_1,\,\alpha_2 \atop 1-\gamma}\Big|\tfrac{b}{b-1}\Big)\Big]\\ &-\frac{b\alpha_2(\alpha_1-\gamma)}{\gamma(1-\gamma)}\Big[(1-b)^{-1-\alpha_2}{}_2F_1\Big({1-\alpha_1,\,1+\alpha_2 \atop 2-\gamma}\Big|\tfrac{b}{b-1}\Big)\Big]\\ &\times\Big[(1-b)^{-\alpha_1}{}_2F_1\Big({\alpha_1,\,1-\alpha_2 \atop 1+\gamma}\Big|\tfrac{b}{b-1}\Big)\Big],\end{aligned}$$

so on multiplying by $(1-b)^{\alpha_1+\alpha_2}$ and denoting $z := \frac{b}{b-1}$ we arrive at

$$\begin{aligned}1=\;&{}_2F_1\Big({\alpha_1,\,1-\alpha_2 \atop 1+\gamma}\Big|z\Big){}_2F_1\Big({1-\alpha_1,\,\alpha_2 \atop 1-\gamma}\Big|z\Big)-z\tfrac{\alpha_1(\alpha_2+\gamma)}{\gamma(1+\gamma)}{}_2F_1\Big({1+\alpha_1,\,1-\alpha_2 \atop 2+\gamma}\Big|z\Big){}_2F_1\Big({1-\alpha_1,\,\alpha_2 \atop 1-\gamma}\Big|z\Big)\\ &-z\tfrac{\alpha_2(\alpha_1-\gamma)}{-\gamma(1-\gamma)}{}_2F_1\Big({1-\alpha_1,\,1+\alpha_2 \atop 2-\gamma}\Big|z\Big){}_2F_1\Big({\alpha_1,\,1-\alpha_2 \atop 1+\gamma}\Big|z\Big).\end{aligned}$$

The right-hand side of this identity, after plugging the Gauss contiguous relations [9, p. 1059, **S** 9.137 (15,16)] or [1, p. 97, Eq. (2.5.11)] into the last two terms, turns to be equal to

$$
\begin{aligned}
&{}_2F_1\left(\begin{smallmatrix}\alpha_1,\,1-\alpha_2\\ 1+\gamma\end{smallmatrix}\Big|z\right){}_2F_1\left(\begin{smallmatrix}1-\alpha_1,\,\alpha_2\\ 1-\gamma\end{smallmatrix}\Big|z\right)\\
&+\left({}_2F_1\left(\begin{smallmatrix}\alpha_1,\,-\alpha_2\\ \gamma\end{smallmatrix}\Big|z\right)-{}_2F_1\left(\begin{smallmatrix}\alpha_1,\,1-\alpha_2\\ 1+\gamma\end{smallmatrix}\Big|z\right)\right){}_2F_1\left(\begin{smallmatrix}1-\alpha_1,\,\alpha_2\\ 1-\gamma\end{smallmatrix}\Big|z\right)\\
&+\left({}_2F_1\left(\begin{smallmatrix}-\alpha_1,\,\alpha_2\\ -\gamma\end{smallmatrix}\Big|z\right)-{}_2F_1\left(\begin{smallmatrix}1-\alpha_1,\,\alpha_2\\ 1-\gamma\end{smallmatrix}\Big|z\right)\right){}_2F_1\left(\begin{smallmatrix}\alpha_1,\,1-\alpha_2\\ 1+\gamma\end{smallmatrix}\Big|z\right).
\end{aligned}
$$

Finally, reduction of similar terms yields

$$
\begin{aligned}
1={}&{}_2F_1\left(\begin{smallmatrix}\alpha_1,\,-\alpha_2\\ \gamma\end{smallmatrix}\Big|z\right){}_2F_1\left(\begin{smallmatrix}1-\alpha_1,\,\alpha_2\\ 1-\gamma\end{smallmatrix}\Big|z\right)+{}_2F_1\left(\begin{smallmatrix}\alpha_1,\,1-\alpha_2\\ 1+\gamma\end{smallmatrix}\Big|z\right){}_2F_1\left(\begin{smallmatrix}-\alpha_1,\,\alpha_2\\ -\gamma\end{smallmatrix}\Big|z\right)\\
&-{}_2F_1\left(\begin{smallmatrix}\alpha_1,\,1-\alpha_2\\ 1+\gamma\end{smallmatrix}\Big|z\right){}_2F_1\left(\begin{smallmatrix}1-\alpha_1,\,\alpha_2\\ 1-\gamma\end{smallmatrix}\Big|z\right),
\end{aligned}
$$

which is (5.19).

Our aim now is to prove the identity (5.19). Let $1+\sum_{k=1}^{\infty}s_k z^k$ be the Taylor expansion of the right-hand side of (5.19) at the origin, then it is enough to prove that $s_1=s_2=\cdots=0$. Observe that

$$
\begin{aligned}
{}_2F_1\left(\begin{smallmatrix}\alpha_1,\,-\alpha_2\\ \gamma\end{smallmatrix}\Big|z\right){}_2F_1\left(\begin{smallmatrix}1-\alpha_1,\,\alpha_2\\ 1-\gamma\end{smallmatrix}\Big|z\right)&=\sum_{j,m=0}^{\infty}\frac{(\alpha_1)_j(-\alpha_2)_j(1-\alpha_1)_m(\alpha_2)_m}{j!(\gamma)_j m!(1-\gamma)_m}z^{j+m}\\
&=\sum_{k=0}^{\infty}z^k\sum_{j=0}^{k}\frac{(\alpha_1)_j(-\alpha_2)_j(1-\alpha_1)_{k-j}(\alpha_2)_{k-j}}{j!(\gamma)_j(k-j)!(1-\gamma)_{k-j}}.
\end{aligned}
$$

On expanding remaining products of the Gauss hypergeometric functions, one obtains the formula $s_k=\sum_{j=0}^{k}s_{k,j}$, in which the summands satisfy

$$
\begin{aligned}
s_{k,j}={}&\frac{(\alpha_1)_j(1-\alpha_2)_{j-1}(1-\alpha_1)_{k-j-1}(\alpha_2)_{k-j}}{j!(\gamma)_{j+1}(k-j)!(-\gamma)_{k-j+1}}\Big((-\alpha_2)(k-j-\alpha_1)(j+\gamma)(-\gamma)\\
&+(j-\alpha_2)(-\alpha_1)\gamma(k-j-\gamma)-(j-\alpha_2)(k-j-\alpha_1)\gamma(-\gamma)\Big)\\
={}&(-1)^{j-1}\frac{(1-\alpha_1-j)_{k-1}(1+\alpha_2-j)_{k-1}\gamma}{j!(k-j)!(-\gamma-j)_{k+1}\gamma}\Big(\alpha_2(k-j-\alpha_1)(j+\gamma)\\
&-\alpha_1(j-\alpha_2)(k-j-\gamma)+(j-\alpha_2)(k-j-\alpha_1)\gamma\Big),
\end{aligned}
$$

using the standard convention $0!=1$. After collecting terms,

$$
\begin{aligned}
s_{k,j}={}&(-1)^{j-1}\frac{(1-\alpha_1-j)_{k-1}(1+\alpha_2-j)_{k-1}}{j!(k-j)!(-\gamma-j)_{k+1}}\\
&\times\big((k-2j-\gamma)\alpha_1\alpha_2+j(k-j)(\alpha_2-\alpha_1+\gamma)\big).
\end{aligned}
$$

Substitution of $j = 0$ into the last formula shows that

$$s_{k,0} = \frac{(1-\alpha_1)_{k-1}(-\alpha_1)\cdot(1+\alpha_2)_{k-1}\alpha_2\cdot(-\gamma+k)}{k!(-\gamma)_{k+1}} = \frac{(-\alpha_1)_k(\alpha_2)_k}{(k)!(-\gamma)_k}.$$

That is, for $m = 0$ we have

$$\sum\nolimits_{j=0}^{m} s_{k,j} = (-1)^m \frac{k-m}{k}\cdot\frac{(-\alpha_1-m)_k(\alpha_2-m)_k}{m!(k-m)!(-\gamma-m)_k}. \tag{5.26}$$

This representation remains true for $m = 1, 2, \ldots, k$, which we are proving by induction: it is enough to derive (5.26) provided that

$$\sum\nolimits_{j=0}^{m-1} s_{k,j} = (-1)^{m-1}\frac{k-m+1}{k}\cdot\frac{(-\alpha_1-m+1)_k(\alpha_2-m+1)_k}{(m-1)!(k-m+1)!(-\gamma-m+1)_k},$$

which is (5.26) with m replaced by $m-1$. However,

$$\begin{aligned}
&(-1)^m \sum\nolimits_{j=0}^{m} s_{k,j} \\
&= -\frac{k-m+1}{k}\cdot\frac{(-\alpha_1-m+1)_k(\alpha_2-m+1)_k}{(m-1)!(k-m+1)!(-\gamma-m+1)_k} \\
&\quad -\frac{(-\alpha_1-m+1)_{k-1}(\alpha_2-m+1)_{k-1}}{m!(k-m)!(-\gamma-m)_{k+1}} \\
&\quad \times\big((k-2m-\gamma)\alpha_1\alpha_2+m(k-m)(\alpha_2-\alpha_1+\gamma)\big) \\
&= \frac{(-\alpha_1-m+1)_{k-1}(\alpha_2-m+1)_{k-1}}{k\,m!(k-m)!(-\gamma-m)_{k+1}} \\
&\quad \Big(-(-\gamma-m)m(-\alpha_1-m+k)(\alpha_2-m+k) \\
&\quad -k\big((k-2m-\gamma)\alpha_1\alpha_2+m(k-m)(\alpha_2-\alpha_1+\gamma)\big)\Big)
\end{aligned}$$

leading to (5.26). In particular, $s_k = 0$ for all $k \geq 1$. □

Proof of Theorem 5.4 Since $R(x)$ for our choice of the parameters is positive and finite for $x \in (\gamma_1+\mathbb{Z}_+)\cup(\gamma_2+\mathbb{Z}_+)$, both measures μ_1, μ_2 are positive and have infinite supports. Whence, by the standard theory of orthogonal polynomials (e.g. [18, **S**2.3]), it follows that all marginal indices are normal.

Let us show the normality of the index (1, 1) for all m, which is equivalent to that the monic (i.e., having the leading coefficient 1) orthogonal polynomials

$$P_{(1,0)}(x;m) = x - c_{\gamma_1,\gamma_2}^{\alpha_1+m,\alpha_2+m} \quad \text{and} \quad P_{(0,1)}(x;m) = x - c_{\gamma_2,\gamma_1}^{\alpha_2+m,\alpha_1+m}$$

are distinct. Put in other words, we are showing that for all $m \in \mathbb{Z}_+$

$$c_{\gamma_1,\gamma_2}^{\alpha_1+m,\alpha_2+m} - c_{\gamma_2,\gamma_1}^{\alpha_2+m,\alpha_1+m} \neq 0. \tag{5.27}$$

The orthogonality conditions imply (without loss of generality put here $m = 0$)

$$c_{\gamma_1,\gamma_2}^{\alpha_1,\alpha_2} = \gamma_1 + b\alpha_1 \frac{\gamma_1 - \gamma_2 + \alpha_2}{\gamma_1 - \gamma_2 + 1} \cdot \frac{{}_2F_1\left(\begin{smallmatrix}\alpha_1+1,\ \gamma_1-\gamma_2+\alpha_2+1\\ \gamma_1-\gamma_2+2\end{smallmatrix}\middle| b\right)}{{}_2F_1\left(\begin{smallmatrix}\alpha_1,\ \gamma_1-\gamma_2+\alpha_2\\ \gamma_1-\gamma_2+1\end{smallmatrix}\middle| b\right)},$$

so it is enough to substitute $\gamma = \gamma_1 - \gamma_2$ in the identity

$$\frac{(1-b)^{-\alpha_1-\alpha_2}}{{}_2F_1\left(\begin{smallmatrix}\alpha_1,\ \alpha_2+\gamma\\ 1+\gamma\end{smallmatrix}\middle| b\right) {}_2F_1\left(\begin{smallmatrix}\alpha_1-\gamma,\ \alpha_2\\ 1-\gamma\end{smallmatrix}\middle| b\right)}$$

$$= \gamma + \tfrac{b\alpha_1(\alpha_2+\gamma)}{(1+\gamma)} \frac{{}_2F_1\left(\begin{smallmatrix}1+\alpha_1,\ 1+\alpha_2+\gamma\\ 2+\gamma\end{smallmatrix}\middle| b\right)}{{}_2F_1\left(\begin{smallmatrix}\alpha_1,\ \alpha_2+\gamma\\ 1+\gamma\end{smallmatrix}\middle| b\right)} - \tfrac{b\alpha_2(\alpha_1-\gamma)}{(1-\gamma)} \frac{{}_2F_1\left(\begin{smallmatrix}1+\alpha_1-\gamma,\ 1+\alpha_2\\ 2-\gamma\end{smallmatrix}\middle| b\right)}{{}_2F_1\left(\begin{smallmatrix}\alpha_1-\gamma,\ \alpha_2\\ 1-\gamma\end{smallmatrix}\middle| b\right)},$$

which is equivalent to the identity provided by Theorem 5.5. Here, the left-hand side is positive, while the right-hand side equals $c_{\gamma_1,\gamma_2}^{\alpha_1,\alpha_2} - c_{\gamma_2,\gamma_1}^{\alpha_2,\alpha_1}$, whence (5.27) follows.

Now, suppose that $n_1, n_2 \geq 1$ and $n_1 + n_2 \geq 3$. If (n_1, n_2) is the index with minimal value of $n_1 + n_2$, for which there is some $m \in \mathbb{Z}_+$ making the index not normal. Then, by the definition of normality, there exists some nontrivial polynomial $Q(x)$ of degree $< n_1 + n_2$ satisfying the same conditions as $P_{n_1,n_2}(x; m)$ in (5.8). However, these orthogonality conditions for $Q(z)$ include all orthogonality conditions posed on both $P_{n_1-1,n_2}(x; m)$ and $P_{n_1,n_2-1}(x; m)$, and hence,

$$Q(z) = C_1 P_{n_1-1,n_2}(x; m) = C_2 P_{n_1,n_2-1}(x; m),$$

due to the normality of the indices $(n_1 - 1, n_2)$ and $(n_1, n_2 - 1)$ for certain nonzero normalizing constants C_1, C_2.

On the other hand, $n_1 + n_2 \geq 3$, so at least one of the numbers n_1, n_2 is more than or equal to 2; without loss of generality, assume that $n_1 \geq 2$. On the one hand, the polynomial $P_{n_1-1,n_2}(x; m)$ satisfies the orthogonality conditions corresponding to the index (n_1, n_2). On the other hand, $P_{n_1-1,n_2}(x; m)$ can be expressed through (5.12) from $P_{n_1-2,n_2-1}(x; m+1)$, the polynomial $P_{n_1-2,n_2-1}(x; m+1)$ by Lemma 5.1 must satisfy all orthogonality conditions posed on $P_{n_1-1,n_2-1}(x; m+1)$. That is, the index $(n_1 - 1, n_2 - 1)$ is not normal. The latter contradicts to the initial assumption, which implies normality of all indices $(n_1, n_2) \in \mathbb{Z}_+^2$. □

References

1. Andrews, G.E., Askey, R., Roy, R.: Special Functions. Encyclopedia of Mathematics and its Applications, vol. 71. Cambridge University Press, Cambridge (1999)
2. Angelesco, A.: Sur deux extensions des fractions continues algébriques. C. R. Acad. Sci. Paris **168**, 262–265 (1919)
3. Aptekarev, A.I., Denisov, S.A., Yattselev, M.L.: Self-adjoint Jacobi matrices on trees and multiple orthogonal polynomials. Trans. Amer. Math. Soc. **373**(2), 875–917 (2020)
4. Aptekarev, A., Dyachenko, A., Lysov, V.: On perfectness of systems of weights satisfying Pearson's equation with nonstandard parameters. Axioms **12**(1) (2023). https://doi.org/10.3390/axioms12010089
5. Aptekarev, A.I., Dyachenko, A.V., Lysov, V.G.: Meixner multiple orthogonal polynomials on interlacing lattices. Math. Notes **115**(3–4), 642–646 (2024). https://doi.org/10.1134/S0001434624030374
6. Chihara, T.S.: An Introduction to Orthogonal Polynomials. Mathematics and its Applications, vol. 13. Gordon and Breach Science Publishers, London (1978)
7. Dyachenko, A., Lysov, V.: Discrete multiple orthogonal polynomials on shifted lattices (2019) https://arxiv.org/abs/1908.11467. Preprint
8. Filipuk, G., Van Assche, W.: Discrete orthogonal polynomials with hypergeometric weights and Painlevé VI. SIGMA **14**(088) (2018)
9. Gradshteyn, I.S., Ryzhik, I.M.: Table of integrals, series, and products. Translated from the Russian. Translation edited and with a preface by Alan Jeffrey and Daniel Zwillinger, 6th edn. Academic Press, San Diego (2000)
10. Hermite, C.: Sur la fonction exponentielle. C. R. Acad. Sci. Paris Sér. I Math. **77**, 18–24, 74–79, 226–233, 285–293 (1873)
11. Ismail, M.E.H.: Classical and Quantum Orthogonal Polynomials in One Variable. Encyclopedia of Mathematics and its Applications. Cambridge University Press, Cambridge (2005). https://doi.org/10.1017/CBO9781107325982
12. Jovanović, A.N., Stanić, M.P., Tomović, T.V.: Construction of the optimal set of quadrature rules in the sense of Borges. Electron. Trans. Numer. Anal. **50**, 164–181 (2018). https://doi.org/10.1553/etna_vol50s164
13. Kuijlaars, A.B.J.: Multiple orthogonal polynomial ensembles. In: Recent Trends in Orthogonal Polynomials and Approximation Theory. Contemporary Mathematics, vol. 507, pp. 155–176. American Mathematical Society, Providence (2010)
14. Mahler, K.: Perfect systems. Compos. Math. **19**, 95–166 (1968)
15. Milovanović, G.V.: Generalized weighted Birkhoff-Young quadratures with the maximal degree of exactness. Appl. Numer. Math. **116**, 238–255 (2017)
16. Nikishin, E.M., Sorokin, V.N.: Rational approximations and orthogonality, Translations of Mathematical Monographs, vol. 92. American Mathematical Society, Providence (1991). Translated from the Russian by Ralph P. Boas
17. Sorokin, V.N.: On multiple orthogonal polynomials for discrete Meixner measures. Sb. Math. **201**, 1539–1561 (2010). https://doi.org/10.1070/SM2010v201n10ABEH004121
18. Szegő, G.: Orthogonal Polynomials. American Mathematical Society, Providence (1939)
19. Tomović, T.V., Stanić, M.P.: Construction of the optimal set of two or three quadrature rules in the sense of Borges. Numer. Algorithms **78**(4), 1087–1109 (2018)
20. Van Assche, W.: Multiple orthogonal polynomials, irrationality and transcendence. In: Berndt, B.C., Gesztesy, F. (eds.) Continued Fractions: From Analytic Number Theory to Constructive Approximation. Contemporary Mathematics, vol. 236, pp. 325–342. American Mathematical Society, Providence (1999)
21. Van Assche, W.: Nearest neighbor recurrence relations for multiple orthogonal polynomials. J. Approx. Theory **163**(10), 1427–1448 (2011). https://doi.org/10.1016/j.jat.2011.05.003

Chapter 6
On Reverse Triangle Inequality and Some of Its Applications

Ivan D. Aranđelović

Dedicated to Academician Gradimir V. Milovanović on the occasion of his 75th birthday

6.1 Introduction

6.1.1 Triangle Inequality on Euclidean Space

In Euclidean space, the triangle inequality asserts that the sum of any two sides of a triangle is strictly bigger than the remaining third side. The reverse triangle inequality is its equivalent, which says that any side of a triangle is strictly greater than the difference between the other two sides. Both statements define the necessary conditions for the existence of the triangle with given edges, and their superposition is a sufficient condition for existence of such a triangle.

The oldest proof of the triangle inequality can be found in Euclid's elements, Book 1, Proposition 20. As far as we know, the oldest proof of reverse triangle inequality was presented in 19th century.

6.1.2 Triangle Inequality on Pseudometric Space

In [1], M. Fréchet introduced the notion of metric space, by the following way.

Definition 6.1 Let X be a nonempty set and $d : X \times X \to [0, +\infty)$ functions, which satisfies the following conditions:

(M1) $d(x, y) = 0$ if and only if $x = y$;

I. D. Aranđelović (✉)
University of Belgrade - Faculty of Mechanical Engineering, Belgrade, Serbia
e-mail: iarandjelovic@mas.bg.ac.rs

M. P. Stanić et al. (eds.), *Analysis, Approximation, Optimization: Computation and Applications*, Springer Optimization and Its Applications 224,
https://doi.org/10.1007/978-3-031-85743-0_6

(M2) $d(x, y) = d(y, x)$;
(M3) $d(x, y) \leq d(x, z) + d(z, y)$,

for each $x, y, z \in X$, then ordered pair (X, d) is called *metric space*.

If in Definition 6.1, we replace M1) with

(M1') $d(x, x) = 0$ for any $x \in X$;

we obtained definition of *pseudometric space*.

Let (X, d) be a pseudometric space. We defined equivalence relation $\sim$ on X by $x \sim y$ if and only if $d(x, y) = 0$. Then quotient space $\tilde{X} \equiv X/\sim$ is the metric space with quotient metric.

In [2], A. Lindenbaum introduced the following characterization of metric space.

Proposition 6.1 *Let X be a nonempty set and $d : X \times X \to [0, +\infty)$ functions, which satisfies the following conditions:*

(M1) $d(x, y) = 0$ *if and only if* $x = y$*;*
(M3') $d(x, y) \leq d(x, z) + d(y, z)$,

for each $x, y, z \in X$. Then ordered pair (X, d) is metric space.

6.1.3 Adamović's Inequality

In [3], D. S. Mitrinović proposed the following problem.

Problem 6.1 Let a, b, c, and d be positive numbers and $n \geq 1$ a natural number. Prove that

$$\left|\sqrt[n]{a^n + b^n} - \sqrt[n]{c^n + d^n}\right| \leq \min\{|a - c| + |b - d|, |a - d| + |b - c|\}$$

and determine the corresponding estimation for

$$\left|\sqrt[n]{\sum_{i=1}^{N} a_i^n} - \sqrt[n]{\sum_{i=1}^{N} b_i^n}\right|$$

where $a_1, \ldots, a_N, b_1, \ldots, b_N$ are positive numbers $n \geq 1$, a natural number.

As solution of Problem 6.1, in [4] D. D. Adamović (see also [5] or [6]), gave the following result.

Proposition 6.2 *Let $a_1, \ldots, a_N, b_1, \ldots, b_N$ be positive numbers and $p \geq 1$.*

1. *Then,*

$$\left|\left(\sum_{i=1}^{N} a_i^p\right)^{\frac{1}{p}} - \left(\sum_{i=1}^{N} b_i^p\right)^{\frac{1}{p}}\right| \leq \min\left\{\sum_{i=1}^{N} |a_i - b_{p(i)}| : p(i) \in \mathcal{P}\right\},$$

where $\mathcal{P}$ denotes the set of all permutations of the set $\{1, \ldots, N\}$.

2. *Excluding trivial case $p = N = 1$ inequality turns in equality if and only if its right side is equal to zero.*
3. *With additional conditions $(a_n), (b_n) \in \ell^p$ statements* 1. *and* 2. *are also true for $N = +\infty$.*

We suggest that in 3. "min" must be replaced by "inf" because

$$\min\left\{\sum_{i=1}^{N} |a_i - b_{p(i)}| : p(i) \in \mathcal{P}\right\},$$

need not exist, as follows from next example.

Example 6.1 Let sequences (a_n) and (b_n) be defined by $a_i = \frac{1}{i^2}$ and $b_{2i-1} = b_{2i} = \frac{1}{i^2}$. Then,

$$\inf\left\{\sum_{i=1}^{N} |a_i - b_{p(i)}| : p(i) \in \mathcal{P}\right\} = \frac{1}{\pi^2},$$

but minimum does not exist.

6.1.4 Inequality of G. V. Milovanović and I. Ž. Milovanović

A convex function $f : [0, +\infty) \to [0, +\infty)$ belongs to class M if $f(0) = 0$ and if the function F defined by $F(t) = \ln f(\exp(t)$ is convex on $[0, +\infty)$. In [7] the following inequality, which generalizes Proposition 6.2 statement 1, was proved.

Proposition 6.3 *Let $(a_1, \ldots, a_n)$, $(b_1, \ldots, b_n)$ and $(p_1, \ldots, p_n)$ be finite sequences of non-negative real numbers, such that $p_i \geq 1$ for any $1 \leq i \leq n$. If $f \in M$, then*

$$\left|f^{-1}\left(\sum_{i=1}^{n} p_i f(a_i)\right) - f^{-1}\left(\sum_{i=1}^{n} p_i f(b_i)\right)\right| \leq \sum_{i=1}^{n} p_i |a_i - b_i|.$$

Let $p \geq 1$. Then, function $f : [0, +\infty) \to [0, +\infty)$ defined by $f(t) = t^p$ belongs to class M, which implies the following statement.

Corollary 6.1 *Let $(a_1, \ldots, a_n)$, $(b_1, \ldots, b_n)$, and $(p_1, \ldots, p_n)$ be finite sequences of non-negative numbers such that $p_i \geq 1$ for any $1 \leq i \leq n$. Then*

$$\left| \left(\sum_{i=1}^{n} p_i a_i^p \right)^{\frac{1}{p}} - \left(\sum_{i=1}^{n} p_i b_i^p) \right)^{\frac{1}{p}} \right| \leq \sum_{i=1}^{n} p_i |a_i - b_i|.$$

6.1.5 $\ell^{p,q}((p_n), (q_n))$ Spaces

Let w denote the set of all complex sequences. Suppose that (p_n) and (q_n) are two sequences of positive real numbers. Let real numbers $p, q \geq 1$ be arbitrary. Define

$$S_{p,q,(p_n),(q_n)}(x_n) = \sum_{m=0}^{+\infty} q_m \left(\sum_{n \in I(m)} p_n |x_n|^p \right)^{\frac{q}{p}},$$

where $(x_n) \in w$, $I(0) = \{0\}$ and $I(m) = \{n \in \mathbb{N} : 2^{m-1} \leq n < 2^m\}$, for $m > 0$. Then,

$$\ell^{p,q}((p_n), (q_n)) = \{(x_n) \in w : S_{p,q,(p_n),(q_n)}(x_n) < +\infty\}.$$

We say that $\ell^{p,q}((p_n), (q_n))$ is 2-weighted mixed normed sequence space. Define

$$\|(x_n)\|_{\ell^{p,q}} = S^{\frac{1}{q}}_{p,q,(p_n),(q_n)},$$

where $(x_n) \in \ell^{p,q}((p_n), (q_n))$.

If $p = q$ and $p_n = q_n = 1$ for any natural number n, then $\ell^{p,q}((p_n), (q_n))$ reduce to ℓ^p space.

6.2 Basic Results

6.2.1 Characterization of Pseudometric Spaces

Now we present the following characterization of pseudometric spaces.

Theorem 6.1 *Let X be non empty set and $d : X \times X \to [0, +\infty)$, such that for any $x, y \in X$, $x = y$ implies $d(x, y) = 0$. Then, the following statements are equivalent:*

(1) *(X, d) is pseudometric space;*

(2) *for any* $x, y, z \in X$ *holds* $d(x, y) \leq d(x, z) + d(y, z)$*;*
(3) *for any* $x, y, z \in X$ *holds* $|d(x, z) - d(y, z)| \leq d(x, y)$.

Proof [2) implies 1)] It is enough to prove that 2) implies $M2$). It follows from $d(x, y) \leq d(x, x) + d(y, x) = d(y, x)$ and $d(y, x) \leq d(y, y) + d(x, y) = d(x, y)$.

[1) implies 3)] From $d(x, z) \geq d(y, z)$ it follows $|d(x, z) - d(y, z)| = d(x, z) - d(y, z)$. Now, by $d(x, z) \leq d(x, y) + d(y, z)$ we get that

$$|d(x, z) - d(y, z)| = d(x, z) - d(y, z) \leq d(x, y) + d(y, z) - d(y, z) = d(x, y).$$

If $d(x, z) \leq d(y, z)$, then $|d(x, z) - d(y, z)| = d(y, z) - d(x, z)$. By $d(z, y) \leq d(y, x) + d(x, z)$, we obtained

$$\begin{aligned}|d(x, z) - d(y, z)| &= d(y, z) - d(x, z) \leq d(y, x) + d(x, z) - d(x, z) \\ &= d(y, x) = d(x, y).\end{aligned}$$

[3) implies 2)] From $d(x, y) \geq d(z, y)$, it follows

$$|d(x, y) - d(z, y)| = d(x, y) - d(z, y) \leq d(x, z).$$

So, $d(x, y) \leq d(x, z) + d(z, y)$. From $d(z, y) > d(x, y)$, it follows $d(x, y) \leq d(x, z) + d(z, y)$.

Now, it is enough to prove that 2) implies M2). It follows from $d(x, y) = |d(y, y) - d(x, y)| \leq d(y, x)$ and $d(y, x) = |d(x, x) - d(y, x)| \leq d(x, y)$. □

6.2.2 Extended Adamović's Inequality

Now we present our generalization of Adamović's inequality.

Theorem 6.2 *Let* $(a_n), (b_n) \in \ell^p$ *be complex sequences,* $p \geq 1$ *and* $\|(b_n)\|_{\ell^p} = r$*, where* r *is arbitrary positive real number. Then,*

$$\left| \left(\sum_{i=1}^{+\infty} |a_n|^p \right)^{\frac{1}{p}} - \left(\sum_{i=1}^{+\infty} |b_n|^p \right)^{\frac{1}{p}} \right| \leq \inf_{\|(c_n)\|_{\ell^p}=r} \sum_{i=1}^{+\infty} |a_i - c_i|.$$

Proof Let $(a_n), (b_n) \in \ell^p$ be arbitrary. By reverse triangle inequality, we obtained

$$|\|(a_n)\|_{\ell^p} - \|(b_n)\|_{\ell^p}| \leq \|(a_n - b_n)\|_{\ell^p}.$$

So, we get that

$$\left|\left(\sum_{i=1}^{+\infty}|a_i|^p\right)^{\frac{1}{p}}-\left(\sum_{i=1}^{+\infty}|b_i|^p\right)^{\frac{1}{p}}\right|\le\left(\sum_{i=1}^{+\infty}|a_i-b_i|^p\right)^{\frac{1}{p}}\le\sum_{i=1}^{+\infty}|a_i-b_i|,$$

because concave function $x\to x^{\frac{1}{p}}$ is sub-additive.

It follows that, if $\|(b_n)_{\ell^p}\|=r$, then

$$\left|\left(\sum_{i=1}^{+\infty}|a_i|^p\right)^{\frac{1}{p}}-\left(\sum_{i=1}^{+\infty}|b_i|^p\right)^{\frac{1}{p}}\right|\le\inf_{\|(c_n)\|_{\ell^p}=r}\sum_{i=1}^{+\infty}|a_i-c_i|.$$

□

6.2.3 Extended Inequality of G. V. Milovanović and I. Ž. Milovanović

Now we present our infinite-dimensional generalization of Corollary 6.1.

Theorem 6.3 *Let $(p_n)\subseteq[1,+\infty)$ be bounded sequence, $p\ge 1$, $(a_n),(b_n)\in\ell^p$ be complex sequences and $\|(b_n)\}\|_{\ell^p}=r$, where r is arbitrary positive real number. Then,*

$$\left|\left(\sum_{i=1}^{+\infty}p_i|a_i|^p\right)^{\frac{1}{p}}-\left(\sum_{i=1}^{+\infty}p_i|b_i|^p\right)^{\frac{1}{p}}\right|\le\inf_{\|(c_n)\|_{\ell^p}=r}\sum_{i=1}^{+\infty}p_i|a_i-c_i|.$$

Proof Let bounded sequence $(p_n)\subseteq[1,+\infty)$, $p\ge 1$, $(a_n),(b_n)\in\ell^p$, and $\|(b_n)_{\ell^p}\|=r$. We get that $(p_n^{\frac{1}{p}}a_n),(p_n^{\frac{1}{p}}b_n)\in\ell^p$, because (p_n) is bounded.

By reverse triangle inequality, we obtained

$$\begin{aligned}\left|\left(\sum_{i=1}^{+\infty}p_i|a_i|^p\right)^{\frac{1}{p}}-\left(\sum_{i=1}^{+\infty}p_i|b_i|^p\right)^{\frac{1}{p}}\right|&=\left|\left\|\left(p_n^{\frac{1}{p}}a_n\right)\right\|_{\ell^p}-\left\|\left(p_n^{\frac{1}{p}}b_n\right)\right\|_{\ell^p}\right|\\&\le\|(p_n^{\frac{1}{p}}a_n-p_n^{\frac{1}{p}}b_n)\|_{\ell^p}\\&=\left(\sum_{i=1}^{+\infty}|p_i^{\frac{1}{p}}a_i-p_i^{\frac{1}{p}}b_i|^p\right)^{\frac{1}{p}}\le\sum_{i=1}^{+\infty}p_i|a_i-b_i|,\end{aligned}$$

because concave function $x\to x^{\frac{1}{p}}$ is sub-additive and $p_i^{\frac{1}{p}}\le p_i$.

Hence, if $\|(b_n)_{\ell^p}\| = r$, then

$$\left|\left(\sum_{i=1}^{+\infty} p_i|a_i|^p\right)^{\frac{1}{p}} - \left(\sum_{i=1}^{+\infty} p_i|b_i|^p\right)^{\frac{1}{p}}\right| \leq \inf_{\|(c_n)\|_{\ell^p}=r} \sum_{i=1}^{+\infty} p_i|a_i - c_i|.$$

□

If $p_n = 1$ for each non-negative integer n, then Theorem 6.3 reduces to Theorem 6.2.

6.3 Adamović's Inequality for $\ell^{p,q}((p_n), (q_n))$ Spaces

Our main result is the following theorem, which generalizes Theorem 6.2 and Theorem 6.3.

Theorem 6.4 *Let $(p_n), (q_n) \subseteq [1, +\infty)$ be bounded sequence, $p \geq 1$, $(a_n), (b_n) \in \ell^{p,q}((p_n), (q_n))$ be complex sequences, and $\|(b_n)\}\|_{\ell^{p,q}} = r$, where r is arbitrary positive real number. Then,*

$$\left|\left(\sum_{m=1}^{+\infty} q_m \left(\sum_{n\in I(m)} p_n|a_n|^p\right)^{\frac{q}{p}}\right)^{\frac{1}{q}} - \left(\sum_{m=1}^{+\infty} q_m \left(\sum_{n\in I(m)} p_n|b_n|^p\right)^{\frac{q}{p}}\right)^{\frac{1}{q}}\right|$$

$$\leq \inf_{\|(c_n)\|_{\ell^{p,q}}=r} \sum_{m=0}^{+\infty} \sum_{n\in I(m)} q_m p_n|a_n - c_n|.$$

Proof We shall prove that $\|\cdot\|_{l^{p,q}}$ is norm on $\ell^{p,q}((p_n), (q_n))$. It is enough that triangle inequality is proved. Let $(a_n), (b_n) \in \ell^{p,q}((p_n), (q_n))$ be arbitrary. Then,

$$\|(a_n) + (b_n)\|_{l^{p,q}} = \left(\sum_{m=1}^{+\infty} q_m \left(\sum_{n\in I(m)} p_n|a_n + b_n|^p\right)^{\frac{q}{p}}\right)^{\frac{1}{q}}$$

$$= \left(\sum_{m=1}^{+\infty} q_m \left(\sum_{n\in I(m)} |p_n^{\frac{1}{p}} a_n + p_n^{\frac{1}{p}} b_n|^p\right)^{\frac{q}{p}}\right)^{\frac{1}{q}}$$

[inequality of Minkowski implies]

$$= \left(\sum_{m=1}^{+\infty} q_m \left(\left(\sum_{n\in I(m)} |p_n^{\frac{1}{p}} a_n|^p\right)^{\frac{1}{p}} + \left(\sum_{n\in I(m)} |p_n^{\frac{1}{p}} b_n|^p\right)^{\frac{1}{p}}\right)^q\right)^{\frac{1}{q}}$$

[inequality of Minkowski implies]

$$\leq \left(\sum_{m=0}^{+\infty} q_m \left(\sum_{n\in I(m)} p_n|a_n|^p\right)^{\frac{q}{p}}\right)^{\frac{1}{q}}$$

$$+ \left(\sum_{m=0}^{+\infty} q_m \left(\sum_{n\in I(m)} p_n|b_n|^p\right)^{\frac{q}{p}}\right)^{\frac{1}{q}}$$

$$= \|(a_n)\|_{l^{p,q}} + \|(b_n)\|_{l^{p,q}}.$$

By reverse triangle inequality, we obtained

$$\left|\left(\sum_{m=1}^{+\infty} q_m \left(\sum_{n\in I(m)} p_n|a_n|^p\right)^{\frac{q}{p}}\right)^{\frac{1}{q}} - \left(\sum_{m=1}^{+\infty} q_m \left(\sum_{n\in I(m)} p_n|b_n|^p\right)^{\frac{q}{p}}\right)^{\frac{1}{q}}\right|$$

$$= \left|\|(a_n)\|_{l^{p,q}} - \|(b_n)\|_{l^{p,q}}\right| \leq \|(a_n - b_n)\|_{l^{p,q}},$$

because $\ell^{p,q}((p_n), (q_n))$ is normed space. Further, we get that

$$\|(a_n - b_n)\|_{l^{p,q}} = \left(\sum_{m=1}^{+\infty} q_m \left(\sum_{n\in I(m)} p_n|a_n - b_n|^p\right)^{\frac{q}{p}}\right)^{\frac{1}{q}}$$

$$= \left(\sum_{m=1}^{+\infty} q_m \left(\sum_{n\in I(m)} |p_n^{\frac{1}{p}} a_n - p_n^{\frac{1}{p}} b_n|^p\right)^{\frac{q}{p}}\right)^{\frac{1}{q}}$$

$$\leq \left(\sum_{m=1}^{+\infty} \left(q_m^{\frac{1}{q}} \sum_{n\in I(m)} |p_n^{\frac{1}{p}} a_n - p_n^{\frac{1}{p}} b_n|\right)^q\right)^{\frac{1}{q}}$$

$$\le \sum_{m=1}^{+\infty} q_m^{\frac{1}{q}} \sum_{n\in I(m)} p_n^{\frac{1}{p}} |a_n - b_n|$$

$$\le \sum_{m=1}^{+\infty} \sum_{n\in I(m)} q_m p_n |a_n - b_n|,$$

because concave functions $x \to x^{\frac{1}{p}}$, $x \to x^{\frac{1}{q}}$ are sub-additive, $p_n^{\frac{1}{p}} \le p_n$ and $q_m^{\frac{1}{q}} \le q_m$. So, if $\|(b_n)_{\ell^{p,q}}\| = r$, then

$$\left| \left(\sum_{m=1}^{+\infty} q_m \left(\sum_{n\in I(m)} p_n |a_n|^p \right)^{\frac{q}{p}} \right)^{\frac{1}{q}} - \left(\sum_{m=1}^{+\infty} q_m \left(\sum_{n\in I(m)} p_n |b_n|^p \right)^{\frac{q}{p}} \right)^{\frac{1}{q}} \right|$$

$$\le \inf_{\|(c_n)\|_{\ell^{p,q}}=r} \sum_{m=0}^{+\infty} \sum_{n\in I(m)} q_m p_n |a_n - c_n|.$$

□

Acknowledgments The authors would like to thank the reviewers and editors for their valuable suggestions.

This research was supported in part by the Serbian Ministry of Science, Technological Development and Innovations according to Contract 451-03-47/2023-01/200105, dated on February 3, 2023.

References

1. Fréchet, M.: Sur quelques points du calcul fonctionnel. Rend. Circ. Mat. Palermo **22**, 1–74 (1906)
2. Lindenbaum, A.: Contributionsàl'étude de l'espace métrique I. Fund. Math. **8**, 209–222 (1926)
3. Mitrinović, D.S.: Problem no. 109. Mat. Vesnik **4**(19), 453 (1967)
4. Adamović, D.D.: Solution of Problem no. 109. Mat. Vesnik **5**(20), 548–549 (1968)
5. Mitrinović, D.S. (in cooperation with Vasić, P.M.): Analytic Inequalities, p. 281. Springer, Berlin, Heidelberg (1970)
6. Mitrinović, D.S., Pečarić, J.E.: Hölder's and related inequalities. (in Serbian), p. 42. Naučna knjiga, Beograd (1990)
7. Milovanović, G.V., Milovanović, I.Ž.: A generalization of a problem given by D. S. Mitrinović. Publ. Elektroteh. Fak. Univ. Beogr. Ser. Mat. Fiz. **602–633**, 129–132 (1978)

Chapter 7
Classical Karamata Theory of Regular Variability and the Index Function Operator

Danica Fatić and Dragan Djurčić

Dedicated to Academician Gradimir V. Milovanović on the occasion of his 75th birthday

7.1 Introduction

In the past century, in the 1930s, a new area of asymptotic analysis began to develop, which later received the name of Classical Karamata theory of regular variability. Namely, studying the correlation of the limit value and the sequence of Riemann–Stiltjes integrals (especially Dirichlet and power series), Karamata ascertains that this relation is characterized by the asymptotic behavior of the kernel of each of these integrals. Those asymptotic characteristics (serious and essential) for function (and sequence) are: regular variability and rapid variability. Those characteristics very quickly became the subject of study in a qualitative sense and the terms of their applications (see [1]). The mentioned features play an important role in collectivity theory, oscillation theory, the theory of Tauber properties, cosmology, Fourier analysis, number theory, machine learning, differential equations, etc. (see [2–10]). In [11], the authors considered Friedman's equation, which is derived from the Einstein field equation and has a central place in the so-called λ-CDM model of Universe (the Standard model). The complete analysis in the paper is based on Karamata theory. Immediately after the 1930s, due to the needs in other mathematical disciplines, modified models of the classic Karamata theory of regular variables were formed (probability and complex analysis [1] and [12]).

D. Fatić · D. Djurčić (✉)
Department of Mathematics, Faculty of Technical Science, University of Kragujevac, Čačak, Serbia
e-mail: danica.fatic@ftn.kg.ac.rs; dragan.djurcic@ftn.kg.ac.rs

M. P. Stanić et al. (eds.), *Analysis, Approximation, Optimization: Computation and Applications*, Springer Optimization and Its Applications 224,
https://doi.org/10.1007/978-3-031-85743-0_7

7.2 Functional Aspect

The function $f : [a, +\infty) \to (0, +\infty)$ is an O-regularly varying function in the sense of Karamata (see [1] and [13]) if for some fixed $a > 0$ is measurable and

$$\limsup_{x \to +\infty} \frac{f(\lambda x)}{f(x)} = k_f(\lambda) < +\infty \tag{7.1}$$

holds for every $\lambda > 0$.

The function $k_f(\lambda)$, $\lambda > 0$, is called the index function of f, and some of its properties give many asymptotic characteristics of the function f (see [14] and [15]). The function $k_f(\lambda)$, $\lambda > 0$, can be measurable and immeasurable. An example of immeasurable index function was given in [16] by Rubel, who constructed the corresponding function f.

The O-regularly varying functions in the sense of Karamata form the functional class ORV_φ and the elements of that class have a very important convergence condition in the theory of Tauber properties, the so-called Tauber condition (see [1, 6, 17]). Also, the class ORV is important in the structural analysis of functional asymptotic processes (see [14, 18, 19] and [20]).

Let $f \in ORV_\varphi$. With that assumption, we have the following definitions.

(a) A function $f \in IRV_\varphi$ (we can also find in some literature that the class IRV_φ is denoted by CRV_φ) if $k_f(\lambda)$, $\lambda > 0$, is a continuous function. Elements of this class satisfy the famous Schmidt's condition of convergence, very important in collectivity theory (see [1, 21, 22]).

(b) A function $f \in ERV_\varphi$ (the functional class ERV_φ is called the Matuszewska class [1, 23, 24] or the class of extended regularly varying functions in the sense of Karamata [1]) if $k_f(\lambda)$, $\lambda > 0$, has finite one-sided derivatives for every $\lambda > 0$. Elements of this class are very important objects in cosmology (see [1]).

(c) A function $f \in RV_\varphi$ (the functional class RV_φ is the class of regular varying functions in the sense of Karamata [25] and [26]; it represents the basis of the classical Karamata theory of regular variability) if $k_f(\lambda)$, $\lambda > 0$, is a differentiable function. Each element of that class of functions satisfies the famous Karamata convergence condition (see [1, 27]). A particularly important subclass of the class RV_φ is the class SV_φ (of slowly varying functions in the sense of Karamata [1] and [25]). A function $f \in SV_\varphi$ if $k_f(\lambda) = 1$ for every $\lambda > 0$.

(d) A function $f \in SO_\varphi^\beta$ (the class SO_φ^β is the class β-Seneta's functions [28] and [29]), if for some $\beta \geqslant 1$, it holds that $k_f(\lambda) \leqslant \beta$ for every $\lambda > 0$. The class $SO_\varphi = \bigcup_{\beta \geqslant 1} SO_\varphi^\beta$ is the well-known class of Seneta's functions and has a very important role in probability theory and approximation theory (see [1] and [12]).

It is known that

$$SV_\varphi \subsetneq RV_\varphi \subsetneq ERV_\varphi \subsetneq IRV_\varphi \subseteq ORV_\varphi, \tag{7.2}$$

holds.

Also, it holds

$$SV_\varphi \subsetneq SO_\varphi \subsetneq ORV_\varphi, \tag{7.3}$$

and the class SO_φ is incomparable with classes CRV_φ and ERV_φ. Also, we have $SO_\varphi \cap RV_\varphi \setminus SV_\varphi = \emptyset$.

Here are some examples:

i. $f_1(x) = \log x,\ x \geqslant \mathrm{e};\ f_1 \in SV_\varphi$;
ii. $f_2(x) = x^\alpha,\ x \geqslant \mathrm{e}$ and $\alpha \in \mathbb{R} \setminus \{0\}$, $f_2 \in RV_\varphi \setminus SV_\varphi$;
iii. $f_3(x) = x^{\frac{1}{\sqrt{2}}(\sin(\log(\log x)))},\ x \geqslant \mathrm{e};\ f_3 \in ERV_\varphi \setminus RV_\varphi$;
iv. $f_4(x) = \exp\{|\sin(\log x)|\}^{\frac{1}{2}},\ x \geqslant \mathrm{e};\ f_4 \in CRV_\varphi \setminus ERV_\varphi$;
v. $f_5(x) = (2 + \sin x) \cdot \exp\{\int_1^x \frac{2+\sin(\log t)}{t}\,\mathrm{d}t\},\ x \geqslant \mathrm{e};\ f_5 \in ORV_\varphi \setminus CRV_\varphi$;
vi. $f_6(x) = 2 + \sin(\log x),\ x \geqslant \mathrm{e};\ f_6 \in SO_\varphi \setminus SV_\varphi$;
vii. $f_7(x) = f_5(x),\ x \geqslant \mathrm{e};\ f_7 \in ORV_\varphi \setminus SO_\varphi$.

Note In the previous examples, $\log x = \log_{\mathrm{e}} x = \ln x$, for $x \geqslant \mathrm{e}$.

In the paper [30], we can find the next proposition.

Proposition 7.1 *Assume that* $f \in ORV_\varphi$.

(a) *If* $k_f(\lambda)$, $\lambda > 0$, *is a measurable function, then* $k_f \in ORV_\varphi$.
(b) *If* $k_f(\lambda)$, $\lambda > 0$, *is a continuous function, then* $k_f \in IRV_\varphi$.
(c) *If* $k_f(\lambda)$, $\lambda > 0$, *is a measurable and bounded function, then* $k_f \in SO_\varphi$.
(d) *If* $k_f(\lambda)$, $\lambda > 0$, *is a differentiable function, then* $k_f \in RV_\varphi$.
(e) *If* $k_f(\lambda)$, $\lambda > 0$, *is a constant function, then* $k_f \in SV_\varphi$.

Note As we can see, if $k_f(\lambda)$, $\lambda > 0$, is a measurable and bounded function with constant $\beta \geqslant 1$, then $k_f \in SO_\varphi^\beta$.

The next proposition can be found in the paper [30], but here we will use a different way to prove it.

Proposition 7.2 *Assume that* $f \in ORV_\varphi$ *and* $k_f(\lambda)$, $\lambda > 0$, *has finite one-sided derivatives for every* $\lambda > 0$. *Then,* $k_f \in ERV_\varphi$.

Proof From assumptions, we have that $f \in ERV_\varphi$. Hence, there are real numbers c and d (upper and lower Matuszewska index—see [16]) such that $-\infty < c \leqslant d < +\infty$ and

$$\lambda^c \leqslant \frac{f(\lambda x)}{f(x)} \leqslant \lambda^d, \tag{7.4}$$

for every $\lambda > 1$ and every $x \geqslant x(\lambda) \geqslant a$, for some $a > 0$.

So, $\lambda^c \leqslant \liminf_{x\to+\infty} \frac{f(\lambda x)}{f(x)} \leqslant k_f(\lambda) \leqslant \lambda^d$ for every $\lambda > 0$. Hence, for every $\lambda > 1$,

$$k_{k_f} = \limsup_{x\to+\infty} \frac{k_f(\lambda x)}{k_f(x)} \leqslant \limsup_{x\to+\infty} \frac{k_f(\lambda)k_f(x)}{k_f(x)} = k_f(\lambda) \leqslant \lambda^d.$$

Also,

$$\underline{k}_{k_f}(\lambda) = \liminf_{x\to+\infty} \frac{k_f(\lambda x)}{k_f(x)} \geqslant \liminf_{x\to+\infty} \frac{\underline{k}_f(\lambda)\underline{k}_f(x)}{\underline{k}_f(x)} = \underline{k}_f(\lambda) \geqslant \lambda^c,$$

for every $\lambda > 1$. (Here $\underline{k}_f(\lambda) = \liminf_{x\to+\infty} \frac{f(\lambda x)}{f(x)}$, $\lambda > 0$, is the so-called auxiliary index function of the function $f \in ORV_\varphi$ (see [1]).

For the same c and d and every $\lambda > 1$, it holds

$$\lambda^c \leqslant \underline{k}_{k_f}(\lambda) \leqslant k_{k_f}(\lambda) \leqslant \lambda^d. \tag{7.5}$$

Therefore, for every $x \geqslant x(\lambda) \geqslant b > 0$, it is satisfied

$$\lambda^c \leqslant \frac{k_(\lambda x)}{k_f(x)} \leqslant \lambda^d. \tag{7.6}$$

So, we proved that $k_f \in ERV_\varphi$. □

7.2.1 *Index Function Operator*

Let us consider the functional transformation K on the class ORV_φ to the class of positive functions defined on the interval $(0, +\infty)$ given by

$$K(f) = k_f. \tag{7.7}$$

Notice that

$$K(K(f)) = K(f), \tag{7.8}$$

if $f \in RV_\varphi$. It is also valid if $f \in SV_\varphi$.

We call the transformation K from (7.7) the index function operator (see [30]), and its interesting properties are listed in Propositions 7.1 and 7.2.

According to the previous, it makes sense to consider the iterative process given by (see [30])

$$K_n(f) = K(K \cdots (K(f)) \cdots), \tag{7.9}$$

for $n \in \mathbb{N}$ on classes ORV_φ, CRV_φ, ERV_φ, RV_φ and SO_φ.

Let us observe the properties of the mentioned operator on the given function classes [30]. The iterative process for the operator from (7.7) on the class RV_φ is described in (7.8).

Let $f \in ORV_\varphi$, and let $K_n(f)(x)$ be a measurable function for every $x > 0$ and every $n \in \mathbb{N}$. Let us consider the limit

$$\lim_{n \to +\infty} K_n(f)(s), \tag{7.10}$$

for $s > 0$. We see that (7.10) converges to a limit, which is a measurable function for $s > 0$. Let us denote that limit by $p(s)$, $s > 0$. The next proposition is the statement from the paper [30].

Proposition 7.3 *Let $f \in ORV_\varphi$. Then, $p \in ORV_\varphi$ and*

(a) *if $f \in SO_\varphi$, then $p \in SO_\varphi$;*
(b) *if $f \in CRV_\varphi$, then $p \in CVR_\varphi$ (and $p(s)$, $s > 0$, is a continuous function).*

Proof Let us provide the proof for (b); the proof of (a) the reader can find in the paper [30]. Let $f \in CRV_\varphi$. According to Proposition 7.1 (b), the function $K(f) \in CRV_\varphi$. Also, the function $K(f)(s)$ is a continuous function, as well as the function $K_n(f)(s)$, $s > 0$, for every $n \in \mathbb{N}$.

If $s = 1$, then $p(s) = 1$. If $s > 0$, $s \neq 1$, then for every $n \in \mathbb{N}$, we have

$$0 < \frac{1}{K_n(f)(1/s)} \leqslant \frac{1}{K_{n+1}(f)(1/s)} \leqslant K_{n+1}(f)(s) \leqslant K_n(f)(s) < +\infty. \tag{7.11}$$

As we can see, the function $p(s)$, $s > 0$, is finite and positive for every $s > 0$. As $p(s) \leqslant K_1(f)(s)$ for every $s > 0$ and $\lim_{s \to 1} K_1(f)(s) = 1$, then $\limsup_{s \to 1} p(s) \leqslant 1$. It means that the function p is measurable on the interval $(0, +\infty)$ as the limit value of continuous functions and for every s and t greater than zero, it holds

$$p(st) = \lim_{n \to +\infty} K_n(f)(st) \leqslant \lim_{n \to +\infty} K_n(f)(s) \cdot \lim_{n \to +\infty} K_n(f)(t) = p(s)p(t).$$

Based on the previous, the function p is continuous on the interval $(0, +\infty)$. As if $K(p)(s) \leqslant p(s)$ for every $s > 0$ and $\limsup_{s \to 1} K(p)(s) \leqslant 1$, then $K(p)$ is a continuous function on the interval $(0, +\infty)$. Hence, $p \in CRV_\varphi$ holds. □

Corollary 7.1 *Let $f \in ERV_\varphi$. Then, $p \in CRV_\varphi$, where p is given as* (7.10) *for the function f.*

A question was asked in the paper [30]: Does it hold from Corollary 7.1 that $p \in ERV_\varphi$?

The answer is contained in the following statement.

Proposition 7.4 *Let $f \in ERV_\varphi$. Then, $p \in ERV_\varphi$, where p is given as* (7.10) *for the function f.*

Proof Let $\lambda > 1$. As $f \in ERV_\varphi$, there are real numbers c and d such that $c \leqslant d$ and $\lambda^s \leqslant K(f)(\lambda) \leqslant \lambda^d$ for every $\lambda > 1$.

For the same λ, according to Proposition 7.2,

$$\lambda^c \leqslant K_n(f)(s) \leqslant \lambda^d, \tag{7.12}$$

for every $n \in \mathbb{N}$ and for the same n, $K_n(f) \in ERV_\varphi$ holds. Since for each $\lambda \geqslant 1$, there is a limit value $p(\lambda) = \lim_{n \to +\infty} K_n(f)(\lambda)$ and (7.12) holds, then for the same $\lambda > 1$, it is satisfied

$$\lambda^c \leqslant p(\lambda) \leqslant \lambda^d, \tag{7.13}$$

for mentioned numbers c and d. For the same $\lambda > 1$, it holds

$$\lambda^c \leqslant \liminf_{x \to +\infty} \frac{f(\lambda x)}{f(x)} \leqslant \lambda^d. \tag{7.14}$$

If we introduce the auxiliary index function operator as:

$$\underline{K}(f) = \underline{k}_f, \tag{7.15}$$

where $f \in ORV_\varphi$ (and $ERV_\varphi \subsetneq ORV_\varphi$) and $\underline{k}_f(\lambda) = \liminf_{x \to +\infty} \dfrac{f(\lambda x)}{f(x)}$, for $\lambda > 1$, then for $n \in \mathbb{N}$, we can define

$$\underline{K}_n(f) = K(\cdots(K(f))\cdots), \tag{7.16}$$

for every $f \in ORV_\varphi$, also for $f \in ERV_\varphi$. Hence, for $f \in EVR_\varphi$, $\lambda^c \leqslant \underline{K}(f) \leqslant \lambda^d$ for every $\lambda > 1$ holds. So, for $\lambda > 1$,

$$\lambda^c \leqslant \underline{K}_n(f)(\lambda) \leqslant \underline{K}_{n+1}(f)(\lambda) \leqslant K_{n+1}(f)(\lambda) \leqslant K_n(\lambda) \leqslant \lambda^d,$$

for every $n \in \mathbb{N}$ (according to (7.16)).

For $n \in \mathbb{N}$ and $\lambda > 1$, it holds $\underline{K}_n(f) \leqslant \underline{K}_{n+1}(f)$ because: for $f \in ERV_\varphi \subsetneq ORV_\varphi$ is valid that $\lambda^c \leqslant \underline{k}_f(\lambda) \leqslant \lambda^d$, so $\lambda^c \leqslant \underline{K}(\lambda) = \underline{K}_1(\lambda)$, for $\lambda > 1$ and

$$\underline{K}_2(\lambda) = \liminf_{s \to +\infty} \frac{\underline{K}(\lambda s)}{\underline{K}(s)} \geqslant \underline{K}_1(\lambda) = \underline{K}(\lambda) \geqslant \lambda^c. \tag{7.17}$$

We continue in this way and by mathematical induction, we have that for every $n \in \mathbb{N}$, $\underline{K_n}(f) \leqslant K_{n+1}(f)$ for every $\lambda > 1$. We used in the previous that, for every $\lambda > 0$, $\underline{K_n}(f)(\lambda)$ is an element of the class $ERV_\varphi \subsetneq ORV_\varphi$, and so for every $s > 0$

and every $\lambda > 0$, $\underline{K_n}(f)(s\lambda) \geqslant \underline{K_n}(f)(s) \cdot \underline{K_n}(f)(\lambda)$. Hence, for every $\lambda > 1$ and every $x \geqslant x(\lambda) > 0$ it holds that $\lambda^{2c-d} \leqslant \dfrac{p(\lambda x)}{p(x)} \leqslant \lambda^d$, hence $p \in ERV_\varphi$.

Explanation In the previous one, we used the following.

i. According to Proposition 7.1, for every $s > 0$ and every $t > 0$, it holds that $p(st) \leqslant p(s) \cdot p(t)$, because of $p \in CRV_\varphi \subsetneq ORV_\varphi$.
ii. $K(p)(\lambda) \leqslant p(\lambda) \leqslant \lambda^d$ for $\lambda > 1$.
iii. There is a function $g(\lambda)$, $\lambda > 0$, such that $g(\lambda) \leqslant p(\lambda)$, for $\lambda > 0$ and $g(\lambda) = \lim\limits_{n\to+\infty} \underline{K_n}(f)(\lambda)$. The function g satisfies $g(st) \geqslant g(s) \cdot g(t)$, for every $t > 0$ and every $s > 0$.
iv. For $\lambda > 0$, we have $K(g)(\lambda) = \liminf\limits_{s\to+\infty} \dfrac{g(\lambda s)}{g(s)} \geqslant \liminf\limits_{s\to+\infty} \dfrac{g(\lambda)\cdot g(s)}{g(s)} = g(\lambda) \geqslant \lambda^c$.
v. For $\lambda > 1$, it holds

$$\begin{aligned}\lambda^d \geqslant K(p) \geqslant &\limsup_{x\to+\infty}\left(\frac{g(\lambda x)}{g(x)}\cdot\frac{g(x)}{p(x)}\right) \geqslant \liminf_{x\to+\infty}\left(\frac{g(\lambda x)}{g(x)}\cdot\frac{g(x)}{p(x)}\right)\\ \geqslant &\lambda^c \cdot \liminf_{x\to+\infty}\frac{g(x)}{p(x)} \geqslant \lambda^{2c-d}.\end{aligned}$$

□

7.3 Sequence Aspect

Parallel with the functional theory of the classical Karamata regular variation, the sequence theory was also developed (see [25] and [1]). Even earlier (see [31] and [32]), in the collectivity theory, the beginnings of the sequence classical Karamata theory of regular variability appear when considering Tauber's conditions in the theory of Tauber's type properties for series. One of the most interesting papers from that period in the mentioned topic is [31] in which the so-called Schmidt's convergence condition was considered.

Jovan Karamata's (1902–1967) first papers from the theory of regular variation in the 1930s, as well as his doctoral dissertation, equally treat the convergence conditions for summability procedures for series and for integral transformations for functions (see [1] and [26]). Karamata's view on the convergence condition was interesting and perhaps unique at that time, because he believed that continuous and discrete aspects should not be separated. He claimed that the essence is the same and that only the form is in question—testimony of Ranko Bojanic (1925–2017), professor of Columbus University, at a lecture at the Serbian Academy of Science and Arts (Belgrade, 1998).

A sequence $c = (c_n)$ of positive numbers is said to be:

i. Slowly varying in the sense of Karamata, if

$$\lim_{n \to +\infty} \frac{c_{[\lambda n]}}{c_n} = 1, \tag{7.18}$$

for every $\lambda > 0$. We denote the class of all such sequences by SV_s.

ii. Regularly varying in the sense of Karamata, if

$$\lim_{n \to +\infty} \frac{c_{[\lambda n]}}{c_n} = r_c(\lambda) < +\infty, \tag{7.19}$$

for every $\lambda > 0$. We denote the class of all such sequences by RV_s. The inclusion $SV_s \subsetneq RV_s$ holds.

iii. Matuszewska sequence (see [33]) or extended regularly varying in the sense of Karamata, if

$$\lambda^c \leqslant k_c(\lambda) \leqslant \lambda^d, \tag{7.20}$$

for every $\lambda > 1$ and some real numbers c and d such that $s \leqslant d$ is satisfied. We denote the class of all such sequences by ERV_s. Note that $RV_s \subsetneq ERV_s$ holds. (In (7.20), it holds

$$k_c(\lambda) = \limsup_{n \to +\infty} \frac{c[\lambda n]}{c_n}, \tag{7.21}$$

for every $\lambda > 0$.)

iv. Smith's sequence, if

$$\lim_{n \to +\infty, \lambda \to 1} \frac{c_{[\lambda n]}}{c_n} = 1. \tag{7.22}$$

We denote the class of all such sequences by CRV_s. Observe that $ERV_s \subsetneq CRV_s$ holds (see [33] and [34]).

v. Seneta's sequence, if

$$k_c(\lambda) \leqslant \beta, \tag{7.23}$$

for some $\beta \geqslant 1$ and every $\lambda > 0$. We denote the class of all such sequences by SO_s. The class of Seneta's sequences is incomparable with the classes CRV_s and ERV_s (see [29] and [33]). Also, $SV_s \subsetneq SO_s$ and $SO_s \cap RV_s \setminus SV_s \neq \emptyset$ holds (see [29]).

vi. O-regularly varying in the sense of Karamata (see [35]), if

$$k_c(\lambda) < +\infty, \tag{7.24}$$

for every $\lambda > 0$. We denote the class of all such sequences by ORV_s. Notice that $CRV_s \subsetneq ORV_s$ and $SO_s \subsetneq ORV_s$ hold (see [33] and [29]).

The sequence classes given in 1–6 are very important sequence classes in Karamata theory of regular variation (see [29, 33, 35] and [36]). Those classes of sequences are, each individually, important in the collectivity theory as convergence conditions.

From the creation of the discrete classical Karamata theory of regular variation until the 1970s of the past century, it was developed independently of the functional theory, against the recommendations and opinion of Jovan Karamata.

However, in the papers [36] and [37] in the 1970s, the following proposition was proven.

Proposition 7.5 *Let $c = (c_n)$ be a sequence of positive numbers. Then, $c \in RV_s$ if and only if $f \in RV_s$, where $f(x) = c_{[x]}$, $x \geqslant 1$.*

After those papers, there were revolutionary assertions of the unique results of the discrete and continuous classical theory of regular variability.

According to the well-known characterization theorem $r_c(\lambda) = \lambda^q$, for every $\lambda > 0$ and some $\rho \in \mathbb{R}$, $c \in RV_s$ if and only if $f \in RV_\varphi$, and $\lim\limits_{x \in +\infty} \dfrac{f(\lambda x)}{f(x)} = r_f(\lambda) = r_c(\lambda) = \lambda^\rho$, for $\lambda > 0$ and the same $\rho \in \mathbb{R}$, where $f(x) = c_{[x]}$, $x \geqslant 1$.

Then (along with other important results) a unique (sequence and function) classical Karamata theory of regular variation was established. So, RV_s theory and RV_φ theory formed the RV theory—theory of Karamata's regular variation. Ten years later, Slobodan Aljančić proposed a hypothesis: The statement in Proposition 7.5 also holds when $c \in ORV_s$ and $c_{[x]}$, $x \geqslant 1$ is an element of ORV_φ! He did not deal with the proof.

In the 1990s of the last century, many people from the Department of Analysis of the Faculty of Mathematics in Belgrade tried to prove or disprove it. After that, in 1997, the following statement was proved.

Proposition 7.6 (see [35]) *Let $c = (c_n)$ be a sequence of positive numbers. Then, the following statements are equivalent:*

(a) $c \in ORV_s$;
(b) $f(x) = c_{[x]}$, $x \geqslant 1$, *is an element of* ORV_φ.

Proof The proof that (b) implies (a) is trivial. Let us recall only the proof of (a) implies (b) given in the paper [35]. That part is interesting because it relies on the characteristics of the Cantor's sequence of inserted segments.

Let $c \in ORV_s$, then the function $f(x) = c_{[x]}$, $x \geqslant 1$ is positive and piecewise continuous. We will prove that there is an interval $[a, b]$, $0 < a < b < +\infty$ and some $M > 0$ such that for every $\lambda \in [a, b]$ and every $n \in \mathbb{N}$, $\dfrac{c_{[\lambda n]}}{c_n} < M$ holds. Suppose the opposite that: for every M and every positive a and b $(a < b)$, there exist some $\lambda \in [a, b]$ and some $n \in \mathbb{N}$ such that $\dfrac{c_{[\lambda n]}}{c_n} > M$. Now we will prove that from this assumption, it follows that there exists some $\lambda > 0$ such that

$$\limsup_{n\to+\infty} \frac{c_{[\lambda n]}}{c_n} = +\infty. \tag{7.25}$$

Let λ_1 and n_1 be numbers such that $\frac{c_{[\lambda_1 n_1]}}{c_n} > 1$. Then there exists an interval $[\alpha_1, \beta_1]$ containing a λ_1 such that $\frac{c_{[\lambda n_1]}}{c_{n_1}} = \frac{c_{[\lambda_1 n_1]}}{c_{n_1}} > 1$ for every $\lambda \in [\alpha_1, \beta_1]$.

Let $a_1 = \alpha_1$ and $b_1 = \beta_1$. Consider the interval $\left[\frac{2a_1+b_1}{3}, \frac{a_1+2b_1}{3}\right]$. By assumption, there exist some $\lambda_2 \in \left[\frac{2a_1+b_1}{3}, \frac{a_1+2b_1}{3}\right]$ and some $n_2 \in \mathbb{N}$ such that $\frac{c_{[\lambda_2 n_2]}}{c_{n_2}} > 2$. Then there exists some interval $[\alpha_2, \beta_2]$, $\alpha_2 < \beta_2$ containing λ_2 such that $\frac{c_{[\lambda n_2]}}{c_{n_2}} > 2$ for every $\lambda \in [\alpha_2, \beta_2]$. If we put $[a_2, b_2] = [a_1, b_1] \cap [\alpha_2, \beta_2]$, we see that $a_2 < b_2$. We continue this process successively and obtain a sequences of intervals $[a_s, b_s]$ and natural numbers n_s, $s \in \mathbb{N}$ such that $\frac{c_{[\lambda n_s]}}{c_{n_s}} > s$ for every $\lambda \in [a_s, b_s]$ and $[a_{s+1}, b_{s+1}] \subseteq [a_s, b_s]$. It means, there exists $\lambda > 0$, $\lambda \in \cap_{s=1}^{+\infty}[a_s, b_s]$. For that λ and every $s \in \mathbb{N}$ we have that $\frac{c_{[\lambda n_s]}}{c_{n_s}} > s$, and therefore, $\limsup_{n\to+\infty} = +\infty$. This is a contradiction, so the assumption does not apply.

Hence, there exist some $M > 0$ and some interval $[a, b]$, $0 < a < b < +\infty$ such that $\frac{c_{[\lambda n]}}{c_n} < M$ for every $n \in \mathbb{N}$ and every $\lambda \in [a, b]$. Further, let $\lambda > 0$ and $\alpha = \frac{2\lambda}{a+b}$. For sufficiently large x, there exists some $t = t(x) \in [a, b]$ such that

$$\frac{c_{[\lambda x]}}{c_{[x]}} = \frac{c_{[t[\alpha(x)]]}}{c_{[\alpha(x)]}} \cdot \frac{c_{[\alpha(x)]}}{c_{[x]}}. \tag{7.26}$$

Since $\frac{c_{[t[\alpha(x)]]}}{c_{[\alpha(x)]}} < M$ and $c \in ORV_s$, that is for $\alpha = \alpha(\lambda) > 0$

$$\limsup_{x\to+\infty} \frac{c_{[\lambda x]}}{c_{[x]}} \leqslant M \cdot k_c(\alpha) < +\infty, \tag{7.27}$$

so that $f(x) = c_{[x]}$, $x \geqslant 1$, is an element of class ORV_φ. □

The previous propositions (without proof) can be seen in the monograph [32]. The following propositions were proved in papers [33] and [29] and are also listed in the monograph [32].

Proposition 7.7 *Let $c = (c_n)$ be a sequence of positive numbers.*

(a) *If $c \in ERV_s$, then $f(x) = c_{[x]}$, $x \geqslant 1$ is an element of ERV_φ* [33].
(b) *If $c \in CRV_s$, then $f(x) = c_{[x]}$, $x \geqslant 1$ is an element of CRV_φ* [33].

(c) *If* $c \in SO_s$, *then* $f(x) = c_{[x]}$, $x \geqslant 1$ *is an element of* SO_φ [29].

Proof Let us prove the statement (b) because of the interesting method in its proof. Analogous methods were used in the proof of statements (a) and (b) in the paper [29]. Let $c \in SO_s$. Then there exists some $\beta \geqslant 1$ such that

$$k_c(\lambda) \leqslant \beta, \tag{7.28}$$

for every $\lambda > 0$.

If $\epsilon > 0$, prove that there exists an interval $[a, b] \subsetneq (0, +\infty)$, $a < b$, and $n_0 \in \mathbb{N}$, such that $\frac{c_{[\lambda n]}}{c_n} < \beta + \epsilon$ for every $n \geqslant n_0$ and every $\lambda \in [a, b]$.

For some $\lambda > 0$, let us define $n_\lambda \in \mathbb{N}$ in the following way: $n_\lambda = 1$ if $\frac{c_{[\lambda n]}}{c_n} < \beta + \epsilon$ for every $n \in \mathbb{N}$ and $n_\lambda = 1 + \max\left\{n \in \mathbb{N} \mid \frac{c_{[\lambda n]}}{c_n} \geqslant \beta + \epsilon\right\}$ otherwise. We see that $1 \leqslant n_\lambda < +\infty$.

Let us define a sequence of sets (A_s) with $A_s = \{\lambda > 0 \mid n_\lambda > s\}$, $s \in \mathbb{N}$. Then the sequence A_s is a decreasing sequence of sets such that $\cap_{s=1}^{+\infty} A_s = \emptyset$. Now we will prove that not every set A_s, $s \in \mathbb{N}$, is dense in $(0, +\infty)$. If, for fixed $s \in \mathbb{N}$, $\lambda \in A_s$, then $\frac{c_{[(n_\lambda - 1)\lambda]}}{c_{n_\lambda - 1}} \geqslant \beta + \epsilon$, and there exists $\delta_\lambda > 0$ such that $\frac{c_{[(n_\lambda - 1)t]}}{c_{n_\lambda - 1}} = \frac{c_{[(n_\lambda - 1)\lambda]}}{c_{n_\lambda - 1}} \geqslant \beta + \epsilon$ for every $t \in [\lambda, \lambda + \beta_\lambda)$, since $n_t \geqslant (n_\lambda - 1) + 1 > s$ for that s and $t \in A_s$, $s \in \mathbb{N}$. So, if $\lambda \in A_s$, then $(\lambda, \lambda + \delta_\lambda) \subseteq A_s$. That is, if the set A_s is dense in $(0, +\infty)$, then $(Int A_s)$ is a sequence of dense and open sets in the interval $(0, +\infty)$ of the second category. Hence, $\cap_{s=1}^{+\infty} Int A_s$ is dense in $(0, +\infty)$, so it is non-empty. It is a contradiction to the assumption. Therefore, there exists $n_0 \in \mathbb{N}$ such that A_{n_0} is not dense in $(0, +\infty)$. Hence, there exists an interval $[a, b] =\subsetneq (0, +\infty)$, $a < b$ such that $[a, b] \subsetneq (0, +\infty) \setminus A_{n_0} = \{\lambda \in (0, +\infty) \mid n_\lambda \leqslant n_0\}$. That is, for every $\lambda \in [a, b]$, it holds that $n_\lambda \leqslant n_0$. Therefore, $\frac{c_{[\lambda n]}}{c_n} \leqslant \beta + \epsilon$ is for every $n \geqslant n_0 \geqslant n_\lambda$ and every $\lambda \in [a, b]$. Then for every $\lambda > 0$ and sufficiently large $x \geqslant x_0$ it holds that $\frac{c_{[\lambda x]}}{c_{[x]}} = \frac{c_{[\alpha(x)t]}}{c_{[\alpha(x)]}} \cdot \frac{c_{[\alpha(x)]}}{c_{[x]}}$, where $t = t(x) \in [a, b]$ and $\alpha = \alpha(\lambda) = \frac{2\lambda}{a+b}$. So, $k_f(\lambda) \leqslant (\beta + \epsilon) \cdot \beta$, for every $\lambda > 0$ and arbitrary $\epsilon > 0$. That is, $k_f(\lambda) \leqslant \beta^2$ for every $\lambda > 0$ so, $f \in SO_\varphi$. □

Proposition 7.8 *Let* $c \in CRV_s$. *Then,* $k_c(\lambda) = k_f(\lambda)$, *for* $f(x) = c_{[x]}$, $x \geqslant 1$.

The previous proposition can be found as a statement in [33].

Corollary 7.2 *Let* $c = (c_n)$ *be a sequence of positive numbers.*

(a) *If* $c \in ERV_s$, *then* $k_c \in ERV_\varphi$.
(b) *If* $c \in RV_s$, *then* $k_c \in RV_\varphi$, *where* $f(x) = c_{[x]}$, $x \geqslant 1$.
(c) *If* $c \in SV_s$, *then* $k_c \in SV_\varphi$ *where* $f(x) = c_{[x]}$, $x \geqslant 1$ ($k_c(\lambda) = 1$ *for every* $\lambda > 0$).

Proposition 7.9 *Let $c \in ORV_s$. Then, $k_c \in ORV_\varphi$.*

Proof For $c \in ORV_s$ it holds that

$$\limsup_{n\to+\infty} \frac{c_{[stn]}}{c_n} \leqslant \limsup_{n\to+\infty} \frac{c_{[t\frac{sn}{[sn]}[sn]]}}{c_{[sn]}} \cdot \limsup_{n\to+\infty} \frac{c_{[sn]}}{c_{[n]}}, \tag{7.29}$$

for every $\lambda > 0$. Hence, $k_c \in ORV_\varphi$. □

Let $c = (c_n)$ be an element of the class ORV_s. Consider the transformation

$$K^* : ORV_s \to \text{"class of positive and finite functions on an interval}(0, +\infty)\text{"} \tag{7.30}$$

given by

$$K^*(c) = k_c. \tag{7.31}$$

Proposition 7.10 *Let $c = (c_n) \in ORV_s$.*

(a) *If $c \in IRV_s$, then $p^*(\lambda) = \lim_{n\to+\infty} K(\cdots(K^*(c))\cdots)$ is an element of IRV_φ, for $\lambda > 0$.*
(b) *If $c \in ERV_s$, then $p^*(\lambda) = \lim_{n\to+\infty} K(\cdots(K^*(c))\cdots)$ is an element of IRV_φ, for $\lambda > 0$.*

Remark 7.1 The following is true.

(a) $p^*(\lambda)$, $\lambda > 0$, is an element of the class SV_φ, if $c = (c_n) \in SV_s$ and $K(K^*(c)) = K^*(c)$.
(b) $p^*(\lambda)$, $\lambda > 0$, is an element of the class RV_φ, if $c = (c_n) \in RV_s$ and $K(K^*(c)) = K^*(c)$.
(c) For every $n \in \mathbb{N}$ it follows that $K_n' = K(\cdots(K^*(c))\cdots)$ is an element of the class ERV_φ, or IRV_φ, if $c \in IRV_\varphi$ or $c \in ERV_\varphi$, respectively.
(d) If $c \in ORV_s$ and for every $n \in \mathbb{N}$ and every $\lambda > 0$

$$K_n' = K_n'(c) = K(\cdots(K^*(c))\cdots) \tag{7.32}$$

is measurable function, then $K_n' \in ORV_\varphi$ and $p^*(\lambda) = \lim_{n\to+\infty} K_n'(\lambda)$, $\lambda > 0$, is an element of the class ORV_φ
(e) If $c \in SO_s$ and for every $n \in \mathbb{N}$ and for every $\lambda > 0$, $K_n'(\lambda)$ given by (7.32) is a measurable function, then $K_n' \in SO_\varphi$ and $p^*(\lambda) = \lim_{n\to+\infty} K_n'(\lambda)$, $\lambda > 0$, is an element of the class SO_φ.

7.4 The Index Function Operator of Classes ARV_φ and ARV_φ

Consider the functional class (see [15])

$$A = \{f : [a, +\infty) \to (0, +\infty), a > 0 \mid f\}$$
$$\{\text{is non-decreasing and unlimited function}\} \tag{7.33}$$

and the generated inverse on A given by $\overleftarrow{f}(x) = \inf\{y \geqslant a \mid f(y) > x\}$ (see [1, 38–42]), for $x \geqslant f(a)$. Also, let us look at the functional class $A \cap CRV_\varphi$ (see [15]). Then, $\overleftarrow{f} \in A$, whenever $f(x)$, $x \geqslant a$, is an element of A. Also, it is valid that $\overleftarrow{f}(x) = f^{-1}(x)$ for $x \geqslant f(a)$ whenever f is an increasing and continuous function.

The following proposition was proved in the paper [15].

Proposition 7.11

(a) *If $f \in A \cap CRV_\varphi$, then $\overleftarrow{f}(x)$, $x \geqslant f(a)$, is a function such that $\underline{k}_{\overleftarrow{f}}(\lambda) > 1$, for $\lambda > 1$.*
(b) *If $f \in A$ and $\underline{k}_f(\lambda) > 1$, for $\lambda > 1$, then $\overleftarrow{f} \in A \cap CRV_\varphi$.*

Denote by ARV_φ the class of positive and measurable functions defined on the interval $[a, +\infty)$, $a > 0$, for which

$$\underline{k}_f(\lambda) > 1 \tag{7.34}$$

holds for every $\lambda > 1$.

Also, it is valid that

$$R_{\infty,\varphi} \subsetneq ARV_\varphi \subseteq PI_\varphi, \tag{7.35}$$

where $R_{\infty,\varphi}$ are de Hahn rapidly varying functions in the sense of Karamata (see [1, 43, 44]) and PI_φ are the so-called positively increasing functions in the sense of Karamata (see [45, 46]). The letter A in the ARV_φ is dedicated to the famous Belgrade mathematicians (students of Jovan Karamata) V. Avakumovic (1910–1990), S. Aljancic (1922–1993), and D. Adamovic (1928–2008). It can be noticed that the class ARV_φ contains as a proper subclasses (except for the class $R_{\infty,\varphi}$) and the class of all properly varying functions in the sense of Karameta with a positive index of variability (see [47]).

We see that $ARV_\varphi \cap ORV_\varphi \neq \emptyset$ and $ARV_\varphi \setminus ORV_\varphi \neq \emptyset$ (due to functions from the class $R_{\infty,\varphi}$). Also, $ORV_\varphi \setminus ARV_\varphi \neq \emptyset$ because no slowly varying function in the Karamata's sense belongs to the ARV_φ class. Let us introduce the notation ${}^*ARV_\varphi = ARV_\varphi \cap ORV$.

The next proposition is very interesting and important (see [24]).

Proposition 7.12 *It is valid that* $ERV^* = ARV_\varphi \cap CRV_\varphi^*$ *(where* ERV_φ^* *is the class of Matuszewska functions where both Matuszewska indices are positive, and the class* CRV_φ^* *is the class of Smith's functions* $f \in CRV_\varphi$ *such that the right derivative* $k'(1)$ *is finite).*

Consider the transformation $\underline{K}$ from the class $^*ARV_\varphi$ to the class of positive functions defined on the interval $(0, +\infty)$ given by

$$\underline{K}(f)(\lambda) = \underline{k}_f(\lambda), \quad \text{for } \lambda > 0. \tag{7.36}$$

The transformation (7.36) will be called the operator of the auxiliary index function on the function class $^*ARV_\varphi$. That operator can be correctly defined on the entire class ORV_φ, but we will not deal with that here. The following proposition holds.

Proposition 7.13 *Let* $f \in^* ARV_\varphi$*, then* $\underline{K}(f) \in^* ARV_\varphi$ *if it is measurable.*

Proof Let $f \in^* ARV_\varphi$, Then, $\underline{k}_f(\lambda) = \liminf_{x\to+\infty} \dfrac{f(\lambda x)}{f(x)} > 1$ for $\lambda > 1$. Hence, for $s > 1$ and $t > 1$, it is satisfied

$$\begin{aligned}
\underline{K}(f)(st) = \underline{k}_f(st) &= \liminf_{x\to+\infty} \frac{f(stx)}{f(x)} = \liminf_{x\to+\infty} \left(\frac{f(stx)}{f(tx)} \cdot \frac{f(tx)}{f(x)} \right) \\
&\geqslant \liminf_{x\to+\infty} \frac{f(stx)}{f(tx)} \cdot \liminf_{x\to+\infty} \frac{f(tx)}{f(x)} \\
&= \underline{k}_f(s) \cdot \underline{k}_f(t) = \underline{K}(f)(s) \cdot \underline{K}(f)(t),
\end{aligned}$$

for every $s > 1$ and $t > 1$.

Therefore, for $s > 1$,

$$\underline{k}_{\underline{K}_f}(s) = \liminf_{x\to+\infty} \frac{\underline{K}(f)(st)}{\underline{K}(f)(t)} \geqslant \underline{k}_f(s) > 1, \tag{7.37}$$

i.e., $\underline{K}(f) \in^* ARV_\varphi$. □

According to the previous, it makes sense to consider the iterative process on the class $^*ARV_\varphi$ given by

$$\underline{K}_n(f) = \underline{K}(\underline{K} \cdots (\underline{K}(f)) \cdots), \tag{7.38}$$

for some $n \in \mathbb{N}$.

If, for every $n \in \mathbb{N}$, $\underline{K}_n(f)(\lambda)$, $\lambda > 0$, is a measurable function, then the sequence $(\underline{K}_n(f))$ is non-decreasing (with respect to (7.37)) measurable functions on the interval $(0, +\infty)$, and it is valid that the limit process

$$\lim_{n\to+\infty} \underline{K}_n(f)(s), \tag{7.39}$$

for $s > 0$, must converge in $\mathbb{R}$, depending on s, because

$$\underline{K}_n(f)(s) \leqslant \underline{k}_f\left(\frac{1}{s}\right) = \frac{1}{\underline{K}_1}(f)\left(\frac{1}{s}\right) = \frac{1}{\underline{K}}(f)\left(\frac{1}{s}\right),$$

for every $n \in \mathbb{N}$. Hence, for $s > 0$, a positive function

$$\overline{p}(s) = \lim_{n\to+\infty} \underline{K}_n(f)(s), \tag{7.40}$$

that is measurable and for which it holds that for $s > 0$, $0 < \underline{k}_f(s) \leqslant \overline{p}(s) < +\infty$ and, for $s > 1$, it holds $1 < \underline{k}_f(s) \leqslant \overline{p}(s) < +\infty$.

The conclusion after (7.39): For $s > 0$ and $n \in \mathbb{N}$, it holds

$$\begin{aligned}\frac{1}{\underline{k}_f(\frac{1}{s})} &= \frac{1}{\underline{K}_1(f)(\frac{1}{s})}\\ &\geqslant \cdots \geqslant \frac{1}{\underline{K}_n(f)(\frac{1}{s})} \geqslant \frac{1}{\underline{K}_{n+1}(f)(\frac{1}{s})} \geqslant \underline{K}_{n+1}(f)(s) \geqslant \underline{K}_n(f)(s)\\ &\geqslant \cdots \geqslant \underline{K}_1(f)(s) = \underline{k}_f(s) > 0 \quad \text{and} \quad \frac{1}{\underline{k}_f(\frac{1}{s})} < +\infty.\end{aligned}$$

For s and t greater than zero, it holds

$$\begin{aligned}\overline{p}(st) &= \lim_{n\to+\infty} \underline{K}_n(f)(st)\\ &\geqslant \lim_{n\to+\infty} \underline{K}_n(f)(s) \cdot \lim_{n\to+\infty} K_n(f)(t)\\ &= \overline{p}(s) \cdot \overline{p}(t).\end{aligned}$$

Therefore, $\underline{k}_{\overline{p}}(s) = \liminf_{t\to+\infty} \dfrac{\overline{p}(st)}{\overline{p}(t)} \geqslant \overline{p}(s) > 1$, for every $s > 1$, so $\overline{p} \in ARV_\varphi$.

For $s > 0$, we have that $\underline{K}_1(f)(s) \leqslant \underline{K}_2(f)(s) \leqslant \cdots \leqslant \underline{K}_n(f)(s) \leqslant \overline{p}(s)$. Also, $\underline{K}_1(f)(s) > 0$, for $s > 0$, hence

$$\liminf_{x\to+\infty} \frac{\underline{K}_1(f)(sx)}{\underline{K}_1(f)(x)} = \underline{k}_{\underline{K}_1(f)}(s) \geqslant \underline{k}_f(s) > 0 \tag{7.41}$$

holds. This means $\underline{K}_1(f) \in ORV_\varphi$.

For $s > 0$ and $n \in \mathbb{N}$, $(n \geqslant 2)$

$$\underline{K}_n(f)(s) = \liminf_{x\to+\infty} \frac{\underline{K}_{n-1}(f)(sx)}{\underline{K}_{n-1}(f)(x)} = \underline{k}_{n-1}(f)(s) = \underline{K}_{n-1}(f)(s)$$
$$\geqslant \cdots \geqslant \underline{K}_1(f)(s) \quad \text{and} \quad \underline{k}_f(s) > 0.$$

Hence, $\underline{K}_n(f) \in ORV_\varphi$.

For $s > 0$, $t > 0$,

$$\overline{p}(st) = \lim_{n\to+\infty} \underline{K}_n(f)(st) \geqslant \lim_{n\to+\infty} \underline{K}_n(f)(s) \cdot \lim_{n\to+\infty} \underline{K}_n(f)(t) = \overline{p}(s) \cdot \overline{p}(t)$$

holds.

For $s > 0$, $\underline{k}_{\overline{p}}(s) = \liminf_{t\to+\infty} \frac{\overline{p}(st)}{\overline{p}(t)} = \overline{p}(s) > 0$, so $\overline{p} \in ORV_\varphi$.

That is, since $\overline{p} \in ARV_\varphi$, we conclude $\overline{p} \in^* ARV_\varphi$.

Let $c = (c_n)$ be a sequence of positive numbers. Let us denote by ARV_s the class of sequences c for which

$$\underline{k}_c(\lambda) = \liminf_{n\to+\infty} \frac{c_{[\lambda n]}}{c_n} > 1, \tag{7.42}$$

hold for $\lambda > 1$.

For the classes ARV_s and ARV_φ, the following proposition (which is a theorem of Galambos–Bojanic–Senata type) was proved in the paper [47] (see [15, 45, 46] and [8] for more details about the importance of the class ARV_s in asymptotic analysis, selection principles theory and game theory).

Proposition 7.14 *Let $c = (c_n)$ be a sequence of positive numbers. Then the following statements are equivalent:*

(a) $c \in ARV_s$;
(b) *the function $f(x) = c_{[x]}$, $x \geqslant 1$, is element of the class ARV.*

Let $c = (c_n)$ be an element of the class ${}^*ARV_s = ARV_s \cap ORV_s$.

Consider the transformation K^* given by (7.30) on the class ${}^*ARV$ in the modification

$$\underline{K}^*(c) = \underline{k}_c, \quad \text{for } c \in^* ARV_s. \tag{7.43}$$

Proposition 7.15 *Let a sequence $c = (c_n) \in^* ARV_s$ be given. Then, the function $\underline{K}^*(c)(\lambda)$, $\lambda > 0$, is an element of the class ${}^*ARV_\varphi$ if it is measurable.*

Proof If $c \in^* ARV_s$, then according to Proposition 7.14, the function $f(x) = c_{[x]}$, $x \geqslant 1$ is an element of class ARV_φ, so

$$\liminf_{n\to+\infty} \frac{c_{[stn]}}{c_n} \geqslant \liminf_{n\to+\infty} \frac{c_{[t\frac{sn}{[sn]}[sn]]}}{c_{[sn]}} \cdot \liminf_{n\to+\infty} \frac{c_{[sn]}}{c_n} \text{ for } \lambda > 1. \tag{7.44}$$

Since $c \in ORV_s$, for every $\lambda > 0$ we have $\underline{k}_c(\lambda) > 0$. Also, $\underline{k}_c(\lambda) \leqslant k_c(\lambda) < +\infty$, for every $\lambda > 0$. As for $s > 0$ and $t > 0$, $\underline{k}_c(st) \geqslant M(t) \cdot \underline{k}_c(s)$, for some $M(t) > 1$, $k_{\underline{k}_c}(t) = \liminf_{s \to +\infty} \dfrac{\underline{k}_c(st)}{\underline{k}_c(s)} \geqslant M(t) > 1$ for $t > 1$, and so $\underline{k}_c \in ARV_\varphi$ if it is measurable. According to Proposition 7.9, $\underline{k}_c \in ORV_\varphi$, i.e., $\underline{k}_c \in^* ARV_\varphi$. □

Remark 7.2 From the aforementioned, it is obvious that $\underline{K'}_n(\lambda) = \underline{K}(\cdots(\underline{K}(\lambda))\cdots)$, for $\lambda > 0$, is an element of the class $^*ARV_\varphi$. Also, $\overline{p}^*(\lambda) = \lim_{n \to +\infty} \underline{K'}_n(\lambda)$, $\lambda > 0$, is an element of the class $^*ARV_\varphi$.

References

1. Bingham, N.H., Goldie, C.M., Teugels, J.L.: Regular Variation. Cambridge University Press, Cambridge (1987)
2. Grow, D.E., Stanojević, Č.V.: Convergence and Fourier character of trigonometric transforms with slowly varying convergence moduli. Math. Ann. **302**(3),433–472 (1995)
3. Stanojević, Č.: Structure of Fourier and Fourier-Stieltjes coefficients of series with slowly varying convergence moduli. Bull. Am. Math. Soc. **19**(1),283–286 (1988)
4. Kusano, T., Manojlović, J.V, Milošević, J.: Intermediate solutions of fourth order quasilinear differential equations in the framework of regular variation. Appl. Math. Comput. **248**, 246–272 (2014)
5. Timotić, V.,Djurčić, D., Žižović, M.: On rapid equivalence and translational rapid equivalence. Kragujevac J. Math. **46**(2), 259–265 (2022)
6. de Haan, L., Statdtmüller, U.: Dominated variation and related concepts and Tauberian theorems for Laplace transforms. J. Math. Anal. Appl. **108**, 344–365 (1985)
7. Timotić, V., Djurčić, D., Nikolić, R.M.: On slowly varying sequences. Filomat **25**(6), 1075–1080 (2012)
8. Kočinac, Lj.D.R., Djurčić, D., Manojlović, V.J.: Regular and rapid variations and some applications. In: Ruzhansky, M., Dutta, H., Agarwal, R.P. (eds.) Mathematical Analysis and Applications: Selected Topics. Chapter "Partitions and Automorphisms of Finite Kurepa Trees", pp. 429–491. John Wiley, & Sons, New York (2018)
9. Fleige, A.: Characterizations of monotone O-regularly varying functions by means of indefinite eigenvalue problems and HELP type inequalities. J. Math. Anal. Appl. **412**, 345–359 (2014)
10. Berman, S.M.: Sojourns and extremes of a diffusion process on a fixed interval. Adv. Appl. Probab. **14**(4), 811–832 (1982)
11. Mijajlović, Ž., Pejović, N., Šegan, S., Damljanovic, G.: On asymptotic solutions of Friedmann equations. Appl. Math. Comput. **219**, 1273–1286 (2012)
12. Feller, W.: One-side analogues of Karamata's regular variation. L'Enseignement Mathématique **15**, 107–121 (1969)
13. Karamata, J.: Bernerkung über die vorstehende Arbeit des Herrn Avakumović, mit näherer Betrachtung einer Klasse von Funktionen, welche bei den Inversionssätzen vorkommen. Bull Int. Acad. Youg. Sci. **29–30**, 117–123 (1936)
14. Djurčić, D.: O-regularly varying functions and strong asymptotic equivalence. J. Math. Anal. Appl. **220**, 451–461 (1998)
15. Djurčić, D., Torgašev, A., Ješić, S.: The strong asymptotic equivalence and the generalized inverse. Sib. Math. J. **49**(4), 786–795 (2008)
16. Arandjelović, D.: O-regular variation and uniform convergence. Publication de l'Institut Mathematique **48**(62), 25–40 (1990)

17. Aljančić, S., Arandjelović, D.: O–regularly varying functions. Publ. Inst. Math. (Beograd) **22**(36), 5–22 (1977)
18. Buldygin, V.V., Klesov, O.I., Steinebach, J.G.: Properties of a subclass of Avakumović functions and their generalized inverses. Ukrainian Math. J. **54(2)**, 179–206 (2002)
19. Buldygin, V.V., Klesov, O.I., Steinebach, J.G.: On some properties of asymptotically quasi-inverse functions. Teory Imovirn. Mat. Stat. **77**, 33–27 (2007) (in Ukrainian; English translation: Theory of Probability and Mathematcal Statistics **77**, 15–30 (2008))
20. Djurčić, D., Torgašev, A.: Strong asymptotic equivalence and inversion of functions in the class K_c. J. Math. Anal. Appl. **255**, 383–390 (2001)
21. Cline, D.B.H. : Intermediate regular and Π variation. Proc. Lond. Math. Soc. **68**(3), 594–616 (1994)
22. Korenblum, B.I.: On the asymptotic behavior of Laplace integrals near the boundary of a region of convergence. Dokl. Akad. Nauk SSSR **104**, 173–176 (1955)
23. Matuszewska, W., Orlicz W.: On some classes of function with regard to their orders of growth. Stud. Math. **26**, 11–24 (1965)
24. Matuszewska, W.: On a generalization of regularly increasing funkctions. Stud. Math. **24**, 271–279 (1964)
25. Karamata, J.: Sur un mode de croissance régulière des functions. Mathematica (Cluj) **4**, 38–53 (1930)
26. Karamata, J.: Sur un mode de croissance régulière. Théorèmes fondamenteaux. Bulletin de la Société. Mathématique se France **61**, 55–62 (1933)
27. Seneta, E.: Regularly Varying Functions, LNM 508. Springer, New York (1976)
28. Drasin, D., Seneta E.: A generalization of slowly varying function. Proc. Am. Math. Soc. **93(3)**, 470–472 (1986)
29. Djurčić, D., Torga šev, A.: On the Seneta sequences. Acta Math. Sin. **22**(3), 689–692 (2006)
30. Djurčić, D., Fatić, D., Elez, N.: The index function operator for O-regularly varying functions. Kragujevac J. Math. **47**(7), 1041–1046 (2023)
31. Schmidt, R.: Uber divergente Folgen and lineare Mittelbildungen. Math. Z. **22**(1), 89–152 (1925)
32. Djurčić, D., Kočinac, Lj.D.R.: On theorems of Galambos-Bojanić-Seneta type. In: Hararika, B., Acharjee, S., Srivastava, H.M. (Eds) Advances in Mathematical Analysis and its Applications. Chapter "On Reverse Triangle Inequality and Some of Its Applications", pp. 95–111. CRC Press, Boca Raton (2022)
33. Djurčić, D., Torga šev, A.: Representation theorems for sequences of classes CR_c and $ER_{c'}$. Sib. Math. J. **45**(5), 834–838 (2004)
34. Djurčić, D.: A theorem on a representation of *-regularly varying sequence. Filomat **16**, 1–6 (2002)
35. Djurčić, D., Božin, V.: A proof of S. Aljančić hypothesis on O-regularly varying sequences. Publications de l'Institut Mathematique **61**(76), 46–52 (1997)
36. Bojanić, R., Seneta E.: A unified theory of regularly varying sequence. Math. Z. **134**, 91–106 (1973)
37. Galambos, J., Seneta E.,: Regularly varying sequence. Proc. Am. Math. Soc. **41**(1), 110–116 (1973)
38. Djurčić, D., Torgašev, A.: Weak asymptotic equivalence and inverse functions in the class OR. Math. Morav. **7**, 1–6 (2003)
39. Djurčić, D., Torgašev, A.: Some asymptotic relations for the generalized inverse. J. Math. Anal. Appl. **325**, 1397–1402 (2007)
40. Djurčić, D., Nikolić, R.M., Torgašev, A.: The weak asymptotic equivalence and generalized inverse. Lith. Math. J. **50**(1), 34–42 (2010)
41. Djurčić, D., Nikolić, R.M., Torgašev, A.: The weak and strong asymptotic equivalence relations and the generalized inverse. Lith. Math. J.**51**(4), 472–476 (2011)
42. Djurčić, D., Konjokrad, V., Nikolić, R.M.: Asymptotic similarity relation and generalized inverse. Filomat **25**(6), 1075–1080 (2012)

43. Djurčić, D., Kočinac, Lj.D.R., Žižović, M.R.: Some properties of rapidly varying sequences. J. Math. Anal. Appl. **327**, 1297–1306 (2007)
44. Timotić, V., Djurčić, D., Kočinac, Lj.D.R.: On rapidly varying sequences. Appl. Anal. Discrete Math. **16**(1), 83–90 (2022)
45. Djurčić, D., Kočinac LJ.D.R, Žižović M.: A few remarks on divergent sequences: Rates of divergence. J. Math. Anal. Appl. **360**, 588–598 (2009)
46. Djurčić, D., Kočinac LJ.D.R., Žižović M.: A few remarks on divergent sequences: rates of divergence II. J. of Math. Anal. Appl. **367**, 705–709 (2010)
47. Djurčić, D., Torgašev A.: A Theorem of Galambos-Bojanić-Seneta type. Abstr. Appl. Anal. Article ID 360794, 6 p. (2009)

Chapter 8
A Note on Laguerre Truncated Polynomials and Quadrature Formula

Juan C. García-Ardila and Francisco Marcellán

Dedicated to Academician Gradimir V. Milovanović on the occasion of his 75th birthday

8.1 Introduction

Integration with respect to a probability measure supported on the real line is a topic that belongs to analysis, and so is the evaluation or approximation of integrals. According to the seminal work by Gauss, one is led to orthogonal polynomials with respect to such a measure. It is well known that the zeros of such polynomials are real, simple and located in the interior of the convex hull of the support of the measure. Another important fact is that the zeros of two consecutive polynomials interlace, i.e., between two zeros of the orthogonal polynomial of degree $n+1$, you have exactly one zero of the polynomial of degree n (see [1], among others).

The key elements in the study of quadrature formulas are the nodes and weights in order to have the optimal degree of exactness. In fact, it is very well known that the Gaussian quadrature formulas provide the optimal degree of exactness for polynomials of degree at most $2n-1$ when you deal with n nodes. Indeed, they are the zeros of the polynomial of degree n associated with the measure appearing in the integral.

The connection between Gaussian quadrature rules and the algebraic eigenvalue problem was, if not discovered, then certainly exploited in the now classical and widely cited paper [14]. Indeed, the connection between the zeros of orthogonal

J. C. García-Ardila
Departamento de Matemática Aplicada a la Ingeniería Industrial, Universidad Politécnica de Madrid, Madrid, Spain
e-mail: juancarlos.garciaa@upm.es

F. Marcellán (✉)
Departamento de Matemáticas, Universidad Carlos III de Madrid, Leganés, Spain
e-mail: pacomarc@ing.uc3m.es

M. P. Stanić et al. (eds.), *Analysis, Approximation, Optimization: Computation and Applications*, Springer Optimization and Its Applications 224,
https://doi.org/10.1007/978-3-031-85743-0_8

polynomials and the eigenvalues of the truncated Jacobi matrices representing the symmetric operator of multiplication by x for the standard inner product defined by the probability measure is the key point. Thus, we have an interplay between classical analysis and numerical linear algebra as pointed out in [9]. For *Gauss–Radau* and *Gauss–Lobatto quadrature formulas,* the nodes are related with zeros of orthogonal polynomials with respect to polynomial perturbations of the probability measure of degrees 1 and 2, respectively. Those zeros are eigenvalues of Jacobi matrices that are defined from Darboux transformations of the Jacobi matrix associated with the initial probability measure. On the other hand, in [8], the entries of Jacobi matrices associated with generalized Christoffel transformations (perturbations of probability measures by positive rational functions) and their numerical implementation have been analyzed, where the connection with the QD-algorithm is also stated.

In the literature, Gaussian quadrature formulas for classical orthogonal polynomials (Hermite, Laguerre and Jacobi) and efficient algorithms to estimate the reminders in the approximation of integral of continuous and analytic functions have been implemented (see [11], among others).

For nonclassical measures, Gaussian quadrature formulas have been exhaustively analyzed in the literature from the pioneering work [4] for an absolutely continuous measure $e^{-tx^2}\,dx$ supported in the interval $(-1, 1)$, where t is a positive real number. They constitute a basic tool in the computation of integrals related to electron repulsion in molecular quantum mechanics [4]. The corresponding sequences of orthogonal polynomials are known in the literature as Rys polynomials [20, 24]. We must point out that such a perturbation of the Lebesgue measure supported in $(-1, 1)$ yields an example of a Volterra integrable system. On the other hand, by using a change of variables, the orthogonality with respect to a truncated Normal distribution supported in a symmetric interval of the real line appears in a natural way. The corresponding sequences of orthogonal polynomials appear in the study of Gaussian unitary ensembles with two discontinuities [18]. Analytic properties of such orthogonal polynomials (called truncated Hermite polynomials) have been studied in [2]. Other examples of Gauss quadrature formulas for nonclassical measures appear in [19, 22].

More recently, in [22], the authors have considered the problem of construction of quadrature formulas of Gaussian type on $[-1, 1]$, concerning the following two-parametric weight function $w(x; z) = e^{-zx^2}(1 - x^2)^{\lambda-1/2}$, $\lambda > -1/2$, $z > 0$. This weight function is a generalization of the *Rys quadrature*, and it can be obtained by transforming quadratures on $[-1, 1]$ with N nodes to quadratures on $(0, 1)$ with only $(N + 1)/2$ nodes. Such an approach provides a stable and very efficient numerical construction. Notice that a Volterra lattice associated with the symmetric Jacobi (Gegenbauer) measure appears in a natural way. It can be also considered as a continuous perturbation of the Gaussian unitary ensemble with two jumps.

In [3], the authors deal with sequences of polynomials orthogonal with respect to the truncated Gamma distribution, i.e., the absolutely continuous measure $x^{\alpha}e^{-x}\,dx$ supported on an interval $(0, z)$, $z > 0$, $\alpha > -1$. Analytic properties of such polynomials (known as truncated Laguerre polynomials) as well as of their zeros have been studied therein. Notice that a symmetrization problem yields the truncated generalized Hermite polynomials, which have been also analyzed in [3]. The aim of this contribution is to consider Gaussian quadrature formulas for the truncated Gamma distribution as well as to provide an efficient algorithm for the accuracy of them in terms of the parameter z. We emphasize our work on the analysis of computational methods to deal with the nodes and weights of such Gaussian quadrature formulas.

The structure of the manuscript is the following. In Sect. 8.2, we summarize the basic background about Gaussian quadrature formulas, and we explain the modified Chebyshev algorithm based on generalized moments of the measure defined by a new sequence of orthogonal polynomials in order to improve the numerical stability of the eigenvalue problem of the Jacobi matrix. In Sect. 8.3, we study the generalized moments of the truncated Gamma distribution taking into account shifted Jacobi polynomials as supporting sequence. Their expressions are deduced. Finally, in Sect. 8.4, some numerical tests are presented in order to get the coefficients of the three-term recurrence relation that the truncated Laguerre polynomials satisfy and their dependence of z. The nodes and weights in the corresponding Gaussian quadrature formulas are described in an illustrative case for choices of α and z.

8.2 Basic Background

Let $\mathbb{P}$ be the set of polynomials with complex coefficients. We will denote by $\mathbb{P}_n$ the linear subspace of polynomials of degree at most n. The algebraic dual of $\mathbb{P}$, denoted by $\mathbb{P}^*$, is defined as the set of all linear mappings from $\mathbb{P}$ into $\mathbb{C}$, that is,

$$\mathbb{P}^* = \{\mathbf{u} : \mathbb{P} \to \mathbb{C} \, : \, \mathbf{u} \text{ is linear}\}.$$

The image of a polynomial p under $\mathbf{u}$ will be expressed using duality brackets as $\langle \mathbf{u}, p \rangle$. The elements of $\mathbb{P}^*$ are usually called *linear functionals*. Any linear functional $\mathbf{u}$ is completely defined by the values $\mathbf{u}_n =: \langle \mathbf{u}, x^n \rangle$, $n \geq 0$, where $\mathbf{u}_n$ is called the n-th *moment* of the linear functional $\mathbf{u}$. It is extended by linearity to all polynomials.

We turn our attention to orthogonality with respect to a linear functional. The linear functional $\mathbf{u}$ is said to be *quasi-definite* (resp. *positive-definite*) if every leading principal submatrix of the Hankel matrix $H = (\mathbf{u}_{i+j})_{i,j=0}^{\infty}$ is nonsingular (resp. positive-definite). In such a situation, there exists a sequence of monic polynomials $(P_n)_{n\geq 0}$ such that $\deg P_n = n$ and $\langle \mathbf{u}, P_n(x)P_m(x) \rangle = K_n \delta_{n,m}$, where $\delta_{n,m}$ is the Kronecker symbol and $K_n \neq 0$ (see [1, 5]). The sequence $(P_n)_{n\geq 0}$ is said to be the *sequence of monic orthogonal polynomials* (SMOP, in short) with respect to the linear functional $\mathbf{u}$.

Given a quasi-definite linear functional $\mathbf{u}$ and $(P_n)_{n\geq 0}$ its corresponding SMOP, there exist two sequences of complex numbers $(b_n)_{n\geq 0}$ and $(a_n)_{n\geq 1}$, with $a_n \neq 0$, such that

$$x\,P_n(x) = P_{n+1}(x) + b_n\,P_n(x) + a_n\,P_{n-1}(x), \quad n \geq 0,$$
$$P_{-1}(x) = 0, \quad P_0(x) = 1, \tag{8.1}$$

where

$$b_n = \frac{\langle \mathbf{u}, x\,P_n^2(x)\rangle}{\langle \mathbf{u}, P_n^2(x)\rangle}, \quad n \geq 0, \quad a_n = \frac{\langle \mathbf{u}, P_n^2(x)\rangle}{\langle \mathbf{u}, P_{n-1}^2(x)\rangle}, \quad n \geq 1. \tag{8.2}$$

If $\mathbf{u}$ is positive-definite, then b_n and a_n are real numbers with $a_n > 0$. Conversely, from Favard's Theorem (see [1, 5]) if $(P_n)_{n\geq 0}$ is a sequence of monic polynomials generated by a three-term recurrence relation as in (8.1) with $a_n \neq 0$, $n \geq 1$, then there exists a unique linear functional $\mathbf{u}$ such that $(P_n)_{n\geq 0}$ is its SMOP. Moreover, in this case, the sequence of polynomials defined by

$$Q_n(x) = \frac{P_n(x)}{\langle \mathbf{u}, P_n^2(x)\rangle^{1/2}}$$

is said to be *the sequence of orthonormal polynomials associated with* $\mathbf{u}$.

Another way to write the recurrence relation (8.1) is in a matrix form. Indeed, if $\mathbf{P} = (P_0, P_1, \dots)^\top$, where $A^\top$ denotes the transposed of the matrix A, then

$$x\mathbf{P} = \mathbf{J}_{\text{mon}}\mathbf{P}, \tag{8.3}$$

where $\mathbf{J}_{\text{mon}}$ is the semi-infinite matrix

$$\mathbf{J}_{\text{mon}} = \begin{pmatrix} b_0 & 1 & & \\ a_1 & b_1 & 1 & \\ & a_2 & b_2 & \ddots \\ & & \ddots & \ddots \end{pmatrix},$$

which is known in the literature as a *monic Jacobi matrix* (see [1, 5]).

Definition 8.1 Let $(P_n)_{n\geq 0}$ be the SMOP with respect to the quasi-definite moment functional $\mathbf{u}$. We define the n-th Christoffel–Darboux (C-D) kernel polynomial as

$$K_n(x, y) = \sum_{k=0}^{n} \frac{P_k(x)P_k(y)}{\langle \mathbf{u}, P_k^2(x)\rangle}.$$

The classical *(monic) Jacobi polynomials* are well-known families of polynomials depending on two parameters (see [1, 23]). Let $(P_n^{(\alpha,\beta)})_{n\geq 0}$ be the monic Jacobi polynomials of parameters (α, β). For $\alpha, \beta > -1$, these polynomials are orthogonal with respect to the positive-definite linear functional $\mathbf{u}_{\alpha,\beta}$ defined by

$$\langle \mathbf{u}_{\alpha,\,\beta},\, p(x)\rangle = \int_{-1}^{1} p(x)(1-x)^{\alpha}(1+x)^{\beta}\,\mathrm{d}x, \quad p(x) \in \mathbb{P}.$$

The explicit expression of these polynomials is

$$P_n^{(\alpha,\beta)}(x) = \frac{1}{S_n(\alpha,\beta)} \sum_{k=0}^{n} \binom{n+\alpha}{n-k}\binom{n+\beta}{k}(x-1)^k(x+1)^{n-k}, \quad n \geq 0,$$

where

$$S_n(\alpha,\beta) = \binom{2n+\alpha+\beta}{n}.$$

Here $\binom{r}{k} = \frac{\Gamma(r+1)}{\Gamma(k+1)\Gamma(r-k+1)}$ is the usual binomial number and

$$\Gamma(z) = \int_{0}^{\infty} t^{z-1}\mathrm{e}^{-t}\,\mathrm{d}t, \quad \mathrm{Re}(z) > 0,$$

is the Gamma function. The explicit orthogonality relation reads

$$\begin{aligned}&\left\langle \mathbf{u}_{\alpha,\,\beta},\, P_n^{(\alpha,\beta)}(x) P_m^{(\alpha,\beta)}(x)\right\rangle \\ &= 2^{2n+\alpha+\beta+1}\frac{\Gamma(n+\alpha+1)\,\Gamma(n+\beta+1)\,\Gamma(n+\alpha+\beta+1)\,n!}{(2n+\alpha+\beta+1)\,(\Gamma(2n+\alpha+\beta+1))^2}\,\delta_{n,m}.\end{aligned}$$

The Jacobi polynomials satisfy the three-term recurrence relation

$$\begin{aligned}&x P_n^{(\alpha,\beta)}(x) = P_{n+1}^{(\alpha,\beta)}(x) + b_n\, P_n^{(\alpha,\beta)}(x) + a_n\, P_{n-1}^{(\alpha,\beta)}(x), \quad n \geq 0,\\ &P_0^{(\alpha,\beta)}(x) = 1, \quad P_{-1}^{(\alpha,\beta)}(x) = 0,\end{aligned}$$

where

$$\begin{aligned}b_n &= \frac{\beta^2-\alpha^2}{(2n+2+\alpha+\beta)\,(2n+\alpha+\beta)},\\ a_n &= \frac{4\,(n+\beta)\,(n+\alpha+\beta)\,(n+\alpha)\,n}{(2n-1+\alpha+\beta)\,(2n+\alpha+\beta)^2\,(2n+\alpha+\beta+1)},\end{aligned}$$

except that when $\alpha = -\beta$, $b_0 = \beta$, and $b_n = 0$, $n \geq 1$.

We also have

$$P_n^{(\alpha,\beta)}(-x) = (-1)^n P_n^{(\beta,\alpha)}(x), \quad P_n^{(\alpha,\beta)}(1) = \frac{2^n}{S_n(\alpha,\beta)}\binom{n+\alpha}{n}.$$

The representation of Jacobi polynomials in terms of a *generalized Hypergeometric function* is

$$P_n^{(\alpha,\beta)}(x) = \frac{2^n(\alpha+1)_n}{(\alpha+\beta+n+1)_n} \times {}_2F_1\left(-n, \alpha+\beta+n+1; \alpha+1; \frac{1-x}{2}\right), \tag{8.4}$$

where $(a)_k$ is the *Pochhammer symbol* defined by $(a)_0 = 1$,

$$(a)_k = a(a+1)\cdots(a+k-1) = \frac{\Gamma(a+k)}{\Gamma(a)}, \tag{8.5}$$

for $k = 1, 2, \ldots$, and [21, p. 352]

$${}_pF_q\left(a_1, \ldots, a_p; b_1, \ldots, b_q, x\right) = \sum_{k=0}^{\infty} \frac{(a_1)_k \cdots (a_p)_k}{(b_1)_k \cdots (b_q)_k} \frac{x^k}{k!}.$$

8.2.1 Gauss Quadrature Formula

Given a continuous function $f(x)$ and a finite or infinite interval I, a problem of great interest is to approximate the integral

$$\int_I f(x)W(x)\,\mathrm{d}x,$$

(where $W(x)$ is a weight function) by means of numerical integration (quadrature rule)

$$\int_I f(x)W(x)\,\mathrm{d}x \approx f(x_1)w_1 + \cdots + f(x_N)w_N.$$

Here, the integral is approximated by a sum of N terms that involves the weights w_k and the nodes x_k, which must be properly chosen.

Definition 8.2 We say that a quadrature rule has degree of exactness k if for every polynomial $q(x) \in \mathbb{P}_k$,

$$\int_I q(x)W(x)\,\mathrm{d}x = q(x_1)w_1 + \cdots + q(x_N)w_N, \tag{8.6}$$

and there exists a polynomial of degree $k+1$ such that (8.6) does not hold.

A simple way to construct quadrature rules is based on Lagrange interpolation formulas. Recall that given a set of N points, $\mathcal{N} = \{(x_1, y_1), (x_2, y_2), \ldots, (x_N, y_N)\}$, where $x_i \neq x_j$, if $i \neq j$, the *Lagrange interpolating polynomial* of degree at most $N - 1$ corresponding to $\mathcal{N}$ is defined as

$$L_N(x) =: \sum_{k=1}^{N} y_k \, \ell_k(x), \quad \text{where } \ell_k(x) =: \prod_{\substack{j=1 \\ j \neq k}}^{N} \frac{(x - x_j)}{(x_k - x_j)}.$$

Hence, when $N = 1$, we take $L_1(x) \equiv y_1$. The numbers x_j, $1 \leq j \leq N$, are called *nodes*, and the polynomials $(\ell_k(x))_{k=1}^{N}$ are known in the literature as the *Lagrange polynomial basis*. The polynomial $L_N(x)$ has the property $L_N(x_j) = y_j$, $1 \leq j \leq N$. Indeed, since $\ell_k(x_j) = \delta_{k,j}$, we have

$$L_N(x_j) = \sum_{k=1}^{N} y_k \, \delta_{k,j} = y_j, \quad 1 \leq j \leq N.$$

The Lagrange polynomial $L_N(x)$ provides a unique solution to the problem of finding a polynomial of degree at most $N - 1$ whose graph contains the points (x_j, y_j), $1 \leq j \leq N$.

Given a continuous function $f(x)$ and N nodes $(x_j, f(x_j))$, $1 \leq j \leq N$, we can write

$$f(x) = L_N(x) + R_N(x),$$

where $R_N(x)$ is the interpolation error. Observe that if $f(x)$ is a polynomial of degree at most $N - 1$, then by the uniqueness of Lagrange polynomial, $R_N(x) = 0$. With this in mind, we have the following quadrature rule

$$\int_I f(x) W(x) \, \mathrm{d}x \approx \sum_{k=1}^{N} f(x_k) \int_I \ell_k(x) W(x) \, \mathrm{d}x.$$

It is well known [1, 11] that in order to a quadrature formula has the maximum possible degree of exactness, the nodes x_k must be chosen as the zeros of the orthogonal polynomials of degree N with respect to the linear functional $\mathbf{u}$ associated with the weight function $W(x)$, i.e., $\langle \mathbf{u}, p \rangle = \int_I p(x) W(x) \, \mathrm{d}x$. Let us remind that such zeros are real, simple and are located in the interior of the convex hull of the support of the measure associated with the linear functional (see [1]). In other words, the following theorem holds.

Theorem 8.1 (Gauss Quadrature Rule) *Let* $\mathbf{u}$ *be a positive-definite linear functional, and let* $(P_n)_{n \geq 0}$ *be its corresponding SMOP. There exist positive real numbers* $A_{N,1}, \ldots, A_{N,N}$ *such that for every polynomial* $q(x)$ *of degree at most*

$2N-1$, we have

$$\langle \mathbf{u}, q(x) \rangle = \sum_{k=1}^{N} A_{N,k}\, q(x_{N,k}),$$

where $(x_{N,k})_{k=1}^{N}$ are the zeros of $P_N(x)$.

The constants $A_{N,k}$ involved in the Gauss quadrature rule are known in the literature as *Christoffel numbers*.

Corollary 8.1 *The Christoffel numbers $A_{N,k}$, $1 \le k \le N$, have the following representation,*

$$A_{N,k} = \frac{1}{K_{N-1}(x_{N,k}, x_{N,k})}, \tag{8.7}$$

where $K_{N-1}(x, y)$ is the $(N-1)$-th C-D kernel polynomial as in Definition 8.1.

Notice that from (8.3), we get the relation

$$x_{N,k} \begin{pmatrix} P_0(x_{N,k}) \\ \vdots \\ P_{N-1}(x_{N,k}) \end{pmatrix} = (\mathbf{J}_{\text{mon}})_{N\times N} \begin{pmatrix} P_0(x_{N,k}) \\ \vdots \\ P_{N-1}(x_{N,k}) \end{pmatrix},$$

where $(\mathbf{J}_{\text{mon}})_{N\times N}$ is the $N \times N$ leading principal submatrix of $\mathbf{J}_{\text{mon}}$. This means that $x_{N,k}$ is an eigenvalue of $(\mathbf{J}_{\text{mon}})_{N\times N}$ with associated eigenvector $(P_0(x_{N,k}), \dots, P_{N-1}(x_{N,k}))^T$. The matrix $(\mathbf{J}_{\text{mon}})_{N\times N}$ can be symmetrized by a diagonal similarity transformation $\mathbf{D}$. In such a way, we obtain

$$\mathbf{J}_{N\times N} = \mathbf{D}(\mathbf{J}_{\text{mon}})_{N\times N}\,\mathbf{D}^{-1} = \begin{pmatrix} b_0 & \sqrt{a_1} & & & \\ \sqrt{a_1} & b_1 & \sqrt{a_2} & & \\ & \ddots & \ddots & \ddots & \\ & & \sqrt{a_{N-2}} & b_{N-2} & \sqrt{a_{N-1}} \\ & & & \sqrt{a_{N-1}} & b_{N-1} \end{pmatrix}.$$

Notice that $\mathbf{J}_{N\times N}$ is the $N \times N$ leading principal submatrix of the Jacobi matrix associated with the orthonormal polynomials. Since the eigenvalues are preserved by a similarity transformation, $x_{N,k}$ is also an eigenvalue of the symmetric tridiagonal matrix $\mathbf{J}_{N\times N}$. Moreover, taking into account (8.7), if v_k is the eigenvector corresponding to $x_{N,k}$ normalized as $\|v_k\|_2 = 1$, then (see [19, Theorem 3.1])

$$A_{N,k} = \mathbf{u}_0\, v_{k,1}^2,$$

where $v_{k,1}$ is the first component of v_k. It is well known that the seeking of eigenvalues and eigenvectors of a symmetric tridiagonal matrix is a relatively efficient and well-conditioned procedure.

Notice that to use the aforementioned method, we need to know the explicit values of $(b_k)_{k=0}^{N-1}$ and $(a_k)_{k=1}^{N-1}$, but it is not always easy from (8.2). Moreover, the usual approach is to represent each $P_k(x)$ explicitly as a polynomial in x and to compute the inner products by multiplying out term by term. This will be feasible if we know the first $2N$ moments

$$\mathbf{u}_k = \int_I x^k W(x)\, \mathrm{d}x, \quad k = 0, \ldots, 2N-1. \tag{8.8}$$

However, the solution of the resulting set of algebraic equations for the coefficients b_k and a_k in terms of the moments $\mathbf{u}_k$ is in general ill-conditioned. Sack and Donovan [25] (see also [6]) showed that the numerical stability is improved if instead of using the classical moments $\mathbf{u}_k$ we use the modified moments

$$m_k = \int_I \pi_k(x) W(x)\, \mathrm{d}x, \tag{8.9}$$

where $(\pi_k(x))_{k=0}^{2N-1}$ is a set of $2N$ monic polynomials satisfying a three-term recurrence relation

$$\begin{aligned} x\pi_k(x) &= \pi_{k+1}(x) + \mathfrak{b}_k \pi_k(x) + \mathfrak{a}_k \pi_{k-1}(x), \quad k \geq 0, \\ \pi_0(x) &= 1, \quad \pi_{-1}(x) = 0, \end{aligned} \tag{8.10}$$

with known coefficients $\mathfrak{b}_k \in \mathbb{R}$ and $\mathfrak{a}_k \geq 0$. Notice that when $\mathfrak{b}_k = \mathfrak{a}_k = 0$, for all $k \geq 0$, the modified moments (8.9) become the standard moments (8.8). Thus, if we get the modified moments (8.9), then we can find the coefficient as follows.

Assume that we know the modified moments

$$m_k = \int_I \pi_k(x) W(x)\, \mathrm{d}x, \quad j = 0, \ldots, 2N-1,$$

where the $\pi_k(x)$s satisfy the recurrence relation (8.10) and the coefficients $\mathfrak{b}_k$, $\mathfrak{a}_k$ are known explicitly. In Sack and Donovan [25], an efficient algorithm to determine b_k and a_k via a set of intermediate quantities is given. The explanation of the following algorithm can be found in [7, 29].

$$\sigma_{k,\ell} = \langle \mathbf{u}, p_k(x)\pi_\ell(x) \rangle, \quad k, \ell \geq -1,$$

where

$$\sigma_{-1,\ell} = 0, \quad \ell = 1, 2, \ldots, 2N-2,$$

$$\sigma_{0,\ell} = m_\ell, \quad \ell = 0, 1, \ldots, 2N-1,$$
$$b_0 = \mathfrak{b}_0 + \frac{m_1}{m_0}, \quad a_0 = 0,$$

and for $k = 1, 2, \ldots, N-1$,

$$\begin{aligned}
\sigma_{k,\ell} &= \sigma_{k-1,\ell+1} - (b_{k-1} - \mathfrak{b}_\ell)\,\sigma_{k-1,\ell} - a_{k-1}\,\sigma_{k-2,\ell} + \mathfrak{a}_\ell\,\sigma_{k-1,\ell-1}, \\
&\quad \ell = k, \ldots, 2N-k-1, \\
b_k &= \mathfrak{b}_k - \frac{\sigma_{k-1,k}}{\sigma_{k-1,k-1}} + \frac{\sigma_{k,k+1}}{\sigma_{k,k}}, \\
a_k &= \frac{\sigma_{k,k}}{\sigma_{k-1,k-1}}.
\end{aligned} \tag{8.11}$$

The aforementioned algorithm is said to be *modified Chebyshev algorithm.*

8.3 Quadrature Formula for the Truncated Laguerre Polynomials

Let $(L_n^\alpha(x; z))_{n\geq 0}$ be the sequence of monic orthogonal polynomials with respect to the linear functional

$$\langle \boldsymbol{\ell}, p(x) \rangle = \int_0^z p(x)\, x^\alpha \mathrm{e}^{-x}\, \mathrm{d}x, \quad p(x) \in \mathbb{P}, \tag{8.12}$$

with $\alpha > -1$ and $z > 0$. This functional is known in the literature as *the truncated Gamma functional* and the polynomials $(L_n^\alpha(x; z))_{n\geq 0}$ are called *truncated Laguerre orthogonal polynomials* (see [3]). Notice that the moments of the linear functional $\boldsymbol{\ell}$ are given by

$$\boldsymbol{\ell}_m = \int_0^z x^{m+\alpha} \mathrm{e}^{-x}\, \mathrm{d}x = \widehat{\gamma}(m + \alpha + 1, z),$$

where $\widehat{\gamma}(a, z)$ is the *incomplete gamma function* defined by [23, 8.2.1]

$$\widehat{\gamma}(a, z) = \int_0^z x^{a-1} \mathrm{e}^{-x}\, \mathrm{d}x.$$

Its series representation, (see [23, 8.5.1]), is

$$\widehat{\gamma}(a, z) = \frac{z^a \mathrm{e}^{-z}}{a} \sum_{k=0}^{\infty} \frac{z^k}{k(a+1)_k}.$$

A change of variables in (8.12) yields

$$\langle \boldsymbol{\ell}, p \rangle = z^{\alpha} \int_0^1 p(zx)x^{\alpha}\mathrm{e}^{-zx}\,\mathrm{d}x.$$

According to [3, Eq. 7.10–7.11], we get that the recurrence coefficients associated with the monic orthogonal polynomials (8.1) with respect to the weight $w(x, z) = x^{\alpha}\mathrm{e}^{-zx}$ supported on $[0, 1]$ satisfy

$$a_n(z) = \frac{1}{16} + \frac{\left(1 - 2\alpha^2\right)}{64}n^{-2} + \alpha\frac{\alpha(z + 2\alpha) - 1}{64}n^{-3} + O\left(n^{-4}\right),$$

$$b_n(z) = \frac{1}{2} - \frac{\alpha^2}{8}n^{-2} + \frac{z - 2\alpha^2(z + 2\alpha + 2)}{32}n^{-3} + O\left(n^{-4}\right),$$

as $n \to +\infty$, or, equivalently,

$$\lim_{n\to+\infty} a_n(z) = \frac{1}{16}, \qquad \lim_{n\to+\infty} b_n(z) = \frac{1}{2}. \tag{8.13}$$

Notice also that $w(x, z) \to x^{\alpha}$ when $z \to 0$. Therefore, a good choice of sequence of monic orthogonal polynomials to use in the modified Chebyshev algorithm is the *shifted Jacobi polynomials* with parameters $(0, \alpha)$, which are orthogonal with respect to the weight function $w(x) = x^{\alpha}$ supported on the interval $[0, 1]$.

Proposition 8.1 *The shifted monic Jacobi polynomials* $(\widetilde{P}_n^{(0,\alpha)})_{n\geq 0}$ *can be written as*

$$\widetilde{P}_n^{(0,\alpha)}(x) = \frac{1}{2^n}P_n^{(0,\alpha)}(2x - 1) = \frac{(-1)^n}{2^n}P_n^{(\alpha,0)}(1 - 2x), \tag{8.14}$$

where $P_n^{(0,\alpha)}(x)$ *is the Jacobi polynomial of parameters* $(0, \alpha)$ *and degree* n*.*

Proposition 8.2 *Let* $(\widetilde{P}_n^{(0,\alpha)})_{n\geq 0}$ *be the sequence of monic shifted Jacobi polynomials orthogonal with respect to the weight function* $\omega(x) = x^{\alpha}$, $\alpha > -1$, *on* $[0, 1]$. *The modified moments*

$$m_n(\alpha, z) = \int_0^1 \widetilde{P}_n^{(0,\alpha)}(x)x^{\alpha}\mathrm{e}^{-zx}\,\mathrm{d}x,$$

where $n = 0, 1, \ldots,$ *can be expressed via a* $(1, 1)$*-Generalized Hypergeometric function as*

$$m_0(\alpha, z) = \begin{cases} \dfrac{1}{\alpha + 1}, & z = 0, \\ \dfrac{\gamma(\alpha + 1, z)}{z^{\alpha+1}}, & z > 0, \end{cases}$$

$$m_n(\alpha, z) = (-1)^n \frac{n!}{(\alpha+n+1)_n^2(\alpha+2n+1)} z^n e^{-z} \times {}_1F_1\left(n+1; \alpha+2n+2; z\right),$$
$$n \geq 1.$$

Proof From [15, Eq. 7.523],

$$\int_0^1 x^{v-1}(1-x)^{\rho-1}e^{-zx} \times {}_2F_1\left(r, q; v; x\right) dx$$
$$= \frac{\Gamma(v)\Gamma(\rho)\Gamma(v+\rho-r-q)}{\Gamma(v+\rho-r)\Gamma(v+\rho-q)} e^{-z} \times {}_2F_2\left(\rho, v+\rho-r-q;\right.$$
$$\left. v+\rho-r, v+\rho-q; z\right)$$

holds for

$$\mathrm{Re}(v) > 0, \quad \mathrm{Re}(\rho) > 0, \quad \text{and} \quad \mathrm{Re}(v+\rho-r-q) > 0.$$

In particular, for $\varepsilon > 0$ and $n \geq 0$

$$\int_0^1 x^{(\alpha+1)-1}e^{-xz} \times {}_2F_1\left(-n-\varepsilon, \alpha+n+\varepsilon+1; \alpha+1; x\right) dx$$
$$= \frac{\Gamma(\alpha+1)}{\Gamma(\alpha+2+n+\varepsilon)\Gamma(1-n-\varepsilon)} e^{-z} \times {}_2F_2\left(1, 1; \alpha+2+n+\varepsilon, 1-n-\varepsilon; z\right)$$
$$= \frac{\Gamma(\alpha+1)}{\Gamma(\alpha+2+n+\varepsilon)\Gamma(1-n-\varepsilon)} e^{-z} \sum_{k=0}^{\infty} \frac{(1)_k^2}{(\alpha+n+\varepsilon+2)_k(1-n-\varepsilon)_k} \cdot \frac{z^k}{k!}$$
$$= \Gamma(\alpha+1)e^{-z} \sum_{k=0}^{\infty} \frac{(1)_k^2}{\Gamma(\alpha+n+\varepsilon+2+k)\Gamma(1-n-\varepsilon+k)} \cdot \frac{z^k}{k!}. \tag{8.15}$$

On the other hand, from (8.4) we get

$$\frac{(-1)^n}{2^n} P_n^{(\beta,\alpha)}(-x) = (-1)^n \frac{(\beta+1)_n}{(\alpha+\beta+n+1)_n} \times {}_2F_1$$
$$\times \left(-n, \alpha+\beta+n+1; \beta+1; \frac{1+x}{2}\right),$$

and by (8.14)

$$\widetilde{P}_n^{(0,\alpha)}(x) = (-1)^n P_n^{(\alpha,0)}(-(2x-1))$$
$$= (-1)^n \frac{(\alpha+1)_n}{(\alpha+n+1)_n} \times {}_2F_1\left(-n, \alpha+n+1; \alpha+1; x\right).$$

Thus, from (8.15) we obtain

$$
\begin{aligned}
&(-1)^n \frac{(\alpha+n+1)_n}{(\alpha+1)_n} \int_0^1 \widetilde{P}_n^{(0,\alpha)}(x) x^\alpha \mathrm{e}^{-zx}\, \mathrm{d}x \\
&= \int_0^1 x^{(\alpha+1)-1} \mathrm{e}^{-zx} \times \lim_{\varepsilon\to 0} {}_2F_1\left(-n-\varepsilon, \alpha+n+1+\varepsilon; \alpha+1; x\right) \mathrm{d}x \\
&= \lim_{\varepsilon\to 0} \int_0^1 x^{(\alpha+1)-1} \mathrm{e}^{-zx} \times {}_2F_1\left(-n-\varepsilon, \alpha+n+1+\varepsilon; \alpha+1; x\right) \mathrm{d}x \\
&= \Gamma(\alpha+1)\mathrm{e}^{-z} \lim_{\varepsilon\to 0} \sum_{k=0}^{\infty} \frac{(1)_k^2}{\Gamma(\alpha+n+\varepsilon+2+k)\Gamma(1-n-\varepsilon+k)} \cdot \frac{z^k}{k!}.
\end{aligned}
$$

Notice that $\frac{1}{\Gamma(1-n-\varepsilon+k)} \to 0$ when $\varepsilon \to 0$ for $k = 0, \ldots, n-1$. Moreover, using Stirling's Formula (see [23, p. 141, Eq. 5.11.7]) for $k \gg 0$, we get

$$
\begin{aligned}
&\left| \frac{(1)_k^2}{\Gamma(\alpha+n+\varepsilon+2+k)\Gamma(1-n-\varepsilon+k)} \cdot \frac{z^k}{k!} \right| \\
&= \left| \frac{k!}{(\alpha+n+\varepsilon+2)_k \Gamma(1-n-\varepsilon+k)} z^k \right| \\
&\approx \left| \frac{\sqrt{k}\left(\frac{k}{\mathrm{e}}\right)^k}{\sqrt{2\pi(\alpha+n+\varepsilon+1+k)(k-n-\varepsilon)} \left(\frac{\alpha+n+\varepsilon+1+k}{\mathrm{e}}\right)^{\alpha+n+\varepsilon+1+k} \left(\frac{k-n-\varepsilon}{\mathrm{e}}\right)^{k-n-\varepsilon}} z^k \right|.
\end{aligned}
$$

If we denote by s_k the last expression, then we get

$$
\lim_{k\to+\infty} (s_k)^{1/k} = 0.
$$

Consequently, from Tannery's theorem (see [16, p. 448]), we can interchange the limit and the summatory. Thus, using (8.5), we get

$$
\begin{aligned}
&(-1)^n \frac{(\alpha+n+1)_n}{(\alpha+1)_n} \int_0^1 \widetilde{P}_n^{(0,\alpha)}(x) x^\alpha \mathrm{e}^{-zx}\, \mathrm{d}x \\
&= \Gamma(\alpha+1)\mathrm{e}^{-z} \sum_{k=n}^{\infty} \frac{(1)_k^2}{\Gamma(\alpha+2+n+k)\Gamma(k+1-n)} \cdot \frac{z^k}{k!} \\
&= \frac{\Gamma(\alpha+1)\, n!}{\Gamma(\alpha+2n+2)} z^n \mathrm{e}^{-z} \sum_{t=0}^{\infty} \frac{(n+1)_t}{(\alpha+2n+2)_t} \cdot \frac{z^t}{t!} \\
&= \frac{\Gamma(\alpha+1)\, n!}{\Gamma(\alpha+2n+2)} z^n \mathrm{e}^{-z} \times {}_1F_1\left(n+1; \alpha+2n+2; z\right).
\end{aligned}
$$

□

8.4 Numerical Experiments

By default, `MATLAB` works with 16 digits of precision. For higher precision, we can use the function `vpa(expr,digits)` in `Symbolic Math Toolbox`, where `expr` is the numerical variable, and `digits` is the number of precision digits which can be increased from 1 until $2^{29}+1$. By default, the variable-precision arithmetic of `vpa` uses 32 significant decimal digits of precision. For our propose, we take `digits=100`, and we assume that the values obtained with this precision are "exact."

In this section, we will use the *Modified Chebyshev algorithm and the quadrature algorithm*, which are collected in the package `OPQ` (Orthogonal Polynomials and Quadrature) in MATLAB provided by Walter Gautschi. All pieces of software into the package `OPQ` can be downloaded from [13]

www.siam.org/books/se26

The explanation of these routines is illustrated in [10–12]. In particular, the modified Chebyshev moment routine (8.11) for generating the N first coefficients b_n and a_n associated with the polynomials $(L_n^\alpha(x;z))_{n=0}^{N-1}$ we use:

ab = chebyshev(2N, mom, abm),

where $\mathtt{mom} = (m_0(\alpha,z), m_1(\alpha,z), \ldots, m_{2N-1}(\alpha,z))$ is the $1\times 2N$ array of modified moments and `abm` the $(2N-1)\times 2$ array of known recurrence coefficients

$$\mathtt{abm} = \begin{pmatrix} \mathfrak{b}_0 & \mathfrak{a}_0 \\ \mathfrak{b}_1 & \mathfrak{a}_1 \\ \vdots & \vdots \\ \mathfrak{b}_{2N-2} & \mathfrak{a}_{2N-2} \end{pmatrix}.$$

The corresponding symbolic routine is provided in the package `SOPQ` [13]

ab = schebyshev(digits, 2N, mom, abm).

Taking into account the aforementioned, we define *the maximal relative error* by

$$\mathcal{E}_n^\alpha(z;\text{digits}) := \max_{0\le k\le n-1} \left\{ \left|\frac{a_k - \mathbf{a}_k}{\mathbf{a}_k}\right|, \left|\frac{b_k - \mathbf{b}_k}{\mathbf{b}_k}\right| \right\}, \tag{8.16}$$

where the "exact values" $\mathbf{a}_k$ and $\mathbf{b}_k$ are obtained using the symbolic routine

`schebyshev(digits, 2N, mom, abm).`

with `digits=100`. For example, for $\alpha = 1$, $z = 1$, $N = 48$ and `digits=30`, we get the first 48 recurrence coefficients $(b_k)_{k=0}^{47}$, $(a_k)_{k=0}^{47}$ (see Tables 8.1 and 8.2).

Notice that from here we infer that the convergence of the previous sequences to the limits (8.13) is satisfied. However, it convergence appears to be "slow" since for $k = 47$, we only have 4 significant digits.

Table 8.1 Sequence $b_k(1)$, for $k = 0, \ldots, 47$

k	$b_k(1)$	k	$b_k(1)$	k	$b_k(1)$
0	0.607788808822667	16	0.500426808722825	32	0.500113960288728
1	0.531655773460623	17	0.500380956440347	33	0.500107376935035
2	0.513427891887575	18	0.500342117324710	34	0.500101348056586
3	0.507534434505874	19	0.500308930359149	35	0.500095813084832
4	0.504835172165686	20	0.500280349444746	36	0.500090719501977
5	0.503368518898357	21	0.500255559818526	37	0.500086021590060
6	0.502482032582844	22	0.500233919258297	38	0.500081679401089
7	0.501905070635327	23	0.500214916014164	39	0.500077657904668
8	0.501508458210362	24	0.500198138240255	40	0.500073926279007
9	0.501224082197933	25	0.500183251468526	41	0.500070457318386
10	0.501013226345908	26	0.500169981794040	42	0.500067226935724
11	0.500852549532634	27	0.500158103174469	43	0.500064213743185
12	0.500727295329595	28	0.500147427732113	44	0.500061398697152
13	0.500627761645442	29	0.500137798273680	45	0.500058764796504
14	0.500547355966703	30	0.500129082466506	46	0.500056296825256
15	0.500481471186175	31	0.500121168264855	47	0.500053981132248

Table 8.2 Sequence $a_k(1)$, for $k = 0, \ldots, 47$

k	$a_k(1)$	k	$a_k(1)$	k	$a_k(1)$
0	0.2642411176571153	16	0.06244605779295135	32	0.06248566001620657
1	0.06174799916059207	17	0.06245187208765875	33	0.06248649068343801
2	0.06110639779446930	18	0.06245679527042237	34	0.06248725122345314
3	0.06159770170459388	19	0.06246100042276222	35	0.06248794931205289
4	0.06190038948667886	20	0.06246462058754074	36	0.06248859160306462
5	0.06207694279947663	21	0.06246775938221266	37	0.06248918388734983
6	0.06218661764444579	22	0.06247049847158962	38	0.06248973122368508
7	0.06225891009402938	23	0.06247290291817096	39	0.06249023804706056
8	0.06230892605966116	24	0.06247502507158938	40	0.06249070825874153
9	0.06234491040894508	25	0.06247690743561846	41	0.06249114530151897
10	0.06237163970905703	26	0.06247858480860449	42	0.06249155222286876
11	0.06239202635370342	27	0.06248008590029081	43	0.06249193172818968
12	0.06240792454776711	28	0.06248143456640384	44	0.06249228622586166
13	0.06242055888438060	29	0.06248265076085398	45	0.06249261786552981
14	0.06243076389952361	30	0.06248375127700079	46	0.06249292857075320
15	0.06243912391137471	31	0.06248475032972568	47	0.06249322006694707

In order to study the stability of our procedure, we will take $\alpha = 1$, and we will use (8.16) with two different arithmetic, the usual `digits=16` and the high precision `digits=30` for different values of z. The loss of digits practically does not exist for any value of $z \in [0, 30]$ and `digits=30`. For `digits=16`, the loss of digits practically does not exist for any value of $z \in [0, 10)$. When $z = 10$, $z = 15$,

Table 8.3 Maximal relative error $\mathcal{E}_{50}^{1}(z;\text{digits})$ for $z = 5,\ 10,\ 15$

$\alpha = 1$			
`digits`	$z = 5$	$z = 10$	$z = 15$
16	7.8056×10^{-16}	8.8163×10^{-15}	6.0238×10^{-13}
30	7.8238×10^{-38}	8.3718×10^{-37}	4.9234×10^{-35}

Table 8.4 Maximal relative error $\mathcal{E}_{50}^{1}(z;\text{digits})$ for $z = 20,\ 25,\ 30$

$\alpha = 1$			
`digit`	$z = 20$	$z = 25$	$z = 30$
16	2.6790×10^{-11}	4.8261×10^{-9}	4.2894×10^{-7}
30	3.6960×10^{-33}	5.9650×10^{-31}	9.9114×10^{-30}

$z = 20$, $z = 25$, and $z = 30$, (see Tables 8.3 and 8.4) we have a loss of at most one, three, five, seven, or nine digits, respectively, regardless of which arithmetic we use. Thus, for $z < 10$, the method is well-conditioned and its condition number is near 1. On the other hand, taking into account [19], if the condition number of a mapping is 10^t, then approximately t decimal digits are lost. Therefore, we can conclude that the condition number is approximately 10, 10^3, 10^5, ,10^7, and 10^9 for $z = 10$, $z = 15$, $z = 20$, $z = 25$, and $z = 30$, respectively. Thus, from the aforementioned, we conclude that if we want a accuracy of d decimal digits in the recurrence coefficients, then we must work with `digit` $= t + d$, [19].

Suppose now that we want to determinate the coefficients b_k and a_k for $k = 0, \ldots, n-1$, where by hypothesis $\mathfrak{a}_0 = m_0(\alpha, z)$. For simplicity, assume that $\alpha = 1$ and $z \in [0, T]$ with $T > 0$. Taking into account [28], we first select a system of points $Q = \{z_k\}$ in the interval $[0, T]$. In these points, we determine the corresponding values of the coefficients $(b_k(z))_{k=0}^{n-1}$, $(a_k(z))_{k=1}^{n-1}$ and then construct the corresponding interpolating functions for each one of these coefficients, using the function `interp1(Q,f_Q,q,“spline”)`, where `f_Q` is the vector of values of the function f corresponding to the partition Q, and q is the interval where the interpolator function is defined. In this case, the interpolation is made using spline's technique.

We illustrate this procedure for $\alpha = 1$, $T = 30$. By simplicity, we take an equidistant partition of points

$$Q = \{z_{k+1} = z_k + 3/4, \quad z_0 = 0, \quad k = 0, \ldots, 29\},$$

or, equivalently in `MATLAB linspace(0,30,90)`. In Figs. 8.1 and 8.2, we present the coefficient $b_k(z)$, $a_k(z)$ for $\alpha = 1$.

Finally, we use the function `gauss(N,ab)` from package `OPQ` to obtain the zeros of the polynomial $L_{50}^{\alpha}(x; z)$ (in increasing order) as well as the corresponding Christoffel numbers (8.7). If we take $N = 50$, $\alpha = 1$, and $z = 1$, we get Table 8.5.

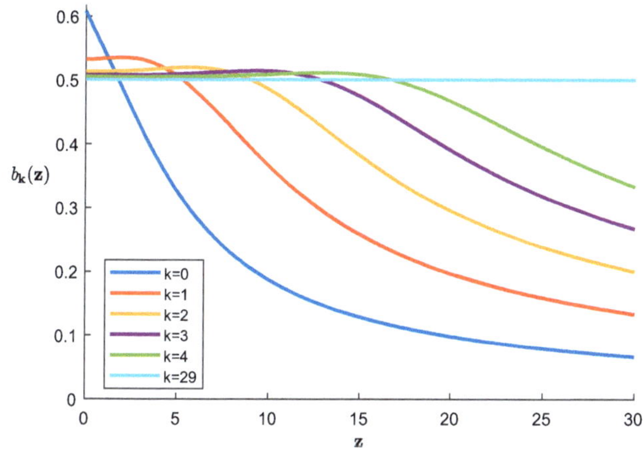

Fig. 8.1 The coefficients $b_k(z)$ for $z \in [0, 30]$ in the case $\alpha = 1$

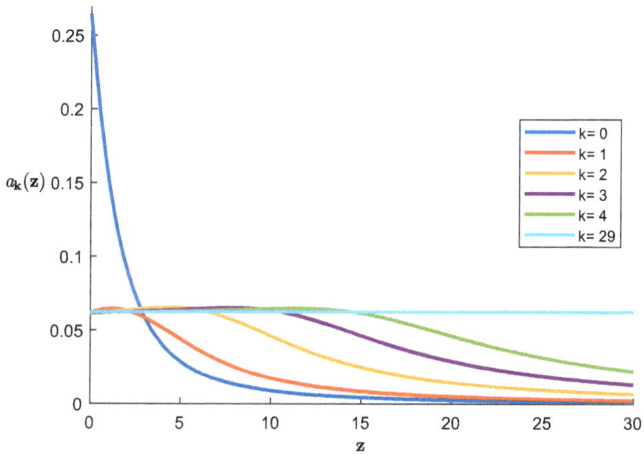

Fig. 8.2 The coefficients $a_k(z)$ for $z \in [0, 30]$ in the case $\alpha = 1$

8.5 Concluding Remarks

The Gauss–Rys quadrature formula with respect to the weight function $w(x) = e^{-tx^2}$ defined on $[-1, 1]$ has been studied in several papers ([26, 27] and references therein) due to its importance to evaluate electron repulsion integrals [28]. Different Rys quadrature algorithms have been proposed in the literature [17, 27, 28]. Recently, G. V. Milovanović and N. Vasović, [22] considered the problem of construction of quadrature formulas of Gaussian type on $[-1, 1]$, concerning the following two-parametric weight function $w(x; z) = e^{-zx^2}(1 - x^2)^{\lambda-1/2}$, $\lambda > -1/2$,, which is a generalization of *Rys quadrature* by a transformation of

Table 8.5 Nodes and weight coefficients. Gauss Quadrature formula

$\lambda_{50,k}$	$A_{50,k}$	$\lambda_{50,k}$	$A_{50,k}$	$\lambda_{50,k}$	$A_{50,k}$
1.0723×10^{-3}	1.8699×10^{-6}	2.2997×10^{-1}	5.5708×10^{-6}	7.2616×10^{-1}	7.3924×10^{-12}
3.5934×10^{-3}	1.0454×10^{-5}	2.5450×10^{-1}	3.0779×10^{-6}	7.5520×10^{-1}	3.1302×10^{-12}
7.5515×10^{-3}	2.8145×10^{-5}	2.8001×10^{-1}	1.6354×10^{-6}	7.8337×10^{-1}	1.3481×10^{-12}
1.2941×10^{-2}	5.3562×10^{-5}	3.0644×10^{-1}	8.3766×10^{-7}	8.1051×10^{-1}	5.9302×10^{-13}
1.9753×10^{-2}	8.2141×10^{-5}	3.3373×10^{-1}	4.1466×10^{-7}	8.3645×10^{-1}	2.6750×10^{-13}
2.7979×10^{-2}	1.0792×10^{-4}	3.6181×10^{-1}	1.9888×10^{-7}	8.6103×10^{-1}	1.2420×10^{-13}
3.7607×10^{-2}	1.2564×10^{-4}	3.9059×10^{-1}	9.2662×10^{-8}	8.8407×10^{-1}	5.9564×10^{-14}
4.8621×10^{-2}	1.3232×10^{-4}	4.2001×10^{-1}	4.2055×10^{-8}	9.0543×10^{-1}	2.9587×10^{-14}
6.1005×10^{-2}	1.2782×10^{-4}	4.4998×10^{-1}	1.8645×10^{-8}	9.2495×10^{-1}	1.5253×10^{-14}
7.4740×10^{-2}	1.1435×10^{-4}	4.8039×10^{-1}	8.0995×10^{-9}	9.4247×10^{-1}	8.1661×10^{-15}
8.9804×10^{-2}	9.5435×10^{-5}	5.1114×10^{-1}	3.4581×10^{-9}	9.5787×10^{-1}	4.5341×10^{-15}
1.0617×10^{-1}	7.4720×10^{-5}	5.4214×10^{-1}	1.4559×10^{-9}	9.7102×10^{-1}	2.5981×10^{-15}
1.2382×10^{-1}	5.5130×10^{-5}	5.7326×10^{-1}	6.0646×10^{-10}	9.8182×10^{-1}	1.5189×10^{-15}
1.4270×10^{-1}	3.8478×10^{-5}	6.0437×10^{-1}	2.5087×10^{-10}	9.9016×10^{-1}	8.8292×10^{-16}
1.6281×10^{-1}	2.5488×10^{-5}	6.3536×10^{-1}	1.0343×10^{-10}	9.9598×10^{-1}	4.7777×10^{-16}
1.8408×10^{-1}	1.6071×10^{-5}	6.6609×10^{-1}	4.2669×10^{-11}	9.9924×10^{-1}	1.8745×10^{-16}
2.0648×10^{-1}	9.6726×10^{-6}	6.9640×10^{-1}	1.7684×10^{-11}		

quadratures on $[-1, 1]$ with N nodes to ones on $(0, 1)$ with only $(N + 1)/2$ nodes. Such an approach provides a stable and very efficient numerical construction.

In this contribution, we explored the quadrature formula with respect to the weight function $w(x; z) = e^{-zx}x^{\alpha}$, $\alpha > -1$, supported on $[0, 1]$. For this propose, we use the *Modified Chebyshev algorithm* using for this end the *shifted monic Jacobi* polynomials. Here we make a computational analysis of our algorithm using for this objective the QPQ and SQPQ package of MATLAB provided by Walter Gautschi [10, 13]. Finally, we want to emphasize that the Truncated Laguerre weight appear in a natural way from the symmetrization process of the weight $w(x; z) = |x|^{2\alpha+1}e^{-zx^2}$, which is other type of generalization of *Rys weights* [3].

Acknowledgments Authors thank the referees for the careful revision of the manuscript. Their suggestions have substantially improved the presentation. The work of J. C. García-Ardila has been supported by the Comunidad de Madrid multiannual agreement with the Universidad Rey Juan Carlos under the grant Proyectos I+D para Jóvenes Doctores, Ref. M2731, project NETA-MM. The work of Francisco Marcellán has been supported by the research project PID2021- 122154NB-I00 Ortogonalidad y Aproximación con Aplicaciones en Machine Learning y Teoría de la Probabilidad funded by MICIU/AEI/10.13039/501100011033 and by "ERDF A Way of making Europe."

References

1. Chihara, T.S.: An introduction to orthogonal polynomials. In: Mathematics and Its Applications Series, vol. 13. Gordon and Breach Science Publishers, New York-London-Paris (1978)
2. Dominici, D., Marcellán, F.: Truncated Hermite polynomials. J. Difference Equ. Appl. **29**(7), 701–732 (2023)

3. Dominici, D., García-Ardila, J.C., Marcellán, F.: Symmetrization process and truncated orthogonal polynomials. Anal. Math. Phys. **14**, 137 (2024)
4. Dupuis, M., Rys, J., King, H.F.: Evaluation of molecular integrals over Gaussian basis functions. J. Chem. Phys. **65**(1), 111–116 (1976)
5. García-Ardila, J.C., Marcellán, F., Marriaga, M.E.: Orthogonal Polynomials and Linear Functionals. An Algebraic Approach and Applications. EMS Series of Lectures in Mathematics. EMS Press, Berlin (2021)
6. Gautschi, W.: On the construction of Gaussian quadrature rules from modified moments. Math. Comp. **24**, 245–260 (1970)
7. Gautschi, W.: Questions of numerical condition related to polynomials. In: de Boor, C., Golub, G.H. (eds.) Symposium on Recent Advances in Numerical Analysis, pp. 45–72. Academic Press, New York (1978)
8. Gautschi, W.: An algorithmic implementation of the generalized Christoffel theorem. In: Hämmerlin, G. (ed.) Numerical Integration International Series of Numerical Mathematics, vol. 57. Birkhäuser Verlag, Basel (1982)
9. Gautschi, W.: The interplay between classical analysis and (numerical) linear algebra—a tribute to Gene H. Golub. ETNA **13**, 119–147 (2002)
10. Gautschi, W.: Orthogonal polynomials (in Matlab). J. Comput. Appl. Math. **178**, 215–234 (2005)
11. Gautschi, W.: Orthogonal polynomials, quadrature, and approximation: computational methods and software (in Matlab). In: Marcellán, F., Van Assche, W. (eds.) Orthogonal Polynomials and Special Functions: Computation and Applications. Lecture Notes in Mathematics 1883, pp. 1–77. Springer, Berlin (2006)
12. Gautschi, W.: Orthogonal Polynomials in Matlab: Exercises and Solutions, Software-Environ. Tools. SIAM, Philadelphia (2016)
13. Gautschi, W.: A Software Repository for Orthogonal Polynomials. Software, Environments and Tools, vol. 28 SIAM, Philadelphia (2018)
14. Golub, G.H., Welsch, J.H.: Calculation of Gauss quadrature rules. Math. Comp. **23**, 221–230 (1969)
15. Gradshteyn, I.S., Ryzhik, I.M.: Table of Integrals, Series, and Products, Corrected and Enlarged Edition. Academic Press, New York, First Printing (1980)
16. Ismail, M.E.H., Koelink, E., Editors: Theory and applications of special functions. A volume dedicated to Mizan Rahman. Papers from the Special Session of the American Mathematical Society Annual Meeting held in Baltimore, MD, January 15–18, 2003. Dev. Math., vol 13. Springer, New York (2005)
17. King, H.F.: Strategies for evaluation of Rys roots and weights. J. Phys. Chem. A **120**, 9348–9351 (2016)
18. Lyu, S., Chen, Y.: Gaussian unitary ensembles with two jump discontinuities, PDEs, and the coupled Painlevé *II* and *IV* systems. Stud. Appl. Math. **146**(1), 118–138 (2021)
19. Milovanović, G.V.: Construction and applications of Gaussian quadratures with nonclassical and exotic weight functions. Stud. Univ. Babeş-Bolyai Math. **60**, 211–233 (2015)
20. Milovanović, G.V.: An efficient computation of parameters in the Rys quadrature formula. Bull. Cl. Sci. Math. Nat. Sci. Math. **43**, 39–64 (2018)
21. Milovanović, G.V., Mitrinović, D.S., Rassias, Th.M.: Topics in Polynomials: Extremal Problems, Inequalities, Zeros. World Scientific Publishing, River Edge (1994)
22. Milovanović, G.V., Vasović, N.: Orthogonal polynomials and generalized Gauss-Rys quadrature formulae. Kuwait J. Sci. **49**(1), 17 (2022)
23. Olver, F.W.J., Lozier, D.W., Boisvert, R.F., Clark, C.W., Editors: NIST Handbook of Mathematical Functions. U. S. Department of Commerce, National Institute of Standards and Technology, Washington, DC. Cambridge University Press, Cambridge (2010)
24. Rys, J., Dupuis, M., King, H.F.: Computation of electron repulsion integrals using Rys quadrature method. J. Comput. Chem. **4**, 154–157 (1983)
25. Sack, R.A., Donovan, A.F.: An algorithm for Gaussian quadrature given modified moments. Numer. Math. **18**, 465–478 (1971)

26. Sagar, R.P., Smith, V.H.: On the calculation of Rys polynomials and quadratures. Int. J. Quant. Chem. **42**(4), 827–836 (1992)
27. Schwenke, D.W.: On the computation of high order Rys quadrature weights and nodes. Comput. Phys. Comm. **185**, 762–763 (2014)
28. Shizgal, B.D.: A novel Rys quadrature algorithm for use in the calculation of electron repulsion integrals. Comput. Theor. Chem. **1074**, 178–184 (2015)
29. Wheeler, J.C.: Modified moments and Gaussian quadratures. Rocky Mountain J. Math. **4**, 287–296 (1974)

Chapter 9
An Optimal Quadrature Formula with Derivative in the Hilbert Space

Abdullo R. Hayotov and Samandar S. Babaev

Dedicated to Academician Gradimir V. Milovanović on the occasion of his 75th birthday

9.1 Introduction

The construction of quadrature formulas involves determining the nodes and weights associated with these nodes to approximate definite integrals. In addition and even more critically, quadrature formulas provide an essential tool for the numerical solution of differential and integral equations. Several methods exist for constructing quadrature formulas, each with its approach. One of these approaches is based on variational methods.

The variational approach for constructing quadrature formulas is based on variational principles and functionals. In this approach, the goal is to find the optimal nodes and weights that minimize or maximize a certain functional associated with the approximation of the integral.

Evidently, the precision of a quadrature formula is contingent upon both its coefficients and the chosen nodes. The challenge of determining the minimum norm of the error functional, considering both coefficients and nodes, is known

A. R. Hayotov (✉)
V.I. Romanovsky Institute of Mathematics, Uzbekistan Academy of Sciences, Tashkent, Uzbekistan

Central Asian University, Tashkent, Uzbekistan
e-mail: hayotov@mail.ru; a.hayotov@centralasian.uz

S. S. Babaev
V.I. Romanovsky Institute of Mathematics, Uzbekistan Academy of Sciences, Tashkent, Uzbekistan

Tashkent International University, Tashkent, Uzbekistan

Bukhara State University, Bukhara, Uzbekistan
e-mail: bssamandar@gmail.com; s.s.boboev@buxdu.uz; samandar.babayev@teacher.tiu.uz

M. P. Stanić et al. (eds.), *Analysis, Approximation, Optimization: Computation and Applications*, Springer Optimization and Its Applications 224,
https://doi.org/10.1007/978-3-031-85743-0_9

as the Nikolsky problem. The resulting formula is recognized as the optimal quadrature formula in the sense of Nikolsky, a problem initially addressed by S.M. Nikolsky [24]. This exploration has been perpetuated by a multitude of scholars, as documented in notable works such as [6, 7, 10, 25], along with additional references.

When the nodes are held fixed, the endeavor to minimize the norm of the error functional solely in terms of coefficients is termed Sard problem. The resulting formula, identified as the optimal quadrature formula in the sense of Sard, was originally scrutinized by A. Sard [26]. This work is also devoted to the construction of an optimal quadrature formula in the Sard sense. In the different spaces, based on these methods, Sard's problem was investigated by many authors (see, [17, 18, 30, 31, 40]).

The papers [11, 14, 20, 33, 39] are devoted to construction of an optimal quadrature formulas for approximation of Fourier integrals in the Hilbert and Sobolev spaces. In these studies, the quadrature sum is a combination of the given function values in a uniform grid. The error functional norm measures the difference between the integral and the quadrature sum. The optimal quadrature formulas were obtained by minimizing the error functional norm with respect to coefficients. Analytical formulas for optimal coefficients can also be obtained using a discrete analog of the differential operator. Additionally, the convergence order of the optimal quadrature formulas was examined.

The filtered back-projection (FBP) algorithm, a renowned image reconstruction technique for computed tomography (CT), is implemented in the capacity of an application, where its accuracy is enhanced through the approximation of Fourier transforms and their inversions [11, 14–16]. The proposed optimal quadrature formula of the first, second, and third orders is employed for this purpose. Through numerical experiments, the quality of the reconstructed image is ascertained using the suggested optimal quadrature formulas in conjunction with the conventional FBP, wherein the fast Fourier transform is employed for Fourier transform calculation and inversion. In the noise test, results indicate that the proposed algorithm yields more reliable outcomes in the face of noise compared to the conventional FBP.

Many problems of science and engineering are naturally reduced to singular integral equations. In the papers [2, 3, 27, 32], the optimal quadrature formulas were constructed for approximate calculation of the Cauchy and Hadamard type singular integrals in the Sobolev space.

These works [9, 23] are devoted to the numerical integration of integrals of fractional order. In article [34], the authors discussed a new algorithm that utilizes optimal quadrature formulas to solve generalized Abel integral equations. The study showed that this new approach is highly effective, extremely accurate, and can be applied to other specific tasks as well.

More specifically, we discuss the numerical integration of the right Riemann–Liouville fractional integral [22, 23]. The right Riemann–Liouville fractional integral with order $\alpha > 0$ of the given function $f(t)$, $t < 1$, are defined as

$$D_{t,1}^{-\alpha} f(t) =_{RL} D_{t,1}^{-\alpha} f(t) = \frac{1}{\Gamma(\alpha)} \int_t^1 (x-t)^{\alpha-1} f(x)\,\mathrm{d}x,$$

where $\Gamma(\cdot)$ is the Euler gamma function.

The Riemann–Liouville fractional integral is a generalization of the standard Riemann integral to non-integer orders. It has practical applications in various fields, particularly in describing complex physical phenomena that exhibit memory effects, long-range dependencies, and anomalous behaviors. Numerous works have been devoted to constructing optimal quadrature formulas on various spaces. These formulas are designed for the numerical integration of Riemann–Liouville fractional integrals.

In [12], an optimal quadrature formula was developed for the numerical integration of right Riemann–Liouville fractional integrals in the space $W_2^{(1,0)}(t,1)$. A sharp upper bound for the error functional associated with the constructed formula was obtained. This quadrature formula was then utilized to calculate the integrals of several functions in the numerical experiments.

Also, optimal quadrature formulas were constructed in $W_2^{(m,m-1)}(t,1)$ space, and the existence and uniqueness of the formulas were proved (see, [5, 13]). In the work [8], the author constructed an optimal quadrature formula for numerically integrating the left Riemann–Liouville integral in $W_2^{(1,0)}(0,t)$ space.

Also, optimal quadrature formulas with derivatives are essential in terms of achieving a high order of approximation. Several scientists have addressed the task of constructing optimal quadrature formulas of this type [19, 21].

Consider a table containing values of functions $\varphi(x_\beta)$ and $\varphi'(x_\beta)$, where $\beta = 0, 1, \ldots, N$ at points $x_\beta \in [t, 1]$ for functions φ. We assume that these functions belong to the Hilbert space

$$W_2^{(2,1)}(t,1) = \{\varphi : [t,1] \to \mathbb{R} \mid \varphi' \text{ is abs. cont. and } \varphi'' \in L_2(t,1)\},$$

equipped with the norm

$$\left\| \varphi | W_2^{(2,1)}(t,1) \right\| = \left\{ \int_t^1 \left(\varphi''(x) + \varphi'(x)\right)^2 \mathrm{d}x \right\}^{1/2},$$

where $\int_t^1 \left(\varphi''(x) + \varphi'(x)\right)^2 \mathrm{d}x < +\infty$.

We consider a quadrature formula in the space $W_2^{(2,1)}(t,1)$, given by

$$\int_t^1 \frac{\varphi(x)}{(x-t)^{1-\alpha}}\,\mathrm{d}x \approx \sum_{\beta=0}^{N} C_{\beta,0}\varphi(x_\beta) + \sum_{\beta=0}^{N} C_{\beta,1}\varphi'(x_\beta). \tag{9.1}$$

The problem of optimizing this quadrature formula for both $C_{\beta,0}$ and $C_{\beta,1}$ coefficients is complicate. In this work, we consider the problem of sequential optimization. In this case, the coefficients of the quadrature formula will be simple and easy to use. Therefore, initially, the optimal quadrature formula of the form

$$\int_t^1 \frac{\varphi(x)}{(x-t)^{1-\alpha}}\,\mathrm{d}x \approx \sum_{\beta=0}^{N} C_{\beta,0}\varphi(x_\beta) \tag{9.2}$$

in the space $L_2^{(1)}(t,1)$ is constructed. Then, the coefficients of the constructed quadrature formula (9.2) are used to construct the optimal quadrature formula (9.1). Such a process is referred to as a sequential optimization process. Consequently, we obtain a quadrature formula of the Euler–Maclaurin type. Here, $x_\beta = h\beta + t$ are equally spaced nodes in the interval $[t, 1]$, $h = \frac{1-t}{N}$, is the distance between adjacent nodes, $0 < \alpha < 1$, and $C_{\beta,1}$ are the unknown coefficients of the quadrature formula (9.1). And, $C_{\beta,0}$ are the known coefficients of optimal quadrature formula (9.2).

The optimal quadrature formula of the form (9.2) was constructed in the space $L_2^{(1)}(t,1)$ in work [38]. The coefficients $C_{\beta,0}$ have the following form ($\beta = \overline{0,N}$):

$$C_{0,0} = \frac{h^\alpha}{\alpha(\alpha+1)},$$

$$C_{\beta,0} = \frac{h^\alpha}{\alpha(\alpha+1)}\left((\beta+1)^{\alpha+1} - 2(\beta)^{\alpha+1} + (\beta-1)^{\alpha+1}\right), \quad \beta = 1,2,\ldots,N-1,$$

$$C_{N,0} = \frac{h^\alpha}{\alpha(\alpha+1)}\left((N-1)^{\alpha+1} - (N)^{\alpha+1}\right) + \frac{(1-t)^\alpha}{\alpha}.$$

The difference

$$(\ell,\varphi) = \int_t^1 \frac{\varphi(x)}{(x-t)^{1-\alpha}}\,\mathrm{d}x - \sum_{\beta=0}^{N} C_{\beta,0}\varphi(x_\beta) - \sum_{\beta=0}^{N} C_{\beta,1}\varphi'(x_\beta), \tag{9.3}$$

is referred to as *the error* of the quadrature formula (9.1). *The error functional* of the quadrature formula (9.1)

$$\ell(x) = \frac{\varepsilon_{[t,1]}(x)}{(x-t)^{1-\alpha}} - \sum_{\beta=0}^{N} C_{\beta,0}\delta(x-x_\beta) + \sum_{\beta=0}^{N} C_{\beta,1}\delta'(x-x_\beta) \tag{9.4}$$

belongs to the space $W_2^{(2,1)*}(t,1)$. The space $W_2^{(2,1)*}(t,1)$ is the dual to the space $W_2^{(2,1)}(t,1)$. Here, $\varepsilon_{[t,1]}(x)$ is the characteristic function of the interval $[t,1]$, and $\delta(x)$ is Dirac's delta-function.

In this work, our initial focus is on calculating the norm of the error functional ℓ of the quadrature formula (9.1). We aim to find the coefficients $C_{\beta,1}$ that minimize the norm of the error functional ℓ. If such coefficients $C_{\beta,1}$ exist, the corresponding quadrature formula is termed the optimal quadrature formula in the space $W_2^{(2,1)}(t,1)$.

It is evident from (9.4) that the error functional ℓ, defined on the space $W_2^{(2,1)}(t,1)$, satisfies the following equalities:

$$\left(\ell, \mathrm{e}^{-x}\right) = 0, \quad (\ell, 1) = 0. \tag{9.5}$$

These equalities indicate that our quadrature formula is exact for the function e^{-x} and the constant function.

Applying the Cauchy–Schwarz inequality, we estimate the absolute value of the error (9.3) as follows:

$$|(\ell, \varphi)| \leq \|\varphi\|_{W_2^{(2,1)}} \cdot \|\ell\|_{W_2^{(2,1)*}},$$

where

$$\|\ell\|_{W_2^{(2,1)*}} = \sup_{\varphi,\ \|\varphi\|\neq 0} \frac{|(\ell, \varphi)|}{\|\varphi\|}.$$

Therefore, to estimate the error of the quadrature formula (9.1) on functions of the space $W_2^{(2,1)}(t,1)$, we need to find the norm of the error functional ℓ in the dual space $W_2^{(2,1)*}(t,1)$. From this, we derive the following problem statement.

Problem 9.1 Find the norm of the error functional ℓ of the quadrature formula (9.1) in the space $W_2^{(2,1)*}(t,1)$.

It is evident that the norm of the error functional ℓ depends on the coefficients $C_{\beta,1}$ and the nodes x_β. The minimization of the quantity $\|\ell\|$ by coefficients $C_{\beta,1}$ is a linear problem, while the minimization by nodes x_β is generally a nonlinear and complicated problem. We aim to minimize the quantity $\|\ell\|$ by finding coefficients $C_{\beta,1}$ for fixed nodes x_β.

The primary objective of this study is outlined as follows.

Problem 9.2 Find the coefficients $\mathring{C}_{\beta,1}$ that minimize the value of $\|\ell\|_{W_2^{(2,1)*}}$, and compute

$$\left\|\mathring{\ell}\right\|_{W_2^{(2,1)*}} = \inf_{C_{\beta,1}} \|\ell\|_{W_2^{(2,1)*}}. \tag{9.6}$$

The functional $\mathring{\ell}$ in (9.4) is termed as the error functional corresponding to *the optimal quadrature formula* in $W_2^{(2,1)}(t,1)$, and $\mathring{C}_{\beta,1}$ are called the optimal coefficients. For convenience, we will remain the optimal coefficients $\mathring{C}_{\beta,1}$ as $C_{\beta,1}$.

The rest of the chapter is organized as follows. In Sect. 9.2, the error of the quadrature formula is estimated; in Sect. 9.3, the extremum of the error functional norm is found; in Sect. 9.4, the analytical forms of optimal coefficients is found; in Sect. 9.5, numerical results are presented.

9.2 Estimation of the Error Functional of the Quadrature Formula

To solve Problem 9.1, our initial step is to determine the norm of the error functional. This requires finding an extremal function for the error functional ℓ. Specifically, we obtain

$$\|\ell\|^2_{W_2^{(2,1)*}} = (\ell, \psi_\ell) = \int_{-\infty}^{+\infty} \ell(x)\psi_\ell(x)\,\mathrm{d}x, \tag{9.7}$$

where ψ_ℓ is the extremal function defined as (see, [4])

$$\psi_\ell(x) = \ell(x) * G_2(x) + d\mathrm{e}^{-x} + p_0, \tag{9.8}$$

where the operator $*$ is convolution and

$$G_2(x) = \frac{\operatorname{sign}(x)}{2}\left(\frac{\mathrm{e}^x - \mathrm{e}^{-x}}{2} - x\right). \tag{9.9}$$

Using (9.8) in (9.7), we derive

$$\|\ell\|^2_{W_2^{(2,1)*}} = (\ell, \psi_\ell) = \int_{-\infty}^{+\infty} \ell(x)\left(\ell(x) * G_2(x) + d\mathrm{e}^{-x} + p_0\right)\mathrm{d}x. \tag{9.10}$$

Substituting (9.10) into the subsequent expression (9.11), taking into account (9.5), yields

$$\begin{aligned}\|\ell\|^2_{W_2^{(2,1)*}} = &\int_t^1\int_t^1 \frac{G_2(x-y)\,\mathrm{d}x\,\mathrm{d}y}{(x-t)^{1-\alpha}(y-t)^{1-\alpha}} - 2\sum_{\beta=0}^{N} C_{\beta,0}\int_t^1 \frac{G_2(x-x_\beta)\,\mathrm{d}x}{(x-t)^{1-\alpha}}\\ &+\sum_{\beta=0}^{N}\sum_{\gamma=0}^{N} C_{\beta,0}C_{\gamma,0}G_2(h\beta-h\gamma) - \sum_{\beta=0}^{N}\sum_{\gamma=0}^{N} C_{\beta,1}C_{\gamma,1}G_2''(h\beta-h\gamma)\\ &+2\sum_{\beta=0}^{N} C_{\beta,1}\left(\int_t^1 \frac{G_2'(x-x_\beta)\,\mathrm{d}x}{(x-t)^{1-\alpha}} + \sum_{\gamma=0}^{N} C_{\gamma,0}G_2'(x_\beta-x_\gamma)\right).\end{aligned} \tag{9.11}$$

9.3 The Extremum of the Error Functional Norm

We consider the Lagrange function

$$\Phi(C_{0,1}, C_{1,1}, \ldots, C_{N,1}, \lambda) = \|\ell\|^2_{W_2^{(2,1)*}} - 2\lambda(\ell, \mathrm{e}^{-x}),$$

to find the minimum of (9.11) under the condition

$$(\ell, \mathrm{e}^{-x}) = \int_t^1 \frac{\mathrm{e}^{-x}\,\mathrm{d}x}{(x-t)^{1-\alpha}} - \sum_{\gamma=0}^{N} C_{\gamma,0}\mathrm{e}^{-x_\gamma} + \sum_{\gamma=0}^{N} C_{\gamma,1}\mathrm{e}^{-x_\gamma} = 0.$$

In that case, equating to zero the partial derivatives of the function Φ by $C_{\beta,1}$, $\beta = 0, 1, \ldots, N$ and λ, we get the following system of linear equations

$$\begin{cases} \sum_{\gamma=0}^{N} C_{\gamma,1} G_2''(h\beta - h\gamma) + \lambda \mathrm{e}^{-x_\beta} = F_2(t + h\beta), \quad \beta = 0, 1, \ldots, N, \\ \sum_{\gamma=0}^{N} C_{\gamma,1}\mathrm{e}^{-x_\gamma} = g(t), \end{cases} \tag{9.12}$$

where

$$F_2(t + h\beta) = \sum_{\gamma=0}^{N} C_{\gamma,0} G_2'(h\beta - h\gamma) + \int_t^1 \frac{G_2'(x - x_\beta)}{(x-t)^{1-\alpha}}\,\mathrm{d}x, \tag{9.13}$$

$$g(t) = \sum_{\gamma=0}^{N} C_{\gamma,0}\mathrm{e}^{-x_\gamma} - \int_t^1 \frac{\mathrm{e}^{-x}}{(x-t)^{1-\alpha}}\,\mathrm{d}x, \tag{9.14}$$

and, function $G_2(x)$ defined in (9.9). Also, generalized derivatives of this function have the following forms:

$$G_2'(x) = \frac{\mathrm{sign}(x)}{2}\left(\frac{\mathrm{e}^{x} + \mathrm{e}^{-x}}{2} - 1\right)$$

and

$$G_2''(x) = \frac{\mathrm{sign}(x)}{2}\left(\frac{\mathrm{e}^{x} - \mathrm{e}^{-x}}{2}\right).$$

It turns out that $G_2''(x)$ is the function $G_1(x)$ which is the fundamental solution of the differential operator $\frac{d^2}{dx^2} - 1$ (see, [28]), that is

$$G_1(x) = \frac{\mathrm{sign}(x)}{2}\left(\frac{e^x - e^{-x}}{2}\right). \tag{9.15}$$

This system (9.12) has a unique solution, as shown in [13], and its solution gives a minimum value to the expression (9.11).

Next, an analytical solution to the system of Eqs. (9.12) is derived.

Assuming $C_{\beta,1} = 0$ for $\beta < 0$ and $\beta > N$, and using the definition of the convolution of discrete functions, we rewrite the system (9.12) as equations in convolution

$$\begin{cases} C_{\beta,1} * G_1(h\beta) + \lambda e^{-x_\beta} = F_2(t + h\beta), & 0 \le \beta \le N, \\ C_{\beta,1} = 0, & \beta < 0 \text{ and } \beta > N, \\ \sum_{\gamma=0}^{N} C_{\gamma,1} e^{-x_\gamma} = g(t), \end{cases} \tag{9.16}$$

where $x_\beta = h\beta + t$.

Here we use the Sobolev method [35–37] for the discrete analogue of the differential operator $\frac{d^2}{dx^2} - 1$. The discrete analog of the operator $\frac{d^2}{dx^2} - 1$ has the following form: [28, 29]:

$$D_1(h\beta) = \frac{1}{1 - e^{2h}} \begin{cases} 0, & |\beta| \ge 2, \\ -2e^h, & |\beta| = 1, \\ 2(1 + e^{2h}), & \beta = 0. \end{cases} \tag{9.17}$$

The solution of the system (9.16) is found by introducing the function $U_2(h\beta)$ in place of $C_{\beta,1}$

$$U_2(h\beta) = C_{\beta,1} * G_1(h\beta) + \lambda e^{-h\beta - t}, \tag{9.18}$$

and the operator $D_1(h\beta)$ satisfying $D_1(h\beta) * G_1(h\beta) = \delta_d(h\beta)$, where $\delta_d(h\beta)$ is the discrete delta-function. The coefficients are expressed as

$$C_{\beta,1} = D_1(h\beta) * U_2(h\beta). \tag{9.19}$$

If we determine the function $U_2(h\beta)$ for all integer values β, we can find coefficients $C_{\beta,1}$ from Eq. (9.19).

From expression (9.18), taking into account (9.15), we have

$$\begin{aligned}U_2(h\beta) &= C_{\beta,1} * G_1(h\beta) + \lambda \mathrm{e}^{-h\beta-t}\\ &= \sum_{\gamma=-\infty}^{+\infty} C_{\gamma,1}\frac{\mathrm{sign}(h\beta-h\gamma)}{4}\left(\mathrm{e}^{h\beta-h\gamma}-\mathrm{e}^{h\gamma-h\beta}\right)+\lambda \mathrm{e}^{-h\beta-t}.\end{aligned}$$

According to the assumption that $C_{\beta,1} = 0$ when $\beta = -1, -2, \ldots,$ and $\beta = N+1, N+2, \ldots$, we get

$$U_2(h\beta) = \sum_{\gamma=0}^{N} C_{\gamma,1}\frac{\mathrm{sign}(h\beta-h\gamma)}{4}\left(\mathrm{e}^{h\beta-h\gamma}-\mathrm{e}^{h\gamma-h\beta}\right)+\lambda \mathrm{e}^{-h\beta-t}. \tag{9.20}$$

From expression (9.20) when $\beta < 0$, taking into account the last equation of the system (9.16), we come

$$U_2(h\beta) = \frac{-\mathrm{e}^{h\beta+t}g(t)}{4} + (\lambda + L)\mathrm{e}^{-h\beta-t},$$

here $L = \frac{1}{4}\sum_{\gamma=0}^{N} C_{\gamma,1}\mathrm{e}^{h\gamma+t}$.

Similarly, when $\beta > N$, we get

$$U_2(h\beta) = \frac{\mathrm{e}^{h\beta+t}g(t)}{4} + (\lambda - L)\mathrm{e}^{-h\beta-t}.$$

Thus, we determined the appearance of the function $U_2(h\beta)$ at all integer values of β

$$U_2(h\beta) = \begin{cases} -\frac{1}{4}\mathrm{e}^{h\beta+t}g(t) + (\lambda+L)\mathrm{e}^{-h\beta-t}, & \beta < 0,\\ F_2(h\beta+t), & 0 \leq \beta \leq N,\\ \frac{1}{4}\mathrm{e}^{h\beta+t}g(t) + (\lambda-L)\mathrm{e}^{-h\beta-t}, & \beta > N. \end{cases} \tag{9.21}$$

Using (9.17) and (9.19), we find the unknowns $\lambda + L$ and $\lambda - L$ when $\beta = -1$ and $\beta = N + 1$ in (9.21), respectively. So, we have

$$\lambda + L = F_2(t)\mathrm{e}^{t} + \frac{1}{4}\mathrm{e}^{2t}g(t) \tag{9.22}$$

and

$$\lambda - L = F_2(1)e - \frac{1}{4}\mathrm{e}^{2}g(t). \tag{9.23}$$

We rewrite the function $U_2(h\beta)$ taking into account equalities (9.22) and (9.23)

$$U_2(h\beta) = \begin{cases} -\frac{1}{4}e^{h\beta+t}g(t) + \left(F_2(t) + \frac{1}{4}e^t g(t)\right)e^{-h\beta}, & \beta < 0, \\ F_2(h\beta + t), & 0 \le \beta \le N, \\ \frac{1}{4}e^{h\beta+t}g(t) + \left(F_2(1) - \frac{1}{4}eg(t)\right)e^{1-h\beta-t}, & \beta > N. \end{cases} \quad (9.24)$$

9.4 The Analytical Forms of Optimal Coefficients

The main result of this work is presented as follows.

Now, utilizing (9.17) and (9.24), we find the optimal coefficients $C_{\beta,1}$ for $\beta = 0, 1, \ldots, N$:

$$C_{\beta,1} = D_1(h\beta) * U_2(h\beta) = \sum_{\gamma=-\infty}^{+\infty} D_1(h\gamma)U_2(h\beta - h\gamma), \quad (9.25)$$

where $D_1(h\beta)$ and $U_2(h\beta)$ are defined by expressions (9.17) and (9.24), respectively.

For $\beta = 0$, the coefficient $C_{0,1}$ is given by:

$$C_{0,1} = \frac{2}{1 - e^{2h}}\left(-e^h(U_2(-h) + F_2(h + t)) + (1 + e^{2h})F_2(t)\right). \quad (9.26)$$

From equality (9.26), and after some simplification, we have

$$C_{0,1} = \frac{2}{1 - e^{2h}}\left(\frac{e^t g(t)}{4}(1 - e^{2h}) + F_2(t) - e^h F_2(h + t)\right).$$

Now we calculate the coefficients $C_{\beta,1}$ from (9.25), when $\beta = 1, 2, \ldots, N - 1$,

$$C_{\beta,1} = \frac{2}{1 - e^{2h}}\left(-e^h(F_2(h(\beta - 1) + t) + F_2(h(\beta + 1) + t)) \right.$$
$$\left. +(1 + e^{2h})F_2(h\beta + t)\right).$$

Similarly, the coefficient $C_{N,1}$ is found as:

$$C_{N,1} = \frac{2}{1 - e^{2h}}\left((1 + e^{2h})F_2(1) - e^h F_2(1 - h) - e^h U_2(h(N + 1))\right).$$

Thus, we get the following expression

$$C_{N,1} = \frac{2}{1-\mathrm{e}^{2h}} \left(\frac{\mathrm{e}g(t)}{4}(1-\mathrm{e}^{2h}) + \mathrm{e}^{h}(\mathrm{e}^{h}F_2(1) - F_2(1-h)) \right).$$

In summary, the coefficients of the optimal quadrature formula (9.1) in the space $W_2^{(2,1)}(t,1)$ are given by the following.

Theorem 9.1 *Coefficients of the optimal quadrature formula* (9.1) *with equidistant nodes in the space* $W_2^{(2,1)}(t,1)$ *have the following forms:*

$$C_{0,1} = \frac{2}{1-\mathrm{e}^{2h}} \left(\frac{\mathrm{e}^{t}g(t)}{4}(1-\mathrm{e}^{2h}) + F_2(t) - \mathrm{e}^{h}F_2(h+t) \right),$$

$$C_{\beta,1} = \frac{2}{1-\mathrm{e}^{2h}} \Big(-\mathrm{e}^{h}(F_2(h(\beta-1)+t) + F_2(h(\beta+1)+t))$$

$$+(1+\mathrm{e}^{2h})F_2(h\beta+t) \Big), \quad \beta = 1, 2, \ldots, N-1,$$

$$C_{N,1} = \frac{2}{1-\mathrm{e}^{2h}} \left(\frac{\mathrm{e}g(t)}{4}(1-\mathrm{e}^{2h}) + \mathrm{e}^{h}(\mathrm{e}^{h}F_2(1) - F_2(1-h)) \right).$$

Remark 9.1 *Considering the expressions* (9.13) *and* (9.14), *we simplify the expressions of the coefficients in Theorem 9.1. So we get the following expressions*

$$C_{0,1} = \frac{h^{\alpha}}{1-\mathrm{e}^{2h}} \left(\sum_{k=0}^{\infty} \frac{h^{k}(1+(-1)^{k}\mathrm{e}^{2h})}{k!(\alpha+k)} - \frac{\mathrm{e}^{2h}+2\mathrm{e}^{h}\alpha+1}{\alpha(\alpha+1)} \right),$$

$$C_{\beta,1} = \frac{h^{\alpha}}{1-\mathrm{e}^{2h}} \Bigg(\frac{(1-\mathrm{e}^{h})^{2}\left((\beta-1)^{\alpha+1} - (\beta+1)^{\alpha+1}\right)}{\alpha(\alpha+1)}$$

$$+2\frac{(1+\mathrm{e}^{2h})\beta^{\alpha} - \mathrm{e}^{h}((\beta+1)^{\alpha} + (\beta-1)^{\alpha})}{\alpha}$$

$$+\mathrm{e}^{-h\beta} \Bigg((\beta+1)^{\alpha} \sum_{k=0}^{\infty} \frac{h^{k}(\beta+1)^{k}}{k!(\alpha+k)}$$

$$+\mathrm{e}^{2h}(\beta-1)^{\alpha} \sum_{k=0}^{\infty} \frac{h^{k}(\beta-1)^{k}}{k!(\alpha+k)} - (1+\mathrm{e}^{2h})\beta^{\alpha} \sum_{k=0}^{\infty} \frac{h^{k}\beta^{k}}{k!(\alpha+k)} \Bigg)$$

$$+\mathrm{e}^{h\beta} \Bigg(\mathrm{e}^{2h}(\beta+1)^{\alpha} \sum_{k=0}^{\infty} \frac{(-1)^{k}h^{k}(\beta+1)^{k}}{k!(\alpha+k)}$$

$$+(\beta-1)^{\alpha} \sum_{k=0}^{\infty} \frac{(-1)^{k}h^{k}(\beta-1)^{k}}{k!(\alpha+k)}$$

$$-(1+e^{2h})\beta^{\alpha}\sum_{k=0}^{\infty}\frac{(-1)^k h^k \beta^k}{k!(\alpha+k)}\Bigg)\Bigg), \quad \beta=1,2,\ldots,N-1,$$

$$C_{N,1} = \frac{h^{\alpha}}{1-e^{2h}}\Bigg(\frac{(1-e^h)^2\left((N-1)^{\alpha+1}-N^{\alpha+1}\right)}{\alpha(\alpha+1)}$$

$$+\frac{(1+e^{2h})N^{\alpha}-2e^h(N-1)^{\alpha}}{\alpha}$$

$$+e^{t-1+2h}\sum_{k=0}^{\infty}\frac{h^k((N-1)^{k+\alpha}-N^{k+\alpha})}{k!(k+\alpha)}$$

$$+e^{1-t}\sum_{k=0}^{\infty}\frac{(-1)^k h^k((N-1)^{k+\alpha}-N^{k+\alpha})}{k!(k+\alpha)}\Bigg).$$

It is known that the series in the last expressions obtained for the coefficients are convergent. We rewrite the series in Remark 9.1 using Kummer's hypergeometric functions.

It is known that Kummer's hypergeometric function is defined as follows (see for instance, [1])

$$M(a,b,z) = \quad 1+\tfrac{az}{b}+\tfrac{(a)_2 z^2}{(b)_2 2!}+\cdots+\tfrac{(a)_n z^n}{(b)_n n!}+\cdots \tag{9.27}$$
$$(a)_n = a(a+1)(a+2)\cdots(a+n-1), \quad (a)_0=1,$$

here n is a positive integer, and $M(a,b,z)$ is a convergent series for all values of a, b, and z.

Using (9.27), we get the following expressions for the coefficients

$$C_{0,1} = \frac{h^{\alpha}}{\alpha(1-e^{2h})}\left(M(\alpha,\alpha+1,h)+e^{2h}M(\alpha,\alpha+1,-h)-\frac{e^{2h}+2e^h\alpha+1}{\alpha+1}\right),$$

$$C_{\beta,1} = \frac{h^{\alpha}}{\alpha(1-e^{2h})}\Bigg(\frac{(1-e^h)^2\left((\beta-1)^{\alpha+1}-(\beta+1)^{\alpha+1}\right)}{\alpha+1}$$

$$+2\left((1+e^{2h})\beta^{\alpha}-e^h((\beta+1)^{\alpha}+(\beta-1)^{\alpha})\right)$$

$$+e^{-h\beta}\left((\beta+1)^{\alpha}M(\alpha,\alpha+1,h(\beta+1))\right.$$

$$\left.+\,e^{2h}(\beta-1)^{\alpha}M(\alpha,\alpha+1,h(\beta-1))-(1+e^{2h})\beta^{\alpha}M(\alpha,\alpha+1,h\beta)\right)$$

$$+e^{h\beta}\left(e^{2h}(\beta+1)^{\alpha}M(\alpha,\alpha+1,-h(\beta+1))\right.$$

$$+(\beta-1)^{\alpha}M(\alpha,\alpha+1,h(-\beta+1))$$

$$- (1+e^{2h})\beta^{\alpha}M(\alpha,\alpha+1,-h\beta)\Big)\Big), \quad \beta = 1,2,\dots,N-1,$$

$$C_{N,1} = \frac{h^{\alpha}}{\alpha(1-e^{2h})}\left(\frac{(1-e^{h})^2\left((N-1)^{\alpha+1}-N^{\alpha+1}\right)}{\alpha+1} + (1+e^{2h})N^{\alpha}\right.$$
$$-2e^{h}(N-1)^{\alpha} + e^{t-1+2h}((N-1)^{\alpha}M(\alpha,\alpha+1,h(N-1))$$
$$+ - N^{\alpha}M(\alpha,\alpha+1,hN))e^{1-t}((N-1)^{\alpha}M(\alpha,\alpha+1,-h(N-1))$$
$$\left. -N^{\alpha}M(\alpha,\alpha+1,-hN))\right),$$

where

$$\frac{1}{\alpha}M(\alpha,\alpha+1,h\beta) = \sum_{k=0}^{+\infty}\frac{h^k\beta^k}{k!(\alpha+k)}.$$

9.5 Numerical Results

Within this section, we shall provide the numerical outcomes that substantiate the theoretical results acquired in the preceding sections.

Now, let us take a look at the following example. We will numerically validate the optimal quadrature formula by using Python. The computations will cover a range of functions with different values of α and N.

Example 9.1 Consider the right Riemann–Liouville fractional integral of the function $f(x) = x\sin(x)$, $x \in [t, 1]$, where $0 < \alpha < 1$

$$\int_t^1 \frac{x\sin(x)}{(x-t)^{1-\alpha}}\,dx \approx \sum_{\beta=0}^{N} C_{\beta,0}x_{\beta}\sin(x_{\beta}) + \sum_{\beta=0}^{N} C_{\beta,1}\left(\sin(x_{\beta}) + x_{\beta}\cos(x_{\beta})\right).$$

We begin by numerically solving the right Riemann–Liouville integral of the function $f(x)$. This is accomplished using the optimal quadrature formula (9.1). We have tabulated the absolute errors of the optimal quadrature formula (ErOQF) in tables, which include t_i values from the interval (0, 1) with a step of 0.2. Each $[t_i; 1]$ interval is divided into N segments. Tables 9.1 and 9.2 illustrate the differences between exact solutions and approximate solutions for various values of α for ErOQF.

The tables illustrate that our optimal quadrature formula, used to approximate the integrals in the provided example, yields good accuracy results even at small values of N.

Table 9.1 The error analysis of OQF at $t = 0.2, 0.4, 0.6, 0.8$, when $\alpha = 0.1$, $N = 10$, and $N = 100$

	Exact value	ErOQF when $N = 10$	ErOQF when $N = 100$
$t = 0.2$	0.449587	$1.423 \cdot 10^{-6}$	$3.134 \cdot 10^{-10}$
$t = 0.4$	0.554900	$6.718 \cdot 10^{-7}$	$1.870 \cdot 10^{-10}$
$t = 0.6$	0.629203	$1.937 \cdot 10^{-7}$	$6.029 \cdot 10^{-11}$
$t = 0.8$	0.591764	$1.923 \cdot 10^{-8}$	$6.354 \cdot 10^{-12}$

Table 9.2 The error analysis of OQF at $t = 0.2, 0.4, 0.6, 0.8$, when $\alpha = 0.5$, $N = 10$, and $N = 100$

	Exact value	ErOQF when $N = 10$	ErOQF when $N = 100$
$t = 0.2$	0.320871	$2.845 \cdot 10^{-7}$	$3.133 \cdot 10^{-11}$
$t = 0.4$	0.315276	$7.856 \cdot 10^{-8}$	$9.667 \cdot 10^{-12}$
$t = 0.6$	0.275814	$1.364 \cdot 10^{-8}$	$1.891 \cdot 10^{-12}$
$t = 0.8$	0.182425	$7.837 \cdot 10^{-10}$	$1.217 \cdot 10^{-13}$

9.6 Conclusions

In this study, we developed an optimal quadrature formula with derivatives for numerically integrating the right Riemann–Liouville integral within the Hilbert space. The construction of this formula involved determining the analytical form of the norm of the error functional, as the error of the quadrature formula is bounded by this norm. We derived a system of linear equations with coefficients for the optimal quadrature formula by identifying the conditional minimum of the norm of the error functional. This process led to the analytical expressions for the coefficients of the optimal quadrature formula.

Acknowledgments We are grateful to the organizing committee for being invited to the International Mathematical Conference dedicated to Academician Gradimir V. Milovanović on the occasion of his 75th anniversary named Analysis, Approximations, and Applications (AAA2023).

References

1. Abramowitz, M., Stegun, I.A.: Handbook of Mathematical Functions. A 10th printing with corrections, National Bureau of Standards, Washington, DC (1972)
2. Akhmedov, D., Shadimetov, K.: Optimal quadrature formulas with derivative for Hadamard type singular integrals. AIP Conf. Proc. **2365**, 020020 (2022)
3. Akhmedov, D.M., Shadimetov, K.M.: Optimal quadrature formulas for approximate solution of the first kind singular integral equation with Cauchy kernel. Stud. Univ. Babeş-Bolyai Math. **67**(3), 633–651 (2022)
4. Babaev, S.S., Hayotov, A.R.: Optimal interpolation formulas in the space $W_2^{(m,m-1)}$". Calcolo (2019). https://doi.org/10.1007/s10092-019-0320-9

5. Babaev, S.S.: Construction of an optimal quadrature formula for the approximation of fractional integrals. AIP Conf. Proc. (2024). https://doi.org/10.1063/5.0199596
6. Blaga, P., Coman, Gh.: Some problems on optimal quadrature. Stud. Univ. Babeş-Bolyai Math. **52**(4), 21–44 (2007)
7. Bojanov, B.: Optimal quadrature formulas. Uspekhi Mat. Nauk. **60/6**(366), 33–52 (2005) (in Russian)
8. Boytillayev, B.A.: Optimal formulas for the approximate solution of the generalized Abel integral equations in the Hilbert space. Probl. Comput. Appl. Math., Tashkent **3/1**(50), 84–91 (2023). https://elibrary.ru/hlgvaq
9. Cai, M., Li, C.: Numerical approaches to fractional integrals and derivatives. A Rev. Math. **8**(43), 1–53 (2020)
10. Cătinaş, T., Coman, Gh.: Optimal quadrature formulas based on the φ-function method. Stud. Univ. Babeş-Bolyai Math. **51**(1), 49–64 (2006)
11. Hayotov, A.R., Babaev, S.S.: Optimal quadrature formulas for computing of Fourier integrals in $W_2^{(m,m-1)}$ space. AIP Conf. Proc. (2021). https://doi.org/10.1063/5.0057127
12. Hayotov, A.R., Babaev, S.S.: An optimal quadrature formula for numerical integration of the right Riemann–Liouville fractional integral. Lobachevskii J. Math. (2023). https://doi.org/10.1134/S1995080223100165
13. Hayotov, A.R., Babaev, S.S.: Optimal quadrature formula for numerical integration of fractional integrals in a Hilbert space. J. Math. Sci. (2023). https://doi.org/10.1007/s10958-023-06844-w
14. Hayotov, A.R., Jeon, S., Lee, C.-O.: On an optimal quadrature formula for approximation of Fourier integrals in the space $L_2^{(1)}$. J. Comput. Appl. Math. **388**, 112713 (2020)
15. Hayotov, A.R., Jeon, S., Shadimetov, Kh.M.: Application of optimal quadrature formulas for reconstruction of CT images. J. Comput. Appl. Math. **388**, 113313 (2021)
16. Hayotov, A.R., Jeon, S., Lee, C-O., Shadimetov, Kh.M.: Optimal quadrature formulas for non-periodic functions in Sobolev space and its application to CT image reconstruction. Filomat **35**, 4177–4195 (2021)
17. Hayotov, A.R., Milovanović, G.V., Shadimetov, Kh.M.: On an optimal quadrature formula in the sense of Sard. Numer. Algor. **57**, 487–510 (2011)
18. Hayotov A.R., Milovanović G.V., Shadimetov Kh.M. Optimal quadratures in the sense of Sard in a Hilbert space. Appl. Math. Comput. **259**, 637–653 (2015). https://doi.org/10.1016/j.amc.2015.02.093
19. Hayotov, A.R., Nuraliev, F.A., Shadimetov, Kh.M.: Optimal quadrature formulas with derivative in the space $L_2^{(m)}(0,1)$. Am. J. Numer. Anal. **2**(4), 115–127 (2014)
20. Hayotov, A.R., Khayriev, U.N.: Construction of an optimal quadrature formula in the hilbert space of periodic functions. Lobachevskii J. Math. **11**(43), 3151–3160 (2022)
21. Hayotov, A.R., Shadimetov, Kh.M., Nuraliev, F.A.: On an optimal quadrature formula in Sobolev space $L_2^{(m)}(0,1)$. J. Comput. Appl. Math. **243**, 91–112 (2013)
22. Kilbas, A.A., Srivastava, H.M., Trujillo, J.J.: Theory and Applications of Fractional Differential Equations, vol. 539. Elsevier, Amsterdam (2006)
23. Li, C., Zeng, F: Numerical Methods for Fractional Calculus, vol. 294. CRC Press Taylor and Francis Group (2015)
24. Nikolsky, S.M.: To question about estimation of approximation by quadrature formulas. Uspekhi Mat. Nauk **5/2**(36), 165–177 (1950) (in Russian)
25. Nikolsky, S.M.: Quadrature Formulas, 4th edn. Nauka, Moscow (1988) (in Russian)
26. Sard, A.: Best approximate integration formulas; best approximation formulas. Am. J. Math. **71**, 80–91 (1949)
27. Shadimetov, K.M., Akhmedov, D.M.: Approximate solution of a singular integral equation using the Sobolev method. Lobachevskii J. Math. **43**(2), 496–505 (2022)
28. Shadimetov, Kh.M., Hayotov, A.R.: Construction of the discrete analogue of the differential operator $d^{2m}/dx^{2m} - d^{2m-2}/dx^{2m-2}$. Uzbek Mat. Jurnal **(2)**, 85–95 (2004) (in Russian)

29. Shadimetov, Kh.M., Hayotov, A.R.: Properties of the discrete analogue of the differential operator $d^{2m}/dx^{2m} - d^{2m-2}/dx^{2m-2}$. Uzbek Mat. Jurnal **(4)**, 72–83 (2004) (in Russian)
30. Shadimetov, Kh.M., Hayotov, A.R.: Optimal quadrature formulas with positive coefficients in $L_2^{(m)}(0, 1)$ space. J. Comput. Appl. Math. **235**, 1114–1128 (2011)
31. Shadimetov, Kh.M., Hayotov, A.R.: Optimal quadrature formulas in the sense of Sard in $W_2^{(m,m-1)}$ space". Calcolo **51**(2), 211–243 (2014)
32. Shadimetov, Kh.M., Hayotov, A.R., Akhmedov, D.M.: Optimal quadrature formulas for Cauchy type singular integrals in Sobolev space. Appl. Math. Comput., **263**, 302–314 (2015). https://doi.org/10.1016/j.amc.2015.04.066
33. Shadimetov, Kh.M., Hayotov, A.R., Bozarov, B.: Optimal quadrature formulas for oscillatory integrals in the Sobolev space. J. Inequal. Appl. (2022). https://doi.org/10.1186/s13660-022-02839-4
34. Shadimetov, Kh.M., Daliev, B.S.: Optimal formulas for the approximate-analytical solution of the general Abel integral equation in the Sobolev space. Results Appl. Math. **15**(100276) (2022). https://doi.org/10.1016/j.rinam.2022.100276
35. Sobolev, S.L.: Introduction to the Theory of Cubature Formulas. Nauka, Moscow (1974) (in Russian)
36. Sobolev, S.L., Vaskevich, V.L.: The Theory of Cubature Formulas, Kluwer Academic Publishers Group, Dordrecht (1997)
37. Sobolev, S. L.: The Coefficients of Optimal Quadrature Formulas. In: Selected Works of S.L. Sobolev, pp. 561–566. Springer, New York (2006)
38. Toshboyev, O.N.: Optimal quadrature formulas for approximate calculation of the Riemann–Liouville fractional order integral. Probl. Comput. Appl. Math. Tashkent **44**(5/1), 171–179 (2022). https://elibrary.ru/mcubra
39. Xu, Z., Milovanović, G.V., Xiang, S.: Efficient computation of highly oscillatory integrals with Henkel kernel. Appl. Math. Comput. **261**, 312–322 (2015)
40. Zhang, S., Novak, E.: Optimal quadrature formulas for the Sobolev space H^1. J. Sci. Comput. **78**, 274–289 (2019)

Chapter 10
Dynamics of Operators on the Space of Radon Measures

Stefan Ivković

Dedicated to Academician Gradimir V. Milovanović on the occasion of his 75th birthday

10.1 Introduction

Linear dynamics of operators have been studied in many articles during several decades; see [15] and [3] as monographs on this topic. Among other concepts, hypercyclicity, topological transitivity, and topological mixing, as important linear dynamical properties of bounded linear operators, have been investigated in many research works; see [2, 5, 10, 12, 22, 24, 25] and their references. Specially, hypercyclic weighted shifts on $\ell^p(\mathbb{Z})$ were characterized in [14, 26], and then C-C. Chen and C-H. Chu, using aperiodic elements of locally compact groups, extended the results in [26] to weighted translations on Lebesgue spaces in the context of a second countable group [11]. Afterward, many other linear dynamics in connection with this theme have been studied; see [8, 9, 23]. Recently, the dynamics of weighted composition operators on solid Banach function spaces has been studied in [20, 21, 27].

Now, for a locally compact Hausdorff space Ω, we can consider $C_0(\Omega)$ as a special kind of a solid Banach function space. Here, as usual, $C_0(\Omega)$ is the space of all continuous functions on Ω vanishing at infinity, which is equipped with the supremum norm. It is well known in functional analysis that the dual of $C_0(\Omega)$ is isometrically isomorphic to the space of Radon measures on Ω equipped with the total variation norm. Hence, the adjoint of a weighted composition operator on $C_0(\Omega)$ can be considered as an operator on the space of Radon measures on Ω. In Sect. 10.3 of this chapter, we study the dynamics of this adjoint operator and provide conditions that ensure topological transitivity and chaos for this operator and for the

S. Ivković (✉)
Mathematical Institute of the Serbian Academy of Sciences and Arts, Belgrade, Serbia
e-mail: stefan.iv10@outlook.com

M. P. Stanić et al. (eds.), *Analysis, Approximation, Optimization: Computation and Applications*, Springer Optimization and Its Applications 224,
https://doi.org/10.1007/978-3-031-85743-0_10

cosine operator function generated by this operator. Moreover, we give concrete examples.

Another essential topic in connection with the dynamics of operators is fixed point theory, which can in fact be viewed as an opposite aspect to hypercyclicity and topological transitivity. An important part of the fixed point theory deals with the convergence of Markov chains, and many papers have been published on this theme, such as [16]. In Sect. 10.4, we consider a Markov integral operator that acts on the real Banach space of all real-valued continuous functions on a compact Hausdorff space Ω equipped with the supremum norm. The adjoint of this operator acts then on the real Banach space of all signed Radon measures on Ω equipped with the total variation norm. Based on the main general result in [16], we provide sufficient conditions on the integral kernel of this operator that ensure the convergence of the Markov chains induced by the adjoint of this operator. Also, we illustrate our result by a concrete example.

Now, the theory of Markov operators relies heavily on the concept of cones in Banach spaces. This fact motivated us to present some structural results concerning the cones in the space of Radon measures. Recall that a subset of a topological vector space is called spaceable if its union with the singleton $\{0\}$ contains a closed infinite-dimensional linear subspace. This notion was introduced in [1, 13] and so far has been considered in several papers, such as [4]. In Sect. 10.5, by applying [4, Theorem 4.2], we characterize a class of cones whose complement is spaceable in the space of Radon measures on locally compact Hausdorff space Ω. As a corollary, we obtain that the complement of the cone consisting of all positive Radon measures on Ω is spaceable when Ω has infinite cardinality.

10.2 Preliminaries

If $\mathcal{X}$ is a Banach space, the set of all bounded linear operators from $\mathcal{X}$ into $\mathcal{X}$ is denoted by $B(\mathcal{X})$. Also, we denote $\mathbb{N}_0 := \mathbb{N} \cup \{0\}$.

We recall now the following definition.

Definition 10.1 ([20, Definition 2.1]) Let $\mathcal{X}$ be a Banach space. A sequence $(T_n)_{n\in\mathbb{N}_0}$ of operators in $B(\mathcal{X})$ is called *topologically transitive* if for each non-empty open subsets U, V of $\mathcal{X}$, $T_n(U) \cap V \neq \varnothing$ for some $n \in \mathbb{N}$. If $T_n(U) \cap V \neq \varnothing$ holds from some n onward, then $(T_n)_{n\in\mathbb{N}_0}$ is called *topologically mixing*.

A single operator T in $B(\mathcal{X})$ is called topologically transitive (respectively topologically mixing) if the sequence $(T^n)_{n\in\mathbb{N}_0}$ is topologically transitive (respectively topologically mixing).

We have the following auxiliary technical lemma concerning open maps.

Lemma 10.1 *Let $T \in B(\mathcal{X})$ and suppose that T is an open map, that is, T maps open sets onto open sets. If T is topologically transitive, then for each non-empty open subsets U, V of $\mathcal{X}$, there exists a strictly increasing sequence $\{n_k\}_k \subseteq \mathbb{N}$ such that $T^{n_k}(U) \cap V \neq \varnothing$ for all $k \in \mathbb{N}$.*

Proof If T is topologically transitive, then given two open non-empty subsets U, V of $\mathcal{X}$, there exists some $n_1 \in \mathbb{N}$ such that $T^{n_1}(U) \cap V \neq \varnothing$. Now, since T is an open map by the assumption, it follows that $T^{n_1}(U)$ is open. Hence, if T is topologically transitive, there exists some $l_1 \in \mathbb{N}$ such that $T^{l_1}(T^{n_1}(U)) \cap V \neq \varnothing$. Put $n_2 = n_1 + l_1$. Proceeding inductively, we can deduce the statement of the lemma. □

Next, we recall the following definition.

Definition 10.2 ([20, Definition 2.3]) Let $\mathcal{X}$ be a Banach space, and $(T_n)_{n\in\mathbb{N}_0}$ be a sequence of operators in $B(\mathcal{X})$. A vector $x \in \mathcal{X}$ is called a *periodic element* of $(T_n)_{n\in\mathbb{N}_0}$ if there exists a constant $N \in \mathbb{N}$ such that for each $k \in \mathbb{N}$, $T_{kN}x = x$. The set of all periodic elements of $(T_n)_{n\in\mathbb{N}_0}$ is denoted by $\mathcal{P}((T_n)_{n\in\mathbb{N}_0})$. The sequence $(T_n)_{n\in\mathbb{N}_0}$ is called *chaotic* if $(T_n)_{n\in\mathbb{N}_0}$ is topologically transitive and $\mathcal{P}((T_n)_{n\in\mathbb{N}_0})$ is dense in $\mathcal{X}$. An operator $T \in B(\mathcal{X})$ is called *chaotic* if the sequence $\{T^n\}_{n\in\mathbb{N}_0}$ is chaotic.

The following concept plays a key role in the proofs.

Definition 10.3 ([20, Definition 2.6]) Let X be a topological space. Let $\alpha : X \longrightarrow X$ be invertible, and α, α^{-1} be Borel measurable. We say that α is *aperiodic* if for each compact subset K of X, there exists a constant $N > 0$ such that for each $n \geq N$, we have $K \cap \alpha^n(K) = \varnothing$, where α^n means the n-fold combination of α.

Throughout this chapter, we let Ω be a locally compact Hausdorff space and α be an aperiodic homeomorphism of Ω in the case when Ω is non-compact. As usual, $C_0(\Omega)$ denotes the space of all continuous functions on Ω vanishing at infinity, $C_b(\Omega)$ denotes the space of bounded, continuous functions on Ω, whereas $C_c(\Omega)$ stands for the set of all continuous, compactly supported functions on Ω. Both $C_0(\Omega)$ and $C_b(\Omega)$ are equipped with the supremum norm. Moreover, we let w be a positive bounded function on Ω such that also $w^{-1} \in C_b(\Omega)$, and we put then $T_{\alpha,w}$ to be the weighted composition operator on $C_0(\Omega)$ with respect to α and w, that is, $T_{\alpha,w}(f) = w \cdot (f \circ \alpha)$ for all $f \in C_0(\Omega)$. Easily, one can see that by the aforementioned assumptions $T_{\alpha,w}$ is well-defined and $\|T_{\alpha,w}\| \leq \|w\|_{\sup}$. Since $\frac{1}{w}$ is also bounded, then $T_{\alpha,w}$ is invertible, and we have

$$T_{\alpha,w}^{-1} f = \frac{f \circ \alpha^{-1}}{w \circ \alpha^{-1}}, \quad f \in C_0(\Omega).$$

Simply we denote $S_{\alpha,w} := T_{\alpha,w}^{-1}$.

Remark 10.1 ([20, Remark 2.5]) If w and $\frac{1}{w}$ are weights, the inverse of a weighted composition operator $T_{\alpha,w}$ is also a weighted composition operator. In fact, $S_{\alpha,w} = T_{\alpha^{-1},\frac{1}{w\circ\alpha^{-1}}}$. Moreover, if T_{α_1,w_1} and T_{α_2,w_2} are two weighted composition operators, then

$$T_{\alpha_2,w_2} \circ T_{\alpha_1,w_1} = T_{\alpha_1\circ\alpha_2, w_2(w_1\circ\alpha_2)},$$

so the composition of two weighted composition operators is again a weighted composition operator. By some calculation, one can see that for each $n \in \mathbb{N}$ and $f \in C_0(\Omega)$,

$$T^n_{\alpha,w} f = \left(\prod_{j=0}^{n-1} (w \circ \alpha^j) \right) \cdot (f \circ \alpha^n) \tag{10.1}$$

and

$$S^n_{\alpha,w} f = \left(\prod_{j=1}^{n} (w \circ \alpha^{-j}) \right)^{-1} \cdot (f \circ \alpha^{-n}). \tag{10.2}$$

The adjoint $T^*_{\alpha,w}$ is a bounded operator on $M(\Omega)$ where $M(\Omega)$ stands for the Banach space of all Radon measures on Ω equipped with the total variation norm. It is straightforward to check that

$$T^*_{\alpha,w}(\mu)(E) = \int_E w \circ \alpha^{-1} \, \mathrm{d}\mu \ \circ \alpha^{-1},$$

for every $\mu \in M(\Omega)$ and every measurable subset E of Ω. Here $\mu \ \circ \alpha^{-1}(E) = \mu(\alpha^{-1}(E))$ for every $\mu \in M(\Omega)$ and every measurable subset E of Ω. By (10.1) and (10.2), it follows then that for every $n \in \mathbb{N}$, $\mu \in M(\Omega)$ and a Borel measurable subset $E \subseteq \Omega$ we have that

$$T^{*n}_{\alpha,w}(\mu)(E) = \int_E \prod_{j=0}^{n-1} w \circ \alpha^{j-n} \, \mathrm{d}\mu \ \circ \alpha^{-n}$$

and

$$S^{*n}_{\alpha,w}(\mu)(E) = \int_E \prod_{j=1}^{n} (w \circ \alpha^{n-j})^{-1} \, \mathrm{d}\mu \ \circ \alpha^{n}.$$

Throughout this chapter, $M_+(\Omega)$ will denote the set of all non-negative Radon measures on Ω. For every Radon measure μ on Ω, we let as usual $|\mu|$ denote the total variation of μ.

10.3 Topologically Transitive Operators on the Space of Radon Measures

We start with the following proposition.

Proposition 10.1 *Under the assumption that α is aperiodic, the following statements are equivalent.*

(i) $T^*_{\alpha,w}$ *is topologically transitive on* $M(\Omega)$.
(ii) *For every compact subset K of Ω and any two measures μ, v in $M(\Omega)$ with $|v|(K^c) = |\mu|(K^c) = 0$, there exist a strictly increasing sequence $\{n_k\}_k \subseteq \mathbb{N}$ and sequences $\{A_k\}$, $\{B_k\}$ of Borel subsets of K such that $\alpha^{n_k}(K) \cap K = \varnothing$ for all $k \in \mathbb{N}$ and*

$$\lim_{k\to+\infty} |\mu|(A_k) = \lim_{k\to+\infty} |v|(B_k) = 0,$$

$$\lim_{k\to+\infty} \sup_{t\in K\cap A_k^c} \left(\prod_{j=0}^{n_k-1} (w\circ\alpha^j)(t) \right) = \lim_{k\to+\infty} \sup_{t\in K\cap B_k^c} \left(\prod_{j=1}^{n_k} (w\circ\alpha^{-j})^{-1}(t) \right) = 0.$$

Proof We prove $(i) \Rightarrow (ii)$ first. Suppose that $T^*_{\alpha,w}$ is topologically transitive on $M(\Omega)$. Notice that since $T^*_{\alpha,w}$ is invertible, it is an open map. For a given compact subset K of Ω,, let $v, \mu \in M(\Omega)$ such that $|v|(K^c) = |\mu|(K^c) = 0$. Since $T^*_{\alpha,w}$ is topologically transitive and an open map, by Lemma 10.1, for each $k \in \mathbb{N}$, we can find some $\eta^{(k)} \in M(\Omega)$ and a strictly increasing sequence $\{n_k\}_{k\subseteq\mathbb{N}}$ such that

$$|\eta^{(k)} - \mu|(\Omega) < \frac{1}{4^k} \quad \text{and} \quad |\gamma^{(k)} - v|(\Omega) < \frac{1}{4^k},$$

where $\gamma^{(k)} = T^{*n_k}_{\alpha,w}(\eta^{(k)})$, that is

$$\gamma^{(k)}(E) = \int_E \prod_{j=0}^{n_k-1} (w\circ\alpha^{j-n_k})\, d\eta^{(k)} \circ \alpha^{-n_k},$$

for every measurable subset E of Ω. Hence, we must have that $|\eta^{(k)}|(K^c) < \frac{1}{4^k}$ and $|\gamma^{(k)}|(K^c) < \frac{1}{4^k}$. Also we have that $S^{*n_k}_{\alpha,w}(\gamma^{(k)}) = \eta^k$, that is

$$\int_E \prod_{j=1}^{n_k} (w\circ\alpha^{n_k-j})^{-1}\, d\gamma^{(k)} \circ \alpha^{n_k} = \eta^{(k)}(E),$$

for every measurable subset E of Ω. Now, since α is aperiodic, we may in fact choose the sequence $\{n_k\}$ in a such way that $\alpha^{n_k}(K)\cap K = \varnothing$ and $\alpha^{-n_k}(K)\cap K = \varnothing$ for all k.

If $\eta^{(k)} = \eta^{(k)}_{1,+} - \eta^{(k)}_{1,-} + i\eta^{(k)}_{2,+} - i\eta^{(k)}_{2,-}$ and $\gamma^{(k)} = \gamma^{(k)}_{1,+} - \gamma^{(k)}_{1,-} + i\gamma^{(k)}_{2,+} - i\gamma^{(k)}_{2,-}$ are Jordan decompositions of $\eta^{(k)}$ and $\gamma^{(k)}$, it follows that $(\eta^{(k)}_{1,+} + \eta^{(k)}_{1,-})(K^c) < \frac{1}{4^k}$,

$(\eta^{(k)}_{2,+} + \eta^{(k)}_{2,-})(K^c) < \frac{1}{4^k}$, $(\gamma^{(k)}_{1,+} + \gamma^{(k)}_{1,-})(K^c) < \frac{1}{4^k}$ and $(\gamma^{(k)}_{2,+} + \gamma^{(k)}_{2,-})(K^c) < \frac{1}{4^k}$. Also, it is easy to see that

$$\gamma^{(k)}_{j,\pm}(E) = \int_E \prod_{i=0}^{n_k-1} (w \circ \alpha^{i-n_k})\, \mathrm{d}\eta^{(k)}_{j,\pm} \circ \alpha^{-n_k}$$

and

$$\eta^{(k)}_{j,\pm}(E) = \int_E \prod_{i=1}^{n_k} (w \circ \alpha^{n_k-i})^{-1}\, \mathrm{d}\gamma^{(k)}_{j,\pm} \circ \alpha^{n_k},$$

for every measurable subset E of Ω and $j \in \{1, 2\}$. This is because α is a homeomorphism of Ω, so taking compositions with α preserves mutual singularity of measures. Hence, for $j \in \{1, 2\}$ we get that

$$\int_{\alpha^{n_k}(K)} \prod_{i=0}^{n_k-1} (w \circ \alpha^{i-n_k})\, \mathrm{d}(\eta^{(k)}_{j,+} + \eta^{(k)}_{j,-}) \circ \alpha^{-n_k} < \frac{1}{4^k}$$

and

$$\int_{\alpha^{-n_k}(K)} \prod_{i=1}^{n_k} (w \circ \alpha^{n_k-i})^{-1}\, \mathrm{d}(\gamma^{(k)}_{j,+} + \gamma^{(k)}_{j,-}) \circ \alpha^{n_k} < \frac{1}{4^k}.$$

By the change of variables, we obtain that

$$\int_K \prod_{i=0}^{n_k-1} (w \circ \alpha^{i})\, \mathrm{d}(\eta^{(k)}_{j,+} + \eta^{(k)}_{j,-}) < \frac{1}{4^k}$$

and

$$\int_K \prod_{i=1}^{n_k} (w \circ \alpha^{-i})^{-1}\, \mathrm{d}(\gamma^{(k)}_{j,+} + \gamma^{(k)}_{j,-}) < \frac{1}{4^k},$$

for $j \in \{1, 2\}$. Let

$$A_k := \left\{ t \in K \mid \prod_{i=0}^{n_k-1} (w \circ \alpha^{i})(t) > \frac{1}{2^k} \right\}$$

and

$$B_k := \left\{ t \in K \mid \prod_{i=1}^{n_k} (w \circ \alpha^{-i})^{-1}(t) > \frac{1}{2^k} \right\}.$$

Then, $(\eta_{j,+}^{(k)} + \eta_{j,-}^{(k)})(A_k) < \frac{1}{2^k}$ and $(\gamma_{j,+}^{(k)} + \gamma_{j,-}^{(k)})(B_k) < \frac{1}{2^k}$ for $j \in \{1, 2\}$. Hence, $|\eta^{(k)}|(A_k) < \frac{1}{2^{k-1}}$ and $|\gamma^{(k)}|(B_k) < \frac{1}{2^{k-1}}$. Therefore, $|\mu|(A_k) < \frac{1}{2^{k-1}} + \frac{1}{4^k}$ and $|v|(B_k) < \frac{1}{2^{k-1}} + \frac{1}{4^k}$.

Next we prove $(ii) \Rightarrow (i)$. Given two open non-empty subsets O_1 and O_2 of $M(\Omega)$, choose some $\mu \in O_1$ and $v \in O_2$. By the regularity of μ and v, there exist some compact subsets K_1 and K_2 of Ω such that $|\mu|(K_1^c) < \frac{\epsilon}{2}$ and $|v|(K_2^c) < \frac{\epsilon}{2}$ for sufficiently small ϵ satisfying that $\epsilon-$neighborhoods of μ and v are contained in O_1 and O_2, respectively. Set $K = K_1 \cup K_2$. Then, K is compact and $|\mu|(K^c), |v|(K^c) < \frac{\epsilon}{2}$. Therefore, we may, without loss of generality, assume that there exists some compact subset $K \subseteq \Omega$ such that $|\mu|(K^c) = |v|(K^c) = 0$.

Choose the strictly increasing sequence $\{n_k\}_k$ and the sequences $\{A_k\}_k$, $\{B_k\}_k$ of Borel subsets of K satisfying the assumptions in (ii) with respect to μ and v. For each $k \in \mathbb{N}$, let $\tilde{\mu}_k$, $\tilde{v}_k$ be the measures in $M(\Omega)$ given by $\tilde{\mu}_k(E) = \mu(E \cap A_k^c \cap K)$ and $\tilde{v}_k(E) = v(E \cap B_k^c \cap K)$ for every measurable subset E of Ω. Then,

$$T_{\alpha,w}^{*n_k}(\tilde{\mu}_k)(E) = \int_E \prod_{j=0}^{n_k-1} (w \circ \alpha^{j-n_k}) \, \mathrm{d}\tilde{\mu}_k \circ \alpha^{-n_k},$$

for every measurable subset E of Ω. However,

$$\begin{aligned} \int_E \prod_{j=0}^{n_k-1} (w \circ \alpha^{j-n_k}) \, \mathrm{d}\tilde{\mu}_k \circ \alpha^{-n_k} &= \int_{E \cap \alpha^{n_k}(A_k^c \cap K)} \prod_{j=0}^{n_k-1} (w \circ \alpha^{j-n_k}) \, \mathrm{d}\tilde{\mu}_k \circ \alpha^{-n_k} \\ &= \int_{\alpha^{-n_k}(E) \cap A_k^c \cap K} \prod_{j=0}^{n_k-1} (w \circ \alpha^{j}) \, \mathrm{d}\tilde{\mu}_k, \end{aligned}$$

so we get that

$$\begin{aligned} |T_{\alpha,w}^{*n_k}(\tilde{\mu}_k)(E)| &\leq \sup_{t \in A_k^c \cap K} \prod_{j=0}^{n_k-1} (w \circ \alpha^j)(t) |\tilde{\mu}_k| (\alpha^{-n_k}(E) \cap A_k^c \cap K) \\ &= \sup_{t \in A_k^c \cap K} \prod_{j=0}^{n_k-1} (w \circ \alpha^j)(t) |\mu| (\alpha^{-n_k}(E) \cap A_k^c \cap K), \end{aligned}$$

for every measurable subset E of Ω. It follows that

$$\|T_{\alpha,w}^{*n_k}(\tilde{\mu}_k)\| \leq \sup_{t\in A_k^c\cap K}\left(\prod_{j=0}^{n_k-1}(w\circ\alpha^j)(t)\right)\|\mu\|, \tag{10.3}$$

for all k, so

$$\lim_{k\to+\infty} T_{\alpha,w}^{*n_k}(\tilde{\mu}_k) = 0.$$

Similarly, we can show that

$$\|S_{\alpha,w}^{*n_k}(\tilde{v}_k)\| \leq \sup_{t\in B_k^c\cap K}\left(\prod_{j=1}^{n_k}(w\circ\alpha^{-j})^{-1}(t)\right)\|v\|, \tag{10.4}$$

for all k, so

$$\lim_{k\to+\infty} S_{\alpha,w}^{*n_k}(\tilde{v}_k) = 0.$$

Moreover, we have that $(\mu-\tilde{\mu}_k)(E) = \mu(E\cap A_k)$ hence $|\mu-\tilde{\mu}_k|(E) \leq |\mu|(E\cap A_k)$ for all k and every measurable subset E of Ω, which gives $|\mu-\tilde{\mu}_k|(\Omega) \leq |\mu|(A_k)$ for all k. Thus, $\lim_{k\to+\infty}\|\mu-\tilde{\mu}_k\| = 0$. Similarly we can show that

$$\lim_{k\to+\infty}\|v-\tilde{v}_k\| = 0.$$

For each $k\in\mathbb{N}$, put $\eta_k = \tilde{\mu}_k + S_{\alpha,w}^{*n_k}(\tilde{v}_k)$. Then we have that $\lim_{k\to+\infty}\eta_k = \mu$ and $\lim_{k\to\infty} T_{\alpha,w}^{*n_k}(\eta_k) = v$, which shows that $T_{\alpha,w}^{*}$ is topologically transitive. □

Remark 10.2 We notice that in the proof of the implication $(ii)\Rightarrow(i)$, we have not used the assumption that α is aperiodic, so this implication holds even if α is not aperiodic.

We have the following corollary.

Corollary 10.1 *We have that* $(ii)\Rightarrow(i)$.

(i) $T_{\alpha,w}^{*}$ *is topologically transitive on* $M(\Omega)$.
(ii) For every compact subset K *of* Ω *we have that*

$$\lim_{n\to+\infty}\sup_{t\in K}\left(\prod_{j=0}^{n-1}(w\circ\alpha^j)(t)\right) = \lim_{n\to\infty}\sup_{t\in K}\left(\prod_{j=1}^{n}(w\circ\alpha^{-j})^{-1}(t)\right) = 0.$$

The cosine operator functions and the dynamics of such mappings have been studied in several papers, such as [6, 7, 17, 28]. In the rest of this section, we shall consider the cosine operator functions generated by $T^{*n}_{\alpha,w}$ for each $n \in \mathbb{N}$. Thus, for each $n \in \mathbb{N}$, we put $C^{(n)^*}_{\alpha,w} = \frac{1}{2}(T^{*n}_{\alpha,w} + S^{*n}_{\alpha,w})$.

The main idea for the proof of the next proposition is inspired by the proof of [17, Theorem 5].

Proposition 10.2 *We have that* $(ii) \Rightarrow (i)$.

(*i*) *The sequence* $(C^{(n)^*}_{\alpha,w})$ *is topologically transitive on* $M(\Omega)$.
(*ii*) *For every compact subset* K *of* Ω *and any two measures* μ, v *in* $M(\Omega)$ *with* $|\mu|(K^c) = |v|(K^c) = 0$, *there exist a strictly increasing sequence* $\{n_k\} \subseteq \mathbb{N}$ *and sequences* $\{A_k\}_k$, $\{F_k\}_k$, $\{D_k\}_k$ *of Borel subsets of* K *such that*

$$\lim_{k\to+\infty} |\mu|(A_k) = \lim_{n\to+\infty} |v|(A_k) = 0,$$

$$\lim_{k\to+\infty} \sup_{t\in K\cap A_k^c} \left(\prod_{j=0}^{n_k-1} (w\circ\alpha^j)(t)\right) = \lim_{k\to+\infty} \sup_{t\in K\cap A_k^c} \left(\prod_{j=1}^{n_k} (w\circ\alpha^{-j})^{-1}(t)\right) = 0,$$

$$\lim_{k\to+\infty} \sup_{t\in F_k} \left(\prod_{j=0}^{2n_k-1} (w\circ\alpha^j)(t)\right) = \lim_{k\to+\infty} \sup_{t\in D_k} \left(\prod_{j=1}^{2n_k} (w\circ\alpha^{-j})^{-1}(t)\right) = 0,$$

where $F_k \cap D_k = \varnothing$ *and* $A_k^c \cap K = F_k \cup D_k$ *for all* k.

Proof As in the proof of Proposition 10.1, given two open subsets O_1 and O_2 of $M(\Omega)$, there exists some compact subset K of Ω and measures $\mu \in O_1, v \in O_2$ such that $|\mu|(K^c) = |v|(K^c) = 0$. Choose the strictly increasing sequence $\{n_k\}_k$ and the sequences of Borel subsets of K, $\{A_k\}_k, \{F_k\}_k, \{D_k\}_k$ satisfying the assumptions of the proposition with respect to μ, v and K. For each $k \in \mathbb{N}$, let $\tilde{\mu}_k$, $\tilde{v}_k$ and $\tilde{\tilde{v}}_k$ be the measures in $M(\Omega)$ defined by $\tilde{v}_k(E) = v(E \cap F_k)$, $\tilde{\tilde{v}}_k(E) = v(E \cap D_k)$ and $\tilde{\mu}_k(E) = \mu(E \cap A_k^c \cap K)$ for every measurable subset E of Ω. Then $(\tilde{v}_k + \tilde{\tilde{v}}_k)(E) = v(E \cap A_k^c \cap K)$ for every measurable subset E of Ω and it is not hard to see that $\|v - \tilde{v}_k - \tilde{\tilde{v}}_k\| \overset{k\to+\infty}{\longrightarrow} 0$ since $|v|(A_k) \to 0$ as $k \to +\infty$ by the assumption. Similarly, since $|\mu|(A_k) \to 0$ as $k \to +\infty$, we have $\|\mu - \mu_k\| \to 0$ as $k \to +\infty$.

By (10.3) we have that

$$\begin{aligned}\|T^{*n_k}_{\alpha,w}(\tilde{v}_k)\| &\le \sup_{t\in F_k} \left(\prod_{j=0}^{n_k-1} (w\circ\alpha^j)(t)\right) \|v\| \\ &\le \sup_{t\in A_k^c\cap K} \left(\prod_{j=0}^{n_k-1} (w\circ\alpha^j)(t)\right) \|v\| \overset{k\to+\infty}{\longrightarrow} 0.\end{aligned}$$

Similarly, we can show that

$$\lim_{k\to+\infty} T^{*^{2n_k}}_{\alpha,w}(\tilde{v}_k) = 0, \quad \lim_{k\to+\infty} S^{*^{n_k}}_{\alpha,w}(\tilde{\tilde{v}}_k) = 0, \quad \lim_{k\to+\infty} S^{*^{2n_k}}_{\alpha,w}(\tilde{\tilde{v}}_k) = 0,$$

$$\lim_{k\to+\infty} T^{*^{n_k}}_{\alpha,w}(\tilde{\mu}_k) = 0 \quad \text{and} \quad \lim_{k\to+\infty} S^{*^{n_k}}_{\alpha,w}(\tilde{\mu}_k) = 0.$$

Set for each $k \in \mathbb{N}$, $\eta_k = \tilde{\mu}_k + 2T^{*^{n_k}}_{\alpha,w}(\tilde{v}_k) + 2S^{*^{n_k}}_{\alpha,w}(\tilde{\tilde{v}}_k)$ and proceed as in the proof of Proposition 10.1 part $(ii) \Rightarrow (i)$. □

Corollary 10.2 *We have that* $(ii) \Rightarrow (i)$.

(*i*) *The sequence* $(C^{(n)^*}_{\alpha,w})$ *is topologically transitive on* $M(\Omega)$.
(*ii*) *For every compact subset* K *of* Ω, *we have that*

$$\lim_{n\to+\infty} \sup_{t\in K} \left(\prod_{j=0}^{n-1} (w \circ \alpha^j)(t) \right) = \lim_{n\to+\infty} \sup_{t\in K} \left(\prod_{j=1}^{n} (w \circ \alpha^{-j})^{-1}(t) \right) = 0,$$

$$\lim_{n\to+\infty} \sup_{t\in K} \left(\prod_{j=0}^{2n-1} (w \circ \alpha^j)(t) \right) = \lim_{n\to\infty} \sup_{t\in K} \left(\prod_{j=1}^{2n} (w \circ \alpha^{-j})^{-1}(t) \right) = 0.$$

In the next proposition, we give sufficient conditions for the operator $T^*_{\alpha,w}$ and for the cosine operator function generated by this operator to be chaotic. The main idea in the proof of this proposition is inspired by the proof of [18, Theorem 5.3].

Proposition 10.3 *We have that* $(ii) \Rightarrow (i)$.

(*i*) *The sequences* $(T^{*n}_{\alpha,w})_n$ *and* $(C^{(n)^*}_{\alpha,w})_n$ *are chaotic on* $M(\Omega)$.
(*ii*) *For any compact subset* K *of* Ω *and any measure* $\mu \in M(\Omega)$ *with* $|\mu|(K^c) = 0$, *there exist a sequence of Borel subsets* $\{A_k\}_k$ *of* K *and a strictly increasing sequence* $\{n_k\}_k \subseteq \mathbb{N}$ *such that* $\lim_{k\to+\infty} |\mu|(A_k) = 0$ *and*

$$\lim_{k\to+\infty} \sum_{l=1}^{+\infty} \sup_{t\in K\cap A_k^c} \left(\prod_{j=0}^{ln_k-1} (w \circ \alpha^j)(t) \right)$$

$$= \lim_{k\to+\infty} \sum_{l=1}^{+\infty} \sup_{t\in K\cap A_k^c} \left(\prod_{j=1}^{ln_k} (w \circ \alpha^{-j})^{-1}(t) \right) = 0,$$

where the corresponding series are convergent for each k.

Proof By Proposition 10.2, it suffices to show that $\mathcal{P}((T^{*n}_{\alpha,w})_n)$ and $\mathcal{P}((C^{(n)^*}_{\alpha,w})_n)$ are dense in $M(\Omega)$. Let O be a non-empty open subset in $M(\Omega)$. Then, as observed earlier, there exists some measure $\mu \in O$ and a compact subset K of Ω such that

$|\mu|(K^c) = 0$. Choose sequence $\{A_k\}$ of Borel subsets and a strictly increasing sequence $\{n_k\}_k \subseteq \mathbb{N}$ satisfying the assumptions in (ii) with respect to μ and K. For each $k \in \mathbb{N}$, let μ_k be the measure given by $\mu_k(E) = \mu(E \cap A_k^c \cap K)$ for every measurable subset E of Ω. Then $\|\mu - \mu_k\| \to 0$ as $k \to +\infty$ since $\lim_{k\to\infty} |\mu|(A_k) = 0$. Moreover, by (10.3), (10.4) and the assumptions in (ii) we have that the series $\sum_{l=1}^{+\infty} T_{\alpha,w}^{*^{ln_k}}(\mu_k)$, $\sum_{l=1}^{+\infty} S_{\alpha,w}^{*^{ln_k}}(\mu_k)$ are absolutely convergent for each k, and moreover,

$$\lim_{k\to+\infty} \sum_{l=1}^{+\infty} \left\| T_{\alpha,w}^{*^{ln_k}}(\mu_k) \right\| = \lim_{k\to+\infty} \sum_{l=1}^{+\infty} \left\| S_{\alpha,w}^{*^{ln_k}}(\mu_k) \right\| = 0.$$

For each $k \in \mathbb{N}$, set

$$v_k = \mu_k + \sum_{l=1}^{+\infty} T_{\alpha,w}^{*^{ln_k}}(\mu_k) + \sum_{l=1}^{+\infty} S_{\alpha,w}^{*^{ln_k}}(\mu_k).$$

It is straightforward to check that $v_k = T_{\alpha,w}^{*ln_k} v_k = C_{\alpha,w}^{(ln_k)^*} v_k$ for all $l, k \in \mathbb{N}$. Moreover, for each $k \in \mathbb{N}$, by relations (10.3) and (10.4), we have

$$\begin{aligned}
\|v_k - \mu\| &\le \|\mu_k - \mu\| + \sum_{l=1}^{+\infty} \|T_{\alpha,w}^{*^{ln_k}}(\mu_k)\| + \sum_{l=1}^{+\infty} \|S_{\alpha,w}^{*^{ln_k}}(\mu_k)\| \\
&\le \|\mu_k - \mu\| + \sum_{l=1}^{+\infty} \sup_{t\in A_k^c\cap K} \left(\prod_{j=0}^{ln_k-1} (w \circ \alpha^j)(t) \right) \|\mu\| \\
&\quad + \sum_{l=1}^{+\infty} \sup_{t\in A_k^c\cap K} \left(\prod_{j=1}^{ln_k} (w \circ \alpha^{-j})^{-1}(t) \right) \|\mu\|.
\end{aligned}$$

By the assumptions in (ii), it follows that $\lim_{k\to\infty} v_k = \mu$, which proves the proposition. □

Corollary 10.3 *We have that $(ii) \Rightarrow (i)$.*

(i) *The sequences $(T_{\alpha,w}^{*n})_n$ and $(C_{\alpha,w}^{(n)^*})_n$ are chaotic on $M(\Omega)$.*
(ii) *For any compact subset K of Ω, we have that*

$$\lim_{n\to+\infty} \sum_{l=1}^{\infty} \sup_{t\in K} \left(\prod_{j=0}^{ln-1} (w \circ \alpha^j)(t) \right) = \lim_{n\to+\infty} \sum_{l=1}^{\infty} \sup_{t\in K} \left(\prod_{j=1}^{ln} (w \circ \alpha^{-j})^{-1}(t) \right) = 0,$$

where the corresponding series are convergent for each n.

As a concrete example, let $\Omega = \mathbb{R}$, $\alpha(t) = t + 1$ for all $t \in \mathbb{R}$ and

$$w(t) = \begin{cases} 2, & \text{for } t \leq -1, \\ \frac{1}{2}, & \text{for } t \geq 1, \\ \text{linear on the segment } [-1, 1]. \end{cases}$$

In this case, the conditions of Corollaries 10.1, 10.2, and 10.3 are satisfied.

In general, if $M, \epsilon > 0$ such that $1 + \epsilon < M$, $1 - \epsilon > \frac{1}{M}$, and $K_1, K_2 > 0$, then if $w \in C_b(\mathbb{R})$ satisfies that $M \geq |w(t)| \geq 1 + \epsilon$ for all $t \leq -K_1$ and $\frac{1}{M} \leq |w(t)| \leq 1 - \epsilon$ for all $t \geq K_2$, then the conditions of Corollaries 10.1, 10.2, and 10.3 are satisfied.

In fact, it is not hard to see from the proof of Proposition 10.1 part $(ii) \Rightarrow (i)$ and from the proof of Proposition 10.2 that the sufficient conditions of Corollaries 10.1 and 10.2 ensure also that the respective operators will be even topologically mixing and not just topologically transitive.

10.4 Convergence of Markov Chains in the Space of Signed Radon Measures

In this section, we shall first recall the notions and definitions from [16] and [19] that are needed to prove the main result denoted as Proposition 10.5.

Let X be a real vector space and $K \subseteq X$, $K \neq \emptyset$. We say that K is a cone in X if K satisfies the following properties:

(i) $K + K \subseteq K$;
(ii) $\lambda K \subseteq K$ for all $\lambda > 0$;
(iii) $K \cap (-K) = \{0\}$.

We let $\leq$ denote the associated order on X, so $x \leq y$ if and only if $(y - x) \in K$. For $e \in K$, set

$$I_e = \{x \in X \mid -e \leq x \leq e\}.$$

Recall that $A \subset X$ is called absorbing if

$$\{t > 0 \mid x \in tA\} \neq \emptyset,$$

for every $x \in X$. An element $e \in K$ is called an order unit for (X, K) if I_e is absorbing.

If Int$K \neq \emptyset$ for $x \in X$ and $y \in \text{Int}(K)$, we define then

$$M(x/y) := \inf\{t \in \mathbb{R} \mid x \leq ty\},$$
$$m(x/y) := \sup\{t \in \mathbb{R} \mid x \geq ty\}.$$

For an order unit $e \in$ K, we define the Thompson's norm with respect to e to be given by $||x||_T := \max\{M(x/e), -m(x/e)\}$ for all $x \in X$.

Let $e \in K\backslash\{0\}$. Then e is an order unit if and only if e is an internal point of K, that is for all $x \in X$ there exists a $\delta > 0$ such that $(e + \lambda x) \in K$ for all $\lambda \in [-\delta, \delta]$.

For $e \in K \setminus \{0\}$, set

$$X_e = \bigcup_{t \geq 0} tI_e, \quad K_e = K \cap X_e.$$

Then X_e is a subspace of X, I_e is an absorbing, absolutely convex subset of X_e, K_e is a cone of X_e, and e is an order unit for (X_e, K_e). If $|\cdot|_e$ denotes the Minkowski functional on X_e associated with I_e, (so $|x|_e = \inf\{t > 0 \mid x \in tI_e\}$), then $|\cdot|_e$ is a seminorm on X_e such that $|e|_e = |-e|_e = 1$. All these facts can be easily checked by elementary calculations. For the details, we refer to [19, Section 2].

Next we recall the following definition.

Definition 10.4 ([16] [19, Definition 1.1]) Let $(X, ||\cdot||)$ be a real normed space, let K be a cone in X. We say that K is normal if there is some constant $M > 0$ such that $||x|| \leq M||y||$ for all $x, y \in K$ satisfying $x \leq y$.

We recall also the following useful proposition.

Proposition 10.4 ([19, Proposition 1.2]) *Assume that $(X, ||\cdot||)$ is a real normed space, K is closed and normal cone in X, and let $e \in K \setminus \{0\}$. Then the following statements hold.*

(i) $|\cdot|_e$ is a norm on X_e and $||\cdot|| \leq \alpha|\cdot|_e$ on X_e for some $\alpha > 0$.
(ii) If $(X, ||\cdot||)$ is a Banach space, then $(X_e, |\cdot|_e)$ is a Banach space.
(iii) If e is an order-unit in (X, K), then $X = X_e$.
(iv) If $e \in \text{Int}(K)$, where $\text{Int}(K)$ is the interior of K in topological sense, then e is order-unit in (X, K) and $||\cdot||$, $|\cdot|_e$ are equivalent norms on X.

Moreover, $|\cdot|_e$ is the same as the Thomson's norm on X defined earlier.

From now on in this section, unless else is specified, we let $(X, ||\cdot||)$ be a real Banach space, $K \subseteq X$ be a normal, closed cone with $\text{Int}(K) \neq \emptyset$, $e \in \text{Int}K$ be an order unit and $||\cdot||_T$ be the Thompson' norm with respect to e. Since $||\cdot||$ and $||\cdot||_T$ are equivalent because K is closed and normal, we have $(X, ||\cdot||)^* = (X, ||\cdot||_T)^*$. This space will be denoted by X^* from now on.

Furthermore, we define the dual cone K^* in X^* by

$$K^* = \{x \in X^* \mid z(x) \geq 0 \text{ for all } x \in K\}$$

and the abstract simplex by

$$P(e) = \{\mu \in K^* \mid \mu(e) = 1\}.$$

Given $(X, e, || \cdot ||_T)$, consider now the quotient space $X/\mathbb{R}e$ and define Hilbert's quotient norm $||| \cdot |||_H : X/\mathbb{R}e \to \mathbb{R}^+$ by

$$|||x + \mathbb{R}e|||_H = 2 \inf_{\lambda \in \mathbb{R}} ||x + \lambda e||_T .$$

Then $||| \cdot |||_H$ is a norm on $X/\mathbb{R}e$.

Moreover, let $M(e) = \{\mu \in X^* \mid \mu(e) = 0\}$ and define Hilbert's dual norm

$$|| \cdot ||_H^* : M(e) \to \mathbb{R}^+ \text{ by } ||\mu||_H^* = \frac{1}{2}||\mu||_T^*, \quad \text{for all } \mu \in M(e).$$

We have then the following lemma.

Lemma 10.2 ([16] [19, Lemma 4.2]) $(X/\mathbb{R}e^*, |||\cdot|||_H^*)$ *is isometrically isomorphic to* $(M(e), || \cdot ||_H^*)$.

We recall that a bounded linear operator T on X is said to be a Markov operator with respect to K and e if $T(K) \subseteq K$ and $T(e) = e$.

We define now the induced linear map $\tilde{T} : X/\mathbb{R}e \to X/\mathbb{R}e$ by

$$\tilde{T}(x + \mathbb{R}e) = T(x) + \mathbb{R}e.$$

It is straightforward to check that $\tilde{T}$ is a well-defined, bounded linear map with respect to $||| \cdot |||_H$. Hence, we may consider $|||\tilde{T}|||_H$, that is the operator norm of $\tilde{T}$ with respect to $||| \cdot |||_H$.

We also define $S^* : M(e) \to M(e)$ by letting $S^* = T^*_{|M(e)}$. One can show that $||S^*||_H^* = |||\tilde{T}|||_H$, for more details, see [19, Section 6].

The crucial tool for proving the main result in this section is the following theorem.

Theorem 10.1 ([16] [19, Theorem 6.1]) *Let* $T : X \to X$ *be a Markov operator with respect to* K *and* e. *If* $|||\tilde{T}|||_H < 1$, *or equivalently if* $\| S^* \|_H^* < 1$, *then there is* $\pi \in P(e)$ *such that for all* $x \in X$ *and* $n \in \mathbb{N}$, *we have*

$$\|T^n(x) - \pi(x)e\|_T \leq (|||\tilde{T}|||_H)^n \|x\|_H$$

$$\text{and } \|(T^*)^n(\mu) - \pi\|_H^* \leq (|||\tilde{T}|||_H)^n,$$

for all $\mu \in P(e)$ *and all* $n \in \mathbb{N}$.

Now we let Ω be a compact Hausdorff space, and we set $X = C_{\mathbb{R}}(\Omega)$, that is, X is the space of all continuous real valued functions on Ω. X is then a Banach space with the supremum norm $\|\cdot\|_\infty$. We put

$$K = \{f \in X \mid f(w) \geq 0 \text{ for all } w \in \Omega\}$$

and we let the order unit e be the constant function 1 on Ω. It is obvious that K closed and normal cone with respect to $||\cdot||_\infty$ and that $1 \in \mathrm{Int}(K)$.

We have the following useful auxiliary lemma, which has also been proved in [19, Example 1.6].

Lemma 10.3 *For all $f \in X$, we have*

$$\|f\|_T = \|f\|_\infty.$$

Proof It holds that

$$\|f\|_T = \inf\{t > 0 \mid f \in tI_e\},$$

since $\|\cdot\|_T$ is equal to the Minkowski functional with respect to I_e by Proposition 10.4. Hence, we obtain that

$$\begin{aligned}\|f\|_T &= \inf\{t > 0 \mid f \in tI_e\} \\ &= \inf\{t > 0 \mid -t \leq f(w) \leq t \text{ for all } w \in \Omega\}, \\ &= \inf\{t > 0 \mid |f(w)| \leq t \text{ for all } w \in \Omega\} = \|f\|_\infty,\end{aligned}$$

which proves the lemma. □

Let $M_r(\Omega)$ denote the space of all signed Radon measures on Ω with the norm $\| v \| = |v|(\Omega)$. For $v \in M_r(\Omega)$, let $\phi_v : C_{\mathbb{R}}(\Omega) \to \mathbb{R}$ be defined by

$$\phi_v(f) = \int_\Omega f \, dv.$$

Then, $v \to \phi_v$ is an isometric isomorphism of $(M_r, (\Omega), \|\cdot\|)$ onto $((C_{\mathbb{R}}(\Omega))^*, \|\cdot\|)$. For the proof, see [19, Remark 2.3].

Since $(C_{\mathbb{R}}(\Omega), ||\cdot||_\infty)^*$ is isometrically isomorphic to $M_r(\Omega)$ equipped with the total variation norm, it follows that

$$||\mu||_T^* = |\mu|(\Omega), \quad \text{for all } \mu \in M_r(\Omega).$$

Also, it is clear that the dual cone and simplex in this case are $K^* = M_+(\Omega)$,

$$P(1) = \{\mu \in M_+(\Omega) \mid \mu(\Omega) = 1\},$$

so the simplex is the set of all probability Radon measures on Ω.

The material in the rest of this section contains the unpublished results from [19, Section 9].

Let $k : \Omega \times \Omega \to \mathbb{R}$ be a continuous non-negative function, and let μ be a positive Radon measure on Ω such that

$$\int_{\Omega} k(x, y)\, \mathrm{d}\mu(y) > 0, \quad \text{for all } x \in \Omega.$$

Put then $\tilde{k} : \Omega * \Omega \to \mathbb{R}$ to be defined as

$$\tilde{k}(x, y) = \frac{k(x, y)}{\int_{\Omega} k(x, y)\, \mathrm{d}\mu(y)}.$$

Then $\tilde{k}$ is continuous, nonnegative and $\int_{\Omega} \tilde{k}(x, y)\, \mathrm{d}\mu(y) = 1$.

We consider now the integral operator T_k on $C_{\mathbb{R}}(\Omega)$ given by

$$T_k(f)(x) = \int_{\Omega} \tilde{k}(x, y) f(y)\, \mathrm{d}\mu(y), \quad \text{for all } x \in \Omega.$$

It is straightforward to check that T_k is a Markov operator with respect to K and the constant function 1. The adjoint of T_k, will be denoted by T_k^*. Since $(C_{\mathbb{R}}(\Omega))^* = M_r(\Omega)$, where $M_r(\Omega)$ is the space of all signed Radon measures on Ω, we have that T_k^* is a bounded linear operator on $M_r(\Omega)$.

Now we are ready to present the main result in this section.

Proposition 10.5 *Under the aforementioned assumptions, if* $||\tilde{k}||_{\infty} < \frac{2}{\mu(\Omega)}$*,, then there exists a unique invariant probability Radon measure* $\tilde{v}$ *on* Ω *such that*

$$|(T_k^*)^n(v) - \tilde{v}|\,(\Omega) \to 0, \quad \text{as } n \to \infty,$$

for all probability Radon measures v *on* Ω.

Proof Let $v \in M_r(\Omega)$ and $w = T_k^*(v)$. Then, we have that

$$\int_{\Omega} f(y)\, \mathrm{d}w(y) = \int_{\Omega} T_k(f)\, \mathrm{d}v(x) = \int_{\Omega} \left(\int_{\Omega} \tilde{k}(x, y) f(y)\, \mathrm{d}\mu(y) \right) \mathrm{d}v(x).$$

Suppose that $v = v_+ - v_-$ is the Jordan decomposition of v. Then

$$\int_{\Omega} \left(\int_{\Omega} |\tilde{k}(x, y) f(y)|\, \mathrm{d}\mu(y) \right) \mathrm{d}v_+(x) \leq ||\tilde{k}||_{\infty}\, ||f||_{\infty}\, \mu(\Omega) v_+(\Omega) < \infty.$$

since u, v_+ are Radon measures and Ω is compact. Similarly,

$$\int_{\Omega} \left(\int_{\Omega} |\tilde{k}(x, y) f(y)|\, \mathrm{d}\mu(y) \right) \mathrm{d}v_-(x) < \infty.$$

By Fubini's theorem, we may change the order of the integration and get

$$\int_\Omega \left(\int_\Omega \tilde{k}(x, y) f(y) \, d\mu(y) \right) dv(x) = \int_\Omega \left(\int_\Omega \tilde{k}(x, y) f(y) \, dv(x) \right) d\mu(y).$$

This gives that

$$\int_\Omega f(y) \, dw(y) = \int_\Omega f(y) \rho(y) \, d\mu(y), \quad \text{where } \rho(y) = \int_\Omega \tilde{k}(x, y) \, dv(x).$$

Since this holds for all $f \in C_{\mathbb{R}}(\Omega)$, we deduce that $dw = \rho \, d\mu$. Then, we get $d|w| = |\rho| \, d\mu$ since μ is a positive measure. So, we obtain that

$$|T_k^*(v)|(\Omega) = |w|(\Omega) = \int_\Omega |\rho(y)| \, d\mu(y) = \int_\Omega \left| \int_\Omega \tilde{k}(x, y) \, dv(x) \right| d\mu(y).$$

Hence, using that $|| \cdot ||_H^* = \frac{1}{2} || \cdot ||_T^*$ by definition and that $||v||_T^* = |v|(\Omega)$ for all $v \in M_r(\Omega)$,, we derive

$$\begin{aligned} ||S_k^*||_H^* = & ||T^*_{k_{|M(1)}}||_H^* = \sup_{v \in M(1),\, v \neq 0} \frac{||T_K^*(v)||_H^*}{||v||_H^*} \\ = & \sup_{v \in M(1),\, v \neq 0} \frac{||T_K^*(v)||_T^*}{||v||_T^*} = \sup_{v \in M(1),\, v \neq 0} \frac{|T_K^*(v)|(\Omega)}{|v|(\Omega)} \\ = & \sup_{v \in M(1),\, v \neq 0} \frac{\int_\Omega \left| \int_\Omega \tilde{k}(x, y) \, dv_+(x) - \int_\Omega \tilde{k}(x, y) \, dv_-(x) \right| d\mu(y)}{v_+(\Omega) + v_-(\Omega)}. \end{aligned}$$

Now, if $v \in M(1)$,, then $v(\Omega) = 0$, hence $v_+(\Omega) = v_-(\Omega) = \frac{1}{2}|v|(\Omega)$. So, we get

$$\begin{aligned} & \left| \int_\Omega \tilde{k}(x, y) \, dv_+(x) - \int_\Omega \tilde{k}(x, y) \, dv_-(x) \right| \\ \leq & \max \left\{ \int_\Omega \tilde{k}(x, y) \, dv_+(x), \ \int_\Omega \tilde{k}(x, y) \, dv_-(x) \right\} \\ \leq & \max \left\{ \int_\Omega ||\tilde{k}||_\infty \, dv_+, \ \int_\Omega ||\tilde{k}||_\infty \, dv_- \right\} \\ = & \max\{ ||\tilde{k}||_\infty v_+(\Omega), \ ||\tilde{k}||_\infty v_-(\Omega) \} \\ = & \frac{1}{2} ||\tilde{k}||_\infty |v|(\Omega), \quad \text{for all } y \in \Omega. \end{aligned}$$

In the first inequality, we have used that $\int_\Omega \tilde{k}(x, y)\, dv_+(x)$, $\int_\Omega \tilde{k}(x, y)\, dv_-(x) \geq 0$ for all $y \in \Omega$ since $\tilde{k} \geq 0$. Hence, we obtain that

$$||T^*_{\tilde{k}_{|M(1)}}||^*_H = \sup_{v \in M(1),\ v \neq 0} \frac{|T^*_K(v)|(\Omega)}{|v|(\Omega)}$$

$$\leq \sup_{v \in M(1),\ v \neq 0} \frac{\int_\Omega \frac{1}{2}||\tilde{k}||_\infty |v|(\Omega)\, d\mu}{|v|(\Omega)} = \frac{1}{2}||\tilde{k}||_\infty \mu(\Omega).$$

Therefore, if $\tilde{k}$ is such that $||\tilde{k}||_\infty < \frac{2}{\mu(\Omega)}$, then $|||\tilde{T}_k|||_H < 1$. By Theorem 10.1, the proposition follows. □

As a concrete example, let now $\Omega = [0, 2\pi]$, μ be the Lebesgue measure on $[0, 2\pi]$ and $k : \Omega \times \Omega \to \mathbb{R}$ be given as $k(x, y) = \frac{1}{4}sin(\frac{1}{4}(x + y))$. Then k is continuous and $k(x + y) \geq 0$ for all $x, y \in [0, 2\pi]$. Furthermore, we have that

$$\int_\Omega k(x, y)\, d\mu(y) = \int_0^{2\pi} k(x, y)\, dy = \cos\left(\frac{1}{4}x\right) + \sin\left(\frac{1}{4}x\right).$$

Using elementary calculus, it can be easily checked that

$$\cos\left(\frac{1}{4}x\right) + \sin\left(\frac{1}{4}x\right) \geq 1,$$

for all $x \in [0, 2\pi]$. In particular,

$$\cos\left(\frac{1}{4}x\right) + \sin\left(\frac{1}{4}x\right) > 0,$$

for all $x \in [0, 2\pi]$, so we can then define $\tilde{k} : \Omega \times \Omega \to \mathbb{R}$ as described earlier by letting

$$\tilde{k}(x, y) = \frac{k(x, y)}{\int_\Omega k(x, y)\, d\mu(y)} = \frac{k(x, y)}{\cos(\frac{1}{4}x) + \sin(\frac{1}{4}x)}.$$

Since $\cos(\frac{1}{4}x) + \sin(\frac{1}{4}x) \geq 1$ for all $x \in [0, 2\pi]$, it follows that

$$||\tilde{k}||_\infty = \sup_{x, y \in [0, 2\pi]} \frac{\frac{1}{4}\sin(\frac{1}{4}(x + y))}{(\cos(\frac{1}{4}x) + \sin(\frac{1}{4}x))} \leq \frac{1}{4},$$

so

$$||\tilde{k}||_\infty \leq \frac{1}{4} < \frac{1}{\pi} = \frac{2}{2\pi} = \frac{2}{\mu(\Omega)},$$

since $\Omega = [0, 2\pi]$ and μ is the Lebesgue measure.

10.5 Spaceability of Cones in the Space of Radon Measures

In this section, for each $\mu, \nu \in M(\Omega)$,, we say that $\mu \sim \nu$ if there are measurable mutually disjoint subsets B_1, B_2 of Ω such that

$$|\mu|(B_1^c) = |\nu|(B_2^c) = 0.$$

Now we recall the following definition.

Definition 10.5 [[4, Definition 4.1]] Let $\mathcal{E}$ be a topological vector space. We say that a relation $\sim$ on $\mathcal{E}$ has property (D) if the following conditions hold.

(i) If (x_n) is a sequence in $\mathcal{E}$ such that $x_n \sim x_m$ for all distinct index m, n, then for each disjoint finite subset A, B of $\mathbb{N}$, we have

$$\sum_{n\in A} \alpha_n x_n \sim \sum_{m\in B} \beta_m x_m,$$

where α_n and β_m's are arbitrary scalars.

(ii) If a sequence (x_n) converges to x in $\mathcal{E}$ and for some $y \in \mathcal{E}$, $x_n \sim y$ for all $n \in \mathbb{N}$, then $x \sim y$.

Then we present the following proposition.

Proposition 10.6 *The relation $\sim$ defined as earlier satisfies the conditions of Definition 10.5.*

Proof Let $\{\mu_n\}_n$ be a sequence in $M(\Omega)$ such that $\mu_n \sim \mu_m$ for all distinct indexes m, n. Let A, B be finite mutually disjoint subsets of $\mathbb{N}$. For any fixed element $j \in A$ and each $m \in B$, there exist disjoint Borel subsets $D_{(j,m)}$ and $D'_{(j,m)}$ of Ω such that $|\mu_j|(D^c_{(j,m)}) = |\mu_m|(D'^c_{(j,m)}) = 0$ and $D_{(j,m)} \cap D'_{(j,m)} = \varnothing$. It follows that

$$|\mu_j| \left(\bigcup_{m\in B} D^c_{(j,m)} \right) \le \sum_{m\in B} |\mu_j|(D^c_{(j,m)}) = 0.$$

Hence, we get for all $i \in A$ that $|\mu_i|(\cap_{j\in A} \cup_{m\in B} D^c_{(j,m)}) = 0$, so

$$\left| \sum_{i\in A} \mu_i \right| (\cap_{j\in A} \cup_{m\in B} D^c_{(j,m)}) \le \sum_{i\in A} |\mu_i|(\cap_{j\in A} \cup_{m\in B} D^c_{(j,m)}) = 0.$$

Similarly,

$$\left| \sum_{i\in B} \mu_i \right| \left(\bigcap_{m\in B} \bigcup_{j\in A} D'^c_{(j,m)} \right) = 0.$$

Since we have that

$$(\cup_{j\in A}\cap_{m\in B} D_{(j,m)})\cap(\cup_{m\in B}\cap_{j\in A} D'_{(j,m)})=\varnothing,$$

because $D_{(j,m)}\cap D'_{(j,m)}=\varnothing$ for all $j\in A$ and $m\in B$, we conclude that $\sum_{i\in A}\mu_i\sim\sum_{m\in B}\mu_m$. By these arguments, it is not hard to see that we actually have $\sum_{i\in A}\alpha_i\mu_i\sim\sum_{m\in B}\beta_m\mu_m$ for all scalars α_i,β_m with $i\in A, m\in B$, since $|\alpha\mu|=|\alpha||\mu|$ for every scalar α and every Radon measure μ on Ω. Thus, the condition (1) of Definition 10.5 is satisfied.

Suppose now that $\{\mu_n\}_n$ is a sequence in $M(\Omega)$, and $\nu\in M(\Omega)$ such that $\mu_n\sim\nu$ for all n. Then, for each $n\in\mathbb{N}$, there exist measurable subsets D_n and D'_n such that $D_n\cap D'_n=\varnothing$ and $|\mu_n|(D_n^c)=|\nu|(D_n'^c)=0$. Hence, for every $m\in\mathbb{N}$, we have $|\nu|(\cup_{n=1}^m D_n'^c)=0$. This implies that

$$|\nu|(\cup_{n=1}^\infty D_n'^c)=\lim_{m\to\infty}|\nu|(\cup_{n=1}^m D_n'^c)=0.$$

Moreover, for each $n\in\mathbb{N}$, we have $|\mu_n|(\cap_{m=1}^\infty D_m^c)=0$. Since $D_n\cap D'_n=\varnothing$ for all n, we get

$$\left(\bigcup_{m=1}^{\infty} D_m\right)\cap(\cap_{n=1}^\infty D'_n)=\varnothing.$$

Now, if $\mu_n\to\mu$ as $n\to\infty$ for some $\mu\in M(\Omega)$, then

$$\begin{aligned}|\mu|(\cap_{n=1}^\infty D_m^c)&\le|\mu-\mu_n|(\cap_{m=1}^\infty D_m^c)+|\mu_n|(\cap_{m=1}^\infty D_m^c)\\&=|\mu-\mu_n|(\cap_{m=1}^\infty D_m^c)\to 0,\end{aligned}$$

as $n\to\infty$. Hence,

$$|\mu|(\cap_{n=1}^\infty D_m^c)=0.$$

Thus, $\mu\sim\nu$, which proves the condition (2) in Definition 10.5. □

We recall that a subset S of a Banach space Y is called *spaceable* in Y if $S\cup\{0\}$ contains a closed infinite-dimensional subspace of Y. In this section, a subset B of a vector space is called a pseudocone if for each scalar c, $cB\subseteq B$. For a Borel measurable subset E and some $\mu\in M(\Omega)$, we let μ_E be the measure given by $\mu_E(B):=\mu(B\cap E)$ for every Borel subset B of Ω. If K is a pseudocone in $M(\Omega)$, we denote

$$\widetilde{K}:=\{\mu_E\mid\mu\in K,\ E\text{ is Borel}\}.$$

Since for every scalar $\lambda\in\mathbb{C}$, we have $(\lambda\mu)_E=\lambda\mu_E$, it follows that $\widetilde{K}$ is a pseudocone since K is a pseudocone. Moreover, $K\subseteq\widetilde{K}$.

Next we recall the following theorem.

Theorem 10.2 ([4, Theorem 4.2]) *Let $(\mathcal{E}, \|\cdot\|)$ be a Banach space, $\sim$ be a relation on $\mathcal{E}$ with property (D), and K be a nonempty subset of $\mathcal{E}$. Assume that:*

(i) there is a constant $k > 0$ such that $\|x + y\| \geq k \|x\|$ for all $x, y \in \mathcal{E}$ with $x \sim y$;
(ii) K is a pseudocone;
(iii) if $x, y \in \mathcal{E}$ such that $x + y \in K$ and $x \sim y$ then $x, y \in K$;
(iv) there is an infinite sequence $\{x_n\}_{n=1}^{\infty} \subseteq \mathcal{E} \setminus K$ such that for each distinct $m, n \in \mathbb{N}$, $x_m \sim x_n$.

Then, $\mathcal{E} \setminus K$ is spaceable in $\mathcal{E}$.

Then we give the following proposition.

Proposition 10.7 *Let K be a pseudocone in $M(\Omega)$. If there exists a sequence of mutually disjoint Borel subsets $\{E_n\}_{n\in\mathbb{N}}$ of Ω such that for all n we have*

$$\{\mu_{E_n} \mid \mu \in \widetilde{K}\} \neq \{\mu_{E_n} \mid \mu \in M(\Omega)\},$$

then $M(\Omega) \setminus \tilde{K}$ (and consequently $M(\Omega) \setminus K$) is spaceable in $M(\Omega)$.

Proof We show that the relation $\sim$ satisfies the conditions of Theorem 10.2. Let $\mu, \nu \in M(\Omega)$, and suppose that $\mu \sim \nu$. Then, there exist Borel subsets B_1, B_2 of Ω such that $B_1 \cap B_2 = \varnothing$ and $|\mu|(B_1^c) = |\nu|(B_2^c) = 0$. Let $\{E_n\}_{n=1}^{N}$ be a partition of Ω. Then, $\{E_n \cap B_1\}_{n=1}^{N} \bigcup \{B_1^c\}$ is also a partition of Ω. Moreover, since $B_1 \cap B_2 = \varnothing$, we have $B_1 \subseteq B_2^c$, hence

$$\begin{aligned}\sum_{n=1}^{N} |(\mu+\nu)(E_n \cap B_1)| + |(\mu+\nu)(B_1^c)| &= \sum_{n=1}^{N} |\mu(E_n \cap B_1)| + |\nu(B_1^c)| \\ &\geq \sum_{n=1}^{N} |\mu(E_n \cap B_1)| = \sum_{n=1}^{N} |\mu(E_n)|.\end{aligned}$$

In particular, $|\mu+\nu|(\Omega) \geq \sum_{n=1}^{N} |\mu(E_n)|$. Since $\{E_n\}_{n=1}^{N}$ was an arbitrary partition of Ω, we deduce that $|\mu + \nu|(\Omega) \geq |\mu|(\Omega)$. So, the condition (i). in Theorem 10.2 holds.

Next, let $\mu, v \in M(\Omega)$ such that $\mu + v \in \tilde{K}$ and $\mu \sim v$. Let B_1, B_2 be Borel subsets of Ω such that $B_1 \cap B_2 = \varnothing$ and $|\mu|(B_1^c) = |\nu|(B_2^c) = 0$. Then, again since $B_1 \subseteq B_2^c$ and $B_2 \subseteq B_1^c$, we get $(\mu + v)_{B_1} = \mu_{B_1} + v_{B_1} = \mu_{B_1} = \mu$ and, similarly, $(\mu + v)_{B_2} = v_{B_2} = v$. However, since $\mu + v \in \tilde{K}$, by the definition of $\tilde{K}$, there exists some measure $\eta \in K$ and some Borel subset E of Ω such that $\eta_E = \mu + v$. Hence, $\mu = (\mu + v)_{B_1} = \eta_{E\cap B_1} \in \tilde{K}$ and $v = (\mu + v)_{B_2} = \eta_{E\cap B_2} \in \tilde{K}$. Thus, $\mu, v \in \tilde{K}$, so the condition (iii) of Theorem 10.2 is satisfied.

By the assumptions in the proposition, the conditions (ii) and (iv) of Theorem 10.2 are also satisfied, so the proposition follows. □

Corollary 10.4 *Let K be the pseudocone of all scalar multiples of positive Radon measures on a locally compact Hausdorff space Ω with infinite cardinality. Then $M(\Omega) \setminus K$ is spaceable in $M(\Omega)$. Consequently, the complement of the cone of all positive Radon measures on Ω is spaceable in $M(\Omega)$.*

Proof Notice first that $K = \tilde{K}$ since $(\lambda\mu)_E = \lambda\mu_E$ for all $\lambda \in \mathbb{C}$, $\mu \in M(\Omega)$ and any Borel subset E. Since Ω consists of infinitely many distinct elements by assumption, we can find two mutually disjoint sequences $(x_n)_n$ and $(y_n)_n$ in Ω where $x_n \neq x_m$ and $y_n \neq y_m$ whenever $n \neq m$.

For each $n \in \mathbb{N}$, set $\mu_n = \delta_{x_n} + i\delta_{y_n}$ where δ denotes the Dirac measure. It is not hard to see that $\mu_n \notin M(\Omega) \setminus K$ for all $n \in \mathbb{N}$, so the sequence $(\mu_n)_n$ satisfies the condition (iv). of Theorem 10.2. Since $K = \tilde{K}$, setting $E_n = \{x_n\} \cup \{y_n\}$ for each $n \in \mathbb{N}$ we can apply Proposition 10.7 to deduce that $M(\Omega) \setminus K$ is spaceable in $M(\Omega)$. □

Acknowledgments The author is very grateful to the reviewers for the careful reading of this chapter and for the useful comments that led to the improved version of the chapter.

The author is supported by the Ministry of Science, Technological Development, and Innovations, Republic of Serbia, grant no. 451-03-66/2024-03/200029.

References

1. Aron, R.M., Gurariy, V.I., Seoane-Sepúlveda, J.B.: Lineability and spaceability of sets of functions on $\mathbb{R}$. Proc. Am. Math. Soc. **133**(3), 795–803 (2005)
2. Ansari, S.I.: Existence of hypercyclic operators on topological vector spaces. J. Funct. Anal. **148**, 384–390 (1997)
3. Bayart, F., Matheron, É.: Dynamics of Linear Operators. Cambridge Tracts in Mathematics, vol. 179. Cambridge University, Cambridge (2009)
4. Bagheri Salec, A., Ivković, S., Tabatabaie, S.M.: Spaceability on some classes of Banach spaces. Math. Inequal. Appl. (2021). https://doi.org/10.7153/mia-2022-25-41
5. Bernal-González, L.: On hypercyclic operators on Banach spaces. Proc. Am. Math. Soc. **127**, 1003–1010 (1999)
6. Bonilla, A., Miana, P.: Hypercyclic and topologically mixing cosine functions on Banach spaces. Proc. Am. Math. Soc. **136**, 519–528 (2008)
7. Butzer, P., Koliha, J.: The a-Drazin inverse and ergodic behaviour of semigroups and cosine operator functions. J. Oper. Theory **62**, 297–326 (2009)
8. Chen, C.-C.: Chaotic weighted translations on groups. Arch. Math. **97**, 61–68 (2011)
9. Chen, C.-C.: Chaos for cosine operator functions generated by shifts. Int. J. Bifurcat. Chaos **24**, 7 pages (2014) (article ID 1450108)
10. Chen, K-Y.: On aperiodicity and hypercyclic weighted translation operators. J. Math. Anal. Appl. **462**, 1669–1678 (2018)
11. Chen, C.-C., Chu, C.-H.: Hypercyclic weighted translations on groups. Proc. Am. Math. Soc. **139**, 2839–2846 (2011)
12. Chen, J.-C., Shaw, S.-Y.: Topological mixing and hypercyclicity criterion for sequences of operators. Proc. Am. Math. Soc. **134**(11), 3171–3179 (2006)

13. Fonf, V.P., Gurariy, V.I., Kadets, M.I.: An infinite dimensional subspace of $C[0, 1]$ consisting of nowhere differentiable functions. C. R. Acad. Bulgare Sci. **52**, 13–16 (1999)
14. Grosse-Erdmann, K.-G.: Hypercyclic and chaotic weighted shifts. Studia Math. **139**, 47–68 (2000)
15. Grosse-Erdmann, K.-G., Peris, A.: Linear Chaos. Universitext, Springer, Berlin (2011)
16. Gaubert, S., Qu, Z.: Dobrushin's ergodicity coefficient for Markov operators on cones. Integr. Equ. Oper. Theory **81**(1), 127–150 (2015). https://doi.org/10.1007/s00020-014-2193-2
17. Kalmes, T.: Hypercyclicity and mixing for cosine operator functions generated by second order partial differential operators. J. Math. Anal. Appl. **365**, 363–375 (2010)
18. Kalmes, T.: Hypercyclic, mixing, and chaotic C0-semigroups induced by semiflows. Ergod. Theory Dyn. Syst. **27**, 1599–1631 (2007)
19. Ivković, S.: On Markov operators and cones. Master thesis, Department of Mathematics, University of Oslo, Oslo (2015). https://www.duo.uio.no/bitstream/handle/10852/49401/Stefan-Ivkovic---Master-thesis.pdf
20. Ivković, S., Tabatabaie, S.M.: Linear dynamics of discrete cosine functions on solid Banach function spaces.. Positivity **25**(4), 1437–1448 (2021). https://doi.org/10.1007/s11117-021-00823-8
21. Ivković, S., Tabatabaie, S.M.: Hypercyclic generalized shift operators. Complex Anal. Oper. Theory **17**, 60 (2023). https://doi.org/10.1007/s11785-023-01376-2
22. Ivković, S., Tabatabaie, S.M.: Hypercyclic translation operators on the algebra of compact operators. Iran. J. Sci. Technol. Trans. A Sci. **45**, 1765–1775 (2021). https://doi.org/10.1007/s40995-021-01186-1
23. Ivković, S., Tabatabaie, S.M.: Disjoint linear dynamical properties of elementary operators. Bull. Iran. Math. Soc. **49**, 63 (2023). https://doi.org/10.1007/s41980-023-00808-1
24. Ivković, S.: Hypercyclic operators on Hilbert C*-modules. Filomat **38**, 1901–1913 (2024)
25. Kostić, M.: Hypercyclic and chaotic integrated C-cosine functions. Filomat **26**, 1–44 (2012)
26. Salas, H.: Hypercyclic weighted shifts. Trans. Am. Math. Soc. **347**, 993–1004 (1995)
27. Sawano, Y., Tabatabaie, S.M., Shahhoseini, F.: Disjoint dynamics of weighted translations on solid spaces. Topology Appl. **298**, 107709 (2021). https://doi.org/10.1016/J.TOPOL.2021.107709
28. Shaw, S.-Y.: Growth order and stability of semigroups and cosine operator functions. J. Math. Anal. Appl. **357**, 340–348 (2009)

Chapter 11
Error Estimates of Gaussian-type Quadratures—A Survey

Davorka R. Jandrlić, Aleksandar V. Pejčev, and Miodrag M. Spalević

Dedicated to Academician Gradimir V. Milovanović on the occasion of his 75th birthday

11.1 Introduction

Numerical integration refers to the procedure of approximating a definite integral of a given function. The n-point quadrature formula (q.f.) is utilized in this context, employing a positive weight function ω on a finite interval that we normalize to be $[-1, 1]$. This formula can be expressed in the following format:

$$\int_{-1}^{1} \omega(t) f(t)\, \mathrm{d}t = \sum_{i=1}^{n} \omega_i f(\xi_i) + R_n(f), \tag{11.1}$$

for some set of nodes ξ_i and weights ω_i. Various options exist for selecting the nodes and weights in numerical integration. The quadrature formula (11.1) is considered to possess an (algebraic) degree of precision d if $R_n(f) = 0$ holds for all $f \in \mathcal{P}_d$, where $\mathcal{P}_d$ represents the set of all (algebraic) polynomials with a degree at most d. The Gauss quadrature formula stands as the unique optimal interpolatory quadrature formula with n nodes, achieving the maximum degree of precision of $2n - 1$. It is named after Gauss, who discovered it in [3].

D. R. Jandrlić (✉) · A. V. Pejčev · M. M. Spalević
University of Belgrade, Faculty of Mechanical Engineering, Department of Mathematics, Belgrade, Serbia
e-mail: djandrlic@mas.bg.ac.rs; apejcev@mas.bg.ac.rs; mspalevic@mas.bg.ac.rs

M. P. Stanić et al. (eds.), *Analysis, Approximation, Optimization: Computation and Applications*, Springer Optimization and Its Applications 224,
https://doi.org/10.1007/978-3-031-85743-0_11

11.2 Progression Through Time: Tracing the Research Path

Let us begin with Jacobi polynomials $P_n^{(\alpha,\beta)}$, which are orthogonal over the interval $[-1, 1]$ with respect to the Jacobi weight function:

$$\omega(t) = (1-t)^\alpha (1+t)^\beta, \quad \alpha, \beta > -1.$$

Initially, we considered the particular case where $\alpha = \beta = 0$ [6], which leads to the Gauss–Legendre quadrature (11.1) with the weight function $\omega(t) = 1$. Subsequently, we proceeded to estimate the remainder term R_n associated with this quadrature formula.

Later, our research expanded to encompass scenarios where α equals β and can represent any natural number. This broader exploration allowed us to delve into a wider range of possibilities within the context of Jacobi polynomials and their associated weight functions. By extending our investigation beyond the confines of specific values to include any natural number for both α and β, we aimed to elucidate the implications and variations in Gaussian-type quadrature formulas associated with the Gegenbauer weight function, with the weight function $\omega(t) = (1-t^2)^\alpha$. This expansion facilitated a more comprehensive understanding of the error estimates for analytic functions on confocal ellipses, enriching our exploration beyond individual cases to generalize our findings for a broader spectrum of polynomial functions.

When f is an analytic function, the remainder term can be expressed as a contour integral with a complex kernel. In our studies, we have specifically investigated kernel on elliptic contours with foci located at the points ∓ 1 and a sum of semi-axes $\rho > 1$, corresponding to the given quadrature formula. We have obtained an explicit expression for the kernel and determined the positions on the ellipses where the maximum modulus of the kernel is attained. Initially, our determination regarding this position was based on an assumption substantiated by numerous numerical examples. This positioning, presented in the form of a conjecture without a formal theoretical proof, was detailed in the paper [6], specifically on pages 429 and 430. This conjecture outlined the specific locations on the ellipses where the maximum modulus of the kernel is anticipated to be achieved for sufficiently large ρ. Building upon this conjecture, we proceeded to derive effective error bounds for this quadrature formula, utilizing the insights gained from our numerical analyses and assumptions regarding the maximal modulus positions on the ellipses.

Later, our efforts culminated in a significant achievement as we successfully provided a theoretical proof confirming our earlier conjecture. This crucial validation of our conjecture was documented in the paper [2]. Through rigorous theoretical analysis, we were able to substantiate and affirm the positions on the ellipses where the maximal modulus of the kernel is attained, thereby solidifying the groundwork laid by our initial conjecture. This theoretical validation added substantial weight to our earlier findings and significantly strengthened the underpinning principles of our research in this domain. The established theorem proved to be applicable not only to our initial research, as outlined in [2], but also to the expanded scope of our investigations concerning Gauss–Gegenbauer quadrature formulas [7].

11.3 Error Bounds of q.f. for Analytic Functions

Let Γ be a simple closed curve in the complex plane surrounding the interval $[-1, 1]$, and let $\mathcal{D}$ be its interior. Suppose f is a function that is analytic in $\mathcal{D}$ and continuous on $\overline{\mathcal{D}}$. It follows from the Lagrange interpolation formula (see, e. g., [16, Chapter 2])

$$r_n(f;t) = f(t) - \sum_{i=1}^{n} l_i(t) f(\xi_i) = \frac{1}{2\pi \mathrm{i}} \oint_\Gamma \frac{f(z)\Omega_n(t)}{(z-t)\Omega_n(z)}\, \mathrm{d}z, \tag{11.2}$$

where

$$\Omega_n(z) = c\prod_{i=1}^{n}(z-\xi_i), \quad c \neq 0,$$

where ξ_i are the zeros of of the corresponding orthogonal polynomial $\Omega_n(t)$ (which is monic when $c = 1$) and l_i are the so-called fundamental functions of Lagrange interpolation. Multiplying (11.2) with the weight function $\omega(t)$ and integrating in t over $(-1, 1)$, we get a contour integral representation for the remainder term $R_n(f)$ in the Gauss q.f. (11.1),

$$R_n(f) = I(f;\omega) - \sum_{i=1}^{n} \omega_i f(\xi_i) = \frac{1}{2\pi \mathrm{i}} \oint_\Gamma K_n(z;\omega) f(z)\, \mathrm{d}z, \tag{11.3}$$

where

$$I(f;\omega) = \int_{-1}^{1} \omega(t)\, f(t)\, \mathrm{d}t, \quad \omega_i = \int_{-1}^{1} \omega(t)\, l_i(t)\, \mathrm{d}t,$$

and the *kernel* $K_n(z) = K_n(z;\omega)$ can be expressed in the form

$$K_n(z;\omega) = \frac{\varrho_n(z;\omega)}{\Omega_n(z)}, \quad \varrho_n(z;\omega) = \int_{-1}^{1} \omega(t)\, \frac{\Omega_n(t)}{z-t}\, \mathrm{d}t, \quad z \in \mathbb{C} \setminus [-1, 1]. \tag{11.4}$$

The integral representation (11.3) leads directly to the error bound

$$|R_n(f)| \leq \frac{l(\Gamma)}{2\pi} \left(\max_{z\in\Gamma} |K_n(z)| \right) \left(\max_{z\in\Gamma} |f(z)| \right), \tag{11.5}$$

where $l(\Gamma)$ is the length of the contour Γ (see [5, 22]).

A common choice for the contour Γ is one of the confocal ellipses, also known as the Bernstein ellipses, with foci at the points ∓ 1, and the sum of semi-axes $\rho > 1$,

$$\mathcal{E}_\rho = \left\{ z \in \mathbb{C} : z = \frac{1}{2}\left(u + u^{-1}\right), u = \rho e^{i\theta},\ 0 \le \theta < 2\pi \right\}. \tag{11.6}$$

The Bernstein ellipse has the major and minor semi-axes given by $\frac{1}{2}(\rho + \rho^{-1})$ and $\frac{1}{2}(\rho - \rho^{-1})$, respectively.

A recent survey on error bounds (11.5) and some others, when $\Gamma = \mathcal{E}_\rho$, for Gaussian type q.f. can be found in [5].

For such Γ, we studied the estimate (11.5) (see [2, 6, 7]) for the Gauss–Legendre and Gauss–Gegenbauer quadrature formulas. Gautschi and Varga [22] (also referenced in [23]) introduced the original method for establishing the error bound of type (11.5). They endeavored to pinpoint the precise location on the confocal ellipses $\mathcal{E}_\rho$ where the kernel of the error functional reaches its maximum modulus. They successfully resolved this inquiry concerning the Chebyshev weight functions and offered empirical findings regarding more extensive Jacobi weight functions.

Rigorous error estimates in the Gaussian quadrature formulas with multiple nodes for functions analytic on confocal ellipses and circles have been developing by Milovanović and his collaborators; see, e.g. [8–14].

11.4 The Modulus of the Kernel

The *kernel* function denoted as $K_n(z)$ is defined as follows: $K_n(z) = \frac{\varrho_n(z)}{\pi_n(z)}$, $z \notin [-1, 1]$ (see (11.4)). In this context, $\pi_n(z) = P_n^{(\alpha,\alpha)}(z)$ denotes the Gegenbauer polynomial of degree n, or in the scenario where α equals 0, it represents the Legendre polynomial. The function $\varrho_n(z)$ is given by the integral

$$\varrho_n(z) = \int_{-1}^{1} \frac{P_n^{(\alpha,\alpha)}(t)}{z - t}(1 - t^2)^\alpha \, dt. \tag{11.7}$$

We will first discuss in some detail the case where α equals 0 and $\omega(t) = 1$. Here one finds [6, Section 2.1], for $z \in \mathcal{E}_\rho$ and $z = \frac{1}{2}(u + u^{-1})$, the explicit formula:

$$K_{2m}(u) = \frac{\dfrac{c_m}{u^{2m+1}} + \dfrac{c_{m+1}}{u^{2m+3}} + \cdots}{d_{2m,2m}u^{2m} + d_{2m-2,2m}u^{2m-2} + \cdots}, \quad |u| \to \infty, \tag{11.8}$$

for even values of n, and similarly for odd values of n

$$K_{2m+1}(u) = \frac{\frac{c_{m+1}}{u^{2m+2}} + \frac{c_{m+2}}{u^{2m+4}} + \cdots}{d_{2m+1,2m+1}u^{2m+1} + d_{2m+1,2m+1}u^{2m-1} + \cdots}, \quad |u| \to \infty. \tag{11.9}$$

The coefficients $d_{k,n}$ are derived from the expression found in [6, Equation 1.8]. The numerator of the kernel follows a specific form for odd values of n:

$$\varrho_n(z) = \sum_{j=[\frac{n+1}{2}]}^{\infty} \frac{c_j}{u^{2j}},$$

with

$$c_j = -16j \sum_{k=1,\ k \text{ odd}}^{n}{}'' \frac{d_{k,n}}{k^2 - (2j)^2},$$

and analogous for n even

$$\varrho_n(z) = \sum_{j=[\frac{n+1}{2}]}^{\infty} \frac{c_j}{u^{2j+1}} = \frac{1}{u}\sum_{j=1}^{\infty} \frac{c_j}{u^{2j}},$$

with

$$c_j = -8(2j+1) \sum_{k=0,\ k \text{ even}}^{n}{}'' \frac{d_{k,n}}{k^2 - (2j+1)^2}.$$

Note In Equations [6, 8. and 9.], an error has been identified. Specifically, these formulas would be accurate if the coefficients in the numerator of the kernel are nonzero. However, it is known that for odd values of n, these coefficients are equal to zero up to the index $[\frac{n}{2}]$, and analogously up to $[\frac{n}{2}] - 1$ for even values of n, as elucidated by Gautschi and Varga [22], and kernel is of the form $O(u^{-2n-1})$. According to these valid expressions for kernels are:

$$K_{2m}(u) = \frac{c_m}{d_{n,n}} \cdot \frac{1}{u^{2n+1}} \cdot \frac{u^2 + \frac{c_{m+1}}{c_m} + o(1) \quad (|u| \to \infty)}{u^2 + \frac{d_{n,n-2}}{d_{n,n}} + o(1) \quad (|u| \to \infty)}. \tag{11.10}$$

Analogous is for odd values of n:

$$K_{2m+1}(u) = \frac{c_{m+1}}{d_{n,n}} \cdot \frac{1}{u^{2n+1}} \cdot \frac{u^2 + \dfrac{c_{m+2}}{c_{m+1}} + o(1) \quad (|u| \to \infty)}{u^2 + \dfrac{d_{n,n-2}}{d_{n,n}} + o(1) \quad (|u| \to \infty)}. \tag{11.11}$$

The remainder of our assertions remains unchanged and does not impact subsequent calculations or outcomes. The result of the given analysis indicates that the value of θ, at which the modulus of the kernel reaches its maximum, is dependent solely on the sign of the difference:

$$D = \frac{c_{m+1}}{c_m} - \frac{d_{2m-2,2m}}{d_{2m,2m}}, \tag{11.12}$$

if n is even, and

$$D = \frac{c_{m+2}}{c_{m+1}} - \frac{d_{2m-1,2m+1}}{d_{2m+1,2m+1}}, \tag{11.13}$$

if n is odd (if $D \neq 0$), i.e.,

(1) if $D < 0$, then the modulus of the kernel should attain its maximum at $\theta = \pi/2$ (and $\theta = 3\pi/2$);
(2) if $D > 0$, then the modulus of the kernel should attain its maximum at $\theta = 0$ (and $\theta = \pi$).

For a comprehensive understanding, please refer to the details provided in [2]. In essence, the numerator is an infinite series represented by u^2 (or, to be more precise, in terms of $\frac{1}{u^2}$), $\sum_{i=1}^{+\infty} \frac{A_i}{u^{2i}}$ ($A_1 = 1$), while the denominator is a polynomial expression in $\frac{1}{u^2}$, $\sum_{i=1}^{n+1} \frac{B_i}{u^{2i}}$ ($B_1 = 1$). In the paper cited as [5], the authors established the corresponding statement (Lemma 4.1). This lemma serves as a foundational tool in establishing the asymptotic behavior of the complex kernel's modulus within the remainder-term of all quadrature formulas outlined in recent papers directly addressing error estimates of Gaussian-type quadrature formulas for analytic functions on ellipses. In this context, the utilization of the corresponding lemma becomes challenging due to the presence of an infinite series within the kernel, necessitating the inclusion of supplementary conditions for its proper application.

In our publication [6], we introduced a conjecture based on multiple numerical examples. This conjecture aimed to elucidate the asymptotic behavior of the modulus of the kernel. However, in the subsequent work referenced as [2], we provided a theoretical proof validating this behavior. Our conjecture pertained to expressions formulated as follows:

$$I(\rho,\theta)=\frac{u^2+A+o\left(\frac{1}{\rho}\right)}{u^2+B+o\left(\frac{1}{\rho}\right)}\quad(\rho\to\infty)=\frac{\frac{1}{u^2}+\frac{A}{u^4}+o\left(\frac{1}{\rho^5}\right)}{\frac{1}{u^2}+\frac{B}{u^4}+o\left(\frac{1}{\rho^5}\right)}\quad(\rho\to\infty),$$

where A and B are distinct real numbers, $u=\rho e^{i\theta}$, and $I(\rho,\theta)\equiv I(\rho,\theta+\pi)$. The claim suggests that, possibly under certain corresponding additional conditions, if $A>B$, then there exists a value $\rho^*>1$ such that $max_{\theta\in[0,\pi)}|I(\rho,\theta)|=|I(\rho,0)|$ for each $\rho>\rho^*$. And if $A<B$, then similarly, there exists a value $\rho^*>1$ such that $max_{\theta\in[0,\pi)}|I(\rho,\theta)|=|I(\rho,\pi/2)|$ for each $\rho>\rho^*$. It appears that "additional conditions" implies a requirement for the coefficients within the series of the numerator to be uniformly bounded.

Since

$$\left|\sum_{j=1}^{\infty}\frac{A_j}{u^{2j}}\right|=\sqrt{\sum_{j=1}^{\infty}A_j^2\rho^{-4j}+2\sum_{i=2}^{\infty}\sum_{j=1}^{i-1}A_iA_j\rho^{-2i-2j}\cos 2(i-j)\theta},$$

after (for the sake of compactness of the record) denoting

$$\pi_n(z)=\sum_{k=0}^{n}d_{|n-2k|,n}u^{n-2k}=u^{n+2}\sum_{j=1}^{n+1}\frac{B_j}{u^{2j}},$$

and further

$$\sum_{j=1}^{\infty}\frac{A_j}{u^{2j}}=A_1\sum_{j=1}^{\infty}\frac{\overline{A}_j}{u^{2j}},\quad \overline{A}_j=\frac{A_j}{A_1},\quad \overline{A}_1=1,$$

$$\sum_{j=1}^{n+1}\frac{B_j}{u^{2j}}=B_1\sum_{j=1}^{n+1}\frac{\overline{B}_j}{u^{2j}},\quad \overline{B}_j=\frac{B_j}{B_1},\quad \overline{B}_1=1,$$

it seems necessary to establish the following:

Theorem 11.1 *For the fixed natural number n and the coefficients* $\overline{A}_1, \overline{A}_2, \ldots$ *and* $\overline{B}_1, \overline{B}_2, \ldots, \overline{B}_{n+1}$ *defined depending the parity of n in the atached manner, if* $\overline{A}_2 > \overline{B}_2$, *then there exists* ρ^* *such that for each* $\rho > \rho*$ *and each* $\theta \in [0, \pi)$ *holds*

$$\frac{\sum_{j_1=1}^{\infty} \overline{A}_{j_1}^2 \rho^{-4j_1} + 2\sum_{i_1=2}^{\infty}\sum_{j_1=1}^{i_1-1} \overline{A}_{i_1}\overline{A}_{j_1}\rho^{-2i_1-2j_1}\cos 2(i_1-j_1)\theta}{\sum_{j_2=1}^{n+1} \overline{B}_{j_2}^2 \rho^{-4j_2} + 2\sum_{i_2=2}^{n+1}\sum_{j_2=1}^{i_2-1} \overline{B}_{i_2}\overline{B}_{j_2}\rho^{-2i_2-2j_2}\cos 2(i_2-j_2)\theta}$$

$$\leq \frac{\sum_{j_1=1}^{\infty} \overline{A}_{j_1}^2 \rho^{-4j_1} + 2\sum_{i_1=2}^{\infty}\sum_{j_1=1}^{i_1-1} \overline{A}_{i_1}\overline{A}_{j_1}\rho^{-2i_1-2j_1}}{\sum_{j_2=1}^{n+1} \overline{B}_{j_2}^2 \rho^{-4j_2} + 2\sum_{i_2=2}^{n+1}\sum_{j_2=1}^{i_2-1} \overline{B}_{i_2}\overline{B}_{j_2}\rho^{-2i_2-2j_2}},$$

and if $\overline{A}_2 < \overline{B}_2$, *then there exists* ρ^* *such that for each* $\rho > \rho*$ *and each* $\theta \in [0, \pi)$ *holds*

$$\frac{\sum_{j_1=1}^{\infty} \overline{A}_{j_1}^2 \rho^{-4j_1} + 2\sum_{i_1=2}^{\infty}\sum_{j_1=1}^{i_1-1} \overline{A}_{i_1}\overline{A}_{j_1}\rho^{-2i_1-2j_1}\cos 2(i_1-j_1)\theta}{\sum_{j_2=1}^{n+1} \overline{B}_{j_2}^2 \rho^{-4j_2} + 2\sum_{i_2=2}^{n+1}\sum_{j_2=1}^{i_2-1} \overline{B}_{i_2}\overline{B}_{j_2}\rho^{-2i_2-2j_2}\cos 2(i_2-j_2)\theta}$$

$$\leq \frac{\sum_{j_1=1}^{\infty} \overline{A}_{j_1}^2 \rho^{-4j_1} + 2\sum_{i_1=2}^{\infty}\sum_{j_1=1}^{i_1-1} \overline{A}_{i_1}\overline{A}_{j_1}(-1)^{i_1-j_1}\rho^{-2i_1-2j_1}}{\sum_{j_2=1}^{n+1} \overline{B}_{j_2}^2 \rho^{-4j_2} + 2\sum_{i_2=2}^{n+1}\sum_{j_2=1}^{i_2-1} \overline{B}_{i_2}\overline{B}_{j_2}(-1)^{i_2-j_2}\rho^{-2i_2-2j_2}}.$$

The proof, while omitted here for brevity, can be found in [2, Theorem 3.1]. Figures 11.1 and 11.2 depict graphs illustrating the modulus of the kernels for $n = 3$ and $n = 4$, respectively, showcasing various values of ρ pertinent to the analyzed scenario ($\alpha = \beta = 0$). To validate the proposed error bounds, we conducted multiple tests, comparing them against the exact errors. These tests involved various special functions found in the literature [6, 7]. The term "Error" denotes the precise (actual) error bound corresponding to each formula. All our experimental results were computed using high-precision arithmetic. Our MATLAB-based program utilizes the codes *sgauss.m* and *sr_jacobi.m* for symbolic calculations (refer to [24] and [25]). The tables labeled as Tables 11.1, 11.2, and 11.3 presented further provide

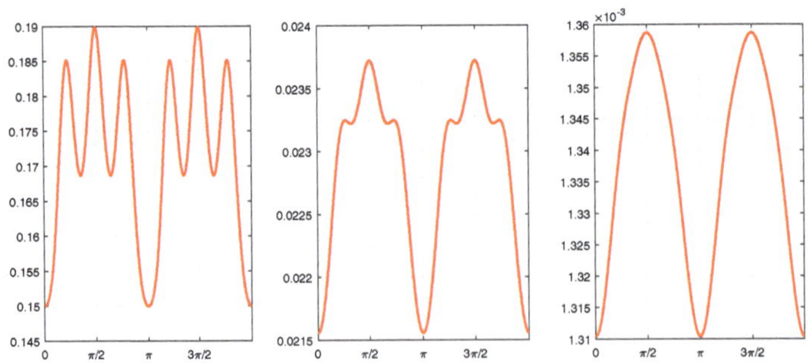

Fig. 11.1 $|K_3(z)|$ in the cases $\rho = 1.5;\ 2;\ 3$, respectively

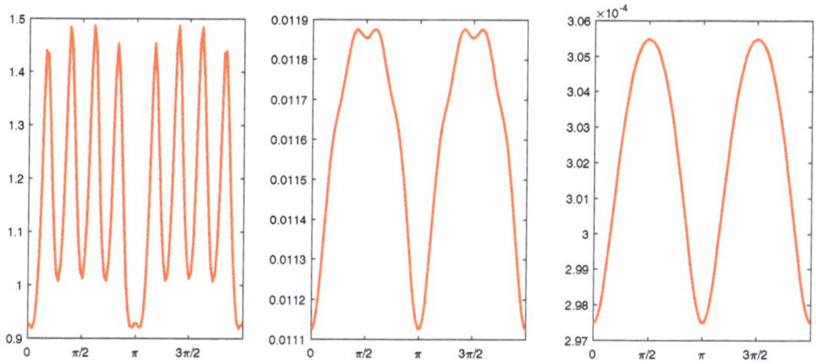

Fig. 11.2 $|K_4(z)|$ in the cases $\rho = 1.5;\ 2;\ 3$, respectively

Table 11.1 Error bounds $r_n(f_1)$, actual errors $(Error)$ and exact value of integral (I_ω)

(n,ω)	$r_n(f_1)$	$Error$	ρ_{opt}	I_ω
(2,1)	7.328 (00)	1.592 (00)	2.071	21.7719...
(4,1)	2.839 (00)	4.175 (-02)	2.117	21.7719...
(5,1)	2.289 (-02)	5.849 (-03)	2.560	21.7719...
(6,1)	6.700 (-03)	7.669(-04)	2.661	21.7719...
(7,1)	4.622 (-04)	9.527(-05)	2.747	21.7719...
(8,1)	1.195(-04)	1.131(-05)	2.821	21.7719...
(9,1)	7.364 (-06)	1.291(-06)	2.884	21.7719...
(10,1)	1.738(-06)	1.423(-07)	2.941	21.7719...
(11,1)	9.894 (-08)	1.523(-08)	2.992	21.7719...
(13,1)	1.166 (-09)	1.611(-10)	3.078	21.7719...
(15,1)	1.237(-11)	1.558(-12)	3.150	21.7719...
(25,1)	5.573(-22)	5.046(-23)	3.410	21.7719...

a condensed overview of the outcomes for specific examples, which constitute a portion of the comprehensive results documented in [6]. The values in parentheses signify decimal exponents. Additionally, the column titled ρ_{opt} presents the value of ρ_{opt} where the infimum of the bound was achieved.

The actual error represents the absolute difference between the integral values:

$$I(f_i) = \int_{-1}^{1} f_i(t)\,\mathrm{d}t, \quad i = 1, 2, 3,$$

and the values computed using the corresponding Gauss quadrature with n nodes, denoted as $I_n(f_i)$. This difference is expressed as $|I(f_i) - I_n(f_i)|$, which can be approximated by

$$Error = Error(n) = |I_{150}(f_i) - I_n(f_i)|.$$

Example 11.1 Let $f_1(z) = \mathrm{e}^{\mathrm{e}^{\cos(\omega z)}}$, $\omega > 0$.

Table 11.2 Error bounds $r_n(f_2)$, actual errors $(Error)$ and exact value of integral (I_ω)

(n, ω)	$r_n(f_2)$	$Error$	ρ_{opt}	I_ω
(2,0.5)	1.068(-01)	2.719(-02)	4.015	2.3899...
(4,0.5)	2.060(-04)	3.785(-05)	6.176	2.3899...
(7,0.5)	1.243(-09)	3.637(-10)	7.481	2.3899...
(9, 0.5)	3.069(-13)	7.974(-14)	8.490	2.3899...
(10, 0.5)	8.098(-15)	1.000(-15)	8.719	2.3899...
(11, 0.5)	4.830(-17)	6.938(-18)	9.382	2.3899...
(15, 0.5)	4.261(-25)	8.566(-26)	10.930	2.3899...
(16, 0.5)	8.438(-27)	6.702(-28)	10.880	2.3899...
(7,1)	2.059(-07)	5.911(-08)	5.291	2.9253...
(9, 1)	2.072(-10)	5.196(-11)	5.961	2.9253...
(11, 1)	1.276(-13)	2.977(-14)	6.640	2.9253...
(13, 1)	1.112(-16)	2.081(-17)	7.211	2.9253...
(15, 1)	1.780(-20)	3.585(-21)	7.750	2.9253...
(16, 1)	5.755(-22)	5.612(-23)	7.699	2.9253...
(19, 1)	2.203(-27)	1.514(-28)	7.810	2.9253...

Table 11.3 Error bounds $r_n(f_3)$, actual errors $(Error)$ and exact value of integral (I_ω)

(n, ω)	$r_n(f_3)$	$Error$	I_ω
(2,0.5)	1.281 (00)	1.031 (00)	4.3181...
(3,0.5)	5.793 (-01)	4.478 (-01)	4.3181...
(4,0.5)	1.913 (-01)	1.643 (-01)	4.3181...
(5,0.5)	7.059 (-02)	6.415 (-02)	4.3181...
(6,0.5)	2.560 (-02)	2.441 (-02)	4.3181...
(7,0.5)	1.139 (-02)	9.370 (-03)	4.3181...
(8,0.5)	5.068 (-03)	3.581 (-03)	4.3181...
(10,0.5)	1.004 (-03)	5.239 (-04)	4.3181...
(2,1)	2.330 (-01)	1.090 (-01)	1.3658...
(3,1)	2.749 (-01)	1.934 (-02)	1.3658...
(4,1)	3.741 (-02)	3.347 (-03)	1.3658...
(5,1)	2.501 (-02)	5.784 (-04)	1.3658...
(6,1)	7.203 (-03)	9.971 (-05)	1.3658...
(7,1)	1.717 (-04)	1.717 (-05)	1.3658...
(8,1)	1.316 (-05)	2.954 (-06)	1.3658...
(10,1)	7.928 (-07)	8.729 (-08)	1.3658...

The function f_1 is entire, and it is known (see [17]) that

$$\max_{z\in\mathcal{E}_\rho} |f_1(z)| = \mathrm{e}^{\mathrm{e}^{\cosh(\omega b_1)}}, \quad b_1 = (\rho - \rho^{-1})/2, \quad \rho_{\max} = +\infty.$$

Example 11.2 Let $f_2(z) = \mathrm{e}^{\omega z^2}$, $\omega > 0$.

The function f_2 is entire, and it is known (see [20]) that

$$\max_{z\in\mathcal{E}_\rho} |f_2(z)| = \mathrm{e}^{\omega a_1^2}, \quad a_1 = (\rho + \rho^{-1})/2, \quad \rho_{\max} = +\infty.$$

Example 11.3 Let $f_3(z) = \frac{\cos(z)}{z^2+\omega^2}$, $\omega > 0$. For function $f_3(z)$ (see [19], [4, p. 25]), it holds that

$$\max_{z\in\mathcal{E}_\rho} |f_3(z)| = \frac{\cosh(b_1)}{-a_1^2 + \omega^2},$$

where $b_1 = (\rho - \rho^{-1})/2$. Here the infimum is calculated with respect to the interval $\rho \in (\rho^*, \rho_{\max})$, where $\rho_{\max} = \omega + \sqrt{1+\omega^2}$. We made a couple of experiments with intervals $(\rho^*, \rho_{\max})$, which exists, and obtained in each case effective bound.

Furthermore, we proceeded by extending this research to establish sufficient conditions for the maximum of the modulus of the kernel, analytically derived, to occur at one of the semi-axes for Gauss–Gegenbauer quadrature (11.1) with the weight function $\omega(t) = (1-t^2)^\alpha$ (see [7]).

In this case, the function $\varrho_n(z)$ is

$$\varrho_n(z) = \int_0^\pi \frac{2\sum_{k=0}^{\prime n} d_{k,n}\cos k\theta}{z-\cos\theta}(\sin\theta)^{2\alpha+1}\, \mathrm{d}\theta. \tag{11.14}$$

By substituting (see [1, Equation 4.1])

$$(\sin\theta)^{2\alpha+1} = 2^{-2\alpha}\sum_{i=0}^{\alpha}(-1)^i\binom{2\alpha+1}{\alpha-i}\sin(2i+1)\theta$$

into (11.14), we got

$$\varrho_n(z) = \int_0^\pi \frac{2\sum_{k=0}^{\prime n} d_{k,n}\cos k\theta\cdot 2^{-2\alpha}\sum_{i=0}^{\alpha}(-1)^i\binom{2\alpha+1}{\alpha-i}\sin(2i+1)\theta}{z-\cos\theta}\,\mathrm{d}\theta \tag{11.15}$$

$$= \frac{1}{2^{2\alpha}}\sum_{k=0}^{n}{}' d_{k,n}\sum_{i=0}^{\alpha}(-1)^i\binom{2\alpha+1}{\alpha-i}\int_0^\pi \frac{\sin(k+1+2i)\theta - \sin(k-2i-1)\theta}{z-\cos\theta}\,\mathrm{d}\theta$$

$$= \frac{1}{2^{2\alpha}}\sum_{k=0}^{n}{}' d_{k,n}\sum_{i=0}^{\alpha}(-1)^i\binom{2\alpha+1}{\alpha-i}h(i,k),$$

where

$$h(i,k) = \begin{cases} I_{k+2i} - I_{k-2i-2}, & k \geq 2i+1, \\ I_{k+2i} + I_{2i-k}, & k < 2i+1. \end{cases} \tag{11.16}$$

The term I_k is obtained using [20, Proposition 2.2, Equation 2.8] and its proof:

$$I_k = \int_0^{\pi} \frac{\sin(k+1)\theta}{z - \cos\theta}\, \mathrm{d}\theta = \int_{-1}^{1} \frac{U_k(t)}{z-t}\, \mathrm{d}t. \tag{11.17}$$

It stands that

$$I_k = U_k(z) \log\frac{z+1}{z-1} - 4 \sum_{j=1}^{\left[\frac{k+1}{2}\right]} \frac{U_{k-2j+1}(z)}{2j-1}, \quad |z| > 1. \tag{11.18}$$

Inserting relation $z = (u+u)^{-1}/2$, $|u| > 1$, in (11.18), it followed that

$$I_k = \frac{u^{k+1} - u^{-(k+1)}}{u - u^{-1}} \log\frac{z+1}{z-1} - 4 \sum_{j=1}^{\left[\frac{k+1}{2}\right]} \frac{u^{k-2j+2} - u^{-(k-2j+2)}}{(2j-1)(u-u^{-1})}. \tag{11.19}$$

We observed that

$$\log\frac{z+1}{z-1} = \log\frac{(u+u^{-1})/2+1}{(u+u^{-1})/2-1} = \log\frac{(1+u^{-1})^2}{(1-u^{-1})^2} = 2\log\frac{1+u^{-1}}{1-u^{-1}}.$$

With this, we demonstrated how to explicitly calculate the expression for the kernel $K_n(z)$, $z = \frac{1}{2}(u+u^{-1})$ for any given u. Based on Schira's findings, we know that the maximum modulus of the kernel is attained on the imaginary axis. Schira also provided a precise estimation of the value of ρ at which this maximum occurs [21, Theorem 3.2].

Similar to our initial research, in this instance as well, we conducted numerous numerical calculations.

The graphical depiction of the modulus of the kernels for the Gauss–Gegenbauer quadrature formula, considering various choices of n, ρ, and α, is illustrated in Figs. 11.3 and 11.4.

In this scenario, once more, to reaffirm the proposed error bounds, we revisited our previous tests, comparing them against the exact errors. Our focus remained on special functions extensively deliberated in the literature sources such as [6] and [18].

For functions f_1 and f_2, previously discussed within Gauss–Legendre quadrature contexts, we now provide concise summaries for Gauss–Gegenbauer quadrature

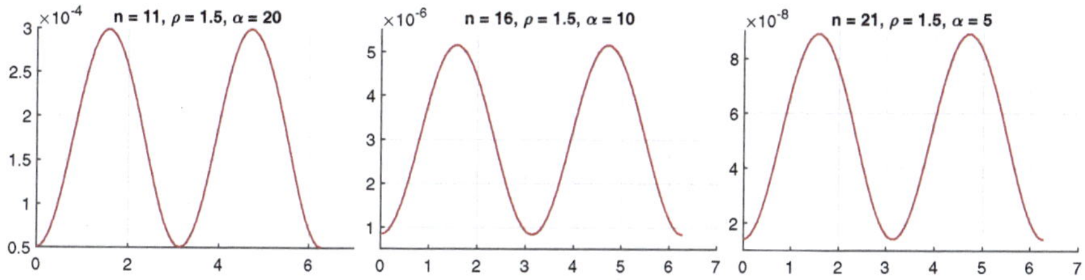

Fig. 11.3 The graphical representation of the modulus of the kernels for various selections of n, ρ, and α

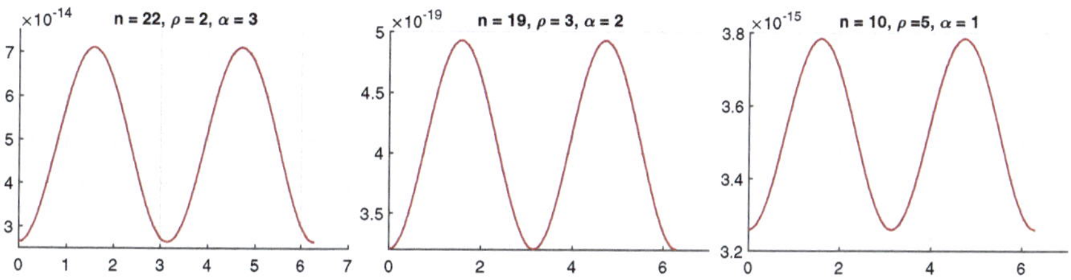

Fig. 11.4 The graphical representation of the modulus of the kernels for various selections of n, ρ, and α

Table 11.4 Error bounds $r_n(f_1)$, actual errors ($Error$) and exact value of integral $I(f_1)$ considering the case where α equals 5 and 10, respectively

(n, ω, α)	$r_n(f_1)(\alpha = 2)$	$Error$	ρ_{opt}	$I(f_1)$
(5,1,5)	1.155(-03)	1.333 (-04)	2.601	1.021...
(10,1,5)	1.869(-08)	1.447 (-09)	3.001	1.021...
(15,1,5)	1.731(-13)	1.047(-14)	3.201	1.021...
(20,1,5)	1.075(-18)	5.474(-20)	3.301	1.021...
(25,1,5)	4.990(-24)	2.204(-25)	3.401	1.021...
(5,1,10)	1.377(-04)	1.531 (-05)	2.701	7.758...
(10,1,10)	8.518(-10)	6.446 (-11)	3.001	7.758...
(15,1,10)	4.337(-15)	2.583(-16)	3.201	7.758...
(20,1,10)	1.804(-20)	8.978(-21)	3.301	7.758...
(25,1,10)	6.194(-25)	2.679(-26)	3.401	7.758...

formulas. Specifically, we cover unexplored α values, detailed in Tables 11.4 and 11.5.

11.5 Conclusion

Here, we've outlined our research chronologically, delving into Gauss–Legendre and Gauss–Gegenbauer quadrature formulae. Our successful derivation of error bounds for these quadratures reveals their significant efficacy, notably in scenarios where the integrand remains an analytic function within the region delineated by

Table 11.5 Error bounds $r_n(f_2)$, actual errors $(Error)$, and exact value of integral $I(f_2)$ considering the case where $\alpha = 3$ and $\alpha = 6$

(n, ω)	$r_n(f_2)(\alpha = 2)$	$Error$	ρ_{opt}	$I(f_2)$
(5, 2, 3)	8.605(-04)	6.745(-05)	4.015	1.191...
(10, 2, 3)	7.110(-10)	5.413(-11)	4.501	1.191...
(15, 2, 3)	7.688(-17)	4.195(-18)	5.001	1.191...
(20, 2, 3)	9.878(-25)	6.611(-26)	6.301	1.191...
(25, 2, 3)	6.493(-31)	3.109(-34)	6.601	1.191...
(5, 2, 6)	1.461(-04)	8.230(-06)	3.301	0.792...
(10, 2, 3)	6.125(-11)	3.516(-12)	4.601	0.792...
(15, 2, 3)	3.459(-18)	1.958(-19)	5.501	0.792...
(20, 2, 3)	4.573(-26)	2.514(-27)	5.601	0.792...
(25, 2, 3)	1.409(-33)	1.028(-35)	5.801	0.792...

confocal ellipses surrounding the integration interval. Moreover, we've developed an asymptotic estimation of the kernel, providing a comprehensive characterization of kernel behavior.

To corroborate our theoretical findings, we conducted numerical experiments. The results serve as validation, affirming the effectiveness of the obtained error bounds in comparison to the actual errors.

Acknowledgments Research of the authors is supported in part by the Serbian Ministry of Science, Technological Development, and Innovations, according to Contract 451-03-47/2023-01/200105 dated on February 3, 2023.

References

1. Ossicini, A., Rosati, F.: Funzioni caratteristiche nelle formule di quadratura gaussiane con nodi multipli. Boll. Unione Mat. Ital. **(4)11**(Suppl. fasc. 3), 224–237 (1975) (Italian)
2. Pejčev, A.V.: A note on "Error bounds of Gaussian quadrature formulae with Legendre weight function for analytic integrands by M. M. Spalević et al.". Electron. Trans. Numer. Anal. **59**, 89–98 (2023)
3. Gauss, C.F.: Methodus nova integralium valores per approximationem inveniendi. Commentationes Societatis Regiae Scientiarum Göttingensis Recentiores **3** (1814) (also in Werke III, pp. 163–196)
4. Djukić, D.Lj., Pejčev, A.V., Spalević, M.M.: The error bounds of Gauss-Kronrod quadrature formulae for weight functions of Bernstein-Szegő type. Numer. Algor. **77**, 1003–1028 (2018)
5. Djukić, D.Lj., Mutavdžić Djukić, R.M., Pejčev, A.V., Spalević, M.M.: Error estimates of Gaussian-type quadrature formulae for analytic functions on ellipses—a survey of recent results. Electron. Trans. Numer. Anal. **53**, 352–382 (2020)
6. Jandrlić, D.R., Krtinić, Dj.M., Mihić, Lj.V., Pejčev, A.V., Spalević, M.M.: Error bounds of Gaussian quadrature formulae with Legendre weight function for analytic integrands. Electron. Trans. Numer. Anal. **55**, 424–437 (2022)
7. Jandrlić, D.R., Pejčev, A.V., Spalević, M.M.: Error bound of Gaussian quadrature rules for certain Gegenbauer weight functions. J. Comput. Appl. Math. **440**, 115661 (2024) (art. 115661)

8. Milovanović, G.V., Spalević, M.M.: Error bounds for Gauss-Turán quadrature formulas of analytic functions. Math. Comp. **72**, 1855–1872 (2003)
9. Milovanović, G.V., Spalević, M.M.: An error expansion for some Gauss-Turán quadratures and L^1-estimates of the remainder term. BIT Numer. Math. **45**, 117–136 (2005)
10. Milovanović, G.V., Spalević, M.M.: Bounds of the error of Gauss–Turán-type quadratures. J. Comput. Appl. Math. **178**, 333–346 (2005)
11. Milovanović, G.V., Pejčev, A.V., Spalević, M.M.: A note on an error bound of Gauss-Turán quadrature with the Chebyshev weight function of the first kind. Filomat **27**, 1037–1042 (2013)
12. Milovanović, G.V., Spalević, M.M., Pranić, M.S.: Maximum of the modulus of kernels in Gauss-Turán quadratures. Math. Comp. **77**, 985–994 (2008)
13. Milovanović, G.V., Spalević, M.M., Pranić, M.S.: Error estimates for Gauss-Turán quadratures and their Kronrod extensions. IMA J. Numer. Anal. **29**, 486–507 (2009)
14. Milovanović, G.V., Spalević, M.M., Pranić, M.S.: Quadrature Formulae with Multiple Nodes. World Scientific Publishing Company, Singapore–New Jersey–London–Hong Kong, 400 (2024) (in preparation)
15. Wang, H., Zhang, L.: Jacobi polynomials on the Bernstein ellipse. J. Sci. Comput. **75**, 457–477 (2018)
16. Mysovskih, I.P.: Lectures of Numerical Methods. Wolters-Noordhoff Publishing Groningen, The Netherlands (1969)
17. Spalević, M.M., Pranić, M.S., Pejčev, A.V.: Maximum of the modulus of kernels of Gaussian quadrature formulae for one class of Bernstein-Szegő weight functions. Appl. Math. Comput. **218**, 5746–5756 (2012)
18. Orive, R., Pejčev, A., Spalević, M., Mihić, Lj.: On the Gauss-Kronrod quadrature formula for a modified weight function of Chebyshev type. Numer. Algor. **11**(91), 1855–1877 (2022)
19. Notaris, S.E.: Error bounds for Gauss-Kronrod quadrature of analytic functions. Numer. Math. **64**, 371–380 (1993)
20. Notaris, S.E.: Integral formulas for Chebyshev polynomials and the error term of interpolatory quadrature formulae for analytic functions. Math. Comp. **75**, 1217–1231 (2006)
21. Schira, T.: The remainder term for analytic functions of symmetric gaussian quadratures. Math. Comp. **66**, 297–380 (1997)
22. Gautschi, W., Varga, R.S.: Error bounds for Gaussian quadrature of analytic functions. SIAM J. Numer. Anal. **20**, 1170–1186 (1983)
23. Gautschi, W., Tychopoupos, E., Varga, R.S.: A note on the contour integral representation of the remainder term for a Gauss-Chebyshev quadrature rule. SIAM J. Numer. Anal. **27**, 219–224 (1990)
24. Gautschi, W. In: OPQ suite. http://www.cs.purdue.edu/archives/2001/wxg/codes
25. Gautschi, W.: Orthogonal Polynomials: Computation and Approximation. Oxford University, Oxford (2004)

Chapter 12
Partitions and Automorphisms of Finite Kurepa Trees

Žarko Mijajlović

Dedicated to Academician Gradimir V. Milovanović on the occasion of his 75th birthday

12.1 Introduction

Đuro Kurepa defined in [17] the left factorial function over nonnegative integers as

$$!n = 0! + 1! + \cdots + (n-1)!, \quad n \geq 1. \tag{12.1}$$

He also posed there the left factorial hypothesis KHn, which states that for no integer $n > 2$, $n \mid !n$. Kurepa also introduced in [19] a finite partially ordered set $\mathbf{T}_n = (T_n, \leq, 0)$, a rooted tree of height n with root 0 of height 0 in which each node of height $k < n$ splits into $k+1$ successors and showed that $|T_n| = 0! + 1! + \cdots + n!$, hence $|T_{n-1}| = !n$. The number of derangements of a set of n elements is also denoted by $!n$, so to avoid confusion, we shall in most cases use the symbol t_n for $|T_n| = !(n+1)$. Obviously, KHn is equivalent to the combinatorial statement:

KHP For no integer $n > 2$, the tree T_{n-1} has a partition into n subsets of the same size.

It suffices to prove KHn for prime numbers p, and this version of the hypothesis is denoted by KHp. This problem is listed as a Problem B44 in R. Guy's book [11], a Problem 37 in Konick-Mercier book [15] and in Sandor–Crstici book [25]. Many mathematicians tried to solve KHn, but it is still an open problem. There were number-theoretical attempts to prove it and many computational efforts to find a counterexample to KHp, a prime p such that $!p = 0 \pmod{p}$. Professor Gradimir Milovanović gave important contributions concerning KHn, in particular

Ž. Mijajlović (✉)
University of Belgrade, Faculty of Mathematics, Belgrade, Serbia
e-mail: zarkom@matf.bg.ac.rs

M. P. Stanić et al. (eds.), *Analysis, Approximation, Optimization: Computation and Applications*, Springer Optimization and Its Applications 224,
https://doi.org/10.1007/978-3-031-85743-0_12

to the theory of integral

$$K(z) = \int_0^\infty \frac{t^z - 1}{t - 1} e^{-t} \, dt, \quad \text{Re}(z) > 0, \tag{12.2}$$

an analytic extension of $!n$, defined by Kurepa in [18]. For example, Milovanović analyzed in [22] asymptotic behavior of this function and gave an approximation of it which can be used for fast and very precise calculations of values of $K(z)$. He also introduced and studied in [23] a sequence of functions defined by the recurrence relation

$$K_m(z+1) = K_{m-1}(z+1) + K_m(z), \quad m = 0, 1, 2, \ldots, \tag{12.3}$$

where $K_{-1}(z) = \Gamma(z)$ and $K_0(z) = K(z)$. We found particularly interesting the question if this sequence of functions is differentially algebraically independent; for details, see [20].

KHp is numerically checked by many authors, with the last achievement for $p \leq 2^{40}$, see [1]. Assuming some heuristic probability models, many authors, e.g., [7], believe that KHp is not true, and therefore, there were so many computational attempts to find an example that violates KHp. Several authors, e.g., [2, 14] and [10], used combinatorial functions related to partition calculus. History of the problem and related results are presented in [12] and [21].

Instead of KHn, we consider here its combinatorial reformulation KHP with the aim to show that if there are such examples, then they are combinatorially complicated. We proved that there are no partitions $\mathcal{P}$ involved in KHP if all components in $\mathcal{P}$ are connected subsets of T_{n-1}, or there is a component invariant under all automorphisms of $\mathbf{T}_{n-1}$. Along the way, we found that Aut($\mathbf{T}_n$) is the wreath product of factorial powers of permutation groups $\mathbf{S}_k$, $k \leq n$. Using Ramsey theory, we also proved that in every partition $\mathcal{P}$ of Kurepa tree, which violates KHn, each component of $\mathcal{P}$ has a large scattered subset.

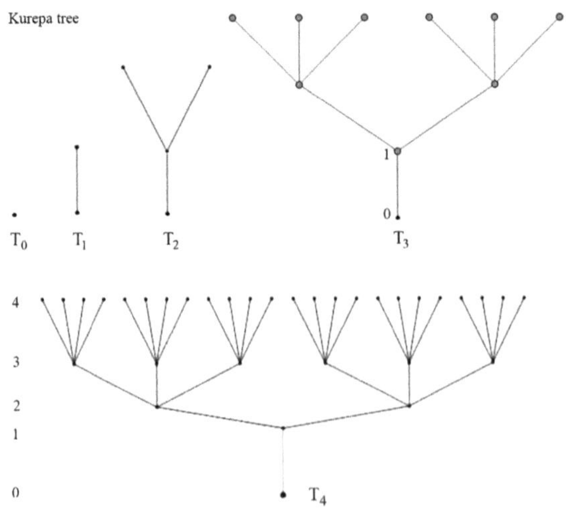

We follow notation and terminology as used in [21]. If X is a set, then $\mathrm{Sym}(X)$ denotes the set of all permutations over X, or symmetric group of X. If f is a function and $X \subseteq \mathrm{dom}(f)$, then $f|X$ denotes the restriction of f to X. By $\mathbf{S}_n$, we denote the permutation group over $\{1, 2, \ldots, n\}$. The identity function of certain domain is denoted by id. We also use the standard notation from group theory. For example, if G is a group, then $K < G$ means that K is a subgroup of G, while $K \lhd G$ denotes that K is a normal subgroup of G. If $K, L \subseteq G$, then $KL = \{ab\colon a \in K, b \in L\}$, while for $a, b \in G$, $b^a = a^{-1}ba$, and $K^a = a^{-1}Ka$.

Notion of a tree somewhat differs in set theory and combinatorics even if they are often graphically presented in the same way. Due to diversity of terminology in the literature concerning trees and related combinatorial objects, we first fixate terminology and notation.

In set theory, rooted tree is a partially ordered set $(T, \leq, 0)$ such that for each $t \in T$, the set $S_a = \{s \in T : s < a\}$ is well-ordered by the relation $<$, and 0 is the least element with respect to the order $\leq$. Here we shall deal exclusively with finite trees; hence, for our purpose, it suffices to assume that S_a is linearly ordered by $\leq$. The order type of S_a is called the height of a, denoted by $\mathrm{ht}(a) = \mathrm{ht}_T(a)$. For example, if b is connected to a and $\mathrm{ht}(b) = \mathrm{ht}(a) + 1$, then b is called the successor of a, while a is the (unique) predecessor of b. We depict the trees by Hasse diagrams from bottom to top, i.e., smaller elements relative to $\leq$ are at the bottom, greater at the top. For other notions such as chain, branch, and level, one may consult, [13] and [16]. Studies of finite trees as set-theoretical or algebraic and topological objects are popular; see, for example, [3].

In some parts of this chapter, we shall also use notions and techniques from graph theory. However, we shall deal only with undirected rooted graphs. We adopt usual definitions from graph theory. For example, the notion of a path in an undirected graph G is a sequence of edges, which joins a sequence of vertices. If there is a path at which lay $a, b \in G$, then a path between a, b is the shortest (with the least number of edges) path joining a, b. If for any $a, b \in G$, there is a path connecting them, then G is connected graph by definition. If this path is unique, it makes sense to introduce the segment $[a, b] = [a, b]_G$, which consists of all vertices in G laying at that path, including a and b. Obviously, $[a, b] = [b, a]$. A combinatorial tree is an undirected graph in which any two vertices are connected by exactly one path. Hence, combinatorial tree is connected graph, and we shall occasionally denote the unique path between vertices a, b by $[a, b]$. Finally, we remind that Hasse diagrams translate set-theoretical trees into combinatorial ones. We must be careful here as the graph of set-theoretical tree $\mathbf{T} = (T, \leq, 0)$ is the set $g(\mathbf{T}) = \{(a, b) \in T^2\colon a \leq b\}$, while the Hasse translation $\mathbf{T}^H$ of $\mathbf{T}$ consists of the set of vertices $V = T$ and the set of edges

$$E = \{\{a, b\}\colon b \text{ is a successor of } a, \text{ or } a \text{ is the predecessor of } b,\ a, b \in T\}$$

Hence, $g(\mathbf{T})$ is directed graph, and its undirected version, when the orientation is omitted, is not at all a tree in the sense of graph theory. For example, if $a < c < b$ in $\mathbf{T} = (T, \leq)$, there is an edge that connects nodes a and b in $g(\mathbf{T})$, while this is not true in $\mathbf{T}^H$. Whenever we mention T as a combinatorial tree in the context of a given set-theoretical tree $\mathbf{T}$, we assume that T is $\mathbf{T}^H$, i.e., T is only typographical relaxation of $\mathbf{T}^H$.

As a set-theoretical tree $\mathbf{T}$ is a lower semilattice, we see that a path between a and b is $[a, c] \cup [c, b]$, where $c = a \wedge b$ is the greatest lower bound for $\{a, b\}$ in $\mathbf{T}$. Observe that c is the greatest element in $[0, a] \cap [0, b]$.

An induced subgraph of a graph G is another graph formed from a subset of the vertices of the graph and all of the edges from the original graph connecting pairs of vertices in that subset. Hence, if $X \subseteq G$, G produces unique induced subgraph on X, so we shall identify X with its induced graph counterpart. It is said that a subset $X \subseteq G$ is connected if for any $a, b \in X$, there is a path p in the induced graph on X connecting a, b. Observe that p is a path in X if and only if p is a path in G. The notion of induced graph differs from the general notion of a subgraph in which the requirement is that the set of edges is only a subset of the original graph. Here we shall work exclusively with induced subgraphs.

We also use some basic concepts from model theory, in particular the part that deals with the first-order definability notion. For example, the sign $\models$ will denote the satisfaction relation. We shall often distinguish a structure (model) $\mathbf{A}$ and its underlying domain A. For model-theoretic notions and their properties, one may consult, for example [4].

12.2 Definability and Automorphisms

In this section, we analyze and describe definable subsets of $\mathbf{T}_n$ and their relations to automorphisms of $\mathbf{T}_n$. Obtained results here will be used in the third section to show that certain partitions of Kurepa tree T_n with equal-sized components are not possible. In the second part, we give a detailed description of $\mathrm{Aut}(\mathbf{T}_n)$.

12.2.1 Definability

We remind that for any structure $\mathbf{A}$, $X \subseteq A$ is definable in $\mathbf{A}$ if there is a formula $\varphi(x)$ in the language $\mathcal{L}$ of $\mathbf{A}$, such that $X = \{a \in A : \mathbf{A} \models \varphi(a)\}$. In case of $\mathbf{T}_n$, the language is $\mathcal{L} = \{\leq, 0\}$ and, for example, "y is a successor of x" is represented by the formula $\mathrm{Succ}(x, y) \equiv x < y \wedge \neg\exists z(x < z < y)$. Then for each k, the level L_k of $\mathbf{T}_n$ is definable as $L_k = \{a \in T_n : \mathbf{T}_n \models \varphi(a)\}$, where $\varphi(x)$ is:

$$\exists x_0 \exists x_1 \cdots \exists x_k \left(x_0 = 0 \wedge x_k = x \wedge \bigwedge_{i<k} \mathrm{Succ}(x_i, x_{i+1}) \right). \tag{12.4}$$

Observe that $|L_k| = k!$ and that T_n is a disjoint union of all its levels. As elements of L_k are incomparable, for each chain $C \subseteq T_n$, we also have $|L_k \cap C| \leq 1$.

For any structure $\mathbf{A}$, $g \in \mathrm{Aut}(\mathbf{A})$, and a formula $\varphi(x)$ in the language of $\mathbf{A}$, for $a \in A$ we have: if $\mathbf{A} \models \varphi(a)$, then $\mathbf{A} \models \varphi(g(a))$. Hence, the followng holds.

Proposition 12.1 *Definable subsets X of $\mathbf{A}$ are invariant under automorphisms, i.e., $g(X) \subseteq X$, $g \in \mathrm{Aut}(\mathbf{T}_n)$.*

In particular, levels L_k of T_n are invariant under automorphisms, i.e., $g(L_k) \subset L_k$, where $g \in \mathrm{Aut}(\mathbf{T}_n)$. In order to describe definable subsets of $\mathbf{T}_n$, we prove several propositions.

Obviously, we may assume $\mathbf{T}_0 \subseteq \mathbf{T}_1 \subseteq \cdots \subseteq \mathbf{T}_n$ for any n. Let $G_n = \mathrm{Aut}(\mathbf{T}_n)$. In the following, we fix enumerations of elements of the levels L_{n-1} and $L_n = \bigcup_{1\leq k\leq m} A_k$ taking

$$\begin{array}{lll} L_{n-1} = \{1, 2, \ldots, m\}, & m = (n-1)! \ \text{ and} \\ A_k = \{k1, k2, \ldots, kn\}, & \text{the set of successors of } k,\ k = 1, 2, \ldots, m. \end{array} \tag{12.5}$$

One would use a_k, and a_{ki} instead, but our notation simplifies symbolism in formulas and avoids multilevel indexing. We may think of ki as of the ordered pair (k, i). Observe that A_k are mutually disjoint sets and that the first enumeration induces the natural linear order on L_{n-1}, while the other enumerations induce linear orders on each set A_k.

Lemma 12.1 *Every $h \in \mathrm{Aut}(\mathbf{T}_{n-1})$ can be extended to some $g \in \mathrm{Aut}(\mathbf{T}_n)$.*

Proof Let $h \in \mathrm{Aut}(\mathbf{T}_{n-1})$. Then h is a permutation over L_{n-1}. We take $g(x) = h(x)$ for $x \in T_{n-1}$ and $g(ki) = h(k)i$. Then g is an automorphism of $\mathbf{T}_n$, which extends h. □

Such extension h of g will be called canonical with respect to the enumerations (12.5).

Lemma 12.2 *Group G_{n-1} is embedded into G_n.*

Proof Let G' be the set of all canonical extensions of $h \in G_{n-1}$ to G_n. If h' is a canonical extension of h, then the map $h \mapsto h'$ is an embedding of G_{n-1} into G_n. □

Hence, we may assume $G_{n-1} \subseteq G_n$.

Lemma 12.3 *If $c, d \in T_n$ belong to the same level, then there is $g \in \mathrm{Aut}(\mathbf{T}_n)$ so that $g(c) = d$.*

Proof We prove Lemma by finite induction. If $c, d \in T_k$, $k < n$, by inductive hypothesis, there is $h \in \mathrm{Aut}(\mathbf{T}_{k-1})$ so that $h(c) = d$. By previous lemma, there is an extension $g \in \mathrm{Aut}(\mathbf{T}_n)$ of h, hence $g(c) = d$.

Suppose c, d belong to the last level L_n. If c, d have a common predecessor at level L_{n-1}, then we take transposition $g(c) = d$, $g(d) = c$, while for $x \in T_n \setminus \{c, d\}$

we take $g(x) = x$. Assume k and l, $k \neq l$, are predecessors of c, d respectively, and $c = ki$, $d = lj$, for some i, j. By inductive hypothesis, there is $h \in G_{n-1}$ so that $h(k) = l$ and an extension $h' \in G_n$ of h. Define $g \in \mathrm{Aut}(\mathbf{T}_n)$ as follows. For $x \in T_n \setminus (A_k \cup A_l)$, we take $g(x) = h'(x)$. To define g over $A_k \cup A_l$, we take a transposition, $g(ki) = lj$, $g(lj) = ki$, while for $s \neq i$ we take $g(ks) = ls$, and for $s \neq j$, $g(ls) = ks$. □

Automorphisms preserve the height of elements in a tree. Hence, in the formulation of the last lemma, we had to assume that c, d belong to the same level, as otherwise there is no automorphism, which permutes c, d. Immediate conclusions of the Lemma are the following corollaries.

Corollary 12.1 *If X is invariant (definable) subset of T_n and $X \cap L_k \neq \emptyset$, then $L_k \subseteq X$.*

Proof Let $a \in X \cap L_k$. Then for any $b \in L_k$ by previous Lemma, there is $g \in G_n$ so that $b = g(a)$, so by definability of X, or the invariance of X under automorphisms, it follows $g(a) \in X$, i.e., $b \in X$. Hence, $L_k \subseteq X$. □

Hence, if $X \subseteq T_n$ is invariant (definable), then $X = X \cap T_n = \bigcup_{0 \leq k \leq n}(X \cap L_k)$. As $X \cap L_k$ is empty set or L_k it follows:

Corollary 12.2 *If X is invariant subset of $\mathbf{T}_n$, then there is $I \subseteq \{0, 1, 2, \ldots, n\}$ such that*

$$X = \bigcup_{k \in I} L_k. \tag{12.6}$$

We have seen that X is definable if and only if it is invariant. Hence, we described all invariant (definable) subsets of $\mathbf{T}_n$. If $\theta : G_n \to \mathrm{Sym}(T_n)$ is the natural action, $\theta(g)(a) = g(a)$, $g \in G_n$, $a \in T_n$, then we see that the orbits a^{G_n} of this action are exactly levels of T_n. If $a \in L_k$, then $a^{G_n} = L_k$.

In the previous proofs, we could rely on a general theorem from model theory, but proofs wouldn't be much simpler. Namely, for finite structures, the following property of definability in model theory is well-known.

Theorem 12.1 *A subset X of a finite structure $\mathbf{A}$ is definable in $\mathbf{A}$ if and only if it is invariant under all automorphisms of $\mathbf{A}$.*

12.2.2 Aut($\mathbf{T}_n$)

The structure of the automorphism group of trees is well understood. Namely, already Camille Jordan had shown that they are obtained from symmetric groups by iterated direct products and wreath products [5]. Products of factorials, known as Jordan–Pólya numbers, arise in this way as the orders of automorphisms groups of trees. Here we give somewhat detailed construction and certain properties of groups of automorphisms of Kurepa trees.

In this section, we fix n, and to easy notation, we often use the letter G to denote the group $G_n = \mathrm{Aut}(\mathbf{T}_n)$. Automorphisms of $\mathbf{T}_n$ induce certain permutations of L_n. In fact, G is isomorphic to the group of automorphisms formed by these permutations.

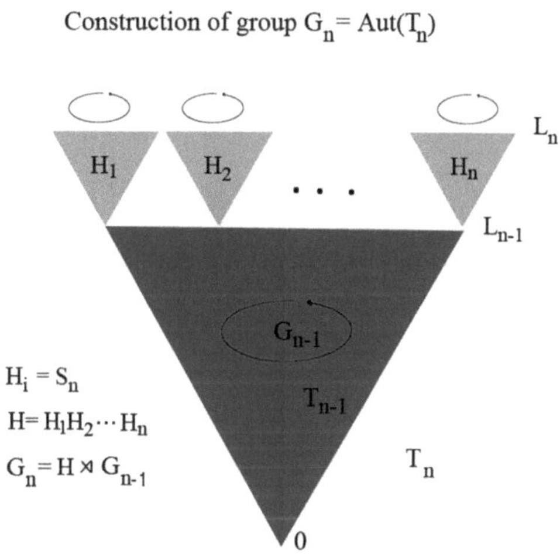

Lemma 12.4 *The mapping* $\theta : G \to \mathrm{Sym}(L_n)$ *defined by* $\theta(f) = f|L_n$, $f \in G$, *is an embedding of* $\mathrm{Aut}(\mathbf{T}_n)$ *into* $S_{n!}$.

Proof We prove here only that

$$\theta \quad \text{is} \quad 1-1 \tag{12.7}$$

as other conditions for embeddability are clearly fulfilled. The condition (12.7) is equivalent to: for $f, g \subset G_n$, $f \circ g^{-1}|L_n = \mathrm{id}|L_n$ implies $f = g$. Hence it is enough to prove for $h \in G_n$ that $h|L_n = \mathrm{id}|L_n$ implies $h = \mathrm{id}|T_n$. So assume $h|L_n = \mathrm{id}|L_n$ and $a \in T_n$. Hence, $a \in L_k$ for some k, so also $h(a) \in L_k$. There is $b \in L_n$ such that $a \le b$, so $h(a) \le h(b)$, i.e., $h(a) \le b$. The set $C = \{x \in T_n : x \le b\}$ is a chain, $a, h(a) \in C$, so a and $h(a)$ are comparable elements at the same level L_k, hence $h(a) = a$. □

Let us define a subgroup $H_k, k = 1, 2, \ldots, m, m = (n-1)!$, of G in the following way. Assume enumeration (12.5). Note that ki, $i = 1, 2, \ldots, n$, are successors of k, hence for any $f \in G$, we have $f(A_k) = A_{f(k)}$, as automorphisms move successors of a node to the successors of the image of that node. In particular, we have $f^{-1}(A_k) = A_{f^{-1}(k)}$. For a permutation $\sigma \in \mathrm{Sym}(A_k)$, define $g_\sigma \in H_k$ taking $g_\sigma(x) = \sigma(x)$ for $x \in A_k$, while $g(x) = x$ for $x \notin A_k$, $x \in T_n$. Here are some properties of H_k.

Proposition 12.2 *For all $k, l = 1, 2, \ldots, m$, $m = (n-1)!$, we have:*

(i) $H_k < G$ *and* $H_k \cong \mathbf{S}_n$;
(ii) *If* $f \in G_n$, *then* $H_k^f = H_{f^{-1}(k)}$;
(iii) $H_k H_l = H_l H_k$;
(iv) $H_k \cap H_1 H_2 \cdots H_{k-1} H_{k+1} \cdots H_m = \mathrm{e}$, *where* e *is the trivial group.*

Proof Property 1 is obvious, so we prove 2. Let $f \in G$, $g \in H_k$, $h = f^{-1} \circ g \circ f$, and $a \in T_n$. If $a \notin A_{f^{-1}(k)}$, then $f(a) \notin A_k$, hence, as $g \in H_k$, $g(f(a)) = f(a)$, so $h(a) = a$. Therefore,

$$a \notin A_{f^{-1}(k)} \Rightarrow h(a) = a. \tag{12.8}$$

If $a \in A_{f^{-1}(k)}$, then $f(a) \in A_k$, so $g(f(a)) \in A_k$. Hence, as $f \circ h = g \circ f$, $f(h(a)) \in A_k$, thus, $h(a) \in f^{-1}(A_k)$, i.e., $h(a) \in A_{f^{-1}(k)}$. Therefore, we proved

$$h|A_{f^{-1}(k)} \quad \text{is a permutation of} \quad A_{f^{-1}(k)}. \tag{12.9}$$

By (12.8) and (12.9), it follows that $h \in H_{f^{-1}(k)}$, so Property 2 is proved.

Now, we prove 3, $H_k H_l = H_l H_k$. If $k = l$, then $H_k H_l = H_k = H_l H_k$, so assume $k \neq l$ and let $f \in H_k H_l$, $g \in H_k$, $h \in H_l$, $f = g \circ h$. As $A_k \cap A_l = \emptyset$, for $x \in T_n$, we have:

$$f|A_k = g, \; f|A_l = h \quad \text{and} \quad f(x) = x \quad \text{if} \quad x \notin A_k \cup A_l. \tag{12.10}$$

Hence, also $f = h \circ g$, i.e., $f \in H_l H_k$, what finishes the proof of 3. Proof of statement 4 is similar, so we omit it. □

Let H be a subgroup of G_n generated by groups $H_1, \ldots, H_m$. By the previous proposition $H = H_1 H_2 \cdots H_m$ and for $f \in G$, we have

$$H^f = H_1^f H_2^f \ldots H_m^f = H_{f^{-1}(1)} H_{f^{-1}(2)} \ldots H_{f^{-1}(m)} = H_1 H_2 \cdots H_m = H \tag{12.11}$$

by Property 3 of the same proposition, hence by (12.11), $H \lhd G$. We also have $H_k \lhd H$, hence H is internal direct product of subgroups $H_1, H_2, \ldots, H_m$. We show that $G_n = H G_{n-1}$. Let $f \in G_n$, $g = f|T_{n-1}$, and $h_k \in H_k$ defined by $h_k = f|A_k$. Then,

$$f = \bar{g} \circ h_{f^{-1}(1)} \circ h_{f^{-1}(2)} \circ \cdots \circ h_{f^{-1}(m)}, \quad m = (n-1)!, \tag{12.12}$$

where $\bar{g}$ is the canonical extension of g defined by Lemma 12.1. As we identified G_{n-1} by canonical extensions of automorphisms of T_{n-1}, we also see that $H \cap G_{n-1} = \mathrm{e}$. Previous derivation shows that G_n is an internal semidirect product of its subgroups H and G_{n-1}. The following theorem summarizes properties of the G_n, its subgroups G_{n-1} and H and recursive representation of the group G_n.

Theorem 12.2 *Let* $m = (n-1)!$. *Then,*

(i) $H \lhd G_n$, $H \cap G_{n-1} = \mathrm{e}$, $G_n = HG_{n-1}$;
(ii) $H \cong \mathbf{S}_n^m$;
(iii) $G_n \cong \mathbf{S}_n^m \rtimes G_{n-1}$.

It is easy to show that the mapping $\theta : g \mapsto \bar{g}H$, $g \in G_{n-1}$, is an isomorphism between groups G_{n-1} and G_n/H, i.e., $\theta : G_{n-1} \cong G_n/H$; hence, the following hold.

Corollary 12.3 *The group* G_n *is an extension of* G_{n-1} *by* H.

By Krasner–Kaloujnine universal embedding theorem, it follows that the group G_n is embedded into the wreath product $H \wr G_{n-1}$. In fact,

Theorem 12.3 $G_n = S_n^m \wr G_{n-1}$.

Proof By Lemma 12.4, the group $G_{n-1} = \mathrm{Aut}(\mathbf{T}_{n-1})$ is acting on the set $\Omega = \{1, 2, \ldots, m\}$, $m = (n-1)!$, the set of terminal nodes (leaves) of T_{n-1}. By Proposition 12.2.2, see also (12.11), this action is extended to the action of G_{n-1} to $H = \prod H_i$ by $g \cdot (a_k)_{k\in\Omega} = (a_{g^{-1}(k)})_{k\in\Omega}$, $g \in G_{n-1}$, $(a_k)_{k\in\Omega} \in \prod H_i$. As $H_i \cong S_n$, we have $G_n = S_n^m \wr_\Omega G_{n-1}$. □

Note that the previous theorem also follows from Theorem 12.2.3.

Let $\gamma_n = |G_n|$, the cardinality of G_n. By the previous theorem, we have at once

$$\gamma_n = \prod_{k=1}^{n} k!^{(k-1)!}. \tag{12.13}$$

From (12.13), we can infer the weak form of KHp. Namely, if p is a prime, then obviously the product $\prod_{k=1}^{p-1}(k!)^{(k-1)!}$ is not divisible by p. Hence, the following theorem holds.

Theorem 12.4 *If* p *is a prime number, then* $p \nmid \gamma_{p-1}$, *i.e.,* p *does not divide* $|\mathrm{Aut}(\mathbf{T}_{p-1})|$.

On the contrary, it is easy to see that for the composite integers n, we have n dividing γ_n. Finally, we note that $\gamma_n | (n!)!$. This follows from the Lagrange theorem and Lemma 12.4. In fact, $|S_{n!} : G_n| = (n!)!/\gamma_n$. For example, $(3!)!/\gamma_3 = 10$, $(4!)!/\gamma_4 = 45092646348750$.

12.3 Partitions of T_n

First, we define some notions related to partitions of finite sets. Suppose S is a finite set. Following [8], we shall say that a partition $\mathcal{P} = \{P_1, P_2, \ldots, P_m\}$ of S is k-balanced if $|P_i| = k$ for all $1 \le i \le m$. Let us note that this notion is related to the so-called k–balanced problem. This problem restricted to trees is thoroughly analyzed in [8].

Let n be a nonnegative integer and $t_n = |T_n|$. Hence, $t_0 = 1$, $t_1 = 2$, $t_2 = 4$, …, and $!n = t_{n-1}$. Assuming $(n+1)|t_n$, a balanced partition $\mathcal{P}$ of T_n such that $|P| = \frac{t_n}{n+1}$ for all $P \in \mathcal{P}$, is called a K-partition. Hence, the existence of K-partition for T_{n-1} violates KHp. Our aim in this section is to exhibit some properties of such partitions. We shall often rely here on the numerical result in [1]:

$$\text{There are no } K \text{ - partitions of } T_{n-1} \text{ for } 2 \le n \le 2^{40}. \tag{12.14}$$

In particular, propositions of the form

$$\text{If } \mathcal{P} \text{ is a } K \text{ - partition of } T_n,\, 2 \le n \le 2^{40}, \text{ then } \varphi, \tag{12.15}$$

φ is some property related to Kurepa trees, hold vacuously. This is useful for a statement φ which holds asymptotically, i.e., of for all $n \ge n_0$ for some $2 \le n_0 \le 2^{40}$, even not knowing the exact value of n_0, to get a simpler form "for all $n_0 \ge 2$, T_n has the property φ." This also allows as not to search for the best (the least) value of $2 \le n_0 \le 2^{40}$, hence, for example many of the inequalities appearing in the rest of the chapter are not best possible, but they serve well to our main goal, to prove main theorems. Next proposition is of this form.

Proposition 12.3 *If $\mathcal{P}$ is a K-partition of T_n and some $P \in \mathcal{P}$ is a chain, then* $n = 0, 1$.

Proof Suppose some $P \in \mathcal{P}$ is a chain. As every chain of T_n can be extended to a maximal chain, we have $|P| \le n+1$. Hence, $t_n/(n+1) \le n+1$, i.e., $t_n \le (n+1)^2$. Therefore,

$$0! + 1! + \cdots + n! \le (n+1)^2. \tag{12.16}$$

But it is easy to see that $0! + 1! + \cdots + n! > (n+1)^2$ for $n \ge 4$, hence $n \le 3$. By (12.14), there are no K partitons for T_2, T_3, hence $n = 0, 1$. □

Corollary 12.4 *If $n \ge 2$ and $\mathcal{P}$ is a K-partition of T_n, then in each $P \in \mathcal{P}$ there are incomparable $a, b \in \mathcal{P}$.*

Elements a, b in the previous corollary make an antichain in T_n. We shall see that components of K-partitions have in fact asymptotically relatively large antichains. First, we prove some technical inequalities we shall use in the rest of the chapter.

Lemma 12.5 *Let T_n be Kurepa tree and $\mathcal{P}$ a K-partition. Then*

$$t_n < (n+1)!, \quad n \ge 2, \tag{12.17a}$$

$$t_n > (n+1) \cdot (n-1)!, \quad n \ge 2, \tag{12.17b}$$

$$t_n > (n+1)t_{n-2}, \quad n \ge 2, \tag{12.17c}$$

$$t_n < (n+1)((n-1)! + (n-3)!), \quad n \geq 4, \tag{12.17d}$$

$$t_n > (n+1)((n-1)! + t_{n-4}), \quad n \geq 4. \tag{12.17e}$$

Proof (12.17a). $t_n = \sum_{i=0}^{n} i! < (n+1) \cdot n!$.

(12.17b). $t_n > (n-1)! + n! = (n+1) \cdot (n-1)!$.

(12.17c). We have the following sequence of equivalent inequalities:

$$\begin{aligned} t_n &> (n+1)t_{n-2}, \\ 0! + 1! + \cdots + n! &> (n+1)(0! + 1! + \cdots + (n-2)!), \\ (n-1)! + n! &> n(0! + 1! + \cdots + (n-2)!), \\ (n+1) \cdot (n-1)! &> n t_{n-2}. \end{aligned}$$

By (12.17a), $(n-1)! > t_{n-2}$, hence the last inequality is true, so the statement (12.17c) follows.

(12.17d). As $t_n = n! + (n-1)! + (n-2)! + (n-3)! + t_{n-4}$ and

$$(n+1)((n-1)! + (n-3)!) = n! + (n-1)! + (n-2)! + (n-3)! + 2(n-3)!,$$

inequality (12.17d) is equivalent to

$$t_{n-4} < 2 \cdot (n-3)!.$$

The last inequality follows from (12.17a), hence inequality (12.17d) holds true.

(12.17e). As in the proof of the previous statement, we have the following sequence of equivalent inequalities:

$$\begin{aligned} \frac{t_n}{n+1} &> (n-1)! + t_{n-4}, \\ 0! + 1! + \cdots + n! &> (n+1)((n-1)! + t_{n-4}) = n! + (n-1)! + (n+1)t_{n-4}, \\ (n-2)! + (n-3)! &> n t_{n-4}, \\ n! + (n-1)! &> (n+2)t_{n-2}, \quad n \geq 2. \end{aligned} \tag{12.18}$$

Using (12.17a) for $n \geq 3$, we also have

$$(n+2)t_{n-2} = (n+2)((n-2)! + t_{n-3}) < (n+2)((n-2)! + (n-2)!) = 2(n+2) \cdot (n-2)!,$$

i.e.,

$$(n+2)t_{n-2} < 2(n+2) \cdot (n-2)! < (n+1) \cdot (n-1)! = n! + (n-1)!, \quad n \geq 4, \tag{12.19}$$

as $n^2 - 1 > 2(n+2)$ for $n \geq 4$. Hence, inequality (12.18) holds true and so (12.17e).

□

Lemma 12.6 $(n-1)! \geq \left(\frac{n}{e}\right)^{n-1}$ *for all integers* $n \geq 1$.

Proof For $1 \leq n \leq 7$, the inequality can be checked by direct calculations. Further, taking the expansion $\mathrm{e}^{n-1} = \sum_{k=0}^{\infty} \frac{(n-1)^k}{k!}$ and three consecutive terms with respect to n, as $n \geq 8$, we have

$$\mathrm{e}^{n-1} > \frac{(n-1)^{n-1}}{(n-1)!} + \frac{(n-1)^n}{n!} + \frac{(n-1)^{n+1}}{(n+1)!} = \frac{(n-1)^{n-1}}{(n-1)!} \cdot \frac{3n-1}{n+1} > \frac{n^{n-1}}{(n-1)!}. \tag{12.20}$$

The lemma follows if we prove the right-hand inequality, which is equivalent to the elementary inequality

$$\left(1 - \frac{1}{n}\right)^{n-1} > \frac{n+1}{3n-1}, \quad \text{for } n \geq 8, \tag{12.21}$$

which is easily proved in many ways. For example, if $a_n = \frac{n+1}{3n-1}$ and $b_n = (1 - \frac{1}{n})^{n-1}$ then $\lim_{n\to\infty} a_n = 1/3$, while $\lim_{n\to\infty} b_n = 1/\mathrm{e}$. As $1/\mathrm{e} > 1/3$, there is n_0, which separates a_n and b_n, i.e., for $k, l \geq n_0$ we have $a_k < b_l$. But a_n is monotonously decreasing, while b_n is increasing, hence any m for which $a_m < b_m$ will separate sequences a_n and b_n. The first such m is $n_0 = 8$. □

Proposition 12.4 *Let* $n \geq 2$, *and assume that* $\mathcal{P}$ *is a* K*-partition of* T_{n-1}. *Then, every* $P \in \mathcal{P}$ *has an antichain of size* $\geq (n-2)!/n$.

Proof As $T_{n-1} = \bigcup_{i=0}^{n-1} L_i$ is a disjoint union and $P = \bigcup_{i=0}^{n-1}(P \cap L_i)$, we have

$$|P| = \sum_{i=0}^{n-1} |P \cap L_i|. \tag{12.22}$$

Let $m = |P \cap L_{i_0}|$ be the greatest of the numbers $|P \cap L_i|$. Then $mn \geq \sum_i |P \cap L_i|$, hence $mn \geq |P|$, so $m \geq t_{n-1}/n^2$. By (12.17b), we have $t_{n-1} \geq n \cdot (n-2)!$, so $m \geq (n-2)!/n$. Hence $P \cap L_{i_0}$ is asymptotically large antichain in P. □

Next proposition states that components of K-partitions spread out high in T_n.

Proposition 12.5 *If* $\mathcal{P}$ *is a* K*-partition of* T_n, $n \geq 2$, *then for each component* $P \in \mathcal{P}$

$$P \cap (L_n \cup L_{n-1}) \neq \emptyset. \tag{12.23}$$

Proof By (12.17c) we have $|P| = \frac{t_n}{n+1} > t_{n-2} = |T_{n-2}|$, hence $P \nsubseteq T_{n-2}$. As $P \subseteq T_n$ and $T_n = T_{n-2} \cup (L_{n-1} \cup L_{n-2})$, it follows $P \cap (L_n \cup L_{n-1}) \neq \emptyset$. □

12.3.1 Invariant Partitions

In this subsection, we consider partitions with invariant (definable) components.

Theorem 12.5 *Assume that $\mathcal{P}$ is a K-partition of T_n, $n \geq 2$. Then, no component $P \in \mathcal{P}$ is invariant subset of T_n.*

Proof The proof consists of cardinal arguments on $|P|$ and cardinalities of subsets of P. Suppose on the contrary that P is invariant. Hence, by Corollary 12.1, for each $k \leq n$, $P \cap L_k \neq \emptyset$ implies $L_k \subseteq P$. First, we prove

$$P \cap L_n = \emptyset. \tag{12.24}$$

Suppose that $P \cap L_n \neq \emptyset$. Then we have $L_n \subseteq P$, so $n! \leq \frac{t_n}{n+1}$, what is a contradiction to (12.17a). Hence, (12.24) follows. Further, by proposition 12.5 and (12.24), it follows $P \cap L_{n-1} \neq \emptyset$, hence

$$L_{n-1} \subseteq P. \tag{12.25}$$

Further, we show

$$P \cap L_{n-2} = \emptyset, \quad P \cap L_{n-3} = \emptyset. \tag{12.26}$$

If $P \cap L_{n-3} \neq \emptyset$, then $L_{n-3} \subseteq P$, so by (12.25) we have $\frac{t_n}{n+1} = |P| \geq (n-1)! + (n-3)!$, contradicting (12.17d). Hence, by 12.1 it follows $P \cap L_{n-3} = \emptyset$. Using similar cardinal argument and $|L_{n-1} \cup L_{n-2}| > |L_{n-1} \cup L_{n-3}|$, one can show $P \cap L_{n-2} = \emptyset$.

By (12.25) and (12.26), it follows

$$P = \bigcup_{i=0}^{n} (P \cap L_i) = L_{n-1} \cup (P \cap T_{n-4}) \tag{12.27}$$

so $|P| = |L_{n-1}| + |P \cap T_{n-4}|$, i.e.,

$$\frac{t_n}{n+1} \leq (n-1)! + t_{n-4}, \tag{12.28}$$

contradicting (12.17e), and this proves the theorem. □

Suppose T_n, $n \geq 2$ has a partition $\mathcal{P}$ with components of equal size $n+1$. Then the existence of such partition obviously also would violate KHP. We show that in that case $\mathcal{P}$ has at most one invariant component. To simplify the notation, we shall prove this statement for T_{n-1} and partitions of T_{n-1} with components of equal size n. In the proof, the central role has corollary 12.1, which states that if $X \subseteq T_n$ is invariant and $X \cap L_k \neq \emptyset$, then $L_k \subseteq X$.

Theorem 12.6 *Let $\mathcal{P}$ be a partition of T_{n-1}, $n \geq 2$, with each component $P \in \mathcal{P}$ having size n. Then, at most one component of $\mathcal{P}$ is invariant.*

Proof Let $P \in \mathcal{P}$ be invariant. Then, $|P| = n$ and by Corollary (12.1) if $L_k \cap P \neq \emptyset$, then $L_k \subseteq P$, so $k! \leq n$. Hence, if $n < k!$, then $L_k \cap P = \emptyset$ and so if p is the least k such that $n < k!$ we have

$$P = P \cap T_n = \bigcup_{k \leq n} (P \cap L_k) = \bigcup_{P \cap L_k \neq \emptyset} (P \cap L_k) \subseteq \bigcup_{k \leq p-1} L_k = T_{p-1}. \quad (12.29)$$

Therefore,

$$(p-1)! \leq n < p! \quad \text{and} \quad P \subseteq T_{p-1}. \quad (12.30)$$

Suppose $P \cap L_{p-1} = \emptyset$. Then, $P \subseteq T_{p-2}$ so $|P| \leq t_{p-2}$, i.e., $n \leq t_{p-2}$. By (12.17a), we have $t_{p-2} < (p-1)!$ and so $n < (p-1)!$. But this is a contradiction to (12.30). Hence, $P \cap L_{p-1} \neq \emptyset$, so

$$L_{p-1} \subseteq P. \quad (12.31)$$

Assume that $Q \in \mathcal{P}$ is invariant and $Q \neq P$. Then $Q \cap P = \emptyset$ and by earlier argument, we also have $L_{p-1} \subseteq Q$, i.e., $Q \cap P \neq \emptyset$, a contradiction. Hence, there is at most one invariant component in $\mathcal{P}$. □

Corollary 12.5 *Let P as in the statement of the previous theorem and assume notation as in the proof of the theorem. Then,*

a. $(p-1)! \leq n < p!$ *and* $L_{p-1} \subseteq P$*;*
b. $P = \bigcup_{i \in I} L_i$ *where I is some subset of* $\{0, 1, 2, \ldots, p-1\}$, $p - 1 \in I$*;*
c. $n = |P| = \sum_{k \in I} k!$.

Numbers that are sums of different factorials we shall call factorial numbers. By the aforementioned corollary, we see that the integer n for which T_{n-1} has n-balanced partition with an invariant component must be a factorial number. On the other hand, if $n | t_{n-1}$ and n is a factorial number, then T_{n-1} has an n-balanced partition $\mathcal{P}$ with an invariant component $\mathcal{P}$. To see this, we take $P = \bigcup_{k \in I} L_k$, where $n = \sum_{k \in I} k!$ for some $I \subseteq \{0, 1, \ldots, p-1\}$. As $n | t_{n-1}$, we have $t_{n-1} = mn$ for some m. Then, other components of $\mathcal{P}$ are obtained by partitioning $T_{n-1} \backslash P$ into $m - 1$ subsets of size n. Hence, we have the following theorem holds.

Theorem 12.7 *Suppose $n | t_{n-1}$. Then T_{n-1} has an n-balanced partition with an invariant component if and only if $n = \sum_{k \in I} k!$ is a factorial number.*

Assuming that $P \in \mathcal{P}$ is invariant, we consider some properties of P.

Proposition 12.6 *Let $P \in \mathcal{P}$ be an invariant component of an n-balanced partition of T_{n-1}, $n \geq 3$. Then*

$$P = S \cup \bigcup_{i \in I} L_i, \tag{12.32}$$

where $I \subseteq \{3, 4, \dots, p-1\}$, and S is exactly one of the sets L_0, L_1, $L_0 \cup L_1$, L_2, and $L_0 \cup L_1 \cup L_2$.

Proof By Corollary 12.5, we have $P = S \cup \bigcup_{i \in I} L_i$, where $S = \emptyset$ or S is a union of some sets L_0, L_1, L_2. Suppose $S = \emptyset$. Then,

$$n = |P| = |\bigcup_{i \in I} L_i| = \sum_{i \in I} |L_i| = \sum_{i \in I} i!, \tag{12.33}$$

so $3|n$ as $i!$ is divisible by 3 for each $i \in I$. As $n|t_{n-1}$, we have $3|t_{n-1}$. But it is easy to see that $t_{n-1} = 1 (\text{mod } 3)$ for $n \geq 3$. Hence, $S \neq \emptyset$. Suppose $S = L_2 \cup L_0$ or $S = L_2 \cup L_1$. Then, again $3|n$ and so $3|t_{n-1}$ what is impossible. □

Let $a_0 = 0$ be the root of T_{n-1} and a_1 the successor of a_0. Then $L_0 = \{a_0\}$ and $L_1 = \{a_1\}$. If partition $\mathcal{P}$ of T_{n-1} has a component $P = L_1 \cup \bigcup_{i \in I} T_i$, $I \subseteq \{3, 4, \dots, p-1\}$, then T_{n-1} also has a partition $\mathcal{P}'$ with component $P' = L_0 \cup \bigcup_{i \in I} T_i$. The partition $\mathcal{P}'$ is obtained from $\mathcal{P}$ by interchanging elements $a_1 \in L_1$ with $a_0 \in Q$ for some $Q \in \mathcal{P}$. Such a Q certainly exists since $\mathcal{P}$ covers T_{n-1}. In a similar way, one can prove that if $P = L_2 \cup \bigcup_{i \in I} T_i$, then T_{n-1} has a partition $\mathcal{P}'$ with component $P' = L_0 \cup L_1 \cup \bigcup_{i \in I} T_i$. Invariant components of $\mathcal{P}$ of the form

$$L_0 \cup \bigcup_{i \in I} T_i, \quad L_0 \cup L_1 \cup \bigcup_{i \in I} T_i, \quad L_0 \cup L_1 \cup L_2 \cup \bigcup_{i \in I} T_i, \quad I \subseteq \{3, 4, \dots, p-1\}, \tag{12.34}$$

will be called, respectively the first, second, and third canonical components. The next proposition states that the existence of invariant component in n-balanced partitions is the unique property of T_n.

Proposition 12.7 *Assume that $\mathcal{P}$ and $\mathcal{Q}$ are two n-balanced partitions of T_{n-1} with invariant canonical components P and Q, respectively. Then, $P = Q$.*

Proof There are sequences of layers L_i and L_j such that $|L_{i_1}| > |L_{i_2}| > \cdots > |L_{i_r}|$ and $|L_{j_1}| > |L_{j_2}| > \cdots > |L_{j_s}|$, where $i_1 < i_2 < \cdots < i_r$ and $j_1 < j_2 < \cdots < j_s$, so that

$$P = L_{i_1} \cup L_{i_2} \cup \cdots \cup L_{i_{r-1}} \cup L_{i_r}, \quad Q = L_{j_1} \cup L_{j_2} \cup \cdots \cup L_{j_{r-1}} \cup L_{j_s}. \tag{12.35}$$

By Theorem 12.7, we have $L_{i_1} = L_{p-1} = L_{j_1}$ for some p. Suppose that i and j sequences are not same. Then $L_{i_k} = L_{j_k}$ for $|Li_k|, |L_{jk}| \geq 3!$ or, as $L_{i_1} = L_{j_1}$, there is the greatest integer $k_0 > 1$ such that $L_{i_{k_0}} = L_{j_{k_0}}$ and $|Li_{k_0}|, |L_{j_{k_0}}| \geq 3!$. Suppose the second case. Then we may suppose $|L_{i_m}| > |L_{j_m}|$, where for $m = k_0 + 1$. As $|P| = |Q|$ and $i_1 = j_1, \ldots, i_{k_0} = j_{k_0}$ we have

$$L_{j_s} \cup L_{j_s-1} \cup \cdots \cup L_{j_m} \subseteq T_{i_m}, \tag{12.36}$$

so by (12.17a), we have $|L_{j_s} \cup L_{j_s-1} \cup \cdots \cup L_{j_{k_0}}| \leq t_{i_m} < i_m! = |L_{i_m}|$, hence $|Q| < |P|$, a contradiction. Hence, we have the first case, so $A = \bigcup_{i_k! \leq 2!} L_{i_k} = \bigcup_{j_k! \leq 2!} L_{i_k} = B$. So A and B are unions of some sets L_0, L_1, L_2. By assumption, A and B are canonical components, and there are three possibilities uniquely corresponding to their cardinal numbers $1, 2, 4$. Hence, parts of i and j sequences for $i_k!$, $j_k! \leq 2!$ are of the same length and consist of the same sets L_i. □

Supposing assumptions of Theorem 12.6, we give an estimate of the position of invariant component $P \in \mathcal{P}$ in the tree T_{n-1}, $\mathcal{P}$ is an n-balanced partition of T_{n-1}. Recall that p is the least k such that $n < k!$ and $P \subseteq T_{p-1}$, see Corollary 12.5.

Theorem 12.8 $p < \ln(n)$.

Proof Let $S = \{k : n < k!\}$. Then $p = \min S$, $(p-1)! \leq n < p!$, and $P \subseteq T_{p-1}$. If $n \leq (\frac{k+1}{\mathrm{e}})^k$, then by Lemma 12.6, we have $n < k!$, so $k \in S$. Therefore, such a k is an upper bound for p, i.e., $p < k$. Our aim now is to find a good upper bound for p having in mind the previous observation. So suppose

$$n \leq \left(\frac{k+1}{\mathrm{e}}\right)^k. \tag{12.37}$$

This inequality is equivalent to

$$\ln(n) \leq k\ln(k+1) - k. \tag{12.38}$$

The function $f(x) = x\ln(x+1) - x$ is monotonously increasing for $x \geq 3$ and $596 > \mathrm{e}^{\mathrm{e}^2-1}$, so for $n \geq n_0 = 596$ and $k \geq \ln(n)$ we have

$$\ln(n) \leq \ln(n)(\ln(n)+1) - \ln(n) \leq k\ln(k+1) - k, \tag{12.39}$$

i.e., $\ln(n) \leq \ln(k)\ln(k+1) - k$. Hence, the inequality (12.37) holds true for $n \geq 596$ and $k \geq \ln(n)$. As the assumption is that $n | t_{n-1}$, by (12.14) we have $n \geq 2^{40} > 596$, so

$$n \leq \left(\frac{k+1}{\mathrm{e}}\right)^k, \quad \text{for } k \geq \ln(n). \tag{12.40}$$

Hence, for $k \geq \ln(n)$, we have $n < k!$. By Lindemann–Weirstrass theorem $\ln(n)$ is transcendental so in fact $k \geq \lfloor \ln(n) \rfloor$, therefore, taking $k = \lfloor \ln(n) \rfloor$

$$p \leq \lfloor \ln(n) \rfloor < \ln(n). \tag{12.41}$$

□

For example, if $n = 2^{40}$, then $p = 27$, so the leaves (maximal elements) of P have height 26, while $\mathrm{ht}(T_n) = 2^{40}$.

12.3.2 Partitions of T_n with Connected Components

In this section, we describe another type of partitions of T_{n-1}, which could not be K-partitions. Namely, we prove that a partition of T_n whose all components are connected subsets of T_n cannot be K-partition of T_n. We remind that connected subsets of T_n are automatically induced subgraphs of T_n. Next lemma is a folklore in graph theory and follows from the fact that trees may be described as connected acyclic graphs. The presented proof might give a new look at this property of trees.

Lemma 12.7 *Connected subset X of T_n is a subtree of T_n.*

Proof Let $X \subseteq T_n$ and assume that there are two distinct minimal elements $a, b \in X$. But then for $c = a \wedge b$, we have $c < a, b$, so $c \notin X$. The path p connecting a, b must contain c, hence $p \nsubseteq X$. Therefore, X has exactly one minimal element, say a_0. As there are no other minimal elements in X, we see that a_0 is in fact the least element in X. Then for each $b \in X$, $\{x \in X : x \leq b\} = [a_0, b]_{T_n}$ is a chain, hence X is a tree with the root a_0. □

Hence, we may identify connected subsets of T_n with subtrees of T_n, and we may introduce the following notion. Suppose $X \subseteq T_n$ and $a \in X$. Then by $S_X(a)$, we denote the set of successors of a in X. We say that a connected subset $X \subseteq T_n$ is a complete subtree of T_n if for every $a \in X$, we have $S_X(a) = S_{T_n}(a)$. We immediately have the following property of a complete subtrees of T_n

$$\text{If } X \text{ is complete subtree of } T_n \text{ with root } a, \text{ then } X = \{x \in T_n : a \leq x\}. \tag{12.42}$$

Hence, if $X \subseteq T_n$ is complete subtree and $a \in X$, then X contains all leaves, i.e., maximal elements b of T_n (or of height n), such that $a \leq b$.

Theorem 12.9 *If $n \geq 2$, there is no K-partition of T_n in which all components are connected subsets of T_n.*

Proof Let $\mathcal{P} = \{P_0, P_1, \ldots, P_n\}$ be an $(n+1)$-partition of T_n, such that each P_i is a connected subset of T_n, $n \geq 2$. Observe that $\mathcal{P}$ has at least three components and each component has at least two elements. By previous lemma, P_i are subtrees of T_n, each having root a_i. Let $Q \in \mathcal{P}$ be such that its root has the highest height k in T_n with respect to the elements $a_0, a_1, \ldots a_n$. By renumbering the components of $\mathcal{P}$, we may assume that $Q = P_0$ with root a_0 and so $\mathrm{ht}(a_0) = k$.

Claim Component P_0 is complete subtree of T_n.

Proof of Claim Assume $a \in P_0$. Obviously, $S_{P_0}(a) \subseteq S_{T_n}(a)$. Let b be a successor of a in T_n. Then $\mathrm{ht}(b) > \mathrm{ht}(a) \geq \mathrm{ht}(a_0) = k$, i.e.

$$\mathrm{ht}(b) > k. \tag{12.43}$$

Suppose $b \notin P_0$. Then for some $P \in \mathcal{P}$, $P \neq P_0, b \in P$. First, suppose for all $c \in P$ that $b \wedge c = b$, hence for all $c \in P$, we have $b \leq c$, i.e., b is the root of tree P. But then (12.43) contradicts to the choice of P_0, that P_0 is a component of $\mathcal{P}$ with the highest root a_0, $\mathrm{ht}(a_0) = k$.

Hence, there is $c \in P$, such that $b \wedge c < b$, and so $c \neq b$. As P is connected, the (unique) path $[b, c]_{T_n} \subseteq P$ connects b and c and $b \wedge c \in [b, c]_{T_n}$. Subset $C = \{x \in P : x \leq b\}$ is a chain and $a, b \wedge c \in C$, so a and $b \wedge c$ are comparable. As b is a successor of a, it follows $b \wedge c \leq a$, so $a \in [b \wedge c, b]_{T_n}$. As $[b \wedge c, b]_{T_n} \subseteq [b, c]_{T_n}$, it follows $a \in [b, c]_{T_n}$, i.e., $a \in P$, and so $P \cap P_0 \neq \emptyset$, a contradiction. □

As an immediate consequence of the claim, we have

$$P_0 = \{x \in T_n : a_0 \leq x\}. \tag{12.44}$$

Claim Assume that b is a maximal element of T_n, and let $P_i \in \mathcal{P}$ be such that $b \in P_i$. Then P_i is a complete subtree of T_n and $\mathrm{ht}(a_i) = k$.

Proof of Claim Since a_i is the root of P_i, we have $a_i \leq b$ and by the choice of component P_0, we have $\mathrm{ht}(a_i) \leq k$. Path $[a_i, b]_{P_i}$ connects a_i and b and $[a_i, b]_{P_i} = [a_i, b]_{T_n}$; hence, there is unique $a' \in L_k \cap [a_i, b]_{P_i}$, and so $\mathrm{ht}(a') = k$. By the argument as in the proof of the previous claim, we infer that $S = \{x \in P_i : a' \leq x\}$ is a complete subtree of T_n, hence $S = \{x \in T_n : a' \leq x\}$, and we have $S \subseteq P_i$. As $\mathrm{ht}(a') = \mathrm{ht}(a_0)$, it follows $|P_i| = |P_0|$, and by (12.44) we have $|S| = |P|$, therefore, $P_i = S$. Obviously, a' is a root of P_i, i.e., $a' = a_i$, so $\mathrm{ht}(a_i) = k$. □

Assume that $a \in L_k$, i.e., $\mathrm{ht}(a) = k$. There is a branch l containing a; hence, there is a maximal element $b \geq a$. By previous claim, there is a component P_i such that $b \in P_i$ and $a = a_i$. Hence, each $a \in L_k$ is a root of some $P \in \mathcal{P}$, i.e., $a = a_i$ for some i. Let $I = \{i \leq n : a_i \in L_k\}$ and $S = \bigcup_{i \in I} P_i$. Then S is a subset of the set of all elements of T_n having height $\geq k$, i.e., $S \subseteq T_n \backslash T_{k-1}$. On the other hand, suppose $\mathrm{ht}(x) \geq k$, $x \in T_n$ and b is a maximal element such that $x \leq b$. Then there is $a_i \in L_k$ such that $b \geq a_i$ and $x \in [a_i, b] \subseteq P_i$. Hence $x \in P_i$. It follows $T_n \backslash T_{k-1} \subseteq S$, so

$$\bigcup_{i \in I} P_i = T_n \backslash T_{k-1}. \tag{12.45}$$

If $k = 0$, then there would be only one component in $\mathcal{P}$ (the case $T_n = T_0$) contradicting that $\mathcal{P}$ is a K partition having at least $n+1 = 3$ components. Similarly, if $k = 1$, then there would be only two components (case $T_n = T_1$), yielding again a contradiction.

Hence, $k \geq 2$, T_{k-1} is uncovered by $\bigcup_{i\in I} P_i$ so there is $P \in \mathcal{P}$ such that $P \cap P_i = \emptyset$ for $i \in I$. Observe that $P \subseteq T_{k-1}$ as $\bigcup_{i\in I} P_i$ covers $T_n \backslash T_{k-1}$. If $k \geq n-1$, then each component P_i would have at most $n+1$ elements, so $t_n = (n+1)|P_i| \leq (n+1)^2$. This is possible only for $n = 0, 1, 2$. By assumption of the theorem $n \geq 2$, so $n = 2, 3$, but this contradicts that T_2 and T_3 have no K-partitions. Hence $k \geq n-2$ and so $P \subseteq T_{k-1} \subseteq T_{n-3}$, i.e., $|P| \leq t_{n-3}$, contradicting (12.17c). This proves the theorem. □

Similarly, as in the case of invariant partitions, we consider n-balanced partitions with connected components.

Theorem 12.10 *There is no $(n+1)$-balanced partition of T_n, all of whose components are connected subsets.*

Proof Due to (12.14), we may assume that $n \geq 5$. Suppose, in contrary, that there is an $(n+1)$-balanced partition $\mathcal{P}$ of T_n so that each $P \in \mathcal{P}$ is a connected subset in T_n. Suppose $P \in \mathcal{P}$. By Lemma 12.7, P is a subtree of T_n and by our assumption $|P| = n+1$. As in the proofs of claims 12.3.2, 12.3.2 one can show that if P contains a maximal element of T_n, i.e., some $b \in L_n$, then P is a complete subtree. Hence, as $a \in L_{n-1}$ has exactly n successors, if $P \cap L_n \neq \emptyset$, then for some $a \in L_{n-1}$

$$P = \{a\} \cup \mathrm{Succ}_{T_n}(a). \tag{12.46}$$

Therefore, for some family P_a of components of $\mathcal{P}$, we have

$$L_n \cup L_{n-1} = \bigcup_{a\in L_{n-1}} P_a. \tag{12.47}$$

So, the rest of components Q of $\mathcal{P}$, i.e., which do not belong to the family $\{P_a : a \in L_{n-1}\}$, covers the tree T_{n-2}. Similarly, as elements of L_{n-2} are maximal in T_{n-2}, if $Q \cap L_{n-2} \neq \emptyset$, then Q is a complete subtree of T_{n-2}. If $b \in L_{n-3} \cap Q$, then b has $n-2$ successors in T_{n-2}, so $\{b\} \cup \mathrm{Succ}_{T_{n-2}}(b)$ has only n elements, not enough to fill out Q. As Q is connected, we see that there is $c \in Q \cap L_{n-3}$, and as it is a complete subtree, for $S = \{x \in T_{n-2} : c \leq x\}$ we have $S \subseteq Q$. Further

$$|S| = 1 + (n-3) + (n-3)(n-2), \tag{12.48}$$

so as $S \subseteq Q$, we have $|S| \leq |P| = n+1$, i.e.,

$$n^2 - 5n + 3 \leq 0. \tag{12.49}$$

But $n^2 - 5n + 3 \geq 0$ for $n \geq 5$, so (12.49) is a contradiction to our assumption $n \geq 5$. □

12.4 Ramsey Theory of $\mathbf{T}_n$

We shall use Ramsey theorem, [24], from combinatorics, which states that there is a monochromatic cliques in any edge labeling (with colours) of a sufficiently large complete graph. We remind that a clique of a graph is a subset of vertices of an undirected graph such that every two distinct vertices in the clique are adjacent.

Theorem 12.11 (F. Ramsey) *Let r and s be positive integers. Then there exists the least positive integer $m = R(r, s)$ for which every blue-red edge coloring of the complete graph on m vertices contains a blue clique on r vertices or a red clique on s vertices.*

The value $R(r, s)$ is called the Ramsey number. It is difficult to compute them, only few of their values are known. The special case $m = n$ is of the particular interest for us. In set theory, a homogeneous set is used for the term clique, while $[X]^2$ denotes partition of a set X into two-element subsets. Then the set-theoretical version of Ramsey theorem is stated in the following way.

Theorem 12.12 (F. Ramsey) *Let X be a finite set and $\pi : [X]^2 \to \{0, 1\}$. Then here is m such that: if $|X| \geq m$ then there is $S \subseteq X$, $|S| \geq n$, so that $\pi|S = const$.*

The next P. Erdös theorem [6], which gives an upper bound for $R(n, n)$, will be particularly useful.

Theorem 12.13 $R(n, n) < 4^n$ *for every* $n \geq 3$.

Both versions of Ramsey theorem, finite or infinite, belong to Ramsey theory, an important branch of combinatorics. Ramsey started this theory, but Erdös further developed and popularized. A good overview of this theory can be found in [9]. We will use also the following technical lemmas.

Lemma 12.8 $n > \mathrm{e}^2 \ln^2 n$ *for all* $n > \mathrm{e}^6$.

Proof It is easy to see that the real function $f(x) = \frac{\ln^2 x}{x}$ has the maximum at $x_0 = \mathrm{e}^2$ and for $x > x_0$ is monotonously decreasing. Hence, if $x > \mathrm{e}^6$, then as $\mathrm{e}^6 > x_0$,

$$\frac{\ln^2 x}{x} < f(\mathrm{e}^6) = \frac{36}{\mathrm{e}^6}, \tag{12.50}$$

i.e., $x > \frac{\mathrm{e}^4}{36}\mathrm{e}^2 \ln^2 x > \mathrm{e}^2 \ln^2 x$ as $\mathrm{e}^4/36 > 1$. □

Lemma 12.9 $\left(\dfrac{n}{\mathrm{e}}\right)^{n-1} > n4^{n \ln \ln n}$ *for all* $n > \mathrm{e}^6$.

Proof Note that $\ln 4 < 2$ and suppose $n > \mathrm{e}^6$. Then we have

$$\frac{2 \ln n}{n} + 1 - \frac{1}{n} < 2. \tag{12.51}$$

By previous Lemma and (12.51), we also have

$$\ln n > 2\ln\ln n + 2 > \ln(4)\ln\ln n + \frac{2\ln n}{n} + 1 - \frac{1}{n}, \tag{12.52}$$

wherefrom

$$n\ln n > n - 1 + 2\ln n + \ln(4)\cdot n\ln\ln n. \tag{12.53}$$

Applying the exponential function e^x on the last inequality, we obtain the statement of the lemma. □

Corollary 12.6 *Assume that $\mathcal{P}$ is a K-partition of T_n and $P \in \mathcal{P}$. Then for $n > \mathrm{e}^6$*

$$|P| > n4^{n\ln\ln n} \quad \textit{and} \quad t_n > n(n+1)n4^{n\ln\ln n}. \tag{12.54}$$

Proof We have $|P| = \frac{t_n}{n+1}$ and $t_n > n! + (n-1)! = (n+1)\cdot(n-1)!$. Hence, by Lemma 12.6 and previous lemma

$$|P| > (n-1)! > \left(\frac{n}{\mathrm{e}}\right)^{n-1} > n4^{n\ln\ln n}. \tag{12.55}$$

□

In graph theory, the distance $d(a, b)$ between two vertices a and b in a graph G is the number of edges in a shortest path connecting them. This is also known as the geodesic distance or shortest-path distance. In trees, there is only one path between a and b; hence, it is also the shortest path connecting them. Also, the number of edges on that path is equal to the number of elements in $[a, b]_G$ minus 1, hence $d(a, b) = |[a, b]_G| - 1$. The distance function $d(x, y)$ defines a metric on G, so it also introduces a kind of geometry on G. For example, a circle with the center a and radius r, r is a non-negative integer, would be $C_a(r) = \{x \in G : d(a, x) = r\}$, while the disk bounded by that circle would be $D_a(r) = \{x \in G : d(a, x) \le r\}$.

Proposition 12.8 *Assume that a is a maximal element of T_n and $n \ge 4$. Then, $|D_a(3)| = 2n + 1$.*

Proof Let b be the predecessor of a, c of b, and d of c. Then the distances of b, c, and d from a are respectively 1, 2, and 3 and $b \in L_{n-1}$, $c \in L_{n-2}$, and $d \in L_{n-3}$. The only element of T_n having distance one from a is b. An element $x \in T_n$ having distance 2 from a must be c, or it is a successor of b. Finally, if x has distance 3 from a, then $x = d$, or x is a successor of c. Therefore, counting the number of elements having distances 0, 1, 2, and 3, respectively, from a, we find

$$|D_3(a)| = 1 + 1 + (1 + (n-1)) + (1 + (n-2)) = 2n + 1. \tag{12.56}$$

□

The next theorem states that for large n, if $\mathcal{P}$ is a partition of T_n, then every component P of $\mathcal{P}$ has a relatively large subset, which is sparse.

Theorem 12.14 *Let $\mathcal{P}$ be a K-partition of T_n and $P \in \mathcal{P}$. Then P has a subset S so that*

1. $|S| \geq \frac{1}{2} \log_2 n + n \ln \ln n$;
2. $d(x, y) \geq 4$ *for all* $x, y \in S$, $x \neq y$.

Proof Due to (12.14), we may assume $n \geq 2^{40}$. Let

$$m = \left\lfloor \frac{1}{2} \log_2 n + n \ln \ln n \right\rfloor. \tag{12.57}$$

Then, by Corollary 12.6 and Theorem 12.13, we have

$$|P| > 4^m > R(m, m). \tag{12.58}$$

Let us define $\pi : [P]^2 \to \{0, 1\}$ for $x, y \in T_n$, $x \neq y$, in the following way:

- If $d(x, y) \leq 3$, then $\pi(\{x, y\}) = 0$;
- If $d(x, y) \geq 4$, then $\pi(\{x, y\}) = 1$.

By (12.58) and Theorem 12.13, which assumes Ramsey theorem, there is homogeneous $S \subseteq P$ of the cardinality

$$|S| = m, \tag{12.59}$$

which is homogeneous for π, hence $\pi|S = const$. We prove that $\pi|S = 1$.

Assume, on the contrary, that $\pi|S = 0$ and let $a \in S$ be of the greatest height $\mathrm{ht}(a) = k$, so $S \subseteq T_k$. As S is homogeneous, we have $\pi(\{a, x\}) = 0$, i.e., $d(a, x) \leq 3$ for all $x \in S$, so $S \subseteq D_3^{T_k}(a)$, the disk of a in T_k of radius 3. The element a is maximal in T_k, therefore by Proposition 12.8, it follows $|S| \leq 2k + 1$, hence as $k \leq n$, we have

$$|S| \leq 2n + 1. \tag{12.60}$$

□

Claim $m \geq 2n + 2$.

Proof of Claim Let

$$f(x) = \frac{1}{2} \log_2 x + x \ln \ln x - 2x - 2 = \frac{1}{\ln(4)} \ln x + x \ln \ln x - 2x - 2. \tag{12.61}$$

Then, $f'(x) = \ln \ln x + \frac{1}{\ln x} + \frac{1}{\ln(4) x} - 2$, so $f'(\mathrm{e}^{\mathrm{e}^2}) = 2 + \mathrm{e}^{-2} + \frac{\mathrm{e}^{-2}}{\ln(4)} - 2 > 0$. As for $x \geq \mathrm{e}^{\mathrm{e}^2}$ $f'(x) \geq f'(\mathrm{e}^{\mathrm{e}^2})$, we have $f'(x) > 0$, so $f(x)$ is monotonously increasing for $x \geq e^{e^2}$. Hence, for $x \geq e^{e^e}$, we have $f(x) > f(e^{e^e}) = \frac{1}{\ln(4)}e^e + e \cdot e^{e^e} - 2e^{e^e} - 2 > 0$, so as as $2^{40} > e^{e^e} > e^{e^2}$

$$\frac{1}{2}\log_2 x + x \ln\ln x > 2x + 2, \quad x > 2^{40}. \tag{12.62}$$

Hence,

$$m = \left\lfloor \frac{1}{2}\log_2 n + n \ln\ln n \right\rfloor > \frac{1}{2}\log_2 n + n \ln\ln n - 1 > (2n+2) - 1 = 2n + 1. \tag{12.63}$$

□

By claim, we have $|S| = m > 2n + 1$ what is a contradiction to (12.60). Hence $\pi|S = 1$, i.e., for all $x, y \in S$, $x \neq y$, we have $d(x, y) \geq 4$. □

In fact by (12.63), we have

Corollary 12.7 *For all K-partitions of T_n, $|S| \geq n \ln\ln n$.*

Example 12.1 Let $n \geq 2^{40}$. Then, assuming notation from the proof of previous theorem, we have $|S| = m \geq \lfloor \frac{1}{2}\log_2(2^{40} + n \ln\ln 2^{40} \rfloor = 20 + 3n > 3n$.

References

1. Andrejić, V., Bostan, A., Tatarević, M.: Improved algorithms for left factorial residues. Inf. Process. Lett. **167**, 106078 (2021)
2. Barsky, D., Benzaghou, B.: Nombres de Bell et somme de factorielles. J. Théor. Nombres Bordx. **16**, 1–17 (2004)
3. Bašić, M.: On the combinatorics of faces of trees and anodyne extensions of dendroidal sets. Glasnik Matematički **53**(2), 403–436 (2018). arXiv:1801.09311v2
4. Chang, C.C., Keisler, J.: Model Theory. Amsterdam, North Holland (1990)
5. Jordan, C.: Sur les assemblages de lignes. J. Reine Angew. Math. **70**, 185–190 (1869)
6. Erdős, P.: Some remarks on the theory of graphs. Bull. Am. Math. Soc. **53**, 292–294 (1947)
7. Fabiano, N., Gardašević-Filipović, M., Mirkov, N., Todorčević, V., Radenović, S.: On the Distribution of Kurepa's Function. Axioms **11**, 388 (2022)
8. Feldmann, A.E., Foschini, L.: Balanced Partitions of Trees and Applications. Algorithmica **71**, 354–376 (2013)
9. Graham, R.L., Nešetřil, J.: Ramsey Theory in the Work of Paul Erdős. In: The Mathematics of Paul Erdős II. Springer, Berlin (1997)
10. Gallardo, R.H.: Artin-Schreier, Erdős, and Kurepa's conjecture. Rad HAZU. Matematičke znanosti **27**(555), 111–121 (2023)
11. Guy, R.: Unsolved problems in number theory. Springer, New York (2004)
12. Ivić, A., Mijajlović, Ž.: On Kurepa problems in number theory. Publ. Inst. Math. **57**(71), 19–28 (1995)
13. Jech, T.: Set Theory. Springer, Berlin (2002)
14. Kellner, B.C.: Some remarks on Kurepa's left factorial (2004). arXiv:math/0410477v1
15. De Koninck, J.M., Mercier, A.: 1001 Problems in Classical Number Theory. AMS, New York (2007)

16. Kunen, K.: Set Theory: An Introduction to Independence Proofs. North-Holland, Amsterdam (1980)
17. Kurepa, Đ.: On the left factorial function !N. Math. Balkan. **1**, 147–153 (1971)
18. Kurepa, Đ.: Left factorial function in complex domain. Math. Balkanica (N.S.) **3**, 297–307 (1973)
19. Kurepa, Đ.: On some new left factorial propositions. Math. Balkanica (N.S.) **4**, 383–386 (1974)
20. Mijajlović, Ž., Malešević, B.: Differentially transcendental functions. Bull. Belg. Math. Soc. Simon Stevin **15**(2), 193–201 (2008)
21. Mijajlović, Ž.: Fifty years of Kurepa's !n hypothesis. Bull. Acad. Serbe Sci. Art. **46**, 169–182 (2021)
22. Milovanović, G.V.: Expansions of the Kurepa function. Publ. Inst. Math. (Beograd) (N.S.) **57**(71), 81–90 (1995)
23. Milovanović, G.V.: A sequence of Kurepa's functions. Scientific Review **19-20**, 137–146 (1996)
24. Ramsey, F.P.: On a Problem in Formal Logic. Proc. Lond. Math. Soc. **30**, 264–286 (1929)
25. Sándor, J., Crstici, B.: Handbook of Number Theory II. Kluwer, Dordrecht (2004)

Chapter 13
Karamata's Theory, Asymptotic Relations, and the Generalized Inverse

Rale M. Nikolić, Valentina Timotić, and Dragan Djurčić

Dedicated to Academician Gradimir V. Milovanović on the occasion of his 75th birthday

13.1 Introduction and Preliminaries

Many problems from mathematical analysis and its applications (e.g., in probability theory) are related to finding an answer to the question under what conditions the following implication

$$\lim_{t\to\infty}\frac{g(t)}{f(t)}=a \quad\Rightarrow\quad \lim_{t\to\infty}\frac{g^{(-1)}(t)}{f^{(-1)}(t)}=b$$

is satisfied, where $f(t)\to+\infty$ and $g(t)\to+\infty$, when $t\to+\infty$, and $f^{(-1)}$ and $g^{(-1)}$ are certain "inverse maps" of f and g, respectively, and $a,b\in[0,+\infty]$. Djurčić and Torgašev [10] studied this problem in the case that functions $f^{(-1)}$ and $g^{(-1)}$ are inverses or generalized inverses of functions f and g, respectively, and

R. M. Nikolić (✉)
Department of Mathematics and Natural Sciences, Military Academy, University of Defence, Belgrade, Serbia
e-mail: rale.nikolic@va.mod.gov.rs

V. Timotić
Department of Mathematics, Faculty of Philosophy, University of East Sarajevo, Pale, Bosnia and Herzegovina
e-mail: valentina.timotic@ffuis.edu.ba

D. Djurčić
Department of Mathematics, Faculty of Technical Sciences, University of Kragujevac, Čačak, Serbia
e-mail: dragan.djurcic@ftn.kg.ac.rs

M. P. Stanić et al. (eds.), *Analysis, Approximation, Optimization: Computation and Applications*, Springer Optimization and Its Applications 224,
https://doi.org/10.1007/978-3-031-85743-0_13

Buldygin et al. [5, 7] in the case that functions $f^{(-1)}$ and $g^{(-1)}$ are inverses, quasi-inverses, or asymptotic quasi-inverses of functions f and g, respectively.

Now, we will consider the next class of functions:

$$\mathcal{A} = \{f : [a, +\infty) \to (0, +\infty)(a > 0) \mid f \text{ is a nondecreasing and unbounded function}\}.$$

We will investigate the properties of generalized inverse of a function which belongs to the class $\mathcal{A}$. Firstly, we will give the definition of generalized inverse.

Definition 13.1 ([4]) Let $f \in \mathcal{A}$ be any function. Then the function

$$f^{\leftarrow}(x) = \inf\{y \geq a \mid f(y) > x\}$$

is called the generalized inverse of the function f, for $x \geq f(a)$.

If $f \in \mathcal{A}$ is a continuous and strictly increasing function, then $f^{\leftarrow}(x) = f^{-1}(x)$, for every $x \geq f(a)$. Besides, $f^{\leftarrow} \in \mathcal{A}$, whenever $f \in \mathcal{A}$. Furthermore, for any right continuous function $g \in \mathcal{A}$, there exists a function $f \in \mathcal{A}$ ($f(x) = g^{\leftarrow}(x)$, for $x \geq g(a)$) such that $g = f^{\leftarrow}$.

Also, an important question is whether a function f that satisfies certain properties and its generalized inverse $f^{\leftarrow}$ (or the inverse of the function in some other sense) belong to the same class of functions studied in Karamata's theory of regular variability or some of its extensions (discussed later). In other words, the question arises as to which properties should be assigned to a function f, so that certain characterizations can be obtained in this sense. The simplest case is when the function f is continuous, is strictly increasing, and belongs to the class of regularly varying functions with a positive index of variability, which will be discussed later in this section. It is well known that then the inverse function f^{-1} of a function f belongs to that same class. More about this can be seen in, e.g., [2, 7, 13].

In the following sections, we will deal with the question of the relationship between the generalized inverse, the weak asymptotic equivalence relation, and the strong asymptotic equivalence relations in certain classes of functions studied in Karamata's theory of regular variability. Therefore, we introduce the notions of weak asymptotic equivalence relation and strong asymptotic equivalence relations.

For $f \in \mathcal{A}$, we will consider the set

$$\{f\} = \{g \in \mathcal{A} \mid f(x) \asymp g(x), \text{ when } x \to +\infty\},$$

where $f(x) \asymp g(x)$, when $x \to +\infty$, denotes the weak asymptotic equivalence relation defined by

$$0 < \underline{\lim}_{x \to +\infty} \frac{f(x)}{g(x)} \leq \overline{\lim}_{x \to +\infty} \frac{f(x)}{g(x)} < +\infty$$

(see, e.g., [4]).

Also, for $f \in \mathcal{A}$, we will consider the set

$$[f]_{\sim} = \{g \in \mathcal{A} \mid f(x) \sim g(x), \text{ when } x \to \infty\},$$

where $f(x) \sim g(x)$, when $x \to +\infty$, denotes the strong asymptotic equivalence relation defined by

$$\lim_{x \to +\infty} \frac{f(x)}{g(x)} = 1.$$

In the following sections, extensions of the next theorem will be given, which is a modified combination of some results from Balkema et al. [3] (also see Bingham et al. [4, p. 190, 14 (ii), (iii))]. Therefore, we state it as follows.

Theorem A *Let $f, g \in \mathcal{A}$, and let f be a regularly varying function with a positive index of variability:*

(a) *If $g \in [f]_{\sim}$, then $g^{\leftarrow} \in [f^{\leftarrow}]_{\sim}$.*
(b) *If $g^{\leftarrow} \in [f^{\leftarrow}]_{\sim}$, then $g \in [f]_{\sim}$.*

At the end of this section, we will give an overview of the classes of regularly varying functions, in which the results in the following sections were obtained. Note that when we use the notion "measurability," we will consider it to be "measurability in Lebesgue's sense." Firstly, for measurable function $f : [a, +\infty) \to (0, +\infty)$ $(a > 0)$, we will introduce the next two functions:

$$\overline{k}_f(l) := \overline{\lim_{x \to +\infty}} \frac{f(lx)}{f(x)} < +\infty, \quad l > 0, \tag{13.1}$$

and

$$\underline{k}_f(l) := \underline{\lim}_{x \to +\infty} \frac{f(lx)}{f(x)} > 0, \quad l > 0. \tag{13.2}$$

$\overline{k}_f(l)$, $l > 0$, is called the index function of f, and $\underline{k}_f(l)$, $l > 0$, is called the auxiliary index function of f. The Tauber condition generated by (13.1) (or (13.2)) is a very important convergence condition in the theory of Tauberian theorems (see, e.g., [2] and [9]), as well as in asymptotic analysis in general (see paper [4]).

13.1.1 RV Class

In 1930 Karamata [20] introduced the notion of regular variability of a function and proved some of the fundamental theorems (see also [21]). These results, along with

with later extensions and generalizations, were suitable for applications in many areas of mathematics (see, e.g., [4, 23, 25]).

A measurable function $f : [a, +\infty) \to (0, +\infty)$, $a > 0$, is called regularly varying if

$$\overline{k}_f(l) = \underline{k}_f(l) = \chi(l) \in (0, +\infty)$$

is satisfied, for every $l > 0$. The class of all such functions is denoted by *RV* (abbreviated from regularly varying). For any regularly varying function *f*, it holds that $\chi(l) = l^\rho$, $l > 0$, for some real number ρ which is called the index of the function *f*. Specially, the class of all regularly varying functions with index ρ is denoted by RV_ρ.

The case when $\rho = 0$ corresponds to the so-called slowly varying functions (some applications for this case can be seen in [26]). The class of all such functions is denoted by *SV* (from the abbreviation for slowly varying). It is obvious that $SV \subsetneq RV$. The classes *RV* (and *SV*) are the most important objects in Karamata's theory of regular variability (about this can be seen, e.g., in [25]) and its applications (see also, e.g., [12] and [4]).

The case when $\rho = +\infty$ corresponds to the so-called rapidly varying functions. Actually, a function f is called rapidly varying if

$$\lim_{x \to \infty} \frac{f(lx)}{f(x)} = +\infty$$

holds, for every $l > 1$. The class of all such functions is denoted by R_∞. Some applications for this case can be seen in [27].

13.1.2 ORV Class

After the mentioned paper of Karamata [20] from 1930, many generalizations of the notion of regular variability were obtained. One of the generalizations was given by Avakumović [2], which was further investigated by Karamata [22], Feller [18], and Aljančić and Arandjelović [1]. The class of functions that was studied by the mentioned, as well as many other authors, is known as the class of *O*-regularly varying functions or the *ORV* class of functions defined on $[a, +\infty)$, $a > 0$.

A measurable function $f : [a, +\infty) \to (0, +\infty)$, $a > 0$, is called *O*-regularly varying if

$$\overline{k}_f(l) < +\infty$$

is satisfied, for every $l > 0$. This class is an important object in asymptotic analysis (see, e.g., [4] and [25]). Connections between regular variation, indefinite operator theory, and HELP-type inequalities can be found in Fleige [19]. A very important subclass of the *ORV* class is the *IRV* class. This class contains all functions from ORV whose index functions are continuous. It holds that $SV \subsetneq RV \subsetneq IRV \subsetneq ORV$ (see [10]).

13.1.3 PI Class

A measurable function $f : [a, +\infty) \to (0, +\infty)$, $a > 0$, is called positively increasing if

$$\underline{k}_f(l_0) > 1 \tag{13.3}$$

holds for some $l_0 > 1$. The class of all such functions is denoted by PI.

A measurable function $f : [a, +\infty) \to (0, +\infty)$, $a > 0$, belongs to the class PI^* if there exists $\lambda_0 \geq 1$ such that

$$\underline{k}_f(\lambda) > 1$$

holds, for every $\lambda > \lambda_0$. The class PI^* is a subclass of PI. In the case that $\lambda_0 = 1$, we get the class ARV, which is a very important strict subclass of the class PI^*. A function $f \in ARV$ is called rapidly varying (in the sense of de Haan) with index of variability $+\infty$ (i.e., it belongs to R_∞) if $\underline{k}_f(\lambda) = +\infty$, for every $\lambda > 1$. The class PI^* also contains, as a strict subclass, the class of regularly varying functions whose index of variability ρ is positive, as well as the class of rapidly varying functions, denoted R_∞, whose de Haan index is $+\infty$, but it does not contain any functions from the class SV. Also, the class $\mathcal{A} \cap PI^*$ is equal to the class $\mathcal{A} \cap PI$.

13.2 The Weak Asymptotic Equivalence and the Generalized Inverse

As already was mentioned in the previous section, we will make certain generalizations of Theorem A. Some extensions of this theorem can be found in Buldygin et al. [5], Djurčić and Torgašev [10], and Djurčić et al. [14]. Also, some modifications of this theorem are also given in Buldygin et al. [6] and Djurčić and Torgašev [13], but in these papers a different asymptotic relation was observed, in the process of inverting functions, in regards to the one that will be observed in this section. One extension of Theorem A for the weak asymptotic equivalence relation for continuous and strictly increasing functions from the class $\mathcal{A}$ was proved in Djurčić and Torgašev [11].

In this section, we will prove the theorems that extend the previously mentioned results and which establish the relationship between the weak asymptotic equivalence relation and the generalized inverse in the functional class $\mathcal{A}$.

Theorem 13.1 ([15]) *Let $f, g \in \mathcal{A}$ and $f^{\leftarrow} \in ORV$. If additionally $g \in \{f\}$, then $g^{\leftarrow} \in \{f^{\leftarrow}\}$.*

Proof If $f, g \in \mathcal{A}$ and $f^{\leftarrow} \in ORV$, $g \in \{f\}$, then there are some constants $m, M \in \mathbb{R}^{+}$ $(m \leq M)$ such that $m \leq g(x)/f(x) \leq M$, $x \geq x_0$. Hence, $g(x) \leq Mf(x)$, for $x \geq x_0$. Therefore, $g^{\leftarrow}(x) \geq f^{\leftarrow}(x/M), x \geq x_0$, i.e., $g^{\leftarrow}(x)/f^{\leftarrow}(x) \geq f^{\leftarrow}(x/M)/f^{\leftarrow}(x), x \geq x_0$. We also have $g(x) \geq mf(x), x \geq x_0$. Hence, $g^{\leftarrow}(x) \leq f^{\leftarrow}(x/m)$ holds for $x \geq x_0$, so that $g^{\leftarrow}(x)/f^{\leftarrow}(x) \leq f^{\leftarrow}(x/m)/f^{\leftarrow}(x)$ for such x. Therefore, for $x \geq x_0$ it holds

$$\frac{f^{\leftarrow}(x/M)}{f^{\leftarrow}(x)} \leq \frac{g^{\leftarrow}(x)}{f^{\leftarrow}(x)} \leq \frac{f^{\leftarrow}(x/m)}{f^{\leftarrow}(x)}.$$

Hence, we have

$$\underline{k}_{f^{\leftarrow}}\left(\frac{1}{M}\right) \leq \underline{\lim}_{x\to+\infty} \frac{g^{\leftarrow}(x)}{f^{\leftarrow}(x)} \leq \underline{k}_{f^{\leftarrow}}\left(\frac{1}{m}\right)$$

and

$$\overline{k}_{f^{\leftarrow}}\left(\frac{1}{M}\right) \leq \overline{\lim}_{x\to+\infty} \frac{g^{\leftarrow}(x)}{f^{\leftarrow}(x)} \leq \overline{k}_{f^{\leftarrow}}\left(\frac{1}{m}\right).$$

Therefore, it follows that

$$0 < \underline{k}_{f^{\leftarrow}}\left(\frac{1}{M}\right) \leq \underline{\lim}_{x\to+\infty} \frac{g^{\leftarrow}(x)}{f^{\leftarrow}(x)} \leq \overline{\lim}_{x\to+\infty} \frac{g^{\leftarrow}(x)}{f^{\leftarrow}(x)} \leq \overline{k}_{f^{\leftarrow}}\left(\frac{1}{m}\right)$$

holds, and hence, $g^{\leftarrow} \in \{f^{\leftarrow}\}$. □

Remark 13.1 ([15]) In general case, if $g^{\leftarrow} \in \{f^{\leftarrow}\}$, then it is not necessarily $g \in \{f\}$. Take, for example, $f(x) = \mathrm{e}^x$, $x \geq 1$, and $g(x) = \sqrt{\mathrm{e}^x}$, $x \geq 1$.

Theorem 13.2 ([15]) *Let $f, g \in \mathcal{A}$. If $g^{\leftarrow} \in \{f^{\leftarrow}\}$ for every $g \in \{f\}$, and then $f^{\leftarrow} \in ORV$ ($g^{\leftarrow} \in ORV$).*

Proof Let $f \in \mathcal{A}$. Then, for every $g \in \mathcal{A}$ it follows that $g^{\leftarrow} \in \{f^{\leftarrow}\}$ whenever $g \in \{f\}$. For arbitrary and fixed $\lambda > 0$, consider the function $g_1(x) = \lambda f(x), x \geq a$. Then $g_1 \in \{f\}$, so it follows that $g_1^{\leftarrow} \in \{f^{\leftarrow}\}$. Since $g_1^{\leftarrow}(x) = f^{\leftarrow}(x/\lambda), x \geq \lambda a$, we have

$$+\infty > M(\lambda) \geq \overline{\lim}_{x\to+\infty} \frac{f^{\leftarrow}(x)}{f^{\leftarrow}(x/\lambda)} = \overline{\lim}_{t\to+\infty} \frac{f^{\leftarrow}(\lambda t)}{f^{\leftarrow}(t)} = \overline{k}_{f^{\leftarrow}}(\lambda).$$

Therefore, $f^{\leftarrow} \in ORV$.

Next, for the same $g \in \mathcal{A}$ we have $g^{\leftarrow} \in \{f^{\leftarrow}\}$, so it follows that $g^{\leftarrow}(x) = h(x)f^{\leftarrow}(x)$, $x \geq f(a)$, where

$$0 < \frac{1}{A(g)} \leq h(x) \leq A(g) < +\infty$$

holds, for every $x \geq f(a)$. Since

$$\overline{k}_{g^{\leftarrow}}(\lambda) \leq \overline{k}_{f^{\leftarrow}}(\lambda)A^2(g) < +\infty, \quad \lambda > 0,$$

is satisfied, we get $g^{\leftarrow} \in ORV$. □

The following Theorem 13.3 and 13.4 have their analogous counterparts, when the asymptotic quasi-inverse is considered instead of the generalized inverse (see, e.g., [8], Theorems 4.1 and 4.2). To prove them we need the following lemma.

Lemma 13.1 ([15]) *Let $f \in PI^*$. Then there is a $\lambda_0 \geq 1$ and at least one function $c(\lambda) > 1$ for every $\lambda > \lambda_0$, depending on f, such that $\lim_{\lambda\to+\infty} c(\lambda) = +\infty$ and $f(\lambda x) \geq c(\lambda)f(x)$ for every $x \geq x_0(\lambda) > 0$.*

Proof Let $\underline{k}_f(\lambda) = \underline{\lim}_{x\to+\infty} \frac{f(\lambda x)}{f(x)}$, where $\lambda > \lambda_0 \geq 1$, and let us define $c(\lambda) = \frac{1}{2}(\underline{k}_f(\lambda) + 1)$ for that λ. Then $c(\lambda) > 1$, $\lambda > \lambda_0$.

Next, let $\lambda_1 > \lambda_0 \geq 1$. If $\lambda \in [\lambda_1^n, \lambda_1^{n+1})$ and $n \geq 2$, we get

$$\begin{aligned}\underline{k}_f(\lambda) &\geq \underline{\lim}_{x\to+\infty} \frac{f(\lambda_1 x)}{f(x)} \cdots \underline{\lim}_{x\to+\infty} \frac{f(\lambda_1^{n-1}x)}{f(\lambda_1^{n-2}x)} \cdot \underline{\lim}_{x\to+\infty} \frac{f(\lambda x)}{f(\lambda_1^{n-1}x)} \\ &= \left(\underline{k}_f(\lambda_1)\right)^{n-1} \cdot \underline{k}_f\left(\frac{\lambda}{\lambda_1^{n-1}}\right),\end{aligned}$$

which gives $\lim_{\lambda\to+\infty} \underline{k}_f(\lambda) = +\infty$, so that $\lim_{\lambda\to+\infty} c(\lambda) = +\infty$. □

Theorem 13.3 ([15]) *Let $f \in \mathcal{A}$. Then $f \in PI^*$ if and only if $f^{\leftarrow} \in ORV$.*

Proof Firstly, assume $f \in \mathcal{A} \cap PI^*$. Then for some $\lambda_0 \geq 1$ and for every $\lambda > \lambda_0$, it holds $f(\lambda x) \geq c(\lambda)f(x)$, $x \geq x_0 = x_0(\lambda)$, where $c(\lambda) = c_f(\lambda) > 1$ for $\lambda > \lambda_0$. Hence, for these λ and x we have $f(\lambda x)/c(\lambda) \geq f(x)$, so it follows that $f^{\leftarrow}(c(\lambda)x)/\lambda \leq f^{\leftarrow}(x)$ for sufficiently large x. Therefore, for this λ we have $\overline{k}_{f^{\leftarrow}}(c(\lambda)) \leq \lambda < +\infty$. Thus, $f^{\leftarrow} \in ORV$.

Next, assume that $f^{\leftarrow} \in ORV \cap \mathcal{A}$. Then by [1] we have

$$\overline{\lim_{x\to+\infty}} \sup_{\lambda\in[1,2]} \frac{f^{\leftarrow}(\lambda x)}{f^{\leftarrow}(x)} = \overline{\lim_{x\to+\infty}} \frac{f^{\leftarrow}(2x)}{f^{\leftarrow}(x)} = \overline{k}_{f^{\leftarrow}}(2) > 1.$$

For every $\varepsilon > 0$ there is $x_0 = x_0(\varepsilon) > 0$ such that

$$\sup_{\lambda\in[1,2]} \frac{f^{\leftarrow}(\lambda x)}{f^{\leftarrow}(x)} \le \overline{k}_{f^{\leftarrow}}(2) + \varepsilon = M(\varepsilon), \quad x \ge x_0,$$

so that for every $x \ge x_0$ and every $\lambda \in [1, 2]$ we have

$$\frac{f^{\leftarrow}(\lambda x)}{f^{\leftarrow}(x)} \le M(\varepsilon).$$

Now it follows that

$$\begin{aligned}
\frac{f^{\leftarrow}(\lambda x)}{M(\varepsilon)} \le f^{\leftarrow}(x) \quad &\Rightarrow \quad \left(\left(\frac{f(M(\varepsilon)x)}{\lambda}\right)^{\leftarrow}\right)^{\leftarrow} \ge (f^{\leftarrow}(x))^{\leftarrow}, \\
&\Rightarrow \quad f(x) \le \frac{f(M^2(\varepsilon)x)}{\lambda}, \\
&\Rightarrow \quad \frac{f(M^2(\varepsilon)x)}{f(x)} \ge \lambda, \\
&\Rightarrow \quad \frac{f(M^2(\varepsilon)x)}{f(x)} \ge 2 > 1, \\
&\Rightarrow \quad \underline{\lim}_{x\to+\infty} \frac{f(M^2(\varepsilon)x)}{f(x)} = \underline{k}_f(M^2(\varepsilon)) \ge 2 > 1.
\end{aligned}$$

Since $\underline{k}_f(s)$ is nondecreasing for $s > 0$, we find that $\underline{k}_f(\lambda) > 1$, for $\lambda > M^2(\varepsilon) > 1$. Hence, $f \in PI^* \cap \mathcal{A}$. □

Corollary 13.1 ([15]) *Let $f \in \mathcal{A}$. Then $f, f^{\leftarrow} \in ORV$ if and only if $f \in ORV \cap PI^*$ ($f^{\leftarrow} \in ORV \cap PI^*$).*

Theorem 13.4 ([15]) *Let $f \in \mathcal{A}$. Then $f \in ORV$ if and only if $f^{\leftarrow} \in PI^*$.*

Proof Firstly, assume that $f \in \mathcal{A} \cap ORV$. Then, by [1],

$$\overline{\lim}_{x\to+\infty} \sup_{\lambda\in[1,2]} \frac{f(\lambda x)}{f(x)} = \overline{\lim}_{x\to+\infty} \frac{f(2x)}{f(x)} = \overline{k}_f(2) \ge 1.$$

For every $\varepsilon > 0$, there is $x_0 = x_0(\varepsilon) > 0$ such that

$$\sup_{\lambda\in[1,2]} \frac{f(\lambda x)}{f(x)} \le \overline{k}_f(2) + \varepsilon = m(\varepsilon), \quad \text{for every } x \ge x_0,$$

so for the same x and for every $\lambda \in [1, 2]$, we have $f(\lambda x)/f(x) \le m(\varepsilon)$. Therefore, $f(\lambda x)/m(\varepsilon) \le f(x)$, and it follows that

$$\begin{aligned}
\frac{f^{\leftarrow}(m(\varepsilon)x)}{\lambda} \ge f^{\leftarrow}(x) \quad &\Rightarrow \quad f^{\leftarrow}(m(\varepsilon)x) \ge \lambda f^{\leftarrow}(x),\\
&\Rightarrow \quad f^{\leftarrow}(m(\varepsilon)x) \ge 2 f^{\leftarrow}(x),\\
&\Rightarrow \quad \frac{f^{\leftarrow}(m(\varepsilon)x)}{f^{\leftarrow}(x)} \ge 2 > 1,\\
&\Rightarrow \quad \underline{\lim}_{x\to+\infty} \frac{f^{\leftarrow}(m(\varepsilon)x)}{f^{\leftarrow}(x)} \ge 2 > 1,\\
&\Rightarrow \quad \underline{k}_{f^{\leftarrow}}(m(\varepsilon)) < 1.
\end{aligned}$$

Hence, $\underline{k}_{f^{\leftarrow}}(\lambda) > 1$ for $\lambda > m(\varepsilon) = \lambda_0 \ge 1$, so it follows that $f^{\leftarrow} \in PI^*$.

Next, assume that $f^{\leftarrow} \in PI^* \cap \mathcal{A}$. Then, for some $\lambda_0 \ge 1$ and every $\lambda > \lambda_0$ we have $f^{\leftarrow}(\lambda x) \ge c(\lambda) f^{\leftarrow}(x)$, for every $x \ge x_0 = x_0(\lambda)$, where $c(\lambda) = c_f(\lambda) > 1$, $\lambda > \lambda_0$. Hence, for those λ and x we have $f^{\leftarrow}(\lambda x)/c(\lambda) \ge f^{\leftarrow}(x)$, so that $\big(f(c(\lambda)x)/\lambda\big)^{\leftarrow} \ge f^{\leftarrow}(x)$. Then, similarly as in the previous proof, we have $f(c(\lambda)x)/\lambda \le f\big(\sqrt{c(\lambda)}x\big)$. Therefore, $f(c(\lambda)x)/f\big(\sqrt{c(\lambda)}x\big) \le \lambda$, and consequently, for a fixed $\lambda > \lambda_0$, we obtain $\overline{k}_f\big(\sqrt{c(\lambda)}\big) \le \lambda < +\infty$. Hence, we get $f \in ORV$. □

Corollary 13.2 ([15]) *Let $f \in \mathcal{A}$. Then $f, f^{\leftarrow} \in PI^*$ if and only if $f \in ORV \cap PI^*$ ($f^{\leftarrow} \in ORV \cap PI^*$).*

Theorem 13.5 ([15]) *If $f \in \mathcal{A} \cap ORV$, then $f(x) \asymp (f^{\leftarrow}(x))^{\leftarrow}$, when $x \to +\infty$, i.e., $f \in \{(f^{\leftarrow})^{\leftarrow}\}$.*

Proof Since $f(x) \le (f^{\leftarrow}(x))^{\leftarrow} \le f(\beta x)$ holds for $x \ge 0$ and $\beta > 1$, we obtain

$$1 \le \frac{(f^{\leftarrow}(x))^{\leftarrow}}{f(x)} \le \frac{f(\beta x)}{f(x)}.$$

Therefore,

$$0 < 1 \le \underline{\lim}_{x\to+\infty} \frac{(f^{\leftarrow}(x))^{\leftarrow}}{f(x)} \le \overline{\lim}_{x\to+\infty} \frac{(f^{\leftarrow}(x))^{\leftarrow}}{f(x)} \le \overline{k}_f(\beta) < +\infty.$$

Hence, we get that $f(x) \asymp (f^{\leftarrow}(x))^{\leftarrow}$, when $x \to +\infty$. □

Theorem 13.6 ([15]) *Let $f \in \mathcal{A} \cap ORV$ and $g \in \mathcal{A}$. Then $f(x) \asymp g(x)$, when $x \to +\infty$, if and only if $(f^{\leftarrow}(x))^{\leftarrow} \asymp (g^{\leftarrow}(x))^{\leftarrow}$, when $x \to +\infty$.*

Proof Let $f \in \mathcal{A} \cap ORV$ and $g \in \mathcal{A}$. If $f(x) \asymp g(x)$, when $x \to +\infty$, it follows that $g \in ORV$, and then by Theorem 13.5 we get $f(x) \asymp (f^{\leftarrow}(x))^{\leftarrow}$, when $x \to +\infty$, and $g(x) \asymp (g^{\leftarrow}(x))^{\leftarrow}$, when $x \to +\infty$. From the previous it is obvious that $(f^{\leftarrow}(x))^{\leftarrow} \asymp (g^{\leftarrow}(x))^{\leftarrow}$ holds, when $x \to +\infty$.

Conversely, assume that $f \in \mathcal{A} \cap ORV$, $g \in \mathcal{A}$, and $(f^{\leftarrow}(x))^{\leftarrow} \asymp (g^{\leftarrow}(x))^{\leftarrow}$, when $x \to +\infty$. Then, by Theorem 13.5, we obtain that $f(x) \asymp (g^{\leftarrow}(x))^{\leftarrow}$ is satisfied, when $x \to +\infty$, because $f(x) \asymp (f^{\leftarrow}(x))^{\leftarrow}$, when $x \to +\infty$. Hence, $(g^{\leftarrow}(x))^{\leftarrow} \in ORV \cap \mathcal{A}$, $x \geq a$. Furthermore, from Theorem 13.3 it follows that $g^{\leftarrow} \in PI^*$, while from Theorem 13.4 it follows that $g \in ORV$. Finally, from Theorem 13.5 we obtain that $g(x) \asymp (g^{\leftarrow}(x))^{\leftarrow}$, when $x \to +\infty$, i.e., $f(x) \asymp g(x)$, when $x \to +\infty$. □

Theorem 13.7 ([15])

(a) *Let $f, g \in \mathcal{A}$ and $f \in PI^* \cap ORV$. Then $g \in \{f\}$ if and only if $g^{\leftarrow} \in \{f^{\leftarrow}\}$.*
(b) *Let $f \in \mathcal{A}$ and $f \notin ORV \cap PI^*$. Then there is a $g \in \mathcal{A}$ such that $g \in \{f\}$ and $g^{\leftarrow} \notin \{f^{\leftarrow}\}$, or there is an $g \in \mathcal{A}$ such that $g^{\leftarrow} \in \{f^{\leftarrow}\}$ and $g \notin \{f\}$.*

Proof

(a) Let $f, g \in \mathcal{A}$ and $f \in PI^* \cap ORV$. Firstly, assume that $g \in \{f\}$. Since $f \in PI^*$, then from Theorem 13.3 we have $f^{\leftarrow} \in ORV$, and from Theorem 13.1 we get $g^{\leftarrow} \in \{f^{\leftarrow}\}$.
   Conversely, assume that $g^{\leftarrow} \in \{f^{\leftarrow}\}$. Since $f \in ORV$, from Theorem 13.5 we get $f(x) \asymp (f^{\leftarrow}(x))^{\leftarrow}$, when $x \to +\infty$. Therefore, $(f^{\leftarrow}(x))^{\leftarrow} \in ORV \cap \mathcal{A}$, $x \geq a$. Now, from Theorem 13.1 we get $(f^{\leftarrow}(x))^{\leftarrow} \asymp (g^{\leftarrow}(x))^{\leftarrow}$, when $x \to +\infty$, and from Theorem 13.6 it follows that $f(x) \asymp g(x)$ holds, when $x \to +\infty$. Thus, $g \in \{f\}$.
(b) Assume that $f \in \mathcal{A} \setminus (ORV \cap PI^*)$. We shall discuss the following three cases:

   (1^0) Let $f \in ORV \setminus ARV$. Then from Corollary 1, we get $f^{\leftarrow} \notin ORV$. Hence, from Theorem 13.2 it follows that there exists a $g \in \mathcal{A}$ such that $g \in \{f\}$ and $g^{\leftarrow} \notin \{f^{\leftarrow}\}$.
   (2^0) Let $f \notin ORV \cup PI^*$. Then from Theorem 13.3 it follows that $f^{\leftarrow} \notin ORV$, and from Theorem 13.4 we obtain $f^{\leftarrow} \notin PI^*$. Finally, from Theorem 13.2 there exists a $g \in \mathcal{A}$ such that $g \in \{f\}$ and $g^{\leftarrow} \notin \{f^{\leftarrow}\}$.
   (3^0) Let $f \in PI^* \setminus ORV$. Then, there is a $p > 1$ such that

$$\overline{k}_f(\lambda) = \overline{\lim_{x \to +\infty}} \frac{f(\lambda x)}{f(x)} = +\infty, \quad \lambda \geq p.$$

   Consider the function $g(x) = f(px)$, for $x \geq a$. Since the next inequalities

$$1 \leq \underline{\lim_{x \to +\infty}} \frac{g(x)}{f(x)} \leq \overline{\lim_{x \to +\infty}} \frac{g(x)}{f(x)} = +\infty$$

hold, it follows that $g \notin \{f\}$. Also, we have that

$$g^{\leftarrow}(x) = (f(px))^{\leftarrow} = \inf\{y > 0 \mid f(py) > x\}$$
$$= \inf\left\{\frac{t}{p} > 0 \mid f(t) > x\right\} = \frac{f^{\leftarrow}(x)}{p}$$

holds, and we get $f^{\leftarrow}(x) \asymp g^{\leftarrow}(x)$, when $x \to +\infty$. Hence, there exists a $g \in \mathcal{A}$ such that $g^{\leftarrow} \in \{f^{\leftarrow}\}$ and $g \notin \{f\}$.

□

Let $a > 0$. Functions $f, g : [a, +\infty) \to (0, +\infty)$ are called mutually "inverse weak asymptotic" in $+\infty$ (which is denoted by $f(x) \overset{*}{\asymp} g(x)$, when $x \to +\infty$) if there is $\lambda_0 \geq 1$ such that for every $\lambda > \lambda_0$ there is $x_0 = x_0(\lambda) > 0$ such that $f(x/\lambda) \leq g(x) \leq f(\lambda x)$, for every $x \geq x_0$ (see, e.g., [3]).

Theorem 13.8 ([15]) *Let $f, g \in \mathcal{A}$. Then $f^{\leftarrow}(x) \asymp g^{\leftarrow}(x)$, when $x \to +\infty$ if and only if $f(x) \overset{*}{\asymp} g(x)$, when $x \to +\infty$.*

Proof Firstly, assume $f, g \in \mathcal{A}$ and $f(x) \overset{*}{\asymp} g(x)$, when $x \to +\infty$. Then there is a fixed $\lambda_0 \geq 1$ such that for every $\lambda > \lambda_0$ we have $f(x/\lambda) \leq g(x) \leq f(\lambda x)$ for $x \geq x_0 = x_0(\lambda) > 0$.

Since for such λ and x we have $g(x) \leq f(\lambda x)$, it follows that $g^{\leftarrow}(x) \geq f^{\leftarrow}(x)/\lambda$. Consequently, we have $g^{\leftarrow}(x)/f^{\leftarrow}(x) \geq 1/\lambda$ for such λ and x.

From the previous, we have that $f(x/\lambda) \leq g(x)$ holds for the mentioned λ and x, i.e., $\lambda f^{\leftarrow}(x) \geq g^{\leftarrow}(x)$ holds for the mentioned λ and sufficiently large x, i.e., $g^{\leftarrow}(x)/f^{\leftarrow}(x) \leq \lambda$ is satisfied. Now, from $0 < 1/\lambda \leq g^{\leftarrow}(x)/f^{\leftarrow}(x) \leq \lambda < +\infty$, it follows that $g^{\leftarrow}(x) \asymp f^{\leftarrow}(x)$, when $x \to +\infty$.

Conversely, assume that $f, g \in \mathcal{A}$ and $f^{\leftarrow}(x) \asymp g^{\leftarrow}(x)$, when $x \to +\infty$. Then there is $M > 1$ such that

$$\frac{1}{M} \leq \frac{g^{\leftarrow}(x)}{f^{\leftarrow}(x)} \leq M, \quad x \geq x_0(M) > 0.$$

Firstly, for such x we have $g^{\leftarrow}(x) \leq Mf^{\leftarrow}(x)$, so it follows that $g^{\leftarrow}(x) \leq (f(x/M))^{\leftarrow}$, i.e., $(g^{\leftarrow}(x))^{\leftarrow} \geq ((f(x/M))^{\leftarrow})^{\leftarrow}$. Therefore, for such x and any $\lambda > 1$ we have $f(x/M) \leq g(\lambda x)$. If $\lambda > 1$ and $x \geq x_0$, put $t = \lambda x$. Then for every $t \geq t_0$, we have $f(t/\lambda M) \leq g(t)$. Furthermore, for the same x, we have $g^{\leftarrow}(x) \geq f^{\leftarrow}(x)/M$, i.e., $g^{\leftarrow}(x) \geq (f(Mx))^{\leftarrow}$. Therefore, we obtain that $(g^{\leftarrow}(x))^{\leftarrow} \leq ((f(Mx))^{\leftarrow})^{\leftarrow}$ holds and it follows that $g(x) \leq f(\lambda Mx)$ for any $\lambda > 1$ and $x \geq x_0$.

By taking $t = x$ we obtain that $g(t) \leq f(\lambda Mt)$ holds, for any $\lambda > 1$ and $t \geq x_0$. Thus, if $\lambda > 1$ and $t \geq t_0 = t_0(\lambda) > 0$, it follows that $f(t/\lambda M) \leq g(t) \leq f(\lambda Mt)$. Taking $s = \lambda M > M > 1$, for every $s > M > 1$, we have $f(t/s) \leq g(t) \leq f(st)$ as $t \geq t_0(s)$. This finally gives $f(x) \overset{*}{\asymp} g(x)$, when $x \to +\infty$. □

Theorem 13.9 ([15]) *Let $f, g : [a, +\infty) \to (0, +\infty)$, $a > 0$, and let $f \in PI^*$. If $g \in \{f\}$, then $f(x) \overset{*}{\asymp} g(x)$, when $x \to +\infty$.*

Proof Since $f \in PI^*$, Lemma 1 implies that there is a function $c(\lambda) > 1$ ($\lambda > \lambda_0 \geq 1$), depending on f, such that $f(\lambda x) \geq c(\lambda) f(x)$ for every $\lambda > \lambda_0$ and every $x \geq x_0(\lambda) \geq a > 0$, such that $\lim_{\lambda \to +\infty} c(\lambda) = +\infty$.

Since $g \in \{f\}$, there is an $M \in \mathbb{R}$, $M > 1$ such that $1/M \leq g(x)/f(x) \leq M$, for every $x \geq x_1(M) = x_1 > 0$. Hence, for $x \geq \max\{x_0, x_1\}$ we have $f(\lambda x) \geq c(\lambda) f(x) \geq c(\lambda) g(x)/M$. Since $c(\lambda)/M \geq 1$ for $\lambda > \lambda'$, we get $f(\lambda x) \geq g(x)$ for such λ and x. Next, since $\overline{k}_f(1/\lambda) = 1/\underline{k}_f(\lambda)$ for $\lambda \geq \lambda_0$, then for such λ we also obtain $\overline{k}_f(1/\lambda) \leq 1/c(\lambda)$. Hence, for such λ and every $x \geq x_2(\lambda) = x_2 > 0$ we have $f(x/\lambda) \leq 2f(x)/(1 + c(\lambda))$.

Therefore, for such λ and $x \geq \max\{x_1, x_2\}$ we have $f(x/\lambda) \leq 2f(x)/(1 + c(\lambda)) \leq 2Mg(x)/(1 + c(\lambda))$.

Since $2M/(1 + c(\lambda)) \leq 1$ for $\lambda \geq \lambda''$, we have that $f(x/\lambda) \leq g(x)$ for these λ and x. Therefore, we obtain that $f(x/\lambda) \leq g(x) \leq f(\lambda x)$ is satisfied, for $\lambda \geq \max\{\lambda', \lambda''\}$ and $x \geq \max\{x_0, x_1, x_2\}$, which means that $f(x) \overset{*}{\asymp} g(x)$, when $x \to +\infty$. □

Theorem 13.10 ([15]) *Let $f, g : [a, +\infty) \to (0, +\infty)$ $(a > 0)$ and let $f \in ORV$. If $f(x) \overset{*}{\asymp} g(x)$, when $x \to +\infty$, then $g \in \{f\}$.*

Proof From assumptions of Theorem 13.10 we have that

$$\frac{f(x/\lambda)}{f(x)} \leq \frac{g(x)}{f(x)} \leq \frac{f(\lambda x)}{f(x)},$$

for every $\lambda > \lambda_0 \geq 1$ and every $x \geq x_0 = x_0(\lambda) \geq a > 0$. Therefore, from $f \in ORV$ for $\lambda > \lambda_0$, we have that

$$0 < \underline{k}_f\left(\frac{1}{\lambda}\right) \leq \varliminf_{x \to +\infty} \frac{g(x)}{f(x)} \leq \varlimsup_{x \to +\infty} \frac{g(x)}{f(x)} \leq \overline{k}_f(\lambda) < +\infty$$

holds, from which it follows that $f(x) \asymp g(x)$ when $x \to \infty$, i.e., $g \in \{f\}$. □

Corollary 13.3 ([15]) *Let $f, g : [a, +\infty) \to (0, +\infty)$, $a > 0$, and let $f \in ORV \cap PI^*$. Then $g \in \{f\}$ if and only if $f(x) \overset{*}{\asymp} g(x)$, when $x \to +\infty$.*

Remark 13.2 The class $ORV \cap PI^*$ contains all regularly varying functions whose index is positive as well as all functions from the class ERV (for the definition see [4]) whose down Matuszewska index is positive (see [24]). More generally, this class contains all functions from the class $IRV \cap ARV$ (see [11]). Also notice that the same class does not contain any slowly varying Karamata functions and any rapidly varying function in the sense of de Haan (see, e.g., [13]).

Remark 13.3 ([15]) The function $f(x) = (2 + \sin x)x$, $x \geq 1$, satisfies

$$f \in (PI^* \cap ORV) \setminus (ARV \cap IRV).$$

We end this section with an open question.

Question 13.1 Is the class $ORV \cap PI^*$ the largest class for which Corollary 3 remains true?

13.3 The Weak and Strong Asymptotic Equivalence Relations and the Generalized Inverse

In this section we will give a complete characterization of the functional class $R_\infty \cap \mathcal{A}$, where R_∞ is the class of rapidly varying functions. Also, the characterization of the functional class $PI^* \cap \mathcal{A}$ will also be given.

In the following theorem, (a) gives one extension of Theorem A by taking the weaker condition $g \in \{f\}$ instead of $g \in [f]_\sim$, while (b) determines the maximal class $R_\infty \cap \mathcal{A}$ for which (a) holds.

Theorem 13.11 ([16])

(a) *Let $f, g \in \mathcal{A}$ and $f \in R_\infty$. If additionally $g \in \{f\}$, then $g^\leftarrow \in [f^\leftarrow]_\sim$.*
(b) *Let $f, g \in \mathcal{A}$. If $g^\leftarrow \in [f^\leftarrow]_\sim$ for every $g \in \{f\}$, then $f \in R_\infty$ ($g \in R_\infty$).*

Proof

(a) From the fact that $f \in \mathcal{A} \cap R_\infty$ and by the result obtained in [4], we have that $f^\leftarrow \in SV$. Since $g \in \{f\}$, there is $m > 0$ such that $g(x) \cdot m \leq f(x)$ for sufficiently large x. Furthermore, for the same m and sufficiently large x, it holds $g^\leftarrow(x) \geq f^\leftarrow(mx)$, and we have

$$\underline{\lim}_{x \to +\infty} \frac{g^\leftarrow(x)}{f^\leftarrow(x)} \geq \lim_{x \to +\infty} \frac{f^\leftarrow(mx)}{f^\leftarrow(x)} = 1.$$

On the other hand, there is an $M > 0$ such that $f(x) \leq g(x) \cdot M$ holds for sufficiently large x. Then, for the same M and sufficiently large x, $g^\leftarrow(x) \leq f^\leftarrow(Mx)$ holds. From the previous inequality it follows that

$$\overline{\lim}_{x \to +\infty} \frac{g^\leftarrow(x)}{f^\leftarrow(x)} \leq \lim_{x \to +\infty} \frac{f^\leftarrow(Mx)}{f^\leftarrow(x)} = 1.$$

Hence, we obtain

$$\lim_{x \to +\infty} \frac{g^\leftarrow(x)}{f^\leftarrow(x)} = \underline{\lim}_{x \to +\infty} \frac{g^\leftarrow(x)}{f^\leftarrow(x)} = \overline{\lim}_{x \to +\infty} \frac{g^\leftarrow(x)}{f^\leftarrow(x)} = 1,$$

i.e., $g^\leftarrow \in [f^\leftarrow]_\sim$.

(b) Let $f \in \mathcal{A}$ and let $g(x) = \lambda \cdot f(x)$ for $x \geq a$, where $\lambda > 0$ is a fixed and arbitrary number. Then, we have $g \in \mathcal{A}$ and $g \in \{f\}$, i.e., $g^{\leftarrow} \in [f^{\leftarrow}]_{\sim}$. From the previous we obtain that $g^{\leftarrow}(x) = f^{\leftarrow}\left(\frac{1}{\lambda}x\right)$ holds, for the same λ and sufficiently large x. Now we get

$$\lim_{x\to+\infty} \frac{f^{\leftarrow}\left(\frac{1}{\lambda}x\right)}{f^{\leftarrow}(x)} = \lim_{x\to+\infty} \frac{g^{\leftarrow}(x)}{f^{\leftarrow}(x)} = 1.$$

Hence, $\overline{k}_{f^{\leftarrow}}(\alpha) = 1$ holds for every $\alpha > 0$ and every fixed and arbitrary $\lambda > 0$, i.e., we have $f^{\leftarrow} \in SV$ and $f^{\leftarrow} \in \mathcal{A}$. According to results from [13] we obtain $f \in R_{\infty}$. Now, if we choose an arbitrary function $g \in \mathcal{A}$ such that $g \in \{f\}$, then there are $m > 0$ and $M > 0$ such that $f(x) = r(x) \cdot g(x)$, for $x \geq a$, where the function $r(x)$ is defined for $x \geq a$, and the condition $m \leq r(x) \leq M$ is satisfied, for sufficiently large x. Therefore, we get

$$\underline{\lim}_{x\to+\infty} \frac{g(\lambda x)}{g(x)} \geq \frac{m}{M} \cdot \underline{\lim}_{x\to+\infty} \frac{f(\lambda x)}{f(x)} = +\infty,$$

for every $\lambda > 1$, i.e., $g \in R_{\infty}$.

□

In the next theorem, (a) yields an extension of Theorem A by considering the weaker condition $g^{\leftarrow} \in \{f^{\leftarrow}\}$ instead of $g^{\leftarrow} \in [f^{\leftarrow}]_{\sim}$, while (b) shows that $PI^* \cap \mathcal{A}$ is the maximal class for which (a) holds.

Theorem 13.12 ([16])

(a) *Let $f, g \in \mathcal{A}$ and $f \in PI^*$. If additionally $g \in [f]_{\sim}$, then $g^{\leftarrow} \in \{f^{\leftarrow}\}$.*
(b) *Let $f, g \in \mathcal{A}$. If $g^{\leftarrow} \in \{f^{\leftarrow}\}$ for every $g \in [f]_{\sim}$, then $f \in PI^*$ ($g \in PI^*$).*

Proof

(a) This statement is a direct corollary of Proposition 1 from [15]. We can obtain the proof of (a) by applying a methodology analogous to the one used in the proof of Proposition 1 (a).
(b) Let $f \in \mathcal{A}$ and let $g_1(x) = \left(1 - \frac{1}{x}\right) \cdot f(x)$, for $x \geq a$ (without loss of generality we will assume that $a > 1$). Then, we obtain that $g_1 \in [f]_{\sim}$ and g_1 is a strictly increasing function from $\mathcal{A}$. This implies that $g_1^{\leftarrow} \in \{f^{\leftarrow}\}$ and $g_1^{\leftarrow}$ is a continuous function from $\mathcal{A}$. Hence, for every strictly increasing function $g \in \mathcal{A}$, for which $g \in [g_1]_{\sim}$, we obtain $g^{\leftarrow} \in \{g_1^{\leftarrow}\}$. Therefore, there is $M_g \in (0, +\infty)$, associated to g, such that

$$\overline{\lim_{x\to+\infty}} \frac{g_1^{\leftarrow}(g(x))}{g_1^{\leftarrow}(g_1(x))} = \overline{\lim_{x\to+\infty}} \frac{g_1^{\leftarrow}(g(x))}{x} = \overline{\lim_{x\to+\infty}} \frac{g_1^{\leftarrow}(g(x))}{g^{\leftarrow}(g(x))}$$
$$\leq \overline{\lim_{x\to+\infty}} \frac{g_1^{\leftarrow}(x)}{g^{\leftarrow}(x)} = M_g < +\infty.$$

Let $\alpha(x)$ be a continuous function, for $x \geq a$, such that $\alpha(x) \geq 1$ for $x \geq a$, and $\alpha(x) \to 1$, for $x \to +\infty$. Consider the function $r(x) = \max_{a \leq t \leq x} h(t)$, for $x \geq a$, where the function $h(x)$ is defined by $h(x) = x \cdot \alpha(x)$, for $x \geq a$. We see that $r(x)$ is continuous, nondecreasing, and $r(x) \to +\infty$, when $x \to +\infty$. The inequality $r(x) \geq \alpha(x) \cdot x$, for $x \geq a$, is also satisfied. Now, we will prove that $r(x) \sim x$. For this purpose, take $\varepsilon > 0$. There is $x_1 = x_1(\varepsilon) \geq a$ such that $1 \leq \dfrac{h(x)}{x} < 1 + \varepsilon$ for every $x \geq x_1$, and there is $x_2 = x_2(\varepsilon) \geq x_1$ such that $h(x) \geq \max_{a \leq u \leq x_1} h(u)$ for every $x \geq x_2$. Hence, for every $x \geq x_2$ there is a function $v(x)$ with values in $[x_1, x]$ such that

$$1 \leq \frac{r(x)}{x} = \frac{1}{x} \max_{a \leq u \leq x} h(u) = \frac{1}{x} \max_{x_1 \leq u \leq x} h(u) = \frac{1}{x} h(v(x)) \leq \frac{h(v(x))}{v(x)} < 1 + \varepsilon,$$

which means $r(x) \sim x$. Define now the function $r_1(x) = 1 - \dfrac{1}{x} + r(x)$, for $x \geq a$ (without loss of generality we will assume that $a > 1$). Then, r_1 is a strictly increasing and continuous function from the class $\mathcal{A}$, such that $r_1(x) \sim x$. From this we get

$$1 \leq \underline{\lim}_{x \to +\infty} \frac{g_1^{\leftarrow}(\alpha(x) \cdot x)}{g_1^{\leftarrow}(x)} \leq \overline{\lim}_{x \to +\infty} \frac{g_1^{\leftarrow}(\alpha(x) \cdot x)}{g_1^{\leftarrow}(x)} \leq \overline{\lim}_{x \to +\infty} \frac{g_1^{\leftarrow}(r_1(x))}{g_1^{\leftarrow}(x)}.$$

From $\overline{\lim}_{x \to +\infty} \dfrac{g_1^{\leftarrow}(g(x))}{g_1^{\leftarrow}(g_1(x))} \leq M_g < +\infty$ (see above), for sufficiently large x we obtain $\dfrac{g_1^{\leftarrow}(r_1(g_1(x)))}{g_1^{\leftarrow}(g_1(x))} \leq M_{r_1 \circ g_1} < +\infty$, where $M_{r_1 \circ g_1}$ is a positive real number that corresponds to the composition $r_1 \circ g_1$ in the same way as M_g was associated to g. Using the previous facts one can prove that $\dfrac{g_1^{\leftarrow}(r_1(x))}{g_1^{\leftarrow}(x)} \leq M_{r_1 \circ g_1} < +\infty$ is satisfied for sufficiently large x. Finally, we get $\overline{\lim}_{x \to +\infty} \dfrac{g_1^{\leftarrow}(\alpha(x) \cdot x)}{g_1^{\leftarrow}(x)} < +\infty$.

Now, we will assume for a moment that the next two sequences exist: (i) a sequence (λ_n) such that $\lambda_n \geq 1$ for every $n \in \mathbb{N}$, $\lambda_n \to 1$ when $n \to +\infty$, and (ii) an increasing sequence (x_n) such that $x_n \geq a$, for every $n \in \mathbb{N}$, $x_n \to +\infty$, when $n \to +\infty$ such that $\lim_{n \to +\infty} \dfrac{g_1^{\leftarrow}(\lambda_n \cdot x_n)}{g_1^{\leftarrow}(x_n)} = +\infty$. Consider a function $\alpha(x)$, $x \geq a$, such that $\alpha(x_n) = \lambda_n$ for $n \in \mathbb{N}$, $\alpha(x)$ is linear and continuous for $x \in [x_n, x_{n+1}]$, $n \in \mathbb{N}$, and $\alpha(x) = \lambda_1$ for $x \in [a, x_1]$. This function $\alpha : [a, +\infty) \to [1, +\infty)$ is continuous and $\lim_{x \to +\infty} \alpha(x) = 1$. From the definition of α it follows that

$$\overline{\lim}_{n \to +\infty} \frac{g_1^{\leftarrow}(\alpha(x_n) \cdot x_n)}{g_1^{\leftarrow}(x_n)} = \lim_{x \to +\infty} \frac{g_1^{\leftarrow}(\lambda_n \cdot x_n)}{g_1^{\leftarrow}(x_n)} = +\infty.$$

This contradicts the fact $\overline{\lim}_{x\to+\infty} \dfrac{g_1^{\leftarrow}(\alpha(x)\cdot x)}{g_1^{\leftarrow}(x)} < +\infty$ shown above.

So, we obtain that $\overline{\lim}_{\substack{x\to+\infty\\ \lambda\to 1}} \dfrac{g_1^{\leftarrow}(\lambda x)}{g_1^{\leftarrow}(x)} = A$ holds for some $A \in (0, +\infty)$, i.e., for every $\varepsilon > 0$ there are $x_0 \geq a$ and $\delta > 0$ such that $1 \leq \dfrac{g_1^{\leftarrow}(\lambda x)}{g_1^{\leftarrow}(x)} \leq A+\varepsilon$, for every $x \geq x_0$ and every $\lambda \in [1, 1+\delta]$. Hence, for every $\lambda \in (0, 1+\delta]$ we have $\overline{k}_{g_1^{\leftarrow}}(\lambda) \leq A + \varepsilon < +\infty$. On the other hand, from the fact that the function $g_1^{\leftarrow}$ belongs to the class $\mathcal{A}$ (and it is nondecreasing), we have that $\overline{k}_{g_1^{\leftarrow}}(\lambda) < +\infty$, for every $\lambda > 0$ (see [25]). Finally, we obtain $g_1^{\leftarrow} \in ORV$, and if we apply a result from [15], it follows that $g_1 \in PI^*$. Furthermore, as $g_1(x) = \left(1 - \dfrac{1}{x}\right) \cdot f(x)$ for every $x \geq a$, we have $\underline{k}_f(\lambda) \geq \underline{k}_{g_1}(\lambda) \cdot \liminf_{x\to+\infty} \dfrac{\lambda x(x-1)}{(\lambda x - 1)x} = \underline{k}_{g_1}(\lambda)$, for every $\lambda > 0$, i.e., we obtain $f \in PI^*$.

Analogously, we can prove that $g \in PI^*$, for every $g \in \mathcal{A}$ such that $g \in [f]_\sim$.

□

13.4 Asymptotic Similarity Relation and Generalized Inverse

A function $g \in \{f\}$ is called asymptotically similar to f if

$$\lim_{x\to+\infty} \frac{f(x)}{g(x)} = \alpha \tag{13.4}$$

holds, for $\alpha \in \mathbb{R} \setminus \{0\}$; it is denoted by $f(x) \asymp g(x)$, when $x \to +\infty$.

Let $f : [a, +\infty) \to (0, +\infty)$, $(a > 0)$, and let

$$\langle f \rangle = \{g \in \{f\} \mid f(x) \asymp g(x), \text{ when } x \to +\infty\}.$$

Notice that we have that $[f]_\sim \subsetneq \langle f \rangle \subsetneq \{f\}$ is satisfied.

The next results are continuation of research published in [3, 10, 14, 15] and [16].

Theorem 13.13 ([17]) *Let $f \in \mathcal{A}$. Also, let $f \in R_\infty$, or f is a regularly varying function in the sense of Karamata with the index of variability $\rho > 0$. Then, $g^{\leftarrow} \in \langle f^{\leftarrow} \rangle$ for every $g \in \mathcal{A} \cap \langle f \rangle$.*

Proof Firstly, assume that $f \in \mathcal{A} \cap RV_\rho$, for some $\rho > 0$, and let $g \in \mathcal{A} \cap \langle f \rangle$. Then, there exists $\alpha = \alpha(g) > 0$ such that $\lim_{x\to+\infty} \dfrac{f(x)}{g(x)} = \alpha$. Now, it follows that

$$\frac{\alpha}{\varepsilon} \leq \frac{f(x)}{g(x)} \leq \varepsilon \cdot \alpha$$

holds, for every $\varepsilon > 1$ and for sufficiently large x. From the previous it holds that

$$f^{\leftarrow}(x) \geq g^{\leftarrow}\left(\frac{x}{\varepsilon \cdot \alpha}\right) \quad \text{and} \quad f^{\leftarrow}(x) \leq g^{\leftarrow}\left(\frac{\varepsilon}{\alpha} \cdot x\right),$$

for sufficiently large x. Hence, we obtain that

$$\frac{g^{\leftarrow}(\frac{x}{\varepsilon \cdot \alpha})}{g^{\leftarrow}(x)} \leq \frac{f^{\leftarrow}(x)}{g^{\leftarrow}(x)} \leq \frac{g^{\leftarrow}(\frac{\varepsilon}{\alpha} \cdot x)}{g^{\leftarrow}(x)}$$

is satisfied for the same x. Since $g \in \langle f \rangle$, it follows that $g \in RV_\rho$ and we get

$$(\varepsilon \cdot \alpha)^{-1/\rho} \leq \varliminf_{x\to+\infty} \frac{f^{\leftarrow}(x)}{g^{\leftarrow}(x)} \leq \varlimsup_{x\to+\infty} \frac{f^{\leftarrow}(x)}{g^{\leftarrow}(x)} \leq \left(\frac{\varepsilon}{\alpha}\right)^{1/\rho}.$$

If in previous inequalities we let $\varepsilon \to 1_+$, it follows that $\lim_{x\to+\infty} \dfrac{f^{\leftarrow}(x)}{g^{\leftarrow}(x)} = \alpha^{-1/\rho}$, i.e., $g^{\leftarrow} \in \langle f^{\leftarrow} \rangle$ is satisfied.

Now, let $f \in \mathcal{A} \cap R_\infty$ and $g \in \mathcal{A} \cap \langle f \rangle$. Then, we get that $g \in \{f\}$ holds and from results obtained in [16] it follows that $\lim_{x\to+\infty} \dfrac{f^{\leftarrow}(x)}{g^{\leftarrow}(x)} = 1$, i.e., $g^{\leftarrow} \in \langle f^{\leftarrow} \rangle$ holds. □

Theorem 13.14 ([17]) *If $f \in \mathcal{A}$ and $g^{\leftarrow} \in \langle f^{\leftarrow} \rangle$, for every $g \in \mathcal{A} \cap \langle f \rangle$, then $f \in R_\infty \cup \bigcup_{\rho>0} RV_\rho$. The same conclusion holds for every $g \subset \langle f \rangle$.*

Proof Let $f \in \mathcal{A}$ and let $g_1(x) = \alpha \cdot f(x)$ for $x \geq a$, where $\alpha > 0$ is an arbitrary fixed number. Since $g_1 \in \mathcal{A} \cap \langle f \rangle$, it follows that $g_1^{\leftarrow} \in \langle f^{\leftarrow} \rangle$, i.e., there is an $\beta = \beta(\alpha) \in (0, +\infty)$ such that

$$\lim_{x\to+\infty} \frac{f^{\leftarrow}(\frac{1}{\alpha} \cdot x)}{f^{\leftarrow}(x)} = \lim_{x\to+\infty} \frac{g_1^{\leftarrow}(x)}{f^{\leftarrow}(x)} = \beta$$

holds.

Since $f^{\leftarrow} \in \mathcal{A}$, it follows that $f^{\leftarrow} \in RV_\rho$ for some $\rho \geq 0$. If $\rho = 0$, then from the result obtained in [13] we get $f \in R_\infty$. If $\rho > 0$, then $f \in RV_{1/\rho}$ (this is well-known result which can be found in [4]). Now, let $g \in \mathcal{A} \cap \langle f \rangle$ be a fixed function. Then, it follows that

$$\lim_{x\to+\infty} \frac{g(\gamma x)}{g(x)} = \lim_{x\to+\infty} \frac{f(\gamma x)}{f(x)}$$

holds, for every $\gamma > 0$, and hence, we obtain $g \in R_\infty \cup RV_{1/\rho}$ for some $\rho \in (0, +\infty)$. □

Theorem 13.15 ([17]) *If $f \in \mathcal{A} \cap RV$ is with index of variability $\rho \geq 0$, then $g \in \langle f \rangle$, whenever $g \in \mathcal{A}$ and $g^{\leftarrow} \in \langle f^{\leftarrow} \rangle$.*

Proof Let $f \in \mathcal{A} \cap RV_\rho$ for some $\rho \geq 0$. Then $f^{\leftarrow} \in R_\infty$, or $f^{\leftarrow}$ is a regularly varying function in the sense of Karamata with positive index of variability (see, e.g., [4]). Next, assume $g \in \mathcal{A}$ and $g^{\leftarrow} \in \langle f^{\leftarrow} \rangle$. Since $g^{\leftarrow} \in \mathcal{A}$ and $f^{\leftarrow} \in \mathcal{A}$, then from Theorem 13.13 it follows that $(f^{\leftarrow}(x))^{\leftarrow} \asymp (g^{\leftarrow}(x))^{\leftarrow}$, when $x \to +\infty$. Since $f \in RV_\rho$, for $\rho \geq 0$ and $f \in \mathcal{A}$, according to the result from [14] we obtain $\lim_{x\to+\infty} \dfrac{(f^{\leftarrow}(x))^{\leftarrow}}{f(x)} = 1$. Since $g^{\leftarrow} \in \langle f^{\leftarrow} \rangle$, we have that $g^{\leftarrow} \in R_\infty$, or $g^{\leftarrow}$ is a regularly varying function in the sense of Karamata with positive index of variability and $g \in RV_\rho$, for some $\rho \geq 0$. Furthermore, $\lim_{x\to+\infty} \dfrac{(g^{\leftarrow}(x))^{\leftarrow}}{g(x)} = 1$ holds, and finally we get $g \in \langle f \rangle$. □

Theorem 13.16 ([17]) *If $f \in \mathcal{A}$ and $g \in \langle f \rangle$, whenever $g \in \mathcal{A}$ and $g^{\leftarrow} \in \langle f^{\leftarrow} \rangle$, then $f \in RV$ with index of variability $\rho \geq 0$. The same conclusion holds for every $g \in \langle f \rangle$.*

Proof Assume that $f \in \mathcal{A}$ and define $g_1(x) = f(\alpha x)$, for $x \geq a$, where $\alpha > 1$ is an arbitrary and fixed number. Then, we have $g_1^{\leftarrow}(x) = \dfrac{1}{\alpha} \cdot f^{\leftarrow}(x)$ for sufficiently large x, and it follows that $g_1^{\leftarrow} \in \langle f^{\leftarrow} \rangle$, i.e., $g_1 \in \langle f \rangle$. Therefore, it holds that

$$\lim_{x\to+\infty} \frac{g_1(x)}{f(x)} = \lim_{x\to+\infty} \frac{f(\alpha x)}{f(x)} = \beta,$$

for some $\beta \in (0, +\infty)$, where β is a function of $\alpha > 1$. Accordingly, $f \in RV_\rho$ for some $\rho \geq 0$. Now, let $g \in \mathcal{A}$ be an arbitrary and fixed function and let $g \in \langle f \rangle$. Then, analogously to the proof of Theorem 13.14, one can prove that $g \in RV_\rho$, for some $\rho \geq 0$. □

Combining results from Theorems 13.13–13.16, the next corollary can be obtained. Notice that it can be also obtained from the results given in [6] and [8].

Corollary 13.4 ([17]) *Let $f \in \mathcal{A} \cap RV$ be a function with index of variability $\rho > 0$. Then $g \in \langle f \rangle$, for every $g \in \mathcal{A}$ if and only if $g^{\leftarrow} \in \langle f^{\leftarrow} \rangle$. If $f \in \mathcal{A}$ and f is not a regularly varying function in the sense of Karamata with index of variability $\rho > 0$, then there exists $g \in \mathcal{A}$ such that $g \in \langle f \rangle$ and $g^{\leftarrow} \notin \langle f^{\leftarrow} \rangle$, or $g^{\leftarrow} \in \langle f^{\leftarrow} \rangle$ and $g \notin \langle f \rangle$.*

Using the previous corollary, we will prove Theorems 13.17, 13.18, Corollaries 13.5, and 13.6.

Theorem 13.17 ([17]) *Let $f \in \mathcal{A}$.*

(a) *If $f \in R_\infty$, then $g^{\leftarrow} \in [f^{\leftarrow}]_\sim \subsetneq \langle f^{\leftarrow} \rangle$ for every $g \in \{f\}$.*
(b) *If $g^{\leftarrow} \in \langle f^{\leftarrow} \rangle$, whenever $g \in \mathcal{A} \cap \{f\}$, then $f \in RV_\rho \cup R_\infty$ for some $\rho > 0$.*
(c) *If $f \in RV_\rho$, $\rho > 0$, then there exists $g \in \mathcal{A}$ such that $g \in \{f\}$ and $g^{\leftarrow} \notin \langle f^{\leftarrow} \rangle$.*

Proof

(a) Assertion follows from the results in [16].
(b) According to the assumptions and Theorem 13.14 we have $f \in RV_\rho \cup R_\infty$, for some $\rho > 0$.
(c) Assume that $f \in \mathcal{A} \cap RV_\rho$, $\rho > 0$. Then there exists $l_1 \in SV$ such that $l_1(x) \sim l(x) = \dfrac{f(x)}{x^\rho}$, when $x \to +\infty$, where l_1 is continuous and strictly increasing (see, e.g., [4]). Let $f_1(x) = l_1(x) \cdot x^\rho$, $x \geq a$. Then $f_1 \in \mathcal{A} \cap RV_\rho$, $\rho > 0$, and $f_1(x) \sim f(x)$, when $x \to +\infty$.

Now, let us assume the opposite, i.e., we will assume that $g^{\leftarrow} \in \langle f^{\leftarrow} \rangle$, whenever $g \in \{f\} \cap \mathcal{A}$. Since, $f^{\leftarrow} \in RV_{1/\rho}$, we have that $g^{\leftarrow} \in RV_{1/\rho}$, and so we obtain $g \in RV_\rho$. Let

$$g(x) = \begin{cases} f_1(a), & \text{for } x \in [a = a_1, a_2) \quad \big(a_2 > a_1;\ \ f_1(a_2) = 2 \cdot f_1(a_1)\big); \\ f_1(a_2), & \text{for } x \in [a_2, a_3) \qquad\quad \big(a_3 > a_2;\ \ f_1(a_3) = 2 \cdot f_1(a_2)\big); \\ \quad\vdots & \qquad\quad \vdots \qquad\qquad\qquad\qquad\qquad\qquad \vdots \end{cases}$$

Hence, it follows that $g \in \mathcal{A}$ and $1 \leq \dfrac{f_1(x)}{g(x)} < 2$ for $x \geq a$, so we get $g \in \{f\}$. However, $g \notin RV$, because $\dfrac{g(x+1)}{g(x)} \nrightarrow 1$, when $x \to +\infty$. Namely, if we consider the sequence (α_n) defined by $\alpha_n = a_{n+1} - \min\{1/2, a_{n+1} - a_n\}$, $n \in \mathbb{N}$, since $\alpha_n \to +\infty$, when $n \to +\infty$ and $\dfrac{g(\alpha_n+1)}{g(\alpha_n)} \geq 2$, for $n \in \mathbb{N}$, it follows that $\overline{\lim}_{n\to+\infty} \dfrac{g(x+1)}{g(x)} \geq 2$. Since $g \in RV \subsetneq IRV$, it must be $\lim_{x\to+\infty} \dfrac{g(x+1)}{g(x)} = 1$, which is a contradiction with $g \in RV_\rho$.

□

Remark 13.4 The characterization obtained in Theorem 13.17 for any function $f \in \mathcal{A}$ remains true if $g \in \{f\}$.

Theorem 13.18 ([17]) *Let $f \in \mathcal{A}$.*

(a) *If $f \in PI$, then $g^{\leftarrow} \in \{f^{\leftarrow}\}$ for every $g \in \langle f \rangle \cap \mathcal{A}$.*
(b) *If $g^{\leftarrow} \in \{f^{\leftarrow}\}$ whenever $g \in \langle f \rangle \cap \mathcal{A}$, then $f \in PI$. The same conclusion holds for every $g \in \langle f \rangle$.*

Proof

(a) Obviously, $f^{\leftarrow} \in ORV$ and $g \in \langle f \rangle$. Then $g \in \{f\}$, and, according to the result from [15], it follows that $g^{\leftarrow} \in \{f^{\leftarrow}\}$.

(b) Let $\alpha > 0$ and let $g(x) = \alpha \cdot f(x)$, for $x \geq a$. Then $g \in \mathcal{A}$, $g \in \langle f \rangle$, and $g^{\leftarrow}(x) = f^{\leftarrow}\left(\frac{1}{\alpha}x\right)$ for sufficiently large x and

$$\begin{aligned} 0 < m(\alpha) &\leq \varliminf_{x \to +\infty} \frac{g^{\leftarrow}(x)}{f^{\leftarrow}(x)} = \varliminf_{x \to +\infty} \frac{f^{\leftarrow}(\frac{1}{\alpha})}{f^{\leftarrow}(x)} \\ &\leq \varlimsup_{x \to +\infty} \frac{g^{\leftarrow}(x)}{f^{\leftarrow}(x)} = \varlimsup_{x \to +\infty} \frac{f^{\leftarrow}(\frac{1}{\alpha}x)}{f^{\leftarrow}(x)} \\ &\leq M(\alpha) < +\infty. \end{aligned}$$

Now, from $f^{\leftarrow} \in ORV$ and from the result given in [15], $f \in PI$.

□

Corollary 13.5 ([17]) *Let* $f \in \mathcal{A}$.

(a) *If* $f \in ORV$, *then* $g \in \{f\}$ *for every* $g^{\leftarrow} \in \langle f^{\leftarrow} \rangle$, $g \in \mathcal{A}$.

(b) *If* $g \in \{f\}$, *whenever* $g^{\leftarrow} \in \langle f^{\leftarrow} \rangle$, $g \in \mathcal{A}$, *then* $f \in ORV$. *The same conclusion holds for every* $g \in \{f\}$.

Corollary 13.6 ([17]) *Let* $f \in \mathcal{A}$.

(a) *If* $f \in SV$, *then* $g \in [f]_{\sim} \subsetneq \langle f \rangle$, *for every* $g \in \mathcal{A}$, $g^{\leftarrow} \in \{f^{\leftarrow}\}$.

(b) *If* $g \in \langle f \rangle$, *whenever* $g^{\leftarrow} \in \{f^{\leftarrow}\}$, *for* $g \in \mathcal{A}$, *then* $f \in SV \cup RV_{\rho}$, $\rho > 0$.

(c) *If* $f \in RV_{\rho}$, $\rho > 0$, *then exists* $g \in \mathcal{A}$, *such that* $g^{\leftarrow} \in \{f^{\leftarrow}\}$ *and* $g \notin \langle f \rangle$.

Remark 13.5 The characterization obtained in Corollary 13.6 for f holds for every $g \in \langle f \rangle$.

Acknowledgments The first author acknowledges the support of the Ministry of Science, Technological Development and Innovation of the Republic of Serbia, institutionally funded through the Military Academy, University of Defence (Grant no. VA-TT/2/20-25). The third author acknowledges the support of the Ministry of Science, Technological Development and Innovation of the Republic of Serbia, institutionally funded through the Faculty of Technical Science, University of Kragujevac.

References

1. Aljančić, S., Arandjelović, D.: O–regularly varying functions. Publ. Inst. Math. (Beograd) **22(36)**, 5–22 (1977)
2. Avakumović, V.G.: Über einen O–inversionssatz. Bull. Int. Acad. Youg. Sci. **29–30**, 107–117 (1936)

3. Balkema, A.A., Geluk, J.L., De Haan, L.: An extension of Karamata's Tauberian theorem and its connection with complementary convex functions. Q. J. Math. Oxford Ser. **30(2)**, 385–416 (1979)
4. Bingham, N.H., Goldie, C.M., Teugels, J.L.: Regular Variation. Cambridge University Press, Cambridge (1987)
5. Buldygin, V.V., Klesov, O.I., Steinebach, J.G.: Properties of a subclass of Avakumović functions and their generalized inverses. Ukr. Math. Zh. **54**(2), 179–206 (2002)
6. Buldygin, V.V., Klesov, O.I., Steinebach, J.G.: Some properties of asymptotic quasi-inverse function and their applications, I. Theory Probab. Math. Stat. **70**, 11–28 (2005)
7. Buldygin, V.V., Klesov, O.I., Steinebach, J.G.: On some properties of asymptotic quasiinverse function and their applications, II. Theory Probab. Math. Stat. **70**, 37–52 (2005)
8. Buldygin, V.V., Klesov, O.I., Steinebach, J.G.: On some properties of asymptotically quasi-inverse function. Teory Imovirn. Mat. Stat. **77**, 13–27 (2007) (in Ukrainian). English transl.: Theory Probab. Math. Stat. **77**, 15–30 (2008)
9. De Haan, L., Statdtmüller, U.: Dominated variation and related concepts and Tauberian theorems for Laplace transforms. J. Math. Anal. Appl. **108**, 344–365 (1985)
10. Djurčić, D., Torgašev, A.: Strong asymptotic equivalence and inversion of functions in the class K_c. J. Math. Anal. Appl. **255**, 383–390 (2001)
11. Djurčić, D., Torgašev, A.: Weak asymptotic equivalence and inverse functions in the class OR. Math. Morav. **7**, 1–6 (2003)
12. Djurčić, D., Kočinac, Lj.D.R., Žižović, M.R.: Some properties of rapidly varying sequences. J. Math. Anal. Appl. **327**, 1297–1306 (2007)
13. Djurčić, D., Torgašev, A.: Some asymptotic relations for the generalized inverse. J. Math. Anal. Appl. **325**, 1397–1402 (2007)
14. Djurčić, D., Torgašev, A., Ješić, S.: The strong asymptotic equivalence and the generalized inverse. Sib. Math. J. **49**(4), 786–795 (2008)
15. Djurčić, D., Nikolić, R.M., Torgašev, A.: The weak asymptotic equivalence and generalized inverse. Lith. Math. J. **50**(1), 34–42 (2010)
16. Djurčić, D., Nikolić, R.M., Torgašev, A.: The weak and strong asymptotic equivalence relations and the generalized inverse. Lith. Math. J. **51**(4), 472–476 (2011)
17. Djurčić, D., Konjokrad, V., Nikolić, R.M.: Asymptotic similarity relation and generalized inverse. Filomat **25**(5), 1075–1080 (2012)
18. Feller, W.: One-side analogues of Karamata's regular variation. L'Enseignement Math. **15**, 107–121 (1969)
19. Fleige, A.: Characterizations of monotone O-regularly varying functions by means of indefinite eigenvalue problems and HELP type inequalities. J. Math. Anal. Appl. **412**, 345–359 (2014)
20. Karamata, J.: Sur un mode de croissance régulière des functions. Mathematica (Cluj) **4**, 38–53 (1930)
21. Karamata, J.: Sur un mode de croissance régulière. Théorèmes fondamenteaux. Bull Soc. Math. France **61**, 55–62 (1933)
22. Karamata, J.: Bernerkung über die vorstehende Arbeit des Herrn Avakumović, mit näherer Betrachtung einer Klasse von Funktionen, welche bei den Inversionssätzen vorkommen. Bull Int. Acad. Youg. Sci. **29–30**, 117–123 (1936)
23. Kočinac, Lj.D.R., Djurčić, D., Manojlović, J.V.: Regular and rapid variations and some applications. In: Ruzhansky, M., Dutta, H., Agarwal, R.P. (eds.) Mathematical Analysis and Applications: Selected Topics, pp. 414–474. John Wiley & Sons, New York (2018)
24. Matuszewska, W.: On a generalization of regularly increasing funkctions. Stud. Math. **24**, 271–279 (1964)
25. Seneta, E.: Regularly Varying Functions. LNM vol. 508. Springer, New York (1976)
26. Timotić, V., Djurčić, D., Nikolić, R.M.: On slowly varying sequences. Filomat **25**(6), 1075–1080 (2012)
27. Timotić, V., Djurčić, D., Kočinac, Lj.D.R.: On rapidly varying sequences. Appl. Anal. Discr. Math. **16**(1), 83–90 (2022)

Chapter 14
Optimal Solutions of System of (j, φ)-Hilfer FDEs via Best Proximity Point Using MNC

Pradip R. Patle, Moosa Gabeleh, D. R. Abed Al-Zuhairi, and Vladimir Rakočević

Dedicated to Academician Gradimir V. Milovanović on the occasion of his 75th birthday

14.1 Introduction

In the current era of research in mathematics, fractional calculus and the differential equations involving fractional derivatives (FDEs) received immense attention from contemporary mathematicians. This field of research reached utmost importance in recent times because of its applications in numerous areas such as engineering, medical sciences, physical sciences, space sciences, etc. It has very significant real-world applications in such situations where the problem involves memory. There are a number of particular applications of FDEs in various scientific fields such as quantum field theory, control theory, nuclear reactor processes, robotics, transport theory, electromagnetic theory, etc.

Because of extensively numerous applications to real-world problems, fractional calculus is an intensively developing area of mathematical analysis. It is observed that in order to preserve various aspects of the classical integer-order

P. R. Patle (✉)
Department of Mathematics, School of Advanced Sciences, VIT-AP University, Amravati, India

M. Gabeleh
Department of Mathematics, Ayatollah Boroujerdi University, Boroujerd, Iran
e-mail: gabeleh@abru.ac.ir

D. R. Abed Al-Zuhairi
Department of Mathematics, College of Basic Education, University of Diyala, Diyala, Iraq
e-mail: basicmathte2@uodiyala.edu.iq

V. Rakočević
Department of Mathematics, Faculty of Sciences and Mathematics, University of Niš, Niš, Serbia
e-mail: vrakoc@sbb.rs

M. P. Stanić et al. (eds.), *Analysis, Approximation, Optimization: Computation and Applications*, Springer Optimization and Its Applications 224,
https://doi.org/10.1007/978-3-031-85743-0_14

derivative, numerous definitions of fractional derivative are defined in the literature [4, 25, 30, 36, 37, 39]. Till 2006, a bunch of the definitions were introduced in the literature. Some of them are due to Riemann-Liouville, Caputo, Fabrizio, Atangana, Katugampola, Weyl, Hadamard, etc. A detailed discussion of these definitions can be found in [25]. Interestingly all these definitions do not coincide with each other in general. Hilfer [24] defined a generalized form of fractional derivative which unified definitions of Riemann-Liouville and Caputo. It is well known that Hilfer derivative interpolates between Riemann-Liouville and Caputo derivatives. Sousa and Oliveira [40] coined a new definition with the help of another function φ with some assumptions and called it as φ-Hilfer fractional derivative. The φ-Hilfer derivative has unified most of the definitions in the literature. The FDEs involving φ-Hilfer fractional derivative are studied [14, 34] by many mathematicians for different qualitative properties of solutions.

In all the mentioned definitions an important role is played by classical Euler gamma (Γ) function. Diaz and Pariguan [10] in 2007 coined a generalized version of Euler gamma function and called it as k-gamma (Γ_k) function. Interestingly, $\Gamma_k(z)$ converges to $\Gamma(z)$ for $z \in \mathbb{C}$ with $Re(z) > 0$ as $k \to 1$. The introduction of k-gamma function in the literature led to redefine the concepts of fractional derivatives and integrals with the additional parameter $k > 0$. Mubeen and Habibullah [32] and Romero et al. [38] introduced generalized versions of Riemann-Liouville fractional integral and derivative called as k-RL fractional integral and derivative, respectively. Dorrego [11] redefined k-RL derivative and studied its several characteristics. The "k-Hilfer fractional derivative" is defined in [12] but it does not generalize the "k-RL fractional derivative." Several researchers investigated different types of k-fractional derivatives and analyzed nonlinear fractional differential equations (FDEs) involving them.

Recently, Kucche and Mali [26] generalized the φ-Hilfer fractional integrals and derivatives and defined the (j, φ)-Hilfer fractional derivative and integrals. This definition is considered to be the most general definition of fractional-order derivative. They presented some properties and the existence and uniqueness of solutions of the Cauchy problem involving (j, φ)-Hilfer fractional derivative.

Motivated by the above works, in this chapter, we consider the following system of nonlinear fractional differential equations involving the (j, φ)-Hilfer fractional derivatives of arbitrary order with initial conditions of the form

$$\begin{cases} {}^{j,H}D_{a^+}^{m,n;\varphi}\psi(x) = \Lambda_1(x, \psi(x)), x \in (a, b], 0 < m < j, n \in [0, 1], \\ {}^{j}I_{a^+}^{(j-\xi_j);\varphi}\psi(a) = \alpha_a, \ \xi_j = m + n(j - m); \end{cases} \tag{14.1}$$

$$\begin{cases} {}^{j,H}D_{a^+}^{m,n;\varphi}\eta(x) = \Lambda_2(x, \eta(x)), x \in (a, b], 0 < m < j, n \in [0, 1], \\ {}^{j}I_{a^+}^{(j-\xi_j);\varphi}\eta(a) = \beta_a, \ \xi_j = m + n(j - m), \end{cases} \tag{14.2}$$

where ${}^{j,H}D_{a^+}^{m,n;\varphi}$ is the (j, φ)-Hilfer fractional differential operator of order m and type n; ${}^{j}I_{a^+}^{(j-\xi_j);\varphi}$ is the (j, φ)-Riemann-Liouville fractional integral of order $(j -$

ξ_j); and $\Lambda_1 : I \times B_1 \to E$ and $\Lambda_2 : I \times B_2 \to E$ are given mappings satisfying some assumptions. The state $\psi(\cdot)$ takes the values from Banach space E and B_1 and B_2 are some subsets in it. α_a and β_a will be specified later. Our primary motive here is to study the existence of optimum solutions of system of fractional differential equation (14.1)–(14.2). In fact the optimum solution that we are going to study is represented nothing but by the concept of best proximity points. To prove the new best proximity point existence theorems are our secondary goal. The new existence theorems generalize the state-of-art results in the literature.

The workflow in this article is as follows. In Sect. 14.2, we present some basic concepts related to the fractional derivatives and differential equations involving those derivatives. Various characteristics and properties of fractional calculus, the concept of measure of noncompactness, and best proximity point are discussed in Sect. 14.2 itself. The main results concerning the existence of optimal solutions of the system of FDEs (14.1)–(14.2) are proved in Sect. 14.3. Also we have proved the new best proximity point and pair theorems in Sect. 14.3.

14.2 Preliminaries and Basic Concepts

We first present notions and outcomes related to fractional derivatives that are essential to study the fractional differential equations. We denote the collection of all continuous mappings on closed interval $[a, b]$ by $C[a, b]$. Whereas $\mathcal{L}^k(a, b)$, $k \geq 1$ denotes the collection of Lebesgue integrable functions on (a, b). It is clear that $C[a, b]$ and $\mathcal{L}^k(a, b)$, $k \geq 1$ are Banach spaces with respect to specific norms which will be defined when needed. Let $\varphi(x)$ be a positive and monotone increasing mapping defined on $[a, b]$, with derivative $\varphi'(x) \neq 0$ and continuous on (a, b). For $j > 0$ $(j \in \mathbb{R})$, Diaz and Pariguan [10] defined j-gamma function for a complex number ω having positive real part as

$$\Gamma_j(\omega) = \int_0^{+\infty} s^{\omega-1} e^{\left(\frac{-s^j}{j}\right)} \, ds.$$

With the help of this j-gamma function (j, φ)-Riemann-Liouville fractional integral of function $\psi \in L^1(a, b)$ on $[a, b]$ of order $n \in \mathbb{R}_+$ and for $j \in \mathbb{R}_+$ is presented in [27] as follows:

$${}^{j}I_{a+}^{n;\varphi}\psi(x) = \frac{1}{j\,\Gamma_j(n)} \int_a^x \varphi'(s)(\varphi(x) - \varphi(s))^{\frac{n}{j}-1} \psi(s) \, ds, \; x > a.$$

Kucche and Mali [26], with the aim to generalize various definitions of fractional derivatives, introduce the concept of (j, φ)-Hilfer fractional derivative as follows.

Definition 14.1 ([26]) Let $j, m \in \mathbb{R}_+$, $n \in [0, 1]$, $\mu \in \mathbb{N}$, $\psi \in \mathbb{C}^\mu([a, b], \mathbb{R})$ and $\varphi(\cdot)$ as defined above. The expression of the form

$$^{j,H}D_{a^+}^{m,n;\varphi}\psi(x) = {}^jI_{a^+}^{(nj-m);\varphi}\Big(\frac{j}{\varphi'(x)}\frac{d}{dx}\Big)^\mu \; {}^jI_{a^+}^{(1-n)(\mu j-m);\varphi}\psi(x),$$
$$x > a,\ \mu = \left\lceil \frac{m}{j} \right\rceil,$$

is called (j, φ)-Hilfer fractional derivative of order m and type n of function ψ provided the right-hand side exists.

Remark 14.1 The (j, φ)-Hilfer fractional derivative is regarded as the most general and unified definition of fractional differential operator. Indeed, selecting different values for $\varphi(x)$, j, and a and varying the parameters m, n over various limits in Definition 14.1, a wide variety of fractional derivatives in the literature can be obtained. We refer to [26, 40, 41] for more information related to this.

The following results are useful in existence theorems.

Lemma 14.1 ([26]) *Let $m, j \in \mathbb{R}_+$ and let $\xi \in \mathbb{R}$ such that $\frac{\xi}{j} > -1$. Then, we have*

$$^jI_{a^+}^{m;\varphi}(\varphi(t) - \varphi(a))^{\frac{\xi}{j}} = \frac{\Gamma_j(\xi + j)}{\Gamma_j(\xi + j + m)}(\varphi(t) - \varphi(a))^{\frac{\xi+m}{j}}.$$

Lemma 14.2 ([26]) *For $m, n, j \in \mathbb{R}_+$, and $\psi \in L^1(a, b)$, we have*

$$^jI_{a^+}^{m;\varphi}\,{}^jI_{a^+}^{n;\varphi}\psi(x) = {}^jI_{a^+}^{m+n;\varphi}\psi(x),\ a.e.,\ x \in [a, b].$$

In [26], the authors proved that the initial value problem (14.1) is equivalent to the integral equation:

$$\begin{aligned}\psi(x) =& \frac{(\varphi(x) - \varphi(a))^{\frac{\xi_j}{j}-1}}{\Gamma_j(\xi_j)}\alpha_a \\ &+ \frac{1}{j\Gamma_j(m)}\int_a^x \varphi'(s)(\varphi(x) - \varphi(s))^{\frac{m}{j}-1}\Lambda_1(s, \psi(s))\, ds,\ x \in [a, b], \\ \xi_j =& m + n(j - m).\end{aligned}$$

In order to prove the existence of optimum solutions of system (14.1)–(14.2), we will use the concept of measure of noncompactness (MNC) and best proximity points.

The first measure of noncompactness, the function α, was defined and studied by Kuratowski [28] in 1930. In 1955, Darbo [7] was the first to use the function α to prove his famous fixed-point theorem. Darbo's fixed-point theorem is a very important generalization of Schauder's fixed-point theorem and includes the existence part of Banach's fixed-point theorem. The most familiar other measures of noncompactness were introduced by Goldenštein, Gohberg, and Markus [22] in 1957, namely, the Hausdorff or ball measure of noncompactness denoted by β, which was later studied by Goldenštein and Markus [23], and the separation or Istrăţesku measure of noncompactness [29] in 1972. The reader interested in the theory of measures of noncompactness is referred to the monographs [2, 5, 6, 9, 31, 42]. Let us denote the collection of all nonempty bounded subset of $\mathcal{X}$ by $\mathbb{B}(\mathcal{X})$.

Definition 14.2 ([2, 5, 6]) A function $\varkappa : \mathbb{B}(\mathcal{X}) \to \mathbb{R}^+$ is said to be an MNC if it satisfies the below listed three axioms for $\mathcal{P}, \mathcal{Q} \in \mathbb{B}(\mathcal{X})$:

i $\varkappa(\mathcal{P}) = 0 \Leftrightarrow \mathcal{P}$ is relatively compact.
ii $\varkappa(\mathcal{P}) = \varkappa(\overline{\mathcal{P}})$, where $\overline{\mathcal{P}}$ is closure of the set $\mathcal{P}$.
iii $\varkappa(\mathcal{P} \cup \mathcal{Q}) = \max\{\varkappa(\mathcal{P}), \varkappa(\mathcal{Q})\}$.

An MNC $\varkappa$ on $\mathbb{B}(\mathcal{X})$ satisfies following properties:

(a) $\mathcal{P} \subset \mathcal{Q}$ implies $\varkappa(\mathcal{P}) \leq \varkappa(\mathcal{Q})$.
(b) $\varkappa(\mathcal{P}) = 0$ if $\mathcal{P}$ is a finite set.
(c) $\varkappa(\mathcal{P} \cap \mathcal{Q}) \leq \min\{\varkappa(\mathcal{P}), \varkappa(\mathcal{Q})\}$.
(d) If $\varkappa(\mathcal{P}_n) \to 0$ as $n \to +\infty$ for a nonincreasing sequence $\{\mathcal{P}_n\}$ of nonempty, bounded, and closed subsets of $\mathcal{X}$, then $\mathcal{P}_\infty = \cap_{n\geq 1}\mathcal{P}_n$ is nonempty and compact.

On a Banach space $\mathcal{X}$, $\varkappa$ has the following properties.

(i) $\varkappa(\overline{con}(\mathcal{Q})) = \varkappa(\mathcal{Q})$, where $\overline{con}(\mathcal{Q})$ is convex and closed hull of $\mathcal{Q} \in \mathbb{B}(\mathcal{X})$.
(ii) $\varkappa(\lambda\mathcal{Q}) = |\lambda|\varkappa(\mathcal{Q})$ for any number λ and $\mathcal{Q} \in \mathbb{B}(\mathcal{X})$.
(iii) $\varkappa(\mathcal{P} + \mathcal{Q}) \leq \varkappa(\mathcal{P}) + \varkappa(\mathcal{Q})$.

Now we revisit the concept of best proximity points and pairs. Let $\mathcal{P}$ and $\mathcal{Q}$ be two nonempty subsets of a normed linear space (in short, $\mathcal{NLS}$) $\mathcal{X}$. It should be assumed that a pair $(\mathcal{P}, \mathcal{Q})$ has a certain property, if the same property holds true for $\mathcal{P}$ and $\mathcal{Q}$ individually. For instance, a pair $(\mathcal{P}, \mathcal{Q})$ is open if and only if $\mathcal{P}$ and $\mathcal{Q}$ both are open sets. We assume the concept of distance between two sets $\mathcal{P}$ and $\mathcal{Q}$ as $dist(\mathcal{P}, \mathcal{Q}) = \inf\{\|p - q\| : p \in \mathcal{P}, q \in \mathcal{Q}\}$. We define: $\mathcal{P}_0$ is a collection of all p in $\mathcal{P}$ such that there exists q' in $\mathcal{Q}$ with $\|p - q'\| = dist(\mathcal{A}, \mathcal{B})$, and $\mathcal{Q}_0$ is the collection of all q in $\mathcal{Q}$ such that there exists $p' \in \mathcal{P}$ with $\|p' - q\| = dist(\mathcal{A}, \mathcal{B})$. It is to be noted that, in a Banach space $\mathcal{X}$, $(\mathcal{P}_0, \mathcal{Q}_0)$ is a nonempty, convex, and weakly compact pair if $(\mathcal{P}, \mathcal{Q})$ is nonempty, convex, and weakly compact [13]. A nonempty pair $(\mathcal{P}, \mathcal{Q})$ in a $\mathcal{NLS}$ $\mathcal{X}$ is said to be proximinal if $\mathcal{P}_0 = \mathcal{P}$ and $\mathcal{Q}_0 = \mathcal{Q}$.

Let us consider a mapping $T : \mathcal{P} \cup \mathcal{Q} \to \mathcal{P} \cup \mathcal{Q}$. This mapping T is called cyclic if T maps $\mathcal{P}$ into $\mathcal{Q}$ and $\mathcal{Q}$ into $\mathcal{P}$, whereas it is called noncyclic if $T(\mathcal{P}) \subseteq \mathcal{P}$ and $T(\mathcal{Q}) \subseteq \mathcal{Q}$. The mapping T is relatively nonexpansive if it satisfies $\|Tz - Tw\| \leq \|z - w\|$ whenever $z \in \mathcal{P}$ and $w \in \mathcal{Q}$. On the other hand, T is called compact if the image set of $\mathcal{P}$ and $\mathcal{Q}$ under T are relatively compact. Here onwards $\mathcal{M}_{CRE}(\mathcal{P}, \mathcal{Q})$ and $\mathcal{M}_{NCRE}(\mathcal{P}, \mathcal{Q})$ denote the collection of all cyclic and noncyclic relatively nonexpansive mappings $T : \mathcal{P} \cup \mathcal{Q} \to \mathcal{P} \cup \mathcal{Q}$.

The best proximity point exists for a cyclic mapping T and is defined as a point $u^* \in \mathcal{P} \cup \mathcal{Q}$ such that $\|u^* - Tu^*\| = dist(\mathcal{P}, \mathcal{Q})$, whereas in case of noncyclic mapping T, a pair $(u, v) \in (\mathcal{P}, \mathcal{Q})$ satisfying the conditions $u = Tu,\ v = Tv$, and $\|u - v\| = dist(\mathcal{P}, \mathcal{Q})$ is called a best proximity pair.

The concept of best proximity point and pairs is introduced in 2005 by Eldred et al. in [13]. They coined the notion of cyclic (noncyclic) relatively nonexpansive mappings and proved the existence of the best proximity point (pair) in Banach spaces. Gabeleh [15] presented the following existence result for best proximity point (pair) of cyclic (noncyclic) relatively nonexpansive mappings.

Theorem 14.1 ([15]) *Let $T \in \mathcal{M}_{CRE}(\mathcal{P}, \mathcal{Q})$ be a compact mapping. Suppose that $(\mathcal{P}, \mathcal{Q})$ is a nonempty, bounded, closed, and convex (in short, $\mathcal{NBCC}$) pair in a Banach space $\mathcal{X}$. Then T admits a best proximity point if $\mathcal{P}_0$ is nonempty.*

By a strictly convex Banach space $\mathcal{X}$, we mean for $p, q, r \in \mathcal{X}$ and $\Lambda > 0$, we have

$$\left[\|p - r\| \leq \Lambda,\ \|q - r\| \leq \Lambda,\ p \neq q\right] \text{implies} \left\| \frac{p+q}{2} - r \right\| < \Lambda.$$

The L^p space, $1 < p < +\infty$, and Hilbert space are examples of strictly convex Banach spaces.

Theorem 14.2 ([15]) *Let $T \in \mathcal{M}_{NCRE}(\mathcal{P}, \mathcal{Q})$ be a compact mapping and $(\mathcal{P}, \mathcal{Q})$ an $\mathcal{NBCC}$ pair in a strictly convex Banach space $\mathcal{X}$. Then T admits a best proximity pair provided it is compact and P_0 is nonempty.*

The concept of compactness or completely continuous mappings used in Theorem 14.1 and 14.2 is a very strong condition. It is not that much applicable in all the situations. The measure of noncompactness served a great purpose by weakening the condition of compactness and allowing us to select the large classes of mappings which are more general than compact mappings. On this line of idea, Gabeleh and Markin in [19] generalized Theorems 14.1 and 14.2 by relaxing the condition of compactness on operator T by using the concept MNC and applied the obtained results to actualize the optimum solutions of system of differential equations. Recently, the results of [19] have been generalized further in different directions in [16–18, 20, 21, 33–35]) in which best proximity point (pairs) results are obtained using measure of noncompactness.

14.3 Main Results

This section presents the main results of our chapter. There is a twofold aim in this manuscript. The first is to prove the new best proximity point and pair results for novel classes of mappings. The second but primary goal is to prove the existence of optimum solutions for the system of FDEs (14.1)–(14.2).

14.3.1 Best Proximity Point and Pair Theorems

First, we consider the following classes of mappings from [8] which will be used in the sequel to define new classes of cyclic and noncyclic mappings.

i Let us denote by $\mathbb{K}$ the collection of all nondecreasing and continuous functions $\kappa : [0, +\infty) \times [0, +\infty) \to [0, +\infty)$ that satisfy the following conditions:

(1) $\max\{a, b\} \leq \kappa(a, b)$ for $a, b \geq 0$.
(2) $\kappa(a + b, c + d) \leq \kappa(a, c) + \kappa(b, d)$.

ii Let $\mathbb{M}$ denote the class of all functions $\mathfrak{N} : [0, +\infty) \to [0, +\infty)$ such that for any sequence $\{\xi_n\}$ in $[0, +\infty)$, $\lim_{n\to+\infty} \mathfrak{N}(\xi_n) = 0$ if $\lim_{n\to+\infty} \xi_n = 0$.

iii Let $\mathbb{U}$ denote a class of all continuous mappings $\mathfrak{U} : [0, +\infty) \to [0, +\infty)$ satisfying:

(1) $\mathfrak{U}(t) = 0$ if and only if $t = 0$
(2) $\lim_{n\to+\infty} \mathfrak{N}(\xi_n) \leq \mathfrak{U}(\xi)$ if $\lim_{n\to+\infty} \xi_n = \xi > 0$ for $\mathfrak{N} \in \mathbb{M}$

The following is the concept of ξ-admissibility.

Definition 14.3 Let S be a nonempty set. Let $\xi : 2^S \to [0, +\infty)$. A function $T : S \to S$ is said to be ξ-admissible if for every $M_1, M_2 \in 2^S$, we have $\xi(M_1 \cup M_2) \geq 1$ implying $\xi(\overline{con}(T(M_1) \cup T(M_2))) \geq 1$.

The following theorem for a new class of cyclic operators is our first main result.

Theorem 14.3 *Let (P, Q) be a nonempty and convex pair in a Banach space $\mathfrak{E}$ with P_0 being nonempty and $\varkappa$ an MNC on $\mathfrak{E}$. Let $T \in \mathcal{M}_{CRE}(P, Q)$ be a ξ-admissible mapping such that for every $\mathcal{NBCC}$, proximinal, and T-invariant pair $(\mathcal{M}_1, \mathcal{M}_2) \subseteq (P, Q)$ with $dist(\mathcal{M}_1, \mathcal{M}_2) = dist(P, Q)$, T satisfies*

$$\xi(M_1 \cup M_2)\, \mathfrak{U}\left[\kappa\left(\varkappa(T(\mathcal{M}_1) \cup T(\mathcal{M}_2)), \phi(\varkappa(T(\mathcal{M}_1) \cup T(\mathcal{M}_2)))\right)\right]$$
$$\leq \mathfrak{N}\left[\kappa\left(\varkappa(\mathcal{M}_1 \cup \mathcal{M}_2), \phi(\varkappa(\mathcal{M}_1 \cup \mathcal{M}_2))\right)\right] \tag{14.3}$$

where $\kappa \in \mathbb{K}$, $\mathfrak{N} \in \mathbb{M}$, $\mathfrak{U} \in \mathbb{U}$, and $\phi : [0, +\infty) \to [0, +\infty)$ is an increasing and continuous function. Then T has at least one best proximity point provided that there exists an $\mathcal{NBCC}$ pair (P_0, Q_0) in $E \times E$ such that $\xi(P_0, Q_0) \geq 1$.

Proof As P_0 is nonempty, (P_0, Q_0) is nonempty. Also one can show that (P_0, Q_0) is a convex, closed, T-invariant, and proximinal pair considering conditions on T (for more details see [21]). For $a \in P_0$, there is a $b \in Q_0$ such that $\|a - b\| = dist(P, Q)$. Since T is a relatively nonexpansive cyclic mapping,

$$\|Ta - Tb\| \leq \|a - b\| = dist(P, Q),$$

which gives $Ta \in Q_0$, that is, $T(P_0) \subseteq Q_0$. Similarly, $T(Q_0) \subseteq P_0$ and so T is cyclic on $P_0 \cup Q_0$.

Let us define a pair $(\mathscr{G}_n, \mathscr{H}_n)$ as $\mathscr{G}_n = \overline{con}(T(\mathscr{G}_{n-1}))$ and $\mathscr{H}_n = \overline{con}(T(\mathscr{H}_{n-1}))$, $n \geq 1$, with $\mathscr{G}_0 = P_0$ and $\mathscr{H}_0 = Q_0$. We claim that $\mathscr{G}_{n+1} \subseteq \mathscr{H}_n$ and $\mathscr{H}_n \subseteq \mathscr{G}_{n-1}$ for all $n \in \mathbb{N}$. We have $\mathscr{H}_1 = \overline{con}(T(\mathscr{H}_0)) = \overline{con}(TQ_0)) = \overline{con}(P_0) \subseteq P_0 = \mathscr{G}_0$. Therefore, $T(\mathscr{H}_1) \subseteq T(\mathscr{G}_0)$. So $\mathscr{H}_2 = \overline{con}(T(\mathscr{H}_1)) \subseteq \overline{con}(T(\mathscr{G}_0)) = \mathscr{G}_1$. Continuing this pattern, we get $\mathscr{H}_n \subseteq \mathscr{G}_{n-1}$ by using induction. Similarly, we can see that $\mathscr{G}_{n+1} \subseteq \mathscr{H}_n$ for all $n \in \mathbb{N}$. Thus, $\mathscr{G}_{n+2} \subseteq \mathscr{H}_{n+1} \subseteq \mathscr{G}_n \subseteq \mathscr{H}_{n-1}$ for all $n \in \mathbb{N}$. He nce, we get a decreasing sequence $\{(\mathscr{G}_{2n}, \mathscr{H}_{2n})\}$ of nonempty, closed, and convex pairs in $P_0 \times Q_0$. Moreover, $T(\mathscr{H}_{2n}) \subseteq T(\mathscr{G}_{2n-1}) \subseteq \overline{con}(T(\mathscr{G}_{2n-1})) = \mathscr{G}_{2n}$ and $T(\mathscr{G}_{2n}) \subseteq T(\mathscr{H}_{2n-1}) \subseteq \overline{con}(T(\mathscr{H}_{2n-1})) = \mathscr{H}_{2n}$. Therefore, for all $n \in \mathbb{N}$, the pair $(\mathscr{G}_{2n}, \mathscr{H}_{2n})$ is T-invariant.

Now if $(u, v) \in P_0 \times Q_0$ is a proximinal pair, then

$$dist(\mathscr{G}_{2n}, \mathscr{H}_{2n}) \leq \|T^{2n}u - T^{2n}v\| \leq \|u - v\| = dist(P, Q).$$

Next, we show that the pair $(\mathscr{G}_n, \mathscr{H}_n)$ is proximinal using mathematical induction. Obviously for $n = 0$, the pair $(\mathscr{G}_0, \mathscr{H}_0)$ is proximinal. Suppose that $(\mathscr{G}_k, \mathscr{H}_k)$ is proximinal. We show that $(\mathscr{G}_{k+1}, \mathscr{H}_{k+1})$ is also proximinal. Let x be an arbitrary member in $\mathscr{G}_{k+1} = \overline{con}(T(\mathscr{G}_k))$. Then it is represented as $x = \sum_{l=1}^{m} \lambda_l T(x_l)$ with $x_l \in \mathscr{G}_k$, $m \in \mathbb{N}$, $\lambda_l \geq 0$, and $\sum_{l=1}^{m} \lambda_l = 1$. Due to proximinality of the pair $(\mathscr{G}_k, \mathscr{H}_k)$, there exists $y_l \in \mathscr{H}_k$ for $1 \leq l \leq m$ such that $\|x_l - y_l\| = dist(\mathscr{G}_k, \mathscr{H}_k) = dist(P, Q)$. Take $y = \sum_{l=1}^{m} \lambda_l T(y_l)$. Then $y \in \overline{con}(T(\mathscr{H}_k)) = \mathscr{H}_{k+1}$ and

$$\|x - y\| = \left\| \sum_{l=1}^{m} \lambda_l T(x_l) - \sum_{l=1}^{m} \lambda_l T(y_l) \right\| \leq \sum_{l=1}^{m} \lambda_l \|x_l - y_l\| = dist(P, Q).$$

This means that the pair $(\mathscr{G}_{k+1}, \mathscr{H}_{k+1})$ is proximinal and induction does the rest to prove $(\mathscr{G}_n, \mathscr{H}_n)$ is proximinal for all $n \in \mathbb{N}$.

Now from the assumed hypothesis on T, we have

$$\begin{aligned}
&\mathfrak{U}\Bigg(\kappa\Bigg(\varkappa(\mathscr{G}_{2n+2} \cup \mathscr{H}_{2n+2}), \phi(\varkappa(\mathscr{G}_{2n+2} \cup \mathscr{H}_{2n+2}))\Bigg)\Bigg) \\
=&\mathfrak{U}\Bigg(\kappa\Big(\max\{\varkappa(\mathscr{G}_{2n+2}), \varkappa(\mathscr{H}_{2n+2})\}, \phi(\max\{\varkappa(\mathscr{G}_{2n+2}), \varkappa(\mathscr{H}_{2n+2})\})\Big)\Bigg) \\
\leq&\mathfrak{U}\Bigg(\kappa\Big(\max\{\varkappa(\mathscr{H}_{2n+1}), \varkappa(\mathscr{G}_{2n+1})\}, \phi(\max\{\varkappa(\mathscr{H}_{2n+1}), \varkappa(\mathscr{G}_{2n+1})\})\Big)\Bigg) \\
=&\mathfrak{U}\Bigg(\kappa\Big(\max\{\varkappa(\overline{con}(T(\mathscr{H}_{2n}))), \varkappa(\overline{con}(T(\mathscr{G}_{2n})))\}, \\
&\phi(\max\{\varkappa(\overline{con}(T(\mathscr{H}_{2n}))), \varkappa(\overline{con}(T(\mathscr{G}_{2n})))\})\Big)\Bigg)
\end{aligned}$$

$$
\begin{aligned}
&=\mathfrak{U}\Big(\kappa\big(\max\{\varkappa(T(\mathscr{H}_{2n})), \varkappa(T(\mathscr{G}_{2n}))\}, \phi(\max\{\varkappa(T(\mathscr{H}_{2n})), \varkappa(T(\mathscr{G}_{2n}))\})\big)\Big)\\
&=\mathfrak{U}\Big(\kappa\big(\varkappa(T(\mathscr{G}_{2n}) \cup T(\mathscr{H}_{2n})), \phi(\varkappa(T(\mathscr{G}_{2n}) \cup T(\mathscr{H}_{2n})))\big)\Big)\\
&\leq \xi(\mathscr{G}_{2n} \cup \mathscr{H}_{2n})\, \mathfrak{U}\Big(\kappa\big(\varkappa(T(\mathscr{G}_{2n}) \cup T(\mathscr{H}_{2n})), \phi(\varkappa(T(\mathscr{G}_{2n}) \cup T(\mathscr{H}_{2n})))\big)\Big)\\
&\leq \mathfrak{N}\Big(\kappa\big(\varkappa(\mathscr{G}_{2n} \cup \mathscr{H}_{2n}), \phi(\varkappa(\mathscr{G}_{2n} \cup \mathscr{H}_{2n}))\big)\Big).
\end{aligned} \tag{14.4}
$$

If $\varkappa(\mathscr{G}_{n_0} \cup \mathscr{H}_{n_0}) = 0$ for some $n_0 \in \mathbb{N}$, then $(\mathscr{G}_{n_0}, \mathscr{H}_{n_0})$ is a compact pair and the required outcome will be achieved by virtue of Theorem 14.1.

Now if $\varkappa(\mathscr{G}_{2n} \cup \mathscr{H}_{2n}) > 0$, then $\{\varkappa(\mathscr{G}_{2n} \cup \mathscr{H}_{2n})\}$ is a nonnegative, decreasing sequence which is bounded below sequence for all $n \in \mathbb{N}$. Consequently, there exists $l \geq 0$ such that $\lim_{n\to+\infty} \varkappa(\mathscr{G}_{2n} \cup \mathscr{H}_{2n}) = l$. We claim that $l = 0$. For this suppose that $l > 0$. Applying limit as $n \to +\infty$ in (14.4), we get

$$
\mathfrak{U}\big(\kappa\big(l, \phi(l)\big)\big) \leq \mathfrak{N}\big(\kappa\big(l, \phi(l)\big)\big) \leq \mathfrak{U}\big(\kappa\big(l, \phi(l)\big)\big),
$$

which is a contradiction. Therefore, $l = 0$. That means $\lim_{n\to+\infty} \varkappa(\mathscr{G}_{2n} \cup \mathscr{H}_{2n}) = 0$. That is, $\lim_{n\to+\infty} \varkappa(\mathscr{G}_{2n} \cup \mathscr{H}_{2n}) = \max\{\lim_{n\to+\infty} \varkappa(\mathscr{G}_{2n}), \lim_{n\to+\infty} \varkappa(\mathscr{H}_{2n})\} = 0$. Now let $\mathscr{G}_\infty = \cap_{n=0}^{+\infty} \mathscr{G}_{2n}$ and $\mathscr{H}_\infty = \cap_{n=0}^{+\infty} \mathscr{H}_{2n}$. By property (d) of MNC, the pair $(\mathscr{G}_\infty, \mathscr{H}_\infty)$ is nonempty, convex, compact, and T-invariant with $dist(\mathscr{G}_\infty, \mathscr{H}_\infty) = dist(P, Q)$. All this is sufficient to ensure that T admits a best proximity point by Theorem 14.1. □

The next result is second main result of the chapter which is analogous of the above theorem for relatively nonexpansive noncyclic mapping.

Theorem 14.4 *Let (P, Q) be a nonempty and convex pair in a strictly convex Banach space E with P_0 being nonempty and $\varkappa$ an MNC on $\mathfrak{E}$. Let $T \in \mathscr{M}_{NCRE}(P, Q)$ be a ξ-admissible mapping such that for every $\mathcal{NBCC}$, proximinal, and T-invariant pair $(\mathcal{M}_1, \mathcal{M}_2) \subseteq (P, Q)$ with $dist(\mathcal{M}_1, \mathcal{M}_2) = dist(P, Q)$, T satisfies*

$$
\begin{aligned}
&\xi(M_1 \cup M_2)\, \mathfrak{U}\Big[\kappa\Big(\varkappa(T(\mathcal{M}_1) \cup T(\mathcal{M}_2)), \phi(\varkappa(T(\mathcal{M}_1) \cup T(\mathcal{M}_2)))\Big)\Big]\\
&\leq \mathfrak{N}\Big[\kappa\big(\varkappa(\mathcal{M}_1 \cup \mathcal{M}_2), \phi(\varkappa(\mathcal{M}_1 \cup \mathcal{M}_2))\big)\Big]
\end{aligned} \tag{14.5}
$$

where $\kappa \in \mathbb{K}$, $\mathfrak{N} \in \mathbb{M}$, $\mathfrak{U} \in \mathbb{U}$ and $\phi : [0, +\infty) \to [0, +\infty)$ is an increasing and continuous function. Then T has at least one best proximity pair provided that there exists an $\mathcal{NBCC}$ pair (P_0, Q_0) in $E \times E$ such that $\xi(P_0, Q_0) \geq 1$.

Proof It is clear that (P_0, Q_0) is an $\mathcal{NBCC}$ pair which is proximinal and T-invariant (see [21] for more details on proof). Let $(a, b) \in P_0 \times Q_0$ be such that $\|a - b\| = dist(P, Q)$. Since T is a relatively nonexpansive noncyclic mapping,

$$\|Ta - Tb\| \leq \|a - b\| = dist(P, Q),$$

which gives $Ta \in P_0$, that is, $T(P_0) \subseteq P_0$. Similarly, $T(Q_0) \subseteq Q_0$ and so T is noncyclic on $P_0 \cup Q_0$.

Let us define a pair $(\mathcal{G}_n, \mathcal{H}_n)$ as $\mathcal{G}_n = \overline{con}(T(\mathcal{G}_{n-1}))$ and $\mathcal{H}_n = \overline{con}(T(\mathcal{H}_{n-1}))$, $n \geq 1$, with $\mathcal{G}_0 = P_0$ and $\mathcal{H}_0 = Q_0$. We have $\mathcal{H}_1 = \overline{con}(T(\mathcal{H}_0)) = \overline{con}(T(Q_0)) \subseteq Q_0 = \mathcal{H}_0$. Therefore, $T(\mathcal{H}_1) \subseteq T(\mathcal{H}_0)$. Thus, $\mathcal{H}_2 = \overline{con}(T(\mathcal{H}_1)) \subseteq \overline{con}(T(\mathcal{H}_0)) = \mathcal{H}_1$. Continuing this pattern, we get $\mathcal{H}_n \subseteq \mathcal{H}_{n-1}$ by using induction. Similarly, we can see that $\mathcal{G}_{n+1} \subseteq \mathcal{G}_n$ for all $n \in \mathbb{N}$. Hence, we get a decreasing sequence $\{(\mathcal{G}_n, \mathcal{H}_n)\}$ of nonempty, closed, and convex pairs in $P_0 \times Q_0$. Also, $T(\mathcal{H}_n) \subseteq T(\mathcal{H}_{n-1}) \subseteq \overline{con}(T(\mathcal{H}_{n-1})) = \mathcal{H}_n$ and $T(\mathcal{G}_n) \subseteq T(\mathcal{G}_{n-1}) \subseteq \overline{con}(T(\mathcal{G}_{n-1})) = \mathcal{G}_n$. Therefore, for all $n \in \mathbb{N}$, the pair $(\mathcal{G}_n, \mathcal{H}_n)$ is T-invariant. From the proof of Theorem 14.3, we have $(\mathcal{G}_n, \mathcal{H}_n)$ being a proximinal pair such that $dist(\mathcal{G}_n, \mathcal{H}_n) = dist(P, Q)$ for all $n \in \mathbb{N} \cup \{0\}$.

Now since $\{\varkappa(\mathcal{G}_n \cup \mathcal{H}_n)\}$ is a positive nonincreasing sequence and following the proof of Theorem 14.3, we can prove that $\{\varkappa(\mathcal{G}_n \cup \mathcal{H}_n)\}$ converges to 0.

Therefore, $\varkappa(\mathcal{G}_n \cup \mathcal{H}_n) \to 0$ as $n \to +\infty$. That is, $\lim_{n\to+\infty} \varkappa(\mathcal{G}_n \cup \mathcal{H}_n) = \max\left\{\lim_{n\to+\infty} \varkappa(\mathcal{G}_n), \lim_{n\to+\infty} \varkappa(\mathcal{H}_n)\right\} = 0$. Now let $\mathcal{G}_\infty = \cap_{n=0}^{\infty} \mathcal{G}_n$ and $\mathcal{H}_\infty = \cap_{n=0}^{+\infty} \mathcal{H}_n$. By property (d) of MNC, $(\mathcal{G}_\infty, \mathcal{H}_\infty)$ is a nonempty, convex, compact, and T-invariant pair with $dist(\mathcal{G}_\infty, \mathcal{H}_\infty) = dist(P, Q)$. All this is sufficient to ensure that T admits a best proximity pair by Theorem 14.2. □

Now, we give some consequences of above theorems as corollaries.

Corollary 14.1 *Let (P, Q) be a nonempty and convex pair in a (strictly convex) Banach space $\mathfrak{E}$ with P_0 being nonempty and $\varkappa$ an MNC on $\mathfrak{E}$. Let $T : P \cup Q \to P \cup Q$ be a cyclic (noncyclic), relatively nonexpansive, and ξ-admissible mapping such that for every $\mathcal{NBCC}$, proximinal, and T-invariant pair $(\mathcal{M}_1, \mathcal{M}_2) \subseteq (P, Q)$ with $dist(\mathcal{M}_1, \mathcal{M}_2) = dist(P, Q)$, T satisfies*

$$\begin{aligned} &\xi(M_1 \cup M_2)\, \mathfrak{U}\big(\varkappa(T(\mathcal{M}_1) \cup T(\mathcal{M}_2)) + \phi(\varkappa(T(\mathcal{M}_1) \cup T(\mathcal{M}_2)))\big) \\ \leq &\mathfrak{N}\big(\varkappa(\mathcal{M}_1 \cup \mathcal{M}_2) + \phi(\varkappa(\mathcal{M}_1 \cup \mathcal{M}_2))\big) \end{aligned} \tag{14.6}$$

where $\mathfrak{N} \in \mathbb{M}$, $\mathfrak{U} \in \mathbb{U}$ and $\phi : [0, +\infty) \to [0, +\infty)$ is an increasing and continuous function. Then T has at least one best proximity point (pair) provided that there exists an $\mathcal{NBCC}$ pair (P_0, Q_0) in $E \times E$ such that $\xi(P_0, Q_0) \geq 1$.

Proof Taking $\kappa(a, b) = a + b$ in Theorem 14.3 (Theorem 14.4), we obtain the desired result. □

Corollary 14.2 *Let (P, Q) be a nonempty and convex pair in a (strictly convex) Banach space $\mathfrak{E}$ with P_0 being nonempty and $\varkappa$ an MNC on $\mathfrak{E}$. Let $T : P \cup Q \to P \cup Q$ be a cyclic (noncyclic) relatively nonexpansive mapping such that for every $\mathcal{NBCC}$, proximinal, and T-invariant pair $(\mathcal{M}_1, \mathcal{M}_2) \subseteq (P, Q)$ with $dist(\mathcal{M}_1, \mathcal{M}_2) = dist(P, Q)$, T satisfies*

$$\mathfrak{U}\left[\kappa\left(\varkappa(T(\mathcal{M}_1) \cup T(\mathcal{M}_2)), \phi(\varkappa(T(\mathcal{M}_1) \cup T(\mathcal{M}_2)))\right)\right]$$
$$\leq \mathfrak{N}\left[\kappa\left(\varkappa(\mathcal{M}_1 \cup \mathcal{M}_2), \phi(\varkappa(\mathcal{M}_1 \cup \mathcal{M}_2))\right)\right]$$

where $\kappa \in \mathbb{K}$, $\mathfrak{N} \in \mathbb{M}$ and $\mathfrak{U} \in \mathbb{U}$. Then T has at least one best proximity point (pair) provided that there exists an $\mathcal{NBCC}$ pair (P_0, Q_0) in $E \times E$ such that $\xi(P_0, Q_0) \geq 1$.

Proof Taking $\xi(M_1 \cup M_2) = 1$ in Theorem 14.3 (Theorem (14.4)), we obtain the desired result. □

Corollary 14.3 *Let (P, Q) be a nonempty and convex pair in a (strictly convex) Banach space $\mathfrak{E}$ with P_0 being nonempty and $\varkappa$ an MNC on $\mathfrak{E}$. A relatively nonexpansive cyclic (noncyclic) mapping $T : P \cup Q \to P \cup Q$ has at least one best proximity point (pair) if for every $\mathcal{NBCC}$, proximinal, and T-invariant pair $(\mathcal{M}_1, \mathcal{M}_2) \subseteq (P, Q)$ with $dist(\mathcal{M}_1, \mathcal{M}_2) = dist(P, Q)$, T satisfies*

$$\varkappa(T(\mathcal{M}_1) \cup T(\mathcal{M}_2))) \leq k\varkappa(\mathcal{M}_1 \cup \mathcal{M}_2)$$

where $k \in [0, 1)$.

Proof Taking $\varkappa(a, b) = a + b$, $\varphi(t) = 0$, $\mathfrak{U}(t) = t$, and $\mathfrak{N}(t) = kt$ where $k \in [0, 1)$ in Corollary 14.2, we obtain the desired result. □

14.3.2 Optimum Solutions of System of FDEs (14.1)–(14.2)

Now we proceed to prove the existence of optimum solutions of the system (14.1)–(14.2). Let $(\mathfrak{E}, \|\cdot\|)$ be any Banach space containing α_a, β_a and a, b and δ be numbers that are positive and real. We denote by $\mathcal{A}_\alpha$ and $\mathcal{B}_\beta$ closed balls in $\mathfrak{E}$ centered at α_a and β_a, respectively, with radius δ. Let $\mathcal{J} \subseteq [a, b]$. We denote the collection of all continuous mappings from $\mathcal{J}$ into $\mathfrak{E}$ by $C(\mathcal{J}, \mathfrak{E})$. In fact, if $\mathfrak{S} = C(\mathcal{J}, \mathfrak{E})$ is endowed with a supremum norm, then it is a Banach space. Let

$$\mathfrak{A}_1 = \{\psi \in C(\mathcal{J}, \mathcal{A}_\alpha) : {}^{j}I^{(j-\xi_j);\varphi}\psi(a) = \alpha_a\},$$
$$\mathfrak{A}_2 = \{\eta \in C(\mathcal{J}, \mathcal{B}_\beta) : {}^{j}I^{(j-\xi_j);\varphi}\eta(a) = \beta_a\}.$$

So, $(\mathfrak{A}_1, \mathfrak{A}_2)$ is an $\mathcal{NBCC}$ pair in $\mathfrak{S} \times \mathfrak{S}$. Here for all $\mu \in \mathfrak{A}_1$ and $\nu \in \mathfrak{B}_2$, we have $\|\mu - \nu\| = \sup \|\mu(s) - \nu(s)\| \geq \|\alpha_a - \beta_a\|$. Therefore, $\text{dist}(\mathfrak{A}_1, \mathfrak{A}_2) = \|\alpha_a - \beta_a\|$ which indicates that $(\mathfrak{A}_1)_0$ is nonempty.

Now, we define an operator $T : \mathfrak{A}_1 \cup \mathfrak{A}_2 \to \mathfrak{S}$ as follows:

$$T\psi(x) = \begin{cases} \dfrac{(\varphi(x)-\varphi(a))^{\frac{\xi_j}{j}-1}}{\Gamma(\xi_j)}\beta_a \\ +\frac{1}{j\Gamma_j(m)}\int_a^x \varphi'(s)(\varphi(x)-\varphi(s))^{\frac{m}{j}-1}\Lambda_1(s,\psi(s))\,\mathrm{d}s, & \psi \in \mathfrak{A}_1, \\ \dfrac{(\varphi(x)-\varphi(a))^{\frac{\xi_j}{j}-1}}{\Gamma(\xi_j)}\alpha_a \\ +\frac{1}{j\Gamma_j(m)}\int_a^x \varphi'(s)(\varphi(x)-\varphi(s))^{\frac{m}{j}-1}\Lambda_2(s,\psi(s))\,\mathrm{d}s, & \psi \in \mathfrak{A}_2. \end{cases} \tag{14.7}$$

Lemma 14.3 *If Λ_i are bounded and continuous functions such that $\Lambda_i \in L^1(a,b)$, $i = 1, 2$, then the operator $T : \mathfrak{A}_1 \cup \mathfrak{A}_2 \to \mathfrak{S}$ defined in* (14.7) *is cyclic.*

Proof We take $\psi \in \mathfrak{A}_1$ and $\xi_j = m + n(j-m)$, and we have

$$\begin{aligned} T\psi(x) =& \frac{(\varphi(x)-\varphi(a))^{\frac{\xi_j}{j}-1}}{\Gamma_j(\xi_j)}\beta_a \\ &+ \frac{1}{j\Gamma_j(m)}\int_a^x \varphi'(s)(\varphi(x)-\varphi(s))^{\frac{m}{j}-1}\Lambda_1(s,\psi(s))\,\mathrm{d}s. \end{aligned}$$

Applying ${}^jI_{a^+}^{j-\xi_j;\varphi}$ on both sides and applying Lemmas 14.1 and 14.2, we get

$$\begin{aligned} {}^jI_{a^+}^{j-\xi_j;\varphi}T\psi(x) =& \frac{\beta_a}{\Gamma_j(\xi_j)}\,{}^jI_{a^+}^{j-\xi_j;\varphi}(\varphi(x)-\varphi(a))^{\frac{\xi_j}{j}-1} \\ &+ {}^jI_{a^+}^{j-\xi_j;\varphi}\,{}^jI_{a^+}^{m;\varphi}\Lambda_1(s,\psi(s))(x) \\ =& \frac{\beta_a}{\Gamma_j(j)} + \left[{}^jI_{a^+}^{j-n(j-m);\varphi}\Lambda_1(s,\psi(s))\right](x). \end{aligned}$$

Using the fact that $\Gamma_j(j) = 1$ and $[I_{a^+}^{j-n(j-m);\varphi}\Lambda_1(s,\psi(s))](x) \to 0$ as $x \to a$ (see the proof of Theorem 7.1 in [26]). Therefore, ${}^jI_{a^+}^{j-\xi_j;\varphi}T\psi(a) = \beta_a$ which means that $T\psi(x) \in \mathfrak{A}_2$. In the same manner, we infer that $T\psi(x) \in \mathfrak{A}_1$ if $\psi \in \mathfrak{A}_2$. This concludes that T is a cyclic operator. □

We recall that $\omega \in \mathfrak{A}_1 \cup \mathfrak{A}_2$ is an optimal solution for the system (14.1) and (14.2) if $\|\omega - T\omega\| = \mathrm{dist}(\mathfrak{A}_1, \mathfrak{A}_2)$. This means we need to prove ω is a best proximity point of the operator T defined in (14.7). We need the following hypotheses to be assumed in order to prove the existence results for optimal solutions.

($\mathscr{H}_1$) Let $\varkappa$ be a MNC on $\mathfrak{E}$ and $\mathscr{L} > 0$. We suppose that for a bounded pair $(M_1, M_2) \subseteq (\mathcal{A}_\alpha, \mathcal{B}_\beta)$ such that

$$\varkappa\big(\Lambda_1(\mathcal{J} \times M_1) \cup \Lambda_2(\mathcal{J} \times M_2)\big) \le \mathscr{L}\varkappa(M_1 \cup M_2).$$

$(\mathscr{H}_2)$

$$\begin{aligned}
&\|\Lambda_1(x, \psi(x)) - \Lambda_2(x, \eta(x))\| \\
&\leq \frac{j\Gamma_j(m)}{(\varphi(b)-\varphi(a))^{\frac{m}{j}}}\left(\|\psi(x)-\eta(x)\| - \frac{(\varphi(b)-\varphi(a))^{\frac{\xi_j}{j}-1}}{\Gamma_j(\xi_j)}\|\beta_a - \alpha_a\|\right),
\end{aligned}$$

for all $(\psi, \eta) \in \mathfrak{A}_1 \times \mathfrak{A}_2$.

The following result is the mean-value theorem for fractional differential.

Theorem 14.5 ([1, 25]) *Let $F : [a, b] \to (-\infty, +\infty)$ be a continuous function, $m > 0$, and $G \in L^1([a, b])$ be a function that is not changing the sign on its domain. Then for almost every $x \in [a, b]$, there exists some $\zeta \in (a, x)$ such that*

$$I_a^m(FG)(x) = F(\zeta)I_a^m G(x). \tag{14.8}$$

Then we present the existence result.

Theorem 14.6 *With the above defined notations, the conditions of Lemma 14.3, and hypotheses $(\mathscr{H}_1)$ and $(\mathscr{H}_2)$, the system of FDEs (14.1)–(14.2) has an optimal solution.*

Proof In order to prove that systems (14.1)–(14.2) have an optimal solution, we prove that the operator T defined in (14.7) has the best proximity point.

From Lemma 14.3, T is a cyclic operator. Also $T(\mathfrak{A}_1)$ is a bounded subset of $\mathfrak{A}_2$. It is our aim to prove that $T(\mathfrak{A}_1)$ is also an equicontinuous subset of $\mathfrak{A}_2$. For $x, y \in \mathscr{J}$ with $x < y$ and $\psi \in \mathfrak{A}_1$, we observe that

$$\begin{aligned}
&\|T\psi(y) - T\psi(x)\| \\
&= \Bigg\| \frac{(\varphi(y)-\varphi(a))^{\frac{\xi_j}{j}-1}}{\Gamma_j(\xi_j)}\beta_a + \frac{1}{j\Gamma_j(m)}\int_a^y \varphi'(s)(\varphi(y)-\varphi(s))^{\frac{m}{j}-1}\Lambda_1(s, \psi(s))\,\mathrm{d}s \\
&\quad - \frac{(\varphi(x)-\varphi(a))^{\frac{\xi_j}{j}-1}}{\Gamma_j(\xi_j)}\beta_a - \frac{1}{j\Gamma_j(m)}\int_a^x \varphi'(s)(\varphi(x)-\varphi(s))^{\frac{m}{j}-1}\Lambda_1(s, \psi(s))\,\mathrm{d}s \Bigg\| \\
&= \Bigg| \frac{\beta_a}{\Gamma_j(\xi_j)}\left((\varphi(y)-\varphi(a))^{\frac{\xi_j}{j}-1} - (\varphi(x)-\varphi(a))^{\frac{\xi_j}{j}-1}\right) \\
&\quad + \frac{1}{j\Gamma_j(m)}\int_a^x \left[\varphi'(s)(\varphi(y)-\varphi(s))^{\frac{m}{j}-1} - \varphi'(s)(\varphi(x)-\varphi(s))^{\frac{m}{j}-1}\right] \\
&\quad \times \Lambda_1(s, \psi(s))\,\mathrm{d}s \\
&\quad + \frac{1}{j\Gamma_j(m)}\int_x^y \varphi'(s)(\varphi(y)-\varphi(s))^{\frac{m}{j}-1}\Lambda_1(s, \psi(s))\,\mathrm{d}s \Bigg|
\end{aligned}$$

$$\leq \frac{\beta_a}{\Gamma_j(\xi_j)} \left|\left((\varphi(y)-\varphi(a))^{\frac{\xi_j}{j}-1} - (\varphi(x)-\varphi(a))^{\frac{\xi_j}{j}-1}\right)\right|$$
$$+ \frac{M}{j\Gamma_j(m)}\left|\int_a^x \left[\varphi'(s)(\varphi(y)-\varphi(s))^{\frac{m}{j}-1} - \varphi'(s)(\varphi(x)-\varphi(s))^{\frac{m}{j}-1}\right] \mathrm{d}s\right|$$
$$+ \frac{M}{j\Gamma_j(m)}\left|\int_x^y \varphi'(s)(\varphi(y)-\varphi(s))^{\frac{m}{j}-1} \mathrm{d}s\right|.$$

As $y \to x$, the right-hand side tends to 0. Thus, $\|T\psi(y) - T\psi(x)\| \to 0$ as $y \to x$. Thus, $T(\mathfrak{A}_1)$ is equicontinuous. Similarly, it can be proved that $T(\mathfrak{A}_2)$ is a bounded and equicontinuous subset of $\mathfrak{A}_1$. Thus, Arzela-Ascoli theorem concludes that $(\mathfrak{A}_1, \mathfrak{A}_2)$ is relatively compact.

Next, we show that T is relatively nonexpansive. For any $(\psi, \eta) \in \mathfrak{A}_1 \times \mathfrak{A}_2$, with assumption $(\mathscr{H}_2)$ we have

$$\|T\psi(x) - T\eta(x)\|$$
$$= \left\| \frac{\beta_a}{\Gamma_j(\xi_j)}(\varphi(x)-\varphi(a))^{\frac{\xi_j}{j}-1} + \frac{1}{j\Gamma_j(m)}\int_a^x \varphi'(s)(\varphi(x)-\varphi(s))^{\frac{m}{j}-1}\Lambda_1(s,\psi(s))\,\mathrm{d}s \right.$$
$$- \left(\frac{\alpha_a}{\Gamma_j(\xi_j)}(\varphi(x)-\varphi(a))^{\frac{\xi_j}{j}-1}\right.$$
$$\left.\left. + \frac{1}{j\Gamma_j(m)}\int_a^x \varphi'(s)(\varphi(x)-\varphi(s))^{\frac{m}{j}-1}\Lambda_2(s,\eta(s))\,\mathrm{d}s\right)\right\|$$
$$= \frac{(\varphi(x)-\varphi(a))^{\frac{\xi_j}{j}-1}}{\Gamma_j(\xi_j)}|\beta_a - \alpha_a|$$
$$+ \left|\frac{1}{j\Gamma_j(m)}\int_a^x \varphi'(s)(\varphi(x)-\varphi(s))^{\frac{m}{j}-1}[\Lambda_1(s,\psi(s)) - \Lambda_2(s,\eta(s))]\,\mathrm{d}s\right|$$
$$\leq \frac{(\varphi(b)-\varphi(a))^{\frac{\xi_j}{j}-1}}{\Gamma_j(\xi_j)}\|\beta_a - \alpha_a\|$$
$$+ \frac{1}{j\Gamma_j(m)}\int_a^t \varphi'(s)(\varphi(x)-\varphi(s))^{\frac{m}{j}-1}\frac{j\Gamma_j(m)}{(\varphi(b)-\varphi(a))^{\frac{m}{j}}}$$
$$\times \left[\|\psi(s)-\eta(s)\| - \frac{(\varphi(b)-\varphi(a))^{\frac{\xi_j}{j}-1}}{\Gamma_j(\xi_j)}\|\beta_a - \alpha_a\|\right]\mathrm{d}s$$
$$= \|\psi - \eta\|,$$

and thereby, $\|Tx - Ty\| \leq \|x - y\|$. Therefore, T is relatively nonexpansive.

We are now in the final step of the proof. Let $(\mathscr{B}_1, \mathscr{B}_2) \subseteq (\mathfrak{A}_1, \mathfrak{A}_2)$ be a nonempty, closed, convex, and proximinal pair which is T-invariant and such that

$\text{dist}(\mathscr{B}_1, \mathscr{B}_2) = \text{dist}(\mathfrak{A}_1, \mathfrak{A}_2)(= \|\alpha_a - \beta_a\|)$. With the help of the generalized version of Arzela-Ascoli theorem (see Ambrosetti [3]) and hypotheses $(\mathscr{H}_1)$, we have

$$\begin{aligned}
&\varkappa(T(\mathscr{B}_1) \cup T(\mathscr{B}_2)) \\
&= \max\{\varkappa(T(\mathscr{B}_1)), \varkappa(T(\mathscr{B}_2))\} \\
&= \max\left\{\sup_{x\in\mathscr{J}}\{\varkappa(\{T\psi(x) : \psi \in \mathscr{B}_1\})\}, \sup_{x\in\mathscr{J}}\{\varkappa(\{T\eta(x) : \eta \in \mathscr{B}_2\})\}\right\} \\
&= \max\left\{\sup_{x\in\mathscr{J}}\left\{\varkappa\left(\left\{\frac{\beta_a\,(\varphi(x)-\varphi(a))^{\frac{\xi_j}{j}-1}}{\Gamma_j(\xi_j)}\right.\right.\right.\right. \\
&\quad\left.\left.\left.+\frac{1}{j\Gamma_j(m)}\int_a^x \varphi'(s)(\varphi(x)-\varphi(s))^{\frac{m}{j}-1}\Lambda_1(s,\psi(s))\,\mathrm{d}s) : \psi \in \mathscr{B}_1\right\}\right)\right\}, \\
&\quad\sup_{x\in\mathscr{J}}\left\{\varkappa\left(\left\{\frac{\alpha_a\,(\varphi(x)-\varphi(a))^{\frac{\xi_j}{j}-1}}{\Gamma_j(\xi_j)}\right.\right.\right. \\
&\quad\left.\left.\left.\left.+\frac{1}{j\Gamma_j(m)}\int_a^x \varphi'(s)(\varphi(x)-\varphi(s))^{\frac{m}{j}-1}\Lambda_2(s,\eta(s))\,\mathrm{d}s) : \eta \in \mathscr{B}_2\}\right)\right\}\right\}.
\end{aligned}$$

So, in view of Theorem (14.5), it follows that

$$\begin{aligned}
&\varkappa(T(\mathscr{B}_1) \cup T(\mathscr{B}_2)) \\
&\le \max\left\{\sup_{x\in\mathscr{J}}\left\{\varkappa\left(\{\frac{\beta_a\,(\varphi(x)-\varphi(a))^{\frac{\xi_j}{j}-1}}{\Gamma_j(\xi_j)}\right.\right.\right. \\
&\quad\left.\left.\left.+\frac{(\varphi(x)-\varphi(a))^{\frac{m}{j}}}{j\Gamma_j(m)}\overline{con}(\{\Lambda_1(\sigma,\psi(\sigma)) : \sigma \in \mathscr{J}\})\right\}\right)\right\}, \\
&\quad\sup_{x\in\mathscr{J}}\left\{\varkappa\left(\left\{\frac{\alpha_a\,(\varphi(x)-\varphi(a))^{\frac{\xi_j}{j}-1}}{\Gamma_j(\xi_j)}\right.\right.\right. \\
&\quad\left.\left.\left.\left.+\frac{(\varphi(x)-\varphi(a))^{\frac{m}{j}}}{j\Gamma_j(m)}\overline{con}(\{\Lambda_2(\sigma,\eta(\sigma)) : \sigma \in \mathscr{J}\})\}\right)\right\}\right\} \\
&= \max\left\{\frac{(\varphi(x)-\varphi(a))^{\frac{m}{j}}}{j\Gamma_j(m)}\varkappa(\Lambda_1(\mathscr{J}\times\mathscr{B}_1)), \frac{(\varphi(x)-\varphi(a))^{\frac{m}{j}}}{j\Gamma_j(m)}\varkappa(\Lambda_2(\mathscr{J}\times\mathscr{B}_2))\right\}
\end{aligned}$$

$$
\begin{aligned}
&=\frac{(\varphi(x)-\varphi(a))^{\frac{m}{j}}}{j\Gamma_j(m)}\varkappa(\Lambda_1(\mathcal{J}\times\mathcal{B}_1))\cup\varkappa(\Lambda_2(\mathcal{J}\times\mathcal{B}_2))\\
&<\frac{\mathcal{L}(\varphi(x)-\varphi(a))^{\frac{m}{j}}}{j\Gamma_j(m)}\varkappa(\mathcal{B}_1\cup\mathcal{B}_2).
\end{aligned}
$$

Since φ is continuous, there exists an $x \in \mathcal{J}$ such that $w = \frac{\Lambda(\varphi(x)-\varphi(a))^{\frac{m}{j}}}{j\Gamma_j(m)} < 1$. Therefore, we conclude that T satisfies all the hypotheses of Corollary 14.3 and so the operator T has the best proximity point $\omega \in \mathfrak{A}_1 \cup \mathfrak{A}_2$ which is an optimal solution for systems (14.1) and (14.2). □

References

1. Angelis, P., Marchis, R., Martire, A.L., Oliva I.: A mean-value Approach to solve fractional differential and integral equations. Chaos Solitons Fractals **138**, 109895 (2020)
2. Akhmerov, A.A, Kamenskii, M.I., Potapov,A.S., Rodkina, A.E, Sadovski, B.N.: Measure of Noncompactness and Condensing Operators. Springer, Basel (1992)
3. Ambrosetti, A.: Un teorema di esistenza per Ie equazioni differenziali negli spazi di Banach. Rend. Sem. Mat. Padova **39**, 349–361 (1967)
4. Baleanu, D., Diethelm, K., Scalas E., Trujillo, J.J.: Fractional calculus models and numerical methods. Series on Complexity, Nonlinearity and Chaos. World Scientific, London (2012)
5. Banas, J., Goebel, K.: Measure of Noncompactness in Banach Spaces. Lecture Notes in Pure and Applied Mathematics, vol. 60. Dekker, New York (1980)
6. Banas, J., Jleli, M., Mursaleen, M., Samet, B., Vetro, C. (eds.): Advances in Nonlinear Analysis via the Concept of measure of Noncompactness. Springer, Singapore (2017)
7. Darbo, G.: Punti uniti in transformazioni a codominio non compatto (Italian). Rend. Sem. Math. Univ. Padova **24**, 84–92 (1955)
8. Das, A., Paunović, M., Parvaneh, V., Mursaleen, M., Bagheri, Z.: Existence of a solution to an infinite system of weighted fractional integral equations of a function with respect to another function via a measure of noncompactness. Demonstratio Math. **56**, 20220192 (2023)
9. De Malafosse, B., Malkowsky, E., Rakočević, V.: Operators Between Sequence Spaces and Applications, pp. 366–373. Springer, Berlin (2021)
10. Diaz, R., Pariguan, E.: On hypergeometric functions and Pochhammer k-symbol, Divulg. Mat. **15**(2), 179–192 (2007)
11. Dorrego, G.A.: An alternative definition for the k-Riemann-Liouville fractional derivative. Appl. Math. Sci. **9**(10), 481–491 (2015)
12. Dorrego, G.A., Cerutti, R.A.: The k-fractional Hilfer derivative. Int. J. Math. Anal. **7**(11), 543–550 (2013)
13. Eldred, A.A., Kirk, W.A., Veeramani, P.: Proximal normal structure and relatively nonexpansive mappings. Studia Math. **171**, 283–293 (2005)
14. Furati, K.M., Kassim, M.D., Tatar, N.e-.: Existence and uniqueness for a problem involving Hilfer fractional derivative. Comput. Math. Appl. **64**, 1616–1626 (2012)
15. Gabeleh, M.: A characterization of proximal normal structures via proximal diametral sequences. J. Fixed Point Theory Appl. **19**, 2909–2925 (2017)
16. Gabeleh, M., Vetro, C.: A best proximity point approach to existence of solutions for a system of ordinary differential equations. Bull. Belgian Math. Soc. **25**(4), 493–503 (2019)
17. Gabeleh, M., Kunzi, H.P.A: Mappings of generalized condensing type in metric spaces with Busemann convex structure. Bull. Iran. Math. Soc. **46**, 1465–1483 (2020)

18. Gabeleh, M., Kunzi, H.P.A :Condensing operators of integral type in Busemann reflexive convex spaces. Bull. Malaysian Math. Soc. **43**(2), 1971–1988 (2020)
19. Gabeleh, M., Markin, J.: Optimum solutions for a system of differential equations via measure of noncompactness. Indag. Math. (N.S.) **29**(3), 895–906 (2018)
20. Gabeleh, M., Patel, D.K., Patle, P.R.: Darbo type best proximity point (pair) results using measure of noncompactness with application. Fixed Point Theory **23**(1), 247–266 (2022)
21. M. Gabeleh, M., Vetro, C. : A new extension of Darbo,s fixed point theorem using relatively Meir-Keeler condensing operators. Bull. Aust. Math. Soc. **98**(2), 286–297 (2018)
22. Goldenštein, L.S., Gohberg, I.T., Markus, A.S.: Investigations of some properties of bounded linear operators with their q-norms. Ucen. Zap. Kishinevsk. Univ. **29**, 29–36 (1957)
23. Goldenštein, L.S. Markus, A.S.: On a measure of noncompactness of bounded sets and linear operators. In: Studies in Algebra and Mathematical Analysis, pp. 45–54. Kishinev (1965)
24. Hilfer, R.: Applications of fractional calculus in Physics. World Scientific, Singapore (2000)
25. Kilbas, A.A., Srivastava, H.M., Trujillo, J.: Theory and Applications of Fractional Differential Equations. North-Holland Mathematics Studies. Elsevier, Amsterdam (2006)
26. Kuchhe, K.D., Mali, A.D.: On the nonlinear (k, ψ)-Hilfer fractional differential equations. Chaos Solitons Fractals **152**, 111335 (2021)
27. Kwun, Y.C., Farid, G., Nazeer, W., Ullah, S., Kang, S.M.: Generalized Riemann-Liouville k-fractional integrals associated with Ostrowski type inequalities and error bounds of Hadamard inequalities. IEEE Access **6**, 64946–64953 (2018)
28. Kuratowski, K.: Sur les espaces complets. Fund. Math. **15**, 301–309 (1930)
29. Istrăţescu, V.: On a measure of noncompactness. Bull. Math. Soc. Sci. Math. R. S. Roumanie (N.S.) **16**, 195–197 (1972)
30. Machado,J.A., Kiryakova, V., Mainardi, F.: Recent history of fractional calculus. Commun. Nonlinear Sci. **16**, 1140–1153 (2011)
31. Malkowsky, E., Rakočević, V.: Advanced Functional Analysis. CRCPress, Taylor and Francis Group, Boca Raton, London, New York (2019)
32. Mubeen, S., Habibullah, G.M.: k-fractional integrals and application. Int. J. Contemp. Math. Sci. **7**(2), 89–94 (2012)
33. Nashine, H.K., Arab, R., Patle, P.R., Patel, D.K.: Best proximity point results via measure of noncompactness and application. Numer. Funct. Anal. Optim. **42**(4), 430–442 (2021)
34. Patle, P.R., Gabeleh, M., Rakočević, V.: Sadovskii type best proximity point (pair) theorems with an application to fractional differential equations. Mediterr. J. Math. **19**, 141 (2022)
35. Patle, P.R.,Patel, D.K., Arab, R.: Darbo type best proximity point results via simulation function with application. Afrika Mat. (3) **31**, 833–845 (2020)
36. Podlubny, I.: Fractional Differential Equations. Academic Press, New York (1999)
37. Podlubny, I.: Matrix approach to discrete fractional calculus. Fract. Calc. Appl. Anal. **3**, 359–386 (2000)
38. Romero, L.G., Luque, L.L., Dorrego, G.A., Cerutti, R.A.: On the k-Riemann-Liouville fractional derivative. Int. J. Contemp. Math. Sci. **8**(1) 41–51 (2013)
39. Samko, S.G., Kilbas, A.A., Marichev, O.I.: Fractional Integral and Derivative. Gordon and Breach, London (1993)
40. Sousa, J., Oliveira, E.: On the ψ-Hilfer fractional derivative. Commun. Nonlinear Sci. Numer. Simul. **60**, 72–91 (2018)
41. Sousa, J., Oliveira, E.: On the Ulama Hyersa Rassiaas stability for nonlinear fractional differential equations using the ψ-Hilfer operator. J. Fixed Point Theory Appl. (2018)
42. Toledano, J.M.A., Benavides, T.D., Acedo, G.L.: Measures of Noncompactness in Metric Fixed Point Theory. Operator Theory Advances and Applications, vol. 99. Birkhauser Verlag, Basel, Boston, Berlin (1997)

Chapter 15
Iterative Methods for the Inclusion of the Inverse Matrix

Marko D. Petković and Miodrag S. Petković

Dedicated to Academician Gradimir V. Milovanović on the occasion of his 75th birthday

15.1 Introduction

A number of tasks in different fields are modeled in the matrix form, which is often reduced to computing the inverse matrix. There is a various literature in this area so that we will not consider all matrix numerical methods of iterative nature. Some recent methods are [10, 11, 13]. Instead, in this chapter we concentrate only on that small branch of matrix iterative analysis concerned with the efficient determination of inverse matrices with upper error bound of the solution using interval arithmetic. The presented study is a two-way connection between linear algebra and computing.

The chapter is divided into four sections and organized as follows. In Sect. 15.2 we give some preliminary matrix properties and definitions and a short study of hyper-power matrix iterations. The main goal of this chapter is to state an efficient iterative method of order six for the inclusion of the inverse of a given regular matrix, which is the subject of Sect. 15.3. This method is constructed by modifying a hyper-power method in such a way that the computational cost is decreased. In order to provide information on the upper error bounds of the approximate interval matrix, interval arithmetic is used. Computational aspects of the considered interval methods and two numerical examples are considered in Sect. 15.4. We show that computational efficiency of the proposed method is equal or higher than the methods of hyper-power's type realized in a Horner scheme fashion. These theoretical considerations are further confirmed by numerical experiments.

M. D. Petković (✉)
University of Niš, Faculty of Sciences and Mathematics, Niš, Serbia

M. S. Petković
University of Niš, Faculty of Electronic Engineering, Niš, Serbia

M. P. Stanić et al. (eds.), *Analysis, Approximation, Optimization: Computation and Applications*, Springer Optimization and Its Applications 224,
https://doi.org/10.1007/978-3-031-85743-0_15

15.2 Hyper-Power Methods

Applying numerical methods on digital computers, one of the most important task is to provide an information on the accuracy of obtained results. The interest of bounding roundoff errors in matrix computations has come from the impossibility of exact representation of elements of matrices in some cases since numbers are represented in the computer by string of bits of fixed, *finite* length. For more details see [4, 7, 8]. Such case also appears in finding inverse matrices, the subject of this chapter. To provide the upper error bound of the outer matrix for the inverse matrix, we will combine point and interval iterations. The essential advantage of the presented interval methods consists of capturing all the roundoff errors automatically, making this approach a useful, elegant, and powerful tool for finding errors in the sought results.

To avoid any confusion, in this chapter interval matrices will be denoted by bold capital letters and real matrices (often called point matrices) by calligraphic letters with a dot below the letter. We use bold small letters to denote real intervals.

The set of $m \times n$ matrices over the real numbers will be denoted by $M_{mn}(\mathbb{R})$. Let $\boldsymbol{C} = [\boldsymbol{c}_{ij}]$ be an $m \times n$ interval matrix, where $\boldsymbol{c}_{ij} = [\underline{c}_{ij}, \overline{c}_{ij}]$, $\overline{c}_{ij} - \underline{c}_{ij} \geq 0$, are real intervals. We will write $M_{mn}(I(\mathbb{R}))$ for the set of $m \times n$ interval matrices over the real intervals. An interval matrix whose all elements are points (real numbers) is called a *point matrix*. Point matrices will be denoted by $\underset{\bullet}{C}$. Basic definitions, operations, and properties of interval matrices can be found in detail in [1, Ch. 10] and [8].

For a given interval matrix $\boldsymbol{C} = [\boldsymbol{c}_{ij}]$ let us define corresponding point matrices, the *midpoint matrix* $m(\boldsymbol{C}) := [m(\boldsymbol{c}_{ij})]$, the *width matrix* $d(\boldsymbol{C}) := [d(\boldsymbol{c}_{ij})]$, and the *absolute value matrix* $|\boldsymbol{C}| := [|\boldsymbol{c}_{ij}|]$, as follows:

$$m(\boldsymbol{c}_{ij}) = \frac{1}{2}(\underline{c}_{ij} + \overline{c}_{ij}), \quad d(\boldsymbol{c}_{ij}) := \overline{c}_{ij} - \underline{c}_{ij}, \quad |\boldsymbol{c}_{ij}| = \max\{|\overline{c}_{ij}|, |\underline{c}_{ij}|\}.$$

If $\boldsymbol{C} = \underset{\bullet}{C} = [c_{ij}]$ is a point matrix, then it is obvious

$$m(\underset{\bullet}{C}) = [c_{ij}], \quad d(\underset{\bullet}{C}) = [0] \text{ (null-matrix)}, \quad |\underset{\bullet}{C}| = [|c_{ij}|].$$

The intersection of two interval matrices $\boldsymbol{A}, \boldsymbol{B} \in M_{mn}(I(\mathbb{R}))$ is defined as a matrix whose elements are intersections of real intervals, that is,

$$\boldsymbol{A} \cap \boldsymbol{B} = [\boldsymbol{a}_{ij} \cap \boldsymbol{b}_{ij}],$$

assuming that the set theoretic intersection is nonempty.

We start with the following matrix identity for an $n \times n$ matrix $\underset{\bullet}{Q}$ and the unity matrix $\underset{\bullet}{I}$:

$$(\underset{\bullet}{I} - \underset{\bullet}{Q})(\underset{\bullet}{I} + \underset{\bullet}{Q} + \cdots + \underset{\bullet}{Q}^{r-2}) = \underset{\bullet}{I} - \underset{\bullet}{Q}^{r-1}.$$

Hence, setting $\underset{\bullet}{Q} = \underset{\bullet}{A}\underset{\bullet}{H}$, where $\underset{\bullet}{H}$ is an $n \times n$ matrix, the following identity is obtained:

$$\underset{\bullet}{H}\sum_{\lambda=0}^{r-2}(\underset{\bullet}{I}-\underset{\bullet}{A}\underset{\bullet}{H})^{\lambda} = \underset{\bullet}{A}^{-1} - \underset{\bullet}{A}^{-1}(\underset{\bullet}{I}-\underset{\bullet}{A}\underset{\bullet}{H})^{r-1}. \tag{15.1}$$

From (15.1) there follows

$$\underset{\bullet}{A}^{-1} = \underset{\bullet}{H}\sum_{\lambda=0}^{r-2}(\underset{\bullet}{I}-\underset{\bullet}{A}\underset{\bullet}{H})^{\lambda} + \underset{\bullet}{A}^{-1}(\underset{\bullet}{I}-\underset{\bullet}{A}\underset{\bullet}{H})^{r-1}. \tag{15.2}$$

This relation will be used for the construction of interval matrix iterations.

Let $\boldsymbol{X}_0$ be an $n \times n$ interval matrix such that $\underset{\bullet}{A}^{-1} \in \boldsymbol{X}_0$, and let the matrix $\underset{\bullet}{H}$ in (15.2) be defined by $\underset{\bullet}{H} = m(\boldsymbol{X}_0)$. Then we obtain from (15.2) using inclusion isotonicity property [4]

$$\underset{\bullet}{A}^{-1} \in \boldsymbol{X}_1 := m(\boldsymbol{X}_0)\sum_{\lambda=0}^{r-2}\Big(\underset{\bullet}{I}-\underset{\bullet}{A}m(\boldsymbol{X}_0)\Big)^{\lambda} + \boldsymbol{X}_0\Big(\underset{\bullet}{I}-\underset{\bullet}{A}m(\boldsymbol{X}_0)\Big)^{r-1}. \tag{15.3}$$

For simplicity, let us introduce $\underset{\bullet}{R}_k = \underset{\bullet}{I} - \underset{\bullet}{A}m(\boldsymbol{X}_k)$. Combining (15.2) and (15.3), it is easy to prove by the set property and mathematical induction that the following is valid for an arbitrary $k \geq 0$:

$$\underset{\bullet}{A}^{-1} \in \boldsymbol{X}_{k+1} := m(\boldsymbol{X}_k)\sum_{\lambda=0}^{r-2}\underset{\bullet}{R}_k^{\lambda} + \boldsymbol{X}_k\underset{\bullet}{R}_k^{r-1}. \tag{15.4}$$

In regard to this property, the following iterative process for finding an inclusion matrix for $\underset{\bullet}{A}^{-1}$ can be stated in a Horner scheme fashion:

$$\begin{cases} \boldsymbol{Y}_k = m(\boldsymbol{X}_k)\Big(\underset{\bullet}{I} + \underbrace{\underset{\bullet}{R}_k(\underset{\bullet}{I} + \underset{\bullet}{R}_k(\underset{\bullet}{I} + \cdots + \underset{\bullet}{R}_k(\underset{\bullet}{I} + \underset{\bullet}{R}_k)}_{r-2 \text{ times}}\cdots\Big) + \boldsymbol{X}_k\underset{\bullet}{R}_k^{r-1}, \\ \boldsymbol{X}_{k+1} = \boldsymbol{Y}_k \cap \boldsymbol{X}_k, \end{cases} \tag{15.5}$$

for $k = 0, 1, \ldots$.

The iterative method (15.5) was considered in detail in the book [1] by Alefeld and Herzberger. As shown in [1, Ch. 18], the most efficient method from the class (15.5) of hyper-power methods is obtained for $r = 3$ and reads

$$\begin{cases} \boldsymbol{Y}_k = m(\boldsymbol{X}^{(k)}) + m(\boldsymbol{X}^{(k)})\underset{\bullet}{R}_k + \boldsymbol{X}_k\underset{\bullet}{R}_k^2, \\ \boldsymbol{X}_{k+1} = \boldsymbol{Y}_k \cap \boldsymbol{X}_k, \end{cases} \quad (k = 0, 1, \ldots). \tag{15.6}$$

The properties of the iterative interval method (15.5) are given in the following theorem proved in [1, Theorem 2, Ch. 18], where $\rho(M)$ denotes the spectral radius of a matrix M.

Theorem 15.1 *Let* $\underset{\bullet}{A}$ *be a nonsingular* $n \times n$ *matrix and* $\boldsymbol{X}_0$ *an* $n \times n$ *interval matrix such that* $\underset{\bullet}{A}^{-1} \in \boldsymbol{X}_0$. *Then*

(a) Each inclusion matrix $\boldsymbol{X}_k$, *calculated by (15.5), contains* A^{-1}.
(b) If $\rho(|\underset{\bullet}{I} - \underset{\bullet}{A}\underset{\bullet}{X}|) < 1$ *for every* $\underset{\bullet}{X} \in \boldsymbol{X}_0$, *then the sequence* $\{\boldsymbol{X}_k\}_{k \geq 0}$ *converges to* A^{-1}.
(c) Using a matrix norm $\|\cdot\|$ *the sequence* $\{d(\boldsymbol{X}_k)\}_{k \geq 0}$ *satisfies*

$$\|d(\boldsymbol{X}_{k+1})\| \leq \gamma \|d(\boldsymbol{X}_k)\|^r, \quad \gamma \geq 0,$$

that is, the R-order of convergence of the method (15.5) is at least r.

Using the iterative formula (15.5) in the Horner form for $r = 6$, we obtain the following iterative method for the inclusion of the inverse matrix:

$$\begin{aligned} \underset{\bullet}{R}_k &= \underset{\bullet}{I} - \underset{\bullet}{A} \odot m(\boldsymbol{X}_k), \\ \underset{\bullet}{S}_k &= \underset{\bullet}{R}_k \odot \underset{\bullet}{R}_k, \\ \underset{\bullet}{M}_k &= \underset{\bullet}{I} + \underset{\bullet}{R}_k \odot (\underset{\bullet}{I} + \underset{\bullet}{R}_k \odot (\underset{\bullet}{I} + \underset{\bullet}{R}_k \odot (\underset{\bullet}{I} + \underset{\bullet}{R}_k))), \qquad (k = 0, 1, \ldots). \\ \boldsymbol{Y}_k &= m(\boldsymbol{X}_k) \odot \underset{\bullet}{M}_k + \boldsymbol{X}_k \otimes (\underset{\bullet}{S}_k \odot \underset{\bullet}{S}_k \odot \underset{\bullet}{R}_k), \\ \boldsymbol{X}_{k+1} &= \boldsymbol{Y}_k \cap \boldsymbol{X}_k, \end{aligned} \tag{15.7}$$

The method (15.7) is a particular case of the general matrix iteration (15.5). According to Theorem 15.1, the method (15.7) has order six and requires 8 multiplication of point matrices (denoted by $\odot$) and one multiplication of interval matrix by point matrix (denoted by $\otimes$).

15.3 New Inclusion Method of High Efficiency

In what follows we are going to show that the computational cost of the interval method (15.7) can be reduced using the identity

$$x^4 + x^3 + x^2 + x + 1 = x^2(x^2 + x + 1) + x + 1 \tag{15.8}$$

and the corresponding matrix relation. Having in mind (15.8) we rewrite (15.7) and state the following algorithm in interval arithmetic for bounding the inverse matrix:

$$\left.\begin{aligned}
\underset{\bullet}{R}_k &= \underset{\bullet}{I} - \underset{\bullet}{A} \odot m(\boldsymbol{X}_k),\\
\underset{\bullet}{S}_k &= \underset{\bullet}{R}_k \odot \underset{\bullet}{R}_k,\\
\underset{\bullet}{T}_k &= \underset{\bullet}{S}_k \odot \underset{\bullet}{S}_k \odot \underset{\bullet}{R}_k,\\
\underset{\bullet}{M}_k &= \underset{\bullet}{I} + \underset{\bullet}{R}_k + \underset{\bullet}{S}_k \odot (\underset{\bullet}{I} + \underset{\bullet}{R}_k + \underset{\bullet}{S}_k),\\
\boldsymbol{Y}_k &= m(\boldsymbol{X}_k) \odot \underset{\bullet}{M}_k + \boldsymbol{X}_k \otimes \underset{\bullet}{T}_k,\\
\boldsymbol{X}_{k+1} &= \boldsymbol{Y}_k \cap \boldsymbol{X}_k,
\end{aligned}\right\} \quad (k = 0, 1, \ldots). \tag{15.9}$$

Compared with the method (15.7), the iterative scheme (15.9) requires 6 multiplications of point matrices (thus, two matrix multiplications less) and still preserves the order six. The above consideration can be summarized in the following theorem.

Theorem 15.2 *Let $\underset{\bullet}{A}$ be a nonsingular $n \times n$ matrix and $\boldsymbol{X}_0$ an $n \times n$ interval matrix such that $\underset{\bullet}{A}^{-1} \in \boldsymbol{X}_0$. Then*

(a) Each inclusion matrix $\boldsymbol{X}_k$, calculated by (15.9), contains $\underset{\bullet}{A}^{-1}$.
(b) If $\rho(|\underset{\bullet}{I} - \underset{\bullet}{A}\underset{\bullet}{X}|) < 1$ holds for all $\underset{\bullet}{X} \in \boldsymbol{X}_0$, then the sequence $\{\boldsymbol{X}_k\}_{k\geq 0}$ converges toward $\underset{\bullet}{A}^{-1}$.
(c) Using a matrix norm $\|\cdot\|$ the sequence $\{d(\boldsymbol{X}_k)\}_{k\geq 0}$ satisfies

$$\|d(\boldsymbol{X}_{k+1})\| \leq \gamma \|d(\boldsymbol{X}_k)\|^6, \quad \gamma \geq 0,$$

that is, the R-order of convergence of the method (15.9) is at least 6.

Proof According to (15.9) we have

$$\underset{\bullet}{M}_k = \underset{\bullet}{I} + \underset{\bullet}{R}_k + \ldots + \underset{\bullet}{R}_k^4, \quad \underset{\bullet}{T}_k = \underset{\bullet}{R}_k^5$$

which implies that the method (15.9) is equivalent with the generic hyper-power method (15.5) of order $r = 6$. Now the result of the theorem holds directly from Theorem 15.1. □

Furthermore, following the results of Zhang et al. [14, Theorem 3.3] we can extend the method (15.9) to the full-rank rectangular matrices $\underset{\bullet}{A}$ and the computation of its Moore–Penrose inverse $\underset{\bullet}{A}^\dagger$.

Theorem 15.3 *If $\underset{\bullet}{A} \in M_{mn}(\mathbb{R})$ is the full-rank matrix and the initial interval $n \times m$ matrix $\boldsymbol{X}_0$ satisfies $m(\boldsymbol{X}_0) = \underset{\bullet}{A}^T \underset{\bullet}{B} \underset{\bullet}{A}^T$ for some matrix $\underset{\bullet}{B} \in M_{mn}(\mathbb{R})$ as well as $\underset{\bullet}{A}^\dagger \in \boldsymbol{X}_0$, then parts (a), (b), and (c) of Theorem 15.2 remain valid when $\underset{\bullet}{A}^{-1}$ is replaced by the Moore–Penrose inverse $\underset{\bullet}{A}^\dagger$.*

Executing iterative interval processes in general, one of the most important but also very difficult task is to find a good initial interval (real interval, complex interval, interval matrix, etc.) that contains the sought result. A similar situation appears in bounding the inverse matrix. We present here an efficient method for constructing an initial matrix $\boldsymbol{X}_0$ that contains the inverse matrix $\underset{\bullet}{A}^{-1}$.

Let $\underset{\bullet}{X} \in \boldsymbol{X}_0$ and let us assume that the matrix $\underset{\bullet}{A}$ can be represented as

$$\underset{\bullet}{A} = \underset{\bullet}{I} - \underset{\bullet}{Y}, \quad \text{where a newly introduced matrix } \underset{\bullet}{Y} \text{ satisfies } \quad \|\underset{\bullet}{Y}\| < 1. \tag{15.10}$$

Such matrices appear (for example) as the design structure matrices (or shorter DSM) where the elements represent the influential parameters of various tasks. This concept was developed in a Massachusetts Institute of Technology project; see, e.g., [3] and [12]. The final conclusions necessary for the improvements/corrections of process or project are provided using the so-called a model of *design iteration*. This model is pretty complex and uses elements of numerical linear algebra so that we refer to the paper [5] for details. Instead, we will only stress that the key task in design iteration is the calculation of the inverse matrix $\underset{\bullet}{A}^{-1} = (\underset{\bullet}{I} - \underset{\bullet}{Y})^{-1}$. The final entries of $\underset{\bullet}{A}^{-1}$ point to further activities in modeling the considered process or project, for example, the schedule of performing particular tasks.

It has been shown in [1, Ch. 18] that the inequality

$$\|\underset{\bullet}{X}\| \le a := \frac{1}{1 - \|\underset{\bullet}{Y}\|}$$

holds. If we use either the row-sum or the column-sum norm, then we find that $-a \le x_{ij} \le a$, $(1 \le i, j \le n)$ holds for all the elements of $\underset{\bullet}{X} = [x_{ij}]$. For the matrix $\boldsymbol{X}_0 = \left[X_{ij}^{(0)}\right]$ with interval coefficients

$$X_{ij}^{(0)} = \begin{cases} [-a, a], & \text{for } i \ne j, \\ [-a, 2 + a], & \text{for } i = j, \end{cases} \tag{15.11}$$

we have $\underset{\bullet}{A}^{-1} \in \boldsymbol{X}_0$ and $m(\boldsymbol{X}_0) = \underset{\bullet}{I}$ (see [1]). If the condition (15.10) is not satisfied, then it is effective to normalize the matrix $\underset{\bullet}{A}$ before running the iterative process.

Having in mind the described procedure of choosing initial inclusion matrix $\boldsymbol{X}_0$, applying point matrix iterations it is convenient to take $\underset{\bullet}{X}_0 = m(\boldsymbol{X}_0) = \underset{\bullet}{I}$. Such choice has already been applied in stating the iterative interval methods (15.7) and (15.9).

15.4 Computational Aspects

Let us compare the computational efficiency of the hybrid methods (15.7) and (15.9). As proved in [6, Ch. 6], the CPU (central processor unit) time necessary for executing an iterative method (IM) can be suitably expressed in a pretty manner in the form

$$CPU_{(IM)} = h \log q \cdot \frac{\theta(IM)}{\log r(IM)}. \tag{15.12}$$

Here $r(IM)$ is the convergence order, $\theta(IM)$ is the computational cost of the iterative method (IM) per iteration, q is the number of significant decimal digits (for example, $q = 15$ or 16 for double-precision arithmetic) and h is a constant that depends on hardware characteristics of the employed digital computer. Assuming that the considered methods are implemented on the same computer, according to (15.12) the comparison of two methods (M_1) and (M_2) is carried out by the *efficiency ratio*:

$$ER_{M_1/M_2}(n) = \frac{CPU_{(M_1)}}{CPU_{(M_2)}} = \frac{\log r(M_2)}{\log r(M_2)} \cdot \frac{\theta(M_1)}{\theta(M_2)}. \tag{15.13}$$

Calculating the computational cost θ, it is necessary to deal with the number of arithmetic operations per iteration taken with certain *weights* depending on the execution times of operations. We assume that floating-point number representation is used, with a binary fraction of b bits, meaning that we deal with " *precision* b" numbers, giving results with a relative error of approximately 2^{-b}. Following results given in [2], the execution time $t_b(A)$ of addition (subtraction) is $O(b)$, where O is the Landau symbol. Using Schönhage–Strassen multiplication (see [2]), often implemented in multi-precision libraries (in the computer algebra systems *Mathematica, Maple, Magma*, for instance), we have $t_b(M) = O\big(b \log b \, \log(\log b)\big)$. For comparison purpose, we chose the weights w_a and w_m proportional to $t_b(A)$ and $t_b(M)$, respectively, for double-precision arithmetic ($b = 64$ bits) and quadruple-precision arithmetic ($b = 128$ bits).

In particular cases, assuming that multiplication of two scalar $n \times n$ matrices requires $n^2(n-1)$ additions and n^3 multiplications and adding combined costs in the iterative formulas (15.7) and (15.9), for the hybrid method (15.7) and (15.9) we have $r(15.7) = r(15.9) = 6$ and, approximately,

$$\theta(15.7) = (9n^3 - 3n^2)b + 10n^3 b \log b \log(\log b),$$
$$\theta(15.9) = (7n^3 - n^2)b + 8n^3 b \log b \log(\log b).$$

In view of this, by (15.13) we determine the efficiency ratio

$$ER_{(15.7)/(15.9)}(n) = \frac{9 - 3/n + 10 \log b \log(\log b)}{7 - 1/n + 8 \log b \log(\log b)}.$$

The graph of the function $ER_{(15.7)/(15.9)}(n)$ for $n \in [2, 40]$ is shown in Fig. 15.1. From this graph we note that the values of $ER(n)$ are grouped about the value 1.25

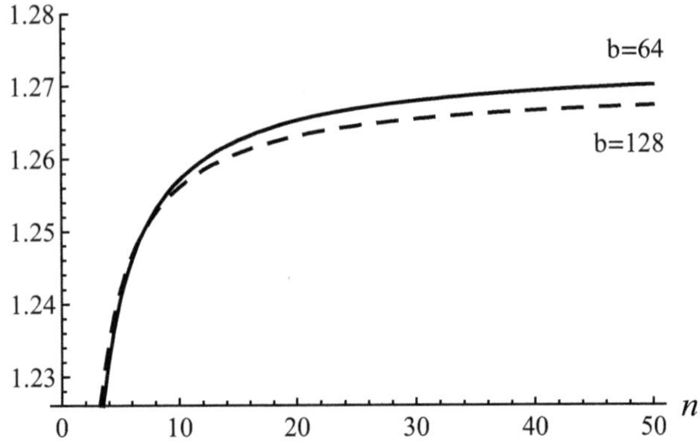

Fig. 15.1 The ratio of CPU times for two different precisions of arithmetical processors

for n in a wide range. This means that the new method (15.9) consumes about 25% less CPU time than the Horner-fashion method (15.7).

A very similar graph is obtained for a lot of computing machines. For example, for double-precision arithmetic and quadruple-precision arithmetic (corresponding approximately to $b = 64$ and $b = 128$, respectively), the values of $ER(n)$ are very close to 1.25 almost independently on the dimension of matrix n. In addition, we find $ER_{(15.5)/(15.9)}(n) > 1$ for every $r \neq 3$ and close to 1 for $r = 3$.

The convergence behavior of the iterative interval method (15.9), together with the choice of initial inclusion matrix $\boldsymbol{X}_0$, will be demonstrated by one simple example. We emphasize that the interval method (15.7) produces the same inclusion matrix, which is obvious since the corresponding iterative formulae are, actually, identical but arranged in different forms. However, as mentioned above, the inclusion method (15.9) has lower computational cost than (15.7).

Example 15.1 We wish to find the inclusion matrix for the inverse of the matrix

$$\underset{\bullet}{A} = \begin{bmatrix} \frac{9}{10} & \frac{1}{5} \\ -\frac{3}{10} & \frac{4}{5} \end{bmatrix}.$$

Note the inverse matrix $\underset{\bullet}{A}^{-1}$ is

$$\underset{\bullet}{A}^{-1} = \begin{bmatrix} \frac{40}{39} & -\frac{10}{39} \\ \frac{5}{13} & \frac{15}{13} \end{bmatrix} = \begin{bmatrix} 1.0\overline{256410} & -0.\overline{256410} \\ 0.\overline{384615} & 1.\overline{153846} \end{bmatrix}.$$

The overlined set of digits indicates that this set of digits repeats periodically.

First, we determine

$$\underset{\bullet}{Y} = \underset{\bullet}{I} - \underset{\bullet}{A} = \begin{bmatrix} 0.1 & -0.2 \\ 0.3 & 0.2 \end{bmatrix} \text{ with } \|\underset{\bullet}{Y}\|_2 = 0.424264 \text{ and } a = \frac{1}{1 - \|\underset{\bullet}{Y}\|_2} = 1.73691.$$

According to (15.11) we form the initial inclusion matrix

$$\boldsymbol{X}_0 = \begin{bmatrix} [-1.73691, 3.73691] & [-1.73691, 1.73691] \\ [-1.73691, 1.73691] & [-1.73691, 3.73691] \end{bmatrix}.$$

Note that the widths of intervals which present the coefficients of the initial inclusion matrix $\boldsymbol{X}_0$ are rather large. We have applied two iterations of (15.9) and obtained the following midpoint matrices (approximations to $\underset{\bullet}{A}^{-1}$) and the width matrices that give the upper error bounds of $\boldsymbol{X}_k$.

$\underline{k = 1}$

$$m(\boldsymbol{X}_1) = \begin{bmatrix} 1.025\ldots & -0.256\ldots \\ 0.384\ldots & 1.153\ldots \end{bmatrix}, \quad d(\boldsymbol{X}_1) = \begin{bmatrix} 1.27 \times 10^{-2} & 8.68 \times 10^{-3} \\ 1.51 \times 10^{-2} & 6.356 \times 10^{-3} \end{bmatrix}.$$

$\underline{k = 2}$

$$m(\boldsymbol{X}_2) = \begin{bmatrix} 1.0256410256410256\ldots & -0.256410256410256\ldots \\ 0.3846153846153846\ldots & 1.153846153846153\ldots \end{bmatrix},$$

$$d(\boldsymbol{X}_2) = \begin{bmatrix} 6.33 \times 10^{-19} & 4.19 \times 10^{-19} \\ 5.99 \times 10^{-19} & 4.54 \times 10^{-19} \end{bmatrix}.$$

All displayed decimal digits of $m(\boldsymbol{X}_1)$ and $m(\boldsymbol{X}_2)$ are correct. The third iteration produces the width matrix $d(\boldsymbol{X}_3)$ with elements in the form of real intervals with widths of order 10^{-99}. We have not listed $m(\boldsymbol{X}_3)$ and $d(\boldsymbol{X}_3)$ to save the space.

We have also tested the interval method (15.6) possessing the highest efficiency among hyper-power methods. Starting with the same initial matrix $\boldsymbol{X}_0$ as above, we obtained the following outcomes:

$\underline{k = 1}$

$$m(\boldsymbol{X}_1) = \begin{bmatrix} 1.05 & -0.26 \\ 0.39 & 1.18 \end{bmatrix}, \quad d(\boldsymbol{X}_1) = \begin{bmatrix} 0.586 & 0.398 \\ 0.666 & 0.318 \end{bmatrix}.$$

$\underline{k = 2}$

$$m(\boldsymbol{X}_2) = \begin{bmatrix} 1.0256\ldots & -0.2564\ldots \\ 0.3846\ldots & 1.1538\ldots \end{bmatrix}, \quad d(\boldsymbol{X}_2) = \begin{bmatrix} 3.60 \times 10^{-4} & 2.43 \times 10^{-4} \\ 3.91 \times 10^{-4} & 2.12 \times 10^{-4} \end{bmatrix}.$$

Table 15.1 Averaged running times of the ordinary hyper-power methods of order $r = 2$ and $r = 6$ and the proposed new method

n	HP2	HP6	HP6imp
300	0.106	0.069	0.059
400	0.239	0.161	0.139
500	0.430	0.292	0.254
600	0.711	0.495	0.423
700	1.078	0.754	0.644
800	1.552	1.093	0.942
900	2.218	1.524	1.327
1000	2.896	2.065	1.741

Method (15.9) produced considerably higher accuracy than (15.6) using only two iterations so that its application is justified in this case. Furthermore, since $ER_{(15.6)/(15.9)}(n)$ is close to 1, which of these two methods will be chosen depends of the nature of solved problem, specific requirements, and available hardware and software (precision of employed computer). For instance, the proposed method (15.9) is more convenient when a high accuracy is requested in a few iterations, which is demonstrated in the following example.

Example 15.2 We compared our method (15.9) with the ordinary hyper-power interval methods of order $r = 2$ and $r = 6$ (given by (15.6) and (15.7)), respectively. These methods will be abbreviated as **HP2**, **HP6** and **HP6imp** respectively. Testings were done on the random matrices of the type (15.10) where the initial values were generated by (15.11). Implementations were carried out in the programming package `Matlab` using the `INTLAB` library (see Rump [9]) for performing the operations in interval arithmetics. We compared the total running times (in seconds) required by each method to obtain the result of the accuracy $\epsilon = 10^{-10}$ (the stopping condition was $d(\boldsymbol{X}_k) < \epsilon$). Matrices are generated by $A = I - Y$ where $Y = \texttt{rand}(n)$ (pseudorandom matrix with the values uniformly distributed on [0, 1]). Results are given in the Table 15.1.

As it can be seen from Table 1, the proposed method **HP6imp** outperformed other two methods for all test matrices.

References

1. Alefeld, G., Herzberger, J.: Introduction to Interval Computations. Academic Press, New York (1983)
2. Brent, R.: Zimmermann, P.: Modern Computer Arithmetic. Cambridge University Press, Cambridge (2011)
3. Eppinger, S.D., Witney, D.E., Smith, R.P., Gebala, D.A.: A model-based method for organizing tasks in product development. Res. Eng. Design **6**, 1–13 (1994)
4. Moore, R.E., Kearfott, R.B., Cloud, M.J.: Introduction to Interval Analysis. SIAM, Philadelphia (2009)
5. Petković, I., Petković, V.: Interval matrix models of design iteration. In: Proceedings of the IEEE International Conference on Electronics and Information Engineering, Kyoto, vol. 1, pp. 20–24 (2010)

6. Petković, M.S.: Iterative Methods for Simultaneous Inclusion of Polynomial Zeros. Springer, Berlin-Heidelberg-New York (1989)
7. Petković, M.S., Herzberger, J.: On the efficiency of a class of combined Schulz's method for bounding the inverse matrix. ZAMM **71**, 181–187 (1991)
8. Petković, M.S., Petković, L.D.: Complex Interval Arithmetic and its Applications. Wiley-VCH, Berlin-Weinheim-New York (1998)
9. Rump, S.M.: INTLAB - Interval Laboratory, the Matlab toolbox for verified computations, Version 5.3 (2006). www.ti3.tu-harburg.de/rump/intlab/
10. Srivastava, S., Gupta, D.K., Stanimirović, P.S., Singh, S., Falguni, R.: A hyperpower iterative method for computing the generalized Drazin inverse of Banach algebra element. Sadhana **42**, 625–630 (2017)
11. Stanimirović, P.S., Srivastava, S., Gupta, D.K.: From Zhang Neural Network to scaled hyperpower iterations. J. Comput. Appl. Math. **331**, 133–155 (2018)
12. Steward, D.V.: The design structure system: a model for managing the design of complex systems. IEEE Trans. Eng. Manag. EM-**28**, 71–74 (1981)
13. Stojanović, I., Stanimirović, P.S., Živković, I.S., Gerontitis, D., Wang, X.Z.: ZNN models for computing matrix inverse based on hyperpower iterative methods. Filomat **31/10**, 2999–3014 (2017)
14. Zhang, X., Cai, J., Wei, Y.: Interval iterative methods for computing Moore-Penrose inverse. Appl. Math. Comput **183**, 522–532 (2006)

Chapter 16
The Optimal Quadrature Formula for the Approximate Calculation of Fourier Coefficients in the Space $\widetilde{W_2}^{(2,1)}$ of Periodic Functions

Kholmat Shadimetov, Abdullo Hayotov, and Umedjon Khayriev

Dedicated to Academician Gradimir V. Milovanović on the occasion of his 75th birthday

16.1 Introduction

Numerical calculation of the strongly oscillating integrals is one of the more critical problems in numerical analysis because such integrals are widely used in science and technology. The following types of Fourier integrals are also examples of strongly oscillating integrals for sufficiently large $\omega \in \mathbb{R}$:

$$I(\omega, \varphi) = \int_{\Omega} \mathrm{e}^{\mathrm{i}\omega g(x)} \varphi(x)\, \mathrm{d}x, \tag{16.1}$$

K. Shadimetov
Tashkent State Transport University, Tashkent, Uzbekistan

V.I. Romanovskiy Institute of Mathematics, Uzbekistan Academy of Sciences, Tashkent, Uzbekistan
e-mail: kholmatshadimetov@mail.ru

A. Hayotov (✉)
V.I. Romanovskiy Institute of Mathematics, Uzbekistan Academy of Sciences, Tashkent, Uzbekistan

Central Asian University, Tashkent, Uzbekistan

Bukhara State University, Bukhara, Uzbekistan
e-mail: hayotov@mail.ru; a.hayotov@centralasian.uz

U. Khayriev
Bukhara State University, Bukhara, Uzbekistan

V.I. Romanovskiy Institute of Mathematics, Uzbekistan Academy of Sciences, Tashkent, Uzbekistan

M. P. Stanić et al. (eds.), *Analysis, Approximation, Optimization: Computation and Applications*, Springer Optimization and Its Applications 224,
https://doi.org/10.1007/978-3-031-85743-0_16

where φ and g are non-oscillating functions, ω is oscillation frequency, and Ω is some piecewise smooth region.

Numerical approximation of these integrals can be challenging, particularly when the oscillations are highly localized and rapidly alternating. However, there are several techniques that can be used to obtain accurate numerical approximations of these integrals. One such technique is the stationary phase method, which involves identifying the stationary points of the integrand and approximating the integral using the properties of the integrand near these points. Another technique is to use a quadrature method that is designed explicitly for oscillatory integrals, such as the Filon, Clenshaw-Curtis, modified Clenshaw-Curtis method, Levin-type methods, Gauss-Laguerre quadrature, and generalized quadrature (see [26] for full details, for instance, [17, 22] and references therein).

These methods involve different approaches to approximating oscillatory integrals, and which approach is best suited to a given problem will depend on the specific properties of the integrand and the desired accuracy of the approximation. Additionally, it is worth noting that an effective approximation of a strongly oscillatory integral may require a large number of samples.

Based on the properties of the functions φ and g, various methods have been developed for the numerical calculation of highly oscillating integrals (16.1). There are *the Filon method, the asymptotic expansion, Levin's collocation method*, and *the steepest descent and optimal quadrature formula* methods of approximate calculation of such integrals.

In recent years, scientists such as A. Iserles [13], S.P. Nørsett [14], S. Olver [21], G.V. Milovanović [18–20], Z. Xu [27], S. Zhang, and E. Novak [28] have been engaged in the approximate solution of integral (16.1). Also, G.V. Milovanović, Kh.M. Shadimetov, and A.R. Hayotov [2, 3] have conducted scientific research on constructing optimal quadrature formulas for calculating Fourier coefficients and integrals in different spaces, including in spaces $L_2^{(m)}$ and $W_2^{(m,m-1)}$. The results of constructing optimal quadrature formulas for the numerical calculation of Fourier integrals in the Sobolev space $\widetilde{L_2^{(m)}}$ of periodic functions and applying the constructed formulas to the reconstruction of computed tomography images were obtained in the researches of A.R. Hayotov, S. Jeon, Ch.-O. Lee, and Kh. M. Shadimetov [5–8].

In this work, the problem of constructing an optimal quadrature formula for the approximate calculation of Fourier coefficients in Hilbert space $\widetilde{W}_2^{(2,1)}(0, 1]$ of periodic functions is studied.

16.2 Statement of the Problem

In this chapter, we are concerned with obtaining optimal quadrature formulas. It is assumed that the integrand belongs to the Hilbert space $W_2^{(2,1)}$. Recall the definition of this Hilbert space, based on the work [1].

$W_2^{(2,1)}(0,1)$ is the Hilbert space of complex-valued functions and it is defined as follows:

$$W_2{}^{(2,1)} = \{\varphi : [0,1] \to \mathbb{C} | \varphi' \text{ is abs. continuous and } \varphi'' \in L_2(0,1)\}.$$

The space $W_2^{(2,1)}$ is a Hilbert space with the inner product

$$\langle \varphi, \psi \rangle_{W_2^{(2,1)}} = \int_0^1 (\varphi''(x) + \varphi'(x))(\overline{\psi}''(x) + \overline{\psi}'(x))\, \mathrm{d}x, \tag{16.2}$$

and the corresponding norm is

$$\|\varphi\|_{W_2^{(2,1)}} = \left(\langle \varphi, \varphi \rangle_{W_2^{(2,1)}} \right)^{1/2}.$$

The last equality is a semi-norm and $\| \varphi \| = 0$ if and only if $\varphi(x) = d_0 + d_1 \mathrm{e}^{-x}$, where d_0 and d_1 are constants. Every element of the space $W_2^{(2,1)}(0,1)$ is a set of functions which differ from each other on a linear combination of any constant and e^{-x}. So, the space $W_2^{(2,1)}(0,1)$ is a factor space.

We denote by $\widetilde{W}_2^{(2,1)}(0,1]$ the subspace of $W_2^{(2,1)}(0,1)$ consisting of 1-periodic functions. Every element of the space $\widetilde{W}_2^{(2,1)}(0,1]$ satisfies the following condition of 1-periodicity:

$$\varphi(x+\beta) = \varphi(x), \quad \text{for } x \in \mathbb{R} \text{ and } \beta \in \mathbb{Z}.$$

Now we consider the following quadrature formula:

$$\int_0^1 \mathrm{e}^{2\pi i w x} \varphi(x)\, \mathrm{d}x \cong \sum_{k=1}^{N} C_k \varphi(hk), \tag{16.3}$$

where ω is a nonzero integer, $\varphi \in \widetilde{W_2}^{(2,1)}(0,1]$, C_k are coefficients to be determined, and $h = \frac{1}{N}$ is a step of the mesh.

The difference

$$(\ell, \varphi) = \int_0^1 \mathrm{e}^{2\pi i w x} \varphi(x)\, \mathrm{d}x - \sum_{k=1}^{N} C_k \varphi(hk) \tag{16.4}$$

between the integral and the quadrature sum is called *the error* of the quadrature formula (16.3) and the corresponding *error functional* is

$$\ell(x) = \mathrm{e}^{2\pi i w x} - \sum_{k=1}^{N} C_k \sum_{\beta=-\infty}^{\infty} \delta(x - hk - \beta), \tag{16.5}$$

where δ is the Dirac delta-function. The error functional $\ell(x)$ is called *the periodic error functional* of the quadrature formula (16.3) and it belongs to the conjugate the space $\widetilde{W}_2^{(2,1)*}(0, 1]$.

Since the error functional ℓ is defined on the space $\widetilde{W}_2^{(2,1)}(0, 1]$, the following equality is valid as in the work [25]:

$$(\ell, 1) = 0\,. \tag{16.6}$$

The error (16.4) of the quadrature formula (16.3) is a linear functional in the space $\widetilde{W}_2^{(2,1)*}$ (0, 1]. The absolute value of the error (16.4) is estimated by the Cauchy-Schwarz inequality as follows:

$$|(\ell, \varphi)| \leq \|\ell\|_{\widetilde{W}_2^{(2,1)*}} \cdot \|\varphi\|_{\widetilde{W}_2^{(2,1)}}\,,$$

where

$$\|\ell\|_{\widetilde{W}_2^{(2,1)*}} = \sup_{\varphi,\ \|\varphi\|_{\widetilde{W}_2^{(2,1)}} \neq 0} \frac{|(\ell, \varphi)|}{\|\varphi\|_{\widetilde{W}_2^{(2,1)}}}$$

is the norm of the error functional (16.5).

Hence, in order to get the minimum of the upper bound of the error for the quadrature formula (16.3) we solve the following.

Problem 16.1 Find the norm of the error functional (16.5) of the quadrature formula (16.3) in the space $\widetilde{W}_2^{(2,1)}$ (0, 1].

Problem 16.2 Find the optimal coefficients C_k that give the minimum value to the norm of the error functional (16.5).

Problem 16.3 Calculate the norm of the error functional for the optimal quadrature formula (16.3) in the following form:

$$\left\|\overset{\circ}{\ell}\right\|_{\widetilde{W}_2^{(2,1)*}} := \inf_{C_k} \|\ell\|_{\widetilde{W}_2^{(2,1)*}}\,.$$

Definition 16.1 The coefficients that give the minimum value to the norm $\|\ell\|_{\widetilde{W}_2^{(2,1)*}}$ are called *optimal coefficients* and are denoted by $\overset{\circ}{C}_k$. The quadrature formula (16.3) with these coefficients is said to be *the optimal quadrature formula.*

A similar problem was first proposed by S.L. Sobolev [25], and later scientists such as Kh.M. Shadimetov [23, pp. 97–104], A.R. Hayotov [9–12], N.D. Boltaev [1], and S.S. Babaev [4] were engaged in solving these problems.

This chapter is composed of the following parts in solving Problems 16.1–16.3:

i The first and second sections are devoted to the introduction and statement of the problem. Also, the work carried out on the approximation of Fourier integrals is studied in it.
ii In the third section, we find the extremal function corresponding to the error functional (16.5) and calculate the norm of the error functional (16.5) using it.
iii In the fourth section, the process of solving the system of linear equations for the coefficients of the optimal quadrature formula is presented.
iv In the fifth section, the sharp upper bound of the error of the constructed optimal quadrature formula is found.
v In the last section some numerical results are presented.

16.3 The Norm of the Error Functional (16.5)

To calculate the norm $\|\ell\|_{\widetilde{W_2}^{(2,1)*}}$, we use *the extremal function* ψ_ℓ for the error functional ℓ (see [25]) that satisfies the following equality:

$$(\ell, \psi_\ell) = \|\ell\|_{\widetilde{W_2}^{(2,1)*}} \cdot \|\psi_\ell\|_{\widetilde{W_2}^{(2,1)}} . \tag{16.7}$$

Since $\widetilde{W_2}^{(2,1)}(0, 1]$ is the Hilbert space, using the Riesz representation theorem, we obtain

$$(\ell, \varphi) = \langle\varphi, \psi_\ell\rangle_{\widetilde{W_2}^{(2,1)}}, \qquad \text{for all } \varphi \in W_2^{(2,1)}, \tag{16.8}$$

where $\langle\psi_\ell, \varphi\rangle_{\widetilde{W_2}^{(2,1)}}$ is the inner product of the functions ψ_ℓ and φ which is defined by expression (16.2). In addition, the equality $\| \ell \|_{\widetilde{W_2}^{(2,1)*}} = \| \psi_\ell \|_{\widetilde{W_2}^{(2,1)}}$ is fulfilled. So, taking into account (16.7), we derive

$$(\ell, \psi_\ell) = \| \ell \|^2_{\widetilde{W_2}^{(2,1)*}} . \tag{16.9}$$

Then from (16.8), integrating by parts, we obtain

$$\int_0^1 \ell(x)\varphi(x)dx = \int_0^1 \left(\overline{\psi_\ell}^{IV}(x) - \overline{\psi_\ell}^{II}(x)\right)\varphi(x)\,dx.$$

From the last equation, we have

$$\overline{\psi_\ell}^{IV}(x) - \overline{\psi_\ell}^{II}(x) = \ell(x). \tag{16.10}$$

For the solution of the last equation the following holds.

Theorem 16.1 *The solution to differential equation (16.10) is the extremal function ψ_ℓ corresponding to the error functional ℓ and it is expressed as*

$$\psi_\ell(x) = d_0 + e^{-2\pi i\omega x} \cdot \kappa(\omega) - \sum_{k=1}^{N} \overline{C_k} \sum_{\beta \neq 0} \kappa(\beta) e^{2\pi i\beta(x-hk)}, \tag{16.11}$$

where d_0 is a complex number and

$$\kappa(\omega) = \frac{1}{(2\pi\omega)^4 + (2\pi\omega)^2}. \tag{16.12}$$

Proof We find the periodic solution of Eq. (16.10) using the following properties of the Fourier transforms (see, for instance, [1, 4]):

$$\begin{aligned} F[\varphi] &= \int_{-\infty}^{\infty} \varphi(x) e^{2\pi i p x}\, dx, \\ F^{-1}[\varphi] &= \int_{-\infty}^{\infty} \varphi(p) e^{-2\pi i p x}\, dp, \\ F[\varphi^{(\alpha)}] &= (-2\pi i p)^\alpha F[\varphi], \quad \alpha \in \mathbb{N}, \\ F^{-1}[F[\varphi(x)]] &= \varphi(x). \end{aligned}$$

We apply the Fourier transform to both sides of Eq. (16.10) and here

$$F[\overline{\psi_\ell}^{IV} - \overline{\psi_\ell}^{II}] = F[\ell].$$

Since the Fourier transform is a linear operator, we have

$$\left((2\pi i p)^4 - (2\pi i p)^2\right) F[\overline{\psi_\ell}] = F\left[e^{2\pi i w x} - \sum_{k=1}^{N} C_k \sum_{\beta=-\infty}^{\infty} \delta(x - hk - \beta)\right]$$

or

$$\left((2\pi i p)^4 - (2\pi i p)^2\right) F[\bar{\psi}_\ell] = F\left[e^{2\pi i\omega x}\right] - \sum_{k=1}^{N} C_k \sum_{\beta=-\infty}^{\infty} F\left[\delta(x - hk - \beta)\right]. \tag{16.13}$$

Now, using the following equalities

$$F\left[e^{2\pi i\omega x}\right] = \delta(p + \omega), \quad \sum_{\beta=-\infty}^{\infty} e^{2\pi i p\beta} = \sum_{\beta=-\infty}^{\infty} \delta(p - \beta)$$

and

$$\sum_{\beta=-\infty}^{\infty} F\left[\delta(x-hk-\beta)\right] = \sum_{\beta=-\infty}^{\infty} \mathrm{e}^{2\pi \mathrm{i} p(hk+\beta)} = \mathrm{e}^{2\pi \mathrm{i} phk} \sum_{\beta=-\infty}^{\infty} \mathrm{e}^{2\pi \mathrm{i} p\beta}$$

$$= \mathrm{e}^{2\pi \mathrm{i} phk} \sum_{\beta=-\infty}^{\infty} \delta(p-\beta),$$

we can rewrite Eq. (16.13) as follows:

$$\left((2\pi \mathrm{i} p)^4 - (2\pi \mathrm{i} p)^2\right) F[\bar{\psi}_\ell] = \delta(p+\omega) - \sum_{k=1}^{N} C_k \sum_{\beta=-\infty}^{\infty} \mathrm{e}^{2\pi \mathrm{i} phk} \delta(p-\beta).$$

We consider that the coefficient on the left-hand side of the last equation is not equal to zero at $p \neq 0$. Consequently, we can multiply both sides of the last equation by $\kappa(p)$ which is defined by (16.12). This multiplication is not uniquely defined. From the last equation the function $F[\bar{\psi}_\ell]$ is defined up to the term of the form $\delta(p)$. Taking into account the above and the properties of Dirac's delta-function, we get

$$F[\bar{\psi}_\ell] = \delta(p+\omega)\kappa(p) - \sum_{k=1}^{N} C_k \sum_{\beta \neq 0} \mathrm{e}^{2\pi \mathrm{i} phk} \delta(p-\beta)\kappa(p) + d_0 \delta(p),$$

where $\kappa(p)$ is defined by formula (16.12) and d_0 is a constant.

Using the property $f(x)\delta(x-a) = f(a)\delta(x-a)$ of the Dirac delta-function, we have the following:

$$F[\bar{\psi}_\ell] = \delta(p+\omega)\kappa(\omega) - \sum_{k=1}^{N} C_k \sum_{\beta \neq 0} \mathrm{e}^{2\pi \mathrm{i} \beta hk} \kappa(\beta)\delta(p-\beta) + d_0 \delta(p).$$

Then, applying the inverse Fourier transform to both sides of the last equation, we have

$$\bar{\psi}_\ell(x) = \mathrm{e}^{2\pi \mathrm{i} \omega x} \kappa(\omega) - \sum_{k=1}^{N} C_k \sum_{\beta \neq 0} \mathrm{e}^{2\pi \mathrm{i} \beta hk} \kappa(\beta) \mathrm{e}^{-2\pi \mathrm{i} \beta x} + d_0.$$

Since $\overline{\overline{\psi}}_\ell$ is equal to ψ_ℓ, we obtain (16.11), that is, Theorem 16.1 is proved. □

The following result is true for the solution of Problem 16.1.

Theorem 16.2 *The square of the norm* $\|\ell\|_{\widetilde{W}_2^{(2,1)*}}$ *for* $\omega \in \mathbb{Z}\backslash\{0\}$ *has the following form:*

$$\|\ell\|^2_{\widetilde{W}_2^{(2,1)*}} = \kappa(\omega) - \kappa(\omega)\left[\sum_{k=1}^{N} C_k e^{-2\pi i\omega hk} + \sum_{k'=1}^{N} \overline{C_{k'}} e^{2\pi i\omega hk'}\right]$$

$$+\sum_{k=1}^{N}\sum_{k'=1}^{N} C_k\overline{C_{k'}}\sum_{\beta\neq 0}\kappa(\beta)e^{2\pi i\beta h(k-k')}, \tag{16.14}$$

where $\kappa(\cdot)$ *is defined by (16.12).*

Proof To prove Theorem 16.2, we calculate the norm $\|\ell\|_{\widetilde{W}_2^{(2,1)*}}$ and use equalities (16.9), (16.14), and (16.11), respectively. As a result, we have the following:

$$\|\ell\|^2_{\widetilde{W}_2^{(2,1)*}} = (\ell, \psi_\ell) = \int_0^1 \ell(x)\psi_\ell(x)\,dx$$

$$= \int_0^1 \ell(x)\left(d_0 + e^{-2\pi i\omega x}\cdot\kappa(\omega) - \sum_{k'=1}^{N}\overline{C_{k'}}\sum_{\gamma\neq 0}\kappa(\gamma)e^{2\pi i\gamma(x-hk')}\right)dx.$$

Taking into account condition (16.6), from the above equation, we get the following:

$$\|\ell\|^2_{\widetilde{W}_2^{(2,1)*}} = \int_0^1 \ell(x)\left(e^{-2\pi i\omega x}\cdot\kappa(\omega) - \sum_{k'=1}^{N}\overline{C}_{k'}\sum_{\gamma\neq 0}\kappa(\gamma)\cdot e^{2\pi i\gamma(x-hk')}\right)dx.$$

Simplifying the last expression, we obtain equality (16.14).

So, Theorem 16.2 is completely proven, that is, Problem 16.2 is solved. □

16.4 Optimal Coefficients of the Optimal Quadrature Formula of the Form (16.3)

In the present section, we find the optimal coefficients of the quadrature formula (16.3) that give the minimum value to the norm for the error functional ℓ. To find the minimum of the norm $\|\ell\|_{\widetilde{W}_2^{(2,1)*}}$ with respect to the coefficients C_k under the condition (16.6), we use the Lagrange method for finding the conditional extremal. We consider the following function :

$$\lambda\left(C_k, \overline{C}_{k'}, \mu\right) = \|\ell\|^2_{\widetilde{W}_2^{(2,1)}} - \mu\cdot(\ell, 1), \quad \text{for } k = 1, 2, \ldots, N,$$

where μ is a constant.

Equating the partial derivatives by all variables C_k, $\overline{C}_{k'}$ and μ of the function $\lambda\left(C_k, \overline{C}_{k'}, \mu\right)$ to zero, we have

$$\frac{\partial \lambda}{\partial C_k} = -\mathrm{e}^{-2\pi \mathrm{i}\omega hk}\kappa(\omega) + \sum_{k'=1}^{N} \overline{C}_{k'} \sum_{\beta\neq 0} \mathrm{e}^{2\pi \mathrm{i}\beta h(k-k')}\kappa(\beta) + \mu = 0,$$

$$\text{for each } k = 1, 2, \dots, N, \tag{16.15}$$

$$\frac{\partial \lambda}{\partial \overline{C}_{k'}} = -\mathrm{e}^{2\pi \mathrm{i}\omega hk'}\kappa(\omega) + \sum_{k=1}^{N} C_k \sum_{\beta\neq 0} \mathrm{e}^{2\pi \mathrm{i}\beta h(k-k')}\kappa(\beta) = 0,$$

$$\text{for each } k' = 1, 2, \dots, N, \tag{16.16}$$

$$\frac{\partial \lambda}{\partial \mu} = \int_0^1 \mathrm{e}^{2\pi \mathrm{i}\omega x}\,\mathrm{d}x - \sum_{k=1}^{N} C_k = 0, \tag{16.17}$$

where $\kappa(\cdot)$ is defined by (16.12).

A solution of this system which we denote by $\{\overset{\circ}{C}_k,\ k = 1, 2, \dots, N\}$ and $\overset{\circ}{\mu}$ is a minimum point for the function $\lambda(C_k, \overline{C}_{k'}, \mu)$. The system of Eqs. (16.15) with (16.17) is equivalent to the system (16.16), (16.17). Therefore, it is sufficient to solve the system (16.16), (16.17).

The following theorem is valid for the solution of the system (16.16), (16.17).

Theorem 16.3 *Among the quadrature formulas of the form (16.3) with $\omega \in \mathbb{Z}\setminus\{0\}$ and $\omega h \notin \mathbb{Z}$ in the space $\widetilde{W}_2^{(2,1)}(0, 1]$ of complex-valued periodic functions, there is a unique quadrature formula with optimal coefficients having the representation*

$$\overset{\circ}{C}_k = \frac{2}{(2\pi\omega)^4 + (2\pi\omega)^2}\left[\frac{h}{1-\cos 2\pi\omega h} - \frac{\mathrm{e}^{2h}-1}{\mathrm{e}^{2h} - 2\mathrm{e}^{h}\cos 2\pi\omega h + 1}\right]^{-1} \cdot \mathrm{e}^{2\pi \mathrm{i}\omega hk}, \tag{16.18}$$

for each $k = 1, 2, \dots, N$.

Proof Let ω be a nonzero integer and $\omega h \notin \mathbb{Z}$. Then we search a solution of the system (16.16), (16.17) in the following form:

$$\overset{\circ}{C}_k = C(\omega, h)\cdot \mathrm{e}^{2\pi \mathrm{i}\omega hk}, \quad \text{for each } k = 1, 2, \dots, N, \tag{16.19}$$

where $C(\omega, h)$ is an unknown function.

Using form (16.19), we rewrite system (16.16) as follows

$$-\mathrm{e}^{2\pi \mathrm{i}\omega hk'}\kappa(\omega) + C(\omega, h)\sum_{\beta\neq 0} \mathrm{e}^{-2\pi \mathrm{i}\beta hk'}\kappa(\beta)\sum_{k=1}^{N} \mathrm{e}^{2\pi \mathrm{i}(\beta+\omega)hk} = 0,$$

for each $k' = 1, 2, \ldots, N$, where

$$\sum_{k=1}^{N} \mathrm{e}^{2\pi \mathrm{i}(\beta+\omega)hk} = \begin{cases} 0, & \text{if } (\beta+\omega)h \neq t \in \mathrm{Z}, \\ N, & \text{if } (\beta+\omega)h = t \in \mathrm{Z}. \end{cases}$$

Simplifying the left-hand side of the last equation, we derive the following equation:

$$-\mathrm{e}^{2\pi \mathrm{i}\omega hk'}\kappa(\omega) + C(\omega,h)\cdot N\sum_{t=-\infty}^{\infty} \mathrm{e}^{2\pi \mathrm{i}\omega hk'}\kappa(t\cdot N-\omega) = 0,$$

$$\text{for each } k' = 1, 2, \ldots, N.$$

Simplifying the last equation, we have

$$C(\omega,h) = h\cdot\kappa(\omega)\left(\sum_{t=-\infty}^{\infty}\kappa(t\cdot N-\omega)\right)^{-1}. \tag{16.20}$$

Calculating the infinity series in the right-hand side of the last expression as in [24, p. 66], we have the following result:

$$\sum_{t=-\infty}^{\infty}\kappa(t\cdot N-\omega) = \frac{h^2}{1-\cos 2\pi\omega h} - \frac{\mathrm{e}^{2h}-1}{\mathrm{e}^{2h}-2\mathrm{e}^{h}\cos 2\pi\omega h+1}\cdot\frac{h}{2}.$$

Considering the last equality and equalities (16.19), (16.20), we get the equality (16.18). Thereby, we have proved Theorem 16.3. □

Thus, we have solved Problem 16.2.

In the theory of quadrature formulas, it is important to find the upper bound their error while constructing the quadrature formulas. Therefore, we find the sharp upper bound of the error of the constructed optimal quadrature formula in the next section.

16.5 The Norm for the Error Functional of the Optimal Quadrature Formula of the Form (16.3)

In the present section, we solve Problem 16.3, that is, we calculate the norm of the error functional (16.5).

The following theorem holds for the norm of the error functional ℓ for the optimal quadrature formula of the form (16.3).

Theorem 16.4 *The norm for the error functional of the optimal quadrature formula (16.3) for $\omega \in \mathbb{Z}\setminus\{0\}$ and $\omega h \notin \mathbb{Z}$ has the following form:*

$$\left\|\overset{\circ}{\ell}\right\|^2_{\widetilde{W}_2^{(2,1)*}} = \frac{1}{(2\pi\omega)^4 + (2\pi\omega)^2}\left[1 - \frac{2N}{(2\pi\omega)^4 + (2\pi\omega)^2}\right.$$
$$\left.\times\left[\frac{h}{1-\cos 2\pi\omega h} - \frac{\mathrm{e}^{2h}-1}{\mathrm{e}^{2h} - 2\mathrm{e}^{h}\cos 2\pi\omega h + 1}\right]^{-1}\right], \quad (16.21)$$

where N is the number of nodes.

Proof We use (16.14):

$$\left\|\overset{\circ}{\ell}\right\|^2_{\widetilde{W}_2^{(2,1)*}} = \kappa(\omega) - \kappa_m(\omega)\sum_{k=1}^{N}\overset{\circ}{C}_k \mathrm{e}^{-2\pi \mathrm{i}\omega hk} - \kappa(\omega)\sum_{k'=1}^{N}\overline{\overset{\circ}{C}_{k'}}\mathrm{e}^{2\pi \mathrm{i}\omega hk'}$$
$$+\sum_{k=1}^{N}\sum_{k'=1}^{N}\overset{\circ}{C}_k\overline{\overset{\circ}{C}_{k'}}\sum_{\beta\neq 0}\kappa_m(\beta)\cdot \mathrm{e}^{2\pi \mathrm{i}\beta h(k-k')}$$
$$= \kappa(\omega) - \kappa(\omega)\sum_{k=1}^{N}\overset{\circ}{C}_k \mathrm{e}^{-2\pi \mathrm{i}\omega hk}$$
$$-\sum_{k'=1}^{N}\overline{\overset{\circ}{C}_{k'}}\left[\kappa(\omega)\mathrm{e}^{2\pi \mathrm{i}\omega hk'} - \sum_{k=1}^{N}\overset{\circ}{C}_k\sum_{\beta\neq 0}\mathrm{e}^{2\pi \mathrm{i}\beta h(k-k')}\kappa(\beta)\right].$$

Considering (16.16) and $\kappa(\omega)$ defined by (16.12), we can rewrite the last expression as

$$\left\|\overset{\circ}{\ell}\right\|^2_{\widetilde{W}_2^{(2,1)*}} = \frac{1}{(2\pi\omega)^4 + (2\pi\omega)^2}\left[1 - \sum_{k=1}^{N}\overset{\circ}{C}_k \mathrm{e}^{-2\pi \mathrm{i}\omega hk}\right].$$

Taking into account the formula (16.18) in the last equality, we get expression (16.21).

Thus, Theorem 16.4 is proved. □

Problem 16.3 is solved.

Next, we present some numerical results for the error of the optimal quadrature formula (16.3).

16.6 Numerical Results

In the present section, we give several numerical results of the sharp upper bounds for the error (16.4) of the optimal quadrature formulas (16.3)in the space $\widetilde{W}_2^{(2,1)}(0,1]$.

According to the Cauchy-Schwarz inequality, in the space $\widetilde{W}_2^{(2,1)}(0,1]$, for the absolute value of the error (16.4), we obtain

$$\left|\left(\overset{\circ}{\ell},\varphi\right)\right| \leq \|\overset{\circ}{\ell}\|_{\widetilde{W}_2^{(2,1)*}} \cdot \|\varphi\|_{\widetilde{W}_2^{(2,1)}},$$

where $\|\overset{\circ}{\ell}\|_{\widetilde{W}_2^{(2,1)*}}$ is defined by equality (16.21).

Using Theorem 16.4 for $\|\overset{\circ}{\ell}\|_{\widetilde{W}_2^{(2,1)*}}$,we obtain the numerical results which are presented in Table 16.1 when $h = 1,\ 10^{-1},\ 10^{-2},\ 10^{-3},\ 10^{-4}$ and $\omega = 1,\ 11,$ $101,\ 1001,\ 10001$. From the first column of this table, we can see that the order of convergence of our optimal quadrature formula is $O(h^2)$ and from the first row of Table 16.1 it is clear that the quantity $\|\overset{\circ}{\ell}\|_{\widetilde{W}_2^{(2,1)*}}$ converges as $O\left(|\omega|^{-2}\right)$. From the other columns and rows of Table 16.1, we conclude that the order of convergence of our optimal quadrature formula is $O\left(\left(\frac{1}{N+|\omega|}\right)^2\right)$ in the space $\widetilde{W}_2^{(2,1)}(0,1]$.

Using the values of Table 16.1, we generate graphs of the norm for the error functional at $N = 10$, $N = 100$ and $N = 1000$, where N is the number of nodes, that is, $N = 1/h$ (Fig. 16.1)

16.7 Conclusion

Herein, the process of constructing the optimal quadrature formula for approximating strongly oscillating integrals in the Hilbert space of complex-valued periodic functions is studied. In this, firstly, to estimate the sharp upper bound of the absolute

Table 16.1 In this graph, the values of the norm of the error function $\|\overset{\circ}{\ell}\|_{\widetilde{W}_2^{(2,1)*}}$ are presented at different h and w

h	$\omega = 1$	$\omega = 11$	$\omega = 101$	$\omega = 1001$	$\omega = 10001$
1	2.5015e-2	2.09319e-4	2.483115e-6	2.5279710e-8	2.53252306e-10
10^{-1}	3.9029e-4	2.09312e-4	2.483115e-6	2.5279710e-8	2.53252306e-10
10^{-2}	3.7285e-6	3.94094e-6	2.483115e-6	2.5279710e-8	2.53252306e-10
10^{-3}	3.7268e-8	3.72890e-8	3.907038e-8	2.5279710e-8	2.53252306e-10
10^{-4}	3.7268e-10	3.72680e-10	3.728566e-10	3.9038127e-10	2.53252306e-10

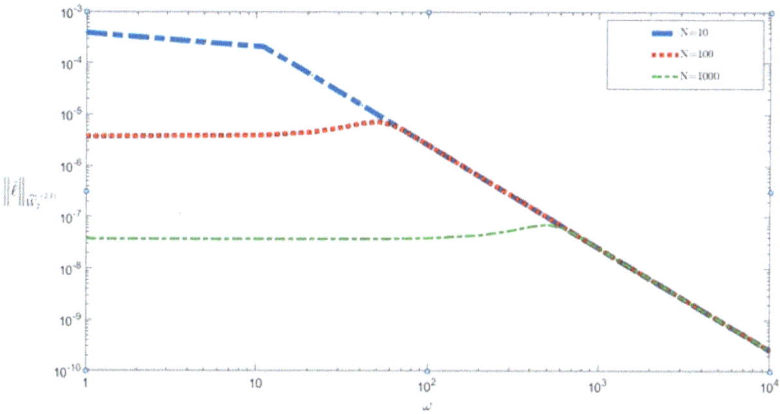

Fig. 16.1 Graph of the norm for the error functional depends on ω at $N = 10$, $N = 100$, and $N = 1000$

value of the error of the quadrature formula, the analytical form of the norm for the error functional is found, in which the extremal function corresponding to the error functional is initially found. In addition, the explicit formulas for optimal coefficients are found in the space $\widetilde{W_2}^{(2,1)}(0, 1]$ of periodic functions. Finally, we presented some numerical results using the optimal coefficient. The numerical results show that the optimal quadrature formula approaches the considered integral by $O(h^2)$when $|\omega| < N$ and by $O\left(|\omega|^{-2}\right)$when $|\omega| \geq N$.

Acknowledgments We are grateful to the organizing committee for being invited to the International Mathematical Conference Analysis, Approximations and Applications (AAA2023), dedicated to Academician Gradimir V. Milovanović on the occasion of his 75th anniversary.

References

1. Boltaev, N.D., Hayotov, A.R., Milovanović, G.V., Shadimetov, Kh.M.: Optimal quadrature formulas for Fourier coefficients in $W_2^{(m,m-1)}$ space. J. Appl. Anal. Comput. **7**, 1233–1266 (2017)
2. Hayotov, A.R., Milovanović, G.V., Shadimetov, Kh.M.: Optimal quadratures in the sense of Sard in a Hilbert space. Appl. Math. Comput. **259**, 637–653 (2015)
3. Hayotov, A.R., Jeon, S., Lee, Ch.O.: On an optimal quadrature formula for approximation of Fourier integrals in the space $L_2^{(1)}$. J. Comput. Appl. Math. **372**, 112713 (2020)
4. Hayotov, A.R., Babaev S.S.: Optimal quadrature formulas for computing of Fourier integrals in $W_2^{(m,m-1)}$ space. AIP Conf. Proc. (2021). https://doi.org/10.1063/5.0057127
5. Hayotov, A.R., Jeon, S., Lee, Ch.O.: On an optimal quadrature formula for approximation of Fourier integrals in the space $L_2^{(1)}$. J. Comput. Appl. Math. **372**, 112713 (2020)
6. Hayotov, A.R., Jeon, S., Lee, Ch.-O. Shadimetov, Kh.M.: Optimal quadrature formulas for non-periodic functions in Sobolev space and its application to CT image reconstruction. Filomat **12**(35), 4177–4195 (2021)

7. Hayotov, A.R., Jeon, S., Shadimetov, Kh.M.: Application of optimal quadrature formulas for reconstruction of CT images. J. Comput. Appl. Math. **388**, 113313 (2021)
8. Hayotov, A.R., Jeon, S., Shadimetov, M.Kh.: Application of optimal quadrature formulas for reconstruction of CT images. J. Comput. Appl. Math. **388**, 13313 (2021)
9. Hayotov, A.R., Khayriev, U.N., Farrukhbek, A.: Exponentially weighted optimal quadrature formula with derivative in the space $L_2^{(2)}$. AIP Conf. Proc. (2023). https://doi.org/10.1063/5.0144753
10. Hayotov, A.R., Khayriev, U.N.: Optimal quadrature formulas in the space $\widetilde{W_2}^{(1,0)}$ of periodic functions. Uzbek Mat. Zh. (2021). https://doi.org/10.29229/uzmj.2021-3-9
11. Hayotov, A.R., Khayriev, U.N.: Construction of an optimal quadrature formula in the Hilbert space of periodic functions. Lobachevskii J. Math. **11**(43), 119–128 (2022). https://doi.org/10.1134/S199508022214013X
12. Hayotov, A.R., Khayriev, U.N.: Optimal quadrature formulas in the space $\widetilde{W_2}^{(m,m-1)}$ of periodic functions. Vestn. KRAUNC (2022). https://doi.org/10.26117/2079-6641-2022-40-3-200-215
13. Iserles, A.: On the numerical quadrature of highly–oscillating integrals II: Irregular oscillators. IMA J. Numer. Anal. **25**, 25–44 (2005)
14. Iserles, A., Nørsett, S.P.: Efficient quadrature of highly–oscillatory integrals using derivatives. Proc. R. Soc. Lond. Ser. A Math. Phys. Eng. Sci. **461**, 1383–1399 (2005)
15. Khayriev, U.N.: An optimal quadrature formula in the space of periodic functions. Probl. Comput. Appl. Math. Tashkent **2/1**(48), 169–179 (2023)
16. Khayriev, U.N.: Construction of the exponentially weighted optimal quadrature formula in a Hilbert space of periodic functions. Probl. Comput. Appl. Math., Tashkent **5/1**(44), 134–142 (2022)
17. Milovanović, G.V., Stanić, M.P.: Numerical integration of highly oscillating functions. In: Analytic Number Theory, Approximation Theory and Special Functions, pp. 613–649. Springer, Berlin (2014)
18. Milovanović, G.V, Cvetković, A.: Special classes of orthogonal polynomials and corresponding quadratures of Gaussian type. Math. Balkanica **26**, 169–184 (2012)
19. Milovanović, G.V.: Generalized Gaussian quadratures for integrals with logarithmic singularity. Filomat **30**, 1111–1126 (2016)
20. Milovanović, G.V., Igić, T. S., Turnić, D.: Generalized quadrature rules of Gaussian type for numerical evaluation of singular integrals. J. Comput. Appl. Math. **278**, 306–325 (2015)
21. Olver, S.: Moment-free numerical integration of highly oscillatory functions. IMA J. Numer. Anal. **26**, 213–227 (2006)
22. Olver, S.: Numerical approximation of highly oscillatory integrals. Ph.D. Thesis, University of Cambridge (2008)
23. Shadimetov, M.Kh.: Optimal Lattice Quadrature and Cubature Formulas in Sobolev Spaces. Monograph, Ministry of Higher and Secondary Specialized Education of the Republic of Uzbekistan, Tashkent (2019)
24. Shadimetov, Kh.M., Hayotov, A.R.: Optimal approximation of error functionals of quadrature and interpolation formulas in spaces of differentiable functions, p. 246. In: Muhr Press, Tashkent (2022) (in Russian)
25. Sobolev, S.L.: Introduction To the Theory of Cubature Formulas. Moscow, Nauka (1974) (in Russian)
26. Xu, Z. Milovanović, G.V., Xiang, S.: Efficient computation of highly oscillatory integrals with Henkel kernel. Appl. Math. Comp. **261**, 312–322 (2015)
27. Xu, Z., Milovanović, G.V.: Efficient method for the computation of oscillatory Bessel transform and Bessel Hilbert transform. J. Comput. Appl.Math. **308** 117–137 (2016)
28. Zhang, S., Novak, E.: Optimal quadrature formulas for the Sobolev space H^1. J. Sci. Comput. **78**, 274–289 (2019)

Chapter 17
On Almost Deferred Sequence Spaces Defined by Infinite Matrix

Ekrem Savas

Dedicated to Academician Gradimir V. Milovanović on the occasion of his 75th birthday

17.1 Definitions

Let S denote the set of all real and complex sequences $x = (x_k)$. By l_∞ and c, we denote the Banach spaces of bounded and convergent sequences $x = (x_k)$ normed by $\|x\| = \sup_n |x_n|$, respectively.

The Hahn-Banach theorem can be considered to introduce methods which are not established by a regular summability matrix. Banach [2] introduced this theorem to contribute that the limit functional can be extended from the set of the convergent sequences to the set of the bounded sequences with the property linearity, positivity, and shift invariance and these extensions were known as Banach limits. If $x \in l_\infty$ has the same value ξ for each Banach limit, we say that the sequence is almost convergent to ξ. Lorentz [9] presented that a sequence $x = (x_n)$ is almost convergent to ξ if and only if

$$\lim_{n\to\infty} \frac{1}{n}\sum_{k=1}^{n} x_{k+m} = \xi,$$

uniformly in m.

$\hat{c}$ will denote the set of all almost convergent sequences.

Different generalizations, extensions of almost convergent sequences have been studied by various authors including Başar and Solak [4], King [8], Nanda[13] and Savaş ([18–20]), and many others.

E. Savas (✉)
Department of Mathematics, Uşak University, Uşak, Turkey

M. P. Stanić et al. (eds.), *Analysis, Approximation, Optimization: Computation and Applications*, Springer Optimization and Its Applications 224,
https://doi.org/10.1007/978-3-031-85743-0_17

Definition 17.1 (see [12]) Let X be a real or complex linear space and g be a function from X to the $\mathbb{R}$. The pair (X, g) is called a paranormed space and g is a paranorm for X, if the following properties are satisfied for all elements $x, y \in X$:

i. $g(\theta) = 0$, where θ is the zero element of X.
ii. $g(x) \geq 0$.
iii. $g(-x) = g(x)$.
iv. $g(x + y) \leq g(x) + g(y)$ (triangle inequality).
v. If (λ_n) is a sequence of scalars with $\lambda_n \to \lambda$ as $n \to \infty$ and $(x_n)_{n=1}^{\infty}$ is a sequence in X with $g(x_n - x) \to 0$ as $n \to \infty$ then $g(\lambda_n x_n - \lambda x) \to 0$ as $n \to \infty$ (continuity of multiplication by scalars).

Observe that, if in addition g is increasing and $g(x) = g(|x|)$, and the sequence (λ_n) converges to λ, then for large $n's$

$$\begin{aligned} g(\lambda_n x_n - \lambda x) &\leq g((\lambda_n - \lambda)(x_n - x)) + g(\lambda(x_n - x)) + g((\lambda_n - \lambda)x) \\ &\leq g(|x_n - x|) + (1 + [|\lambda|])g(|x_n - x|) + g((\lambda_n - \lambda)x) \\ &= (2 + [|\lambda|])g(x_n - x) + g((\lambda_n - \lambda)x), \end{aligned}$$

where $[|\lambda|]$ is the integer part of $|\lambda|$. Hence to prove condition v, it is enough to show that for all fixed $y \in X$ and $\lambda_n \to 0$, $g(\lambda_n y) \to 0$. Throughout this work we introduce the sequence spaces using the following type of transformation:

Definition 17.2 (see [7, 10]) Let $B = (b_{n,k})$ denote a matrix transformation that maps complex sequences x into the sequence Bx where the nth term of Bx is as follows:

$$(Bx)_n = \sum_{k=1}^{\infty} b_{n,k} x_k.$$

Such transformation is said to be nonnegative if $b_{n,k}$ is nonnegative.

Let $p = (p_n)$ and $q = (q_n)$ be sequences of nonnegative integers (see Agnew[1]) satisfying:

(i) $p_n < q_n$ for all $n \in \mathbb{N}_0$
(ii) $\lim_{n \to \infty} q_n = \infty$

Then the deferred Cesàro mean of sequence $x = (x_n)$ is defined (see Nayak et al. [3, 14]) as follows:

$$(D_{p,q}x)_n = \frac{x_{p_{n+1}} + x_{p_{n+2}} + \cdots + x_{q_n}}{q_n - p_n}.$$

In the notation of matrix transformation

$$(B_{p,q}x)_n = \sum_{k=0}^{\infty} b_{nk} x_k,$$

where

$$b_{nk} = \begin{cases} \frac{1}{q_n - p_n}, & p_n < k \leq q_n, \\ 0, & \text{otherwise.} \end{cases}$$

It is clear that (D_{p_n,q_n}) is regular with conditions (i) and (ii). Indeed, this mean generalizes various transforms such as $D_{n-1,n} = I$, the identity transform, and $D_{0,n} = C(0,1)$, the Cesàro transformation.

We now have

Definition 17.3 Let f be a modulus function and $B = (b_{n,k})$ be a nonnegative summability matrix with the property

$$\sup_n \sum_{k=1}^{\infty} b_{n,k} < \infty.$$

We present the new deferred sequence spaces as follows:

$$\hat{c}_0(D, B, f) = \left\{ x \in S : \lim_n \sum_{k=p_n+1}^{q_n} b_{n,k} f(|x_{k+m}|) = 0, \text{ uniformly in } m \right\},$$

$$\hat{c}(D, B, f) = \left\{ x \in S : x - \xi e \in \hat{c}_0(D, B, f) \text{ for some scalar } \xi \right\}$$

and

$$\hat{c}_\infty(D, B, f) = \left\{ x \in S : \sup_{n,m} \sum_{k=p_n+1}^{q_n} b_{n,k} f(|x_{k+m}|) < \infty \right\},$$

where $e = \{1, 1, 1, 1, \ldots\}$ and (p_n), (q_n) are sequences of nonnegative integers satisfying the conditions (i) and (ii).

If $x \in \hat{c}_0(D, B, f)$, then we say that the sequence x is deferred strong B- almost convergent to 0 with respect to a modulus f. It should be note that modulus function was defined by Ruckle [16] as follows.

Definition 17.4 A function $f : [0, \infty) \to [0, \infty)$ is called a modulus function provided that:

i. $f(x) = 0$ if and only if $x = 0$.
ii. $f(x + y) \leq f(x) + f(y)$ for all $x \geq 0$ and $y \geq 0$.
iii. f is increasing.
iv. f is continuous from the right of 0.

Various extensions and applications of modulus function have been considered in [5, 6, 11, 15, 17].

If $f(x) = x$, then the above sequence spaces reduce to the following:

$$\hat{c}_0(D, B) = \left\{ x \in S : \lim_n \sum_{k=p_n+1}^{q_n} b_{n,k}|x_{k+m}| = 0, \text{ uniformly in } m \right\},$$

$$\hat{c}(D, B) = \left\{ x \in S : x - \xi e \in \hat{c}_0(D, B) \text{ for some } \xi \right\}$$

and

$$\hat{c}_\infty(D, B) = \left\{ x \in S : \sup_{n,m} \sum_{k=p_n+1}^{q_n} b_{n,k}|x_{k+m}| < \infty \right\}.$$

If we consider $p_n = 0$ and $q_n = n$ for all $n \in \mathbb{N}$, then $\hat{c}_0(D, B, f)$, $\hat{c}(D, B, f)$ and $\hat{c}_\infty(D, B, f)$ reduce to $\hat{\omega}_0(B, f)$, $\hat{\omega}(B, f)$, and $\hat{c}_\infty(B, f)$, respectively (see [15]).

17.2 Main Results

We now start our main discussions.

Theorem 17.1 *The spaces $\hat{c}_0(D, B, f)$, $\hat{c}(D, B, f)$, and $\hat{c}_\infty(D, B, f)$ are linear space over the complex field $\mathbb{C}$.*

Proof Let us consider that the space $\hat{c}_0(D, B, f)$ is linear. We also prove the linearity of $\hat{c}(D, B, f)$ and $\hat{c}_\infty(D, B, f)$ by proceeding in the similar line. Take $x, y \in \hat{c}_0(D, B, f)$. Let $\phi, \varphi \in \mathbb{C}$, and choose natural numbers K_ϕ and L_φ such that $|\phi| < K_\phi$ and $|\varphi| < L_\varphi$. Using the monotonicity and subadditivity of f we observe, for all m,

$$\begin{aligned}
&\sum_{k=p_n+1}^{q_n} b_{n,k} f(|\phi x_{k+m} + \varphi y_{k+m}|) \\
&\leq \sum_{k=p_n+1}^{q_n} b_{n,k} \left(f(|\phi||x_{k+m}|) + f(|\varphi||y_{k+m}|) \right) \\
&\leq \sum_{k=p_n+1}^{q_n} b_{n,k} f(K_\phi |x_{k+m}|) + \sum_{k=p_n+1}^{q_n} b_{n,k} f(L_\varphi |y_{k+m}|) \\
&\leq K_\phi \sum_{k=p_n+1}^{q_n} b_{n,k} f(|x_{k+m}|) + L_\varphi \sum_{k=p_n+1}^{q_n} b_{n,k} f(|y_{k+m}|).
\end{aligned}$$

Since $x, y \in \hat{c}_0(D, B, f)$, it is obvious that $\lim_{n\to\infty} \sum_{k=p_n+1}^{q_n} b_{n,k} f(|x_{k+m}|) = 0$ and $\lim_{n\to\infty} \sum_{k=p_n+1}^{q_n} b_{n,k} f(|y_{k+m}|) = 0$, uniformly in m . Thus, $\phi x + \varphi y \in \hat{c}_0(D, B, f)$ and finally we can get that the space $\hat{c}_0(D, B, f)$ is linear. □

Theorem 17.2 *For any nonnegative matrix* $B = (b_{nk})$ *with the property* $\sup_n \sum_{k=1}^{\infty} b_{n,k} < \infty$ *and any modulus function* f*, the following inclusions hold:*

i. $\hat{c}(D, B, f) \subset \hat{c}_\infty(D, B, f)$.
ii. $\hat{c}_0(D, B, f) \subset \hat{c}(D, B, f)$.

Proof It may be noted that the second inclusion is obvious. Let $x \in \hat{c}(D, B, f)$. Now, by the definition of a modulus function ii and iii for all m we write

$$\sum_{k=p_n+1}^{q_n} b_{n,k} f(|x_{k+m}|) = \sum_{k=p_n+1}^{q_n} b_{n,k} f(|x_{k+m} - \xi + \xi|)$$
$$\leq \sum_{k=p_n+1}^{q_n} b_{n,k} f(|x_{k+m} - \xi|) + f(|\xi|) \sum_{k=p_n+1}^{q_n} b_{n,k}.$$

Since $\sup_n \sum_{k=p_n+1}^{q_n} b_{n,k} < \infty$ and $x \in \hat{c}(D, B, f)$, it is obvious that $x \in \hat{c}_\infty(D, B, f)$. □

Theorem 17.3 *Let* f *be a modulus function and* $B = (b_{nk})$ *a nonnegative summability matrix with the property* $\sup_n \sum_{k=1}^{\infty} b_{n,k} < \infty$*, then:*

i. $\hat{c}(D, B) \subset \hat{c}(D, B, f)$.
ii. $\hat{c}_0(D, B) \subset \hat{c}_0(D, B, f)$.
iii. $\hat{c}_\infty(D, B) \subset \hat{c}_\infty(D, B, f)$.

Proof We shall give the proof of iii only. The proofs of the remaining parts follow in similar manners. Let $x \in \hat{c}_\infty(D, B)$, that is, $\sup_{nm} \sum_{k=p_n+1}^{q_n} b_{n,k} |x_{k+m}| < \infty$, and f be any modulus function. Also let $\epsilon > 0$ and $\delta \in (0, 1)$ such that $f(t) < \epsilon$ for $0 \leq t \leq \delta$. Consider the following partition of $\sum_{k=p_n+1}^{q_n} b_{n,k} f(|x_{k+m}|)$ and denote each sum by $\sum_1$ and $\sum_2$, respectively:

$$\sum_{k=p_n+1}^{q_n} b_{n,k} f(|x_{k+m}|) = \sum_{\{k=p_n+1 \ \& \ |x_{k+m}|<\delta\}}^{q_n} b_{n,k} f(|x_{k+m}|)$$
$$+ \sum_{\{k=p_n+1 \ \& \ |x_{k+m}|\geq\delta\}}^{q_n} b_{n,k} f(|x_{k+m}|). \tag{17.1}$$

We write

$$|x_{k+m}| < \frac{|x_{k+m}|}{\delta} < 1 + \left[\frac{|x_{k+m}|}{\delta}\right],$$

where $[t]$ denotes the integer part of t. Hence, by properties ii and iii of modulus function, for $|x_{k+m}| \geq \delta$ we have $f(|x_{k+m}|) \leq \left(1 + \left[\frac{|x_{k+m}|}{\delta}\right]\right) f(1) \leq 2f(1)\frac{|x_{k+m}|}{\delta}$. Therefore, the second partition in (17.1) can be bounded as follows, for all m,

$$\sum_{2} \leq 2\frac{f(1)}{\delta} \sum_{k=p_n+1}^{q_n} b_{n,k}|x_{k+m}|.$$

In addition,

$$\sum_{1} \leq \epsilon \sum_{k=p_n+1}^{q_n} b_{n,k},$$

and therefore, for all m

$$\sum_{k=p_n+1}^{q_n} b_{n,k} f(|x_{k+m}|) \leq \epsilon \sum_{k=p_n+1}^{q_n} b_{n,k} + 2\frac{f(1)}{\delta} \sum_{k=p_n+1}^{q_n} b_{n,k}|x_{k+m}|.$$

Since

$$\sup_n \sum_{k=p_n+1}^{q_n} b_{n,k} < \infty$$

and $x \in \hat{c}_\infty(D, B)$, this gives that $x \in \hat{c}_\infty(D, B, f)$. □

Theorem 17.4 *Let f be a modulus function and B a nonnegative summability matrix with the property*

$$\sup_n \sum_{k=p_n+1}^{q_n} b_{n,k} < \infty.$$

If $\beta = \inf_{t>0} \frac{f(t)}{t} > 0$, then

$$\hat{c}(D, B) = \hat{c}(D, B, f).$$

Proof In the previous theorem we know that $\hat{c}(D, B) \subseteq \hat{c}(D, B, f)$. It remains to show that $\hat{c}(D, B, f) \subseteq \hat{c}(D, B)$. Since $f(t) \geq \beta t$ for all $t \geq 0$ and $\beta > 0$, we granted $t \leq \frac{1}{\beta} f(t)$ for all $t \geq 0$. Let $x \in \hat{c}(D, B, f)$, that is, for some scalar ξ, $\lim_{n\to\infty} \sum_{k=p_n+1}^{q_n} f(|x_{k+m} - \xi|) = 0$. Since for every $n \in \mathbb{N}$, for all m,

$$\sum_{k=p_n+1}^{q_n} b_{n,k}|x_{k+m} - \xi| \leq \frac{1}{\beta} \sum_{k=p_n+1}^{q_n} b_{n,k} f(|x_{k+m} - \xi|).$$

We get that $x \in \hat{c}(D, B)$. □

Theorem 17.5 *Let $B = (b_{nk})$ be a nonnegative summability matrix with the property*

$$\sup_n \sum_{k=p_n+1}^{q_n} b_{n,k} < \infty, \quad \inf_n \sum_{k=p_n+1}^{q_n} b_{n,k} > 0,$$

and f is a modulus function and then $\hat{c}_0(D, B, f)$ and $\hat{c}(D, B, f)$ are linear topological spaces with paranorm

$$g(x) = \sup_{n,m} \sum_{k=p_n+1}^{q_n} b_{n,k} f(|x_{k+m}|).$$

Also, if for all $N \in \mathbb{N}$, $\cup_{n=1}^{N} [p_n + 1, q_n) = [1, q_N)$, then $\hat{c}_0(D, B, f)$ and $\hat{c}(D, B, f)$ are complete with respect to the paranorm g.

Proof We shall consider the proof of $\hat{c}(D, B, f)$ only. The case of $\hat{c}_0(D, B, f)$ will follow analogously. For this, we will first show that $\hat{c}(D, B, f)$ with g is a paranormed linear topological space. By Theorem 17.2, for each $x \in \hat{c}(D, B, f)$, $g(x)$ exists. Moreover, note $g(x) = 0$ if and only if $x = 0$, $g(-x) = g(x)$ and $g(x + y) \leq g(x) + g(y)$. Since by the nature of the function f, g is increasing and $g(x) = g(|x|)$, it remains to show that for fixed x taking a sequence (λ_m) of scalars converging to 0, we write that $g(\lambda_m x) \to 0$ (see the remark after definition paranorm).

Fix $\epsilon > 0$ and write $x \in \hat{c}(D, B, f)$ and sequence (λ_m) converging to 0. Then there exists ξ and $N \in \mathbb{N}$ such that for all $n, m \geq N$

$$\sum_{k=p_n+1}^{q_n} b_{n,k} f(|x_{k+m} - \xi|) < \frac{\epsilon}{2}.$$

Denote by $K = \sup_{n,m} \sum_{k=p_n+1}^{q_n} b_{n,k}$ and $r = \max_{k \leq N} |x_{k+m} - \xi|$. Since f is continuous, we can find $M \in \mathbb{N}$ such that $|\lambda_r| < 1$ and $f(|\lambda_r|) < \dfrac{\epsilon}{2K(2 + [|r|] + [|\xi|])}$ for all $r \geq M$, where $[s]$ denotes the integer part of a positive scalar s. Then for all $n, m \in \mathbb{N}$ and $n \geq M$ we have for all m

$$\begin{aligned}\sum_{k=p_n+1}^{q_n} b_{n,k} f(|\lambda_m x_{k+m}|) &= \sum_{k=p_n+1}^{q_n} b_{n,k} f(|\lambda_m x_{k+m} + \lambda_m \xi - \lambda_m \xi|) \\ &\leq \sum_{k=p_n+1 \& n \geq N}^{q_n} b_{n,k} f(|\lambda_m||x_{k+m} - \xi|)\end{aligned}$$

$$+ \sum_{k=p_n+1 \& n<N}^{q_n} b_{n,k} f(|\lambda_m||x_{k+m} - \xi|) + Kf(|\lambda_m \xi|)$$

$$\leq \sum_{k=p_n+1 \& n\geq N}^{q_n} b_{n,k} f(|x_{k+m} - \xi|)$$

$$+ Kf(|\lambda_m|r) + Kf(|\lambda_m \xi|)$$

$$\leq \sum_{k=p_n+1 \& n\geq N}^{q_n} b_{n,k} f(|x_{k+m} - \xi|)$$

$$+ K(2 + [|r|] + [|\xi|])f(|\lambda_m|) < \epsilon.$$

Finally , $g(\lambda_m x) = \sup_{nm} \sum_{k=p_n+1}^{q_n} b_{n,k} f(|\lambda_m x_{k+m}|) \to 0$ as $m \to \infty$, uniformly in m.

We will assume next that

$$\cup_{n=1}^{N} [p_n + 1, q_n) = [1, q_N), \tag{17.2}$$

that is, $p_1 = 0$ and the sums $\sum_{k=p_n+1}^{q_n} y_{k+m}$ taken over all natural numbers n will run through each coordinate y_k of given y at least once. We will establish that under the above condition $\hat{c}(D, B, f)$ is complete with respect to its paranorm g. Let (x_k^s) be a Cauchy sequence in $\big(\hat{c}(D, B, f), g\big)$. Then the convergence $\lim_{s,t\to\infty} g(x^s - x^t) = 0$ means that, as $s, t \to \infty$ for all n, m

$$\sum_{k=p_n+1}^{q_n} b_{n,k} f(|x_{k+m}^s - x_{k+m}^t|) \to 0. \tag{17.3}$$

Hence, by assumption (17.2) for each fixed $k \in \mathbb{N}$, it follows that $b_{n,k} f(|x_{k+m}^s - x_{k+m}^t|) \to 0$, with $s, t \to \infty$, and since $B = (b_{n,k})$ is a nonnegative and f continuous, (x_k^s) is a Cauchy sequence in $\mathbb{C}$ for each fixed k. Since $\mathbb{C}$ is complete, there exists a sequence $x = (x_k) \subseteq \mathbb{C}$ such that as $s \to \infty$, $x_k^s \to x_k$ for each k.

Now from (17.3), given $\epsilon > 0$ there exists a natural number N such that for all $s, t \geq N$ and $n \in \mathbb{N}$, for all m,

$$\sum_{k=p_n+1}^{q_n} b_{n,k} f(|x_{k+m}^s - x_{k+m}^t|) < \frac{\epsilon}{2}. \tag{17.4}$$

Since by the continuity for f we also get that for fixed k, $f(|x_{k+m}^t - x_{k+m}|) \to 0$ as $t \to \infty$, for every $k \in \mathbb{N}$ there is $M_k \in \mathbb{N}$ such that if $t \geq M_k$ then $f(|x_{k+m}^t - x_{k+m}|) < \dfrac{\epsilon}{2C}$, where $C = \sup_n \sum_{k=p_n+1}^{q_n} b_{n,k}$. Write $N_n = \max_{p_n+1\leq k\leq q_n}\{M_k, N\}$. We have for $t \geq N_n$

$$\sum_{k=p_n+1}^{q_n} b_{n,k} f(|x_{k+m}^t - x_{k+m}|) \le \sum_{k=p_n+1}^{q_n} b_{n,k} \frac{\epsilon}{2C} \le C \cdot \frac{\varepsilon}{2C} = \frac{\varepsilon}{2}, \tag{17.5}$$

for all n, m.

Let $n, m \in \mathbb{N}$, and choose $s \ge N$ and $t \ge N_n$. Then from (17.4) and (17.5),

$$\sum_{k=p_n+1}^{q_n} b_{n,k} f(|x_{k+m}^t - x_{k+m}|) \le \sum_{k=p_n+1}^{q_n} b_{n,k} f(|x_{k+m}^s - x_{k+m}^{t_n}|) + \sum_{k=p_n+1}^{q_n} b_{n,k} f(|x_{k+m}^{t_n} - x_{k+m}|) < \varepsilon.$$

Finally , $g(x^s - x) = \sup_{n,m} \sum_{k=p_n+1}^{q_n} b_{n,k} f(|x_{k+m}^t - x_{k+m}|) < \varepsilon$, for all $s \ge N$ and $g(x^s - x) \to 0$ as $n \to \infty$, uniformly in m.

Since all $x^s \in \hat{c}(D, B, f)$, for each s there exists ξ^s with

$$\lim_{n\to\infty} \sum_{k=p_n+1}^{q_n} b_{n,k} f(|x_{k+m}^s - \xi^s|) = 0,$$

uniformly in m. Write by $c = \inf_n \sum_{k=p_n+1}^{q_n} b_{n,k}$, which is assumed to be positive. Write that for all n, m and s, t,

$$\begin{aligned} cf(|\xi^t - \xi^s|) &\le \sum_{k=p_n+1}^{q_n} b_{n,k} f(|\xi^t - \xi^s|) \le \sum_{k=p_n+1}^{q_n} b_{n,k} f(|\xi^t - x_{k+m}^t|) \\ &\quad + \sum_{k=p_n+1}^{q_n} b_{n,k} f(|x_{k+m}^t - x_{k+m}^s|) + \sum_{k=p_n+1}^{q_n} b_{n,k} f(|x_{k+m}^s - \xi^s|) \\ &\le \sum_{k=p_n+1}^{q_n} b_{n,k} f(|\xi^t - x_{k+m}^t|) + g(x^t - x^s) \\ &\quad + \sum_{k=p_n+1}^{q_n} b_{n,k} f(|x_{k+m}^s - \xi^s|). \end{aligned}$$

Since the outer sums for fixed t and s converge to 0 as $n \to \infty$, uniformly in m and $g(x^s - x^t) \to 0$ as $s, t \to \infty$, we have $f(|\xi^s - \xi^t|) \to 0$ as $s, t \to \infty$. By the continuity of f, (ξ^s) is Cauchy in $\mathbb{C}$, and so its limit ξ exists. Using the assumption that $C = \sup_n \sum_{k=p_n+1}^{q_n} b_{n,k} < \infty$ and the estimate, for all m,

$$\sum_{k=p_n+1}^{q_n} b_{n,k} f(|x - \xi|) \le g(x - x^s) + \sum_{k=p_n+1}^{q_n} b_{n,k} f(|x_{k+m}^s - \xi^s|) + Cf(|\xi^s - \xi|),$$

it is clear that

$$\lim_{n\to\infty} \sum_{k=p_n+1}^{q_n} b_{n,k} f(|x_{k+m} - \xi|) = 0,$$

uniformly in m. Finally, $x \in \hat{c}(D, B, f)$. □

Acknowledgments The author thanks the reviewers for the careful reading of the article and for their valuable comments.

References

1. Agnew, R.P.: On deferred Cesaro means. Ann. Math. **33**(3), 413–421 (1932)
2. Banach, S.: Theorie des Operations linearies. Warszawa (1932)
3. Baliarsingh, P.: On statistical deferred A-convergence of uncertain sequences. Inter. J. Uncertain. **29**(4), 499–515 (2021)
4. Başar, F., Solak, I.: Almost-coercive matrix transformations. Rend. Mat. Appl. **7/11**(2), 249–256 (1991)
5. Bhardwaj, V.K., Singh, N.: On some sequence spaces defined by a modulus. Indian J. Pure Appl. Math. **30**(8), 809–817 (1999)
6. Connor, J.: On strong matrix summability with respect to a modulus and statistical convergent. Can. Math. Bull. **(32)**, 194–198 (1989)
7. Hardy, G.H.: Divergent Series. Oxford University Press, London (1949)
8. King, J.P.: Almost summable sequences. Proc. Am. Math. Soc. **17**, 1219–25 (1966)
9. Lorentz, G.G.: A contribution to the theory of divergent sequences. Acta math. **80**, 167–190 (1948)
10. Maddox, I.J.: Elements of Functional Analysis. Cambridge . University Press, Cambridge (1970)
11. Maddox, I.J.: Sequences spaces defined by a modulus. Math. Proc. Camb. Philos. Soc. **(100)**, 161–166 (1986)
12. Malkowsky, E., Basar, F.: A survey on some paranormed sequence spaces. Filomat **31/4**, 1099–1122 (2017)
13. Nanda, S.: Strongly almost summable and strongly almost convergent sequences. Acta Math. Hung. **49**(1–2), 71–76 (1987)
14. Nayak, L., Ray, G., Das B.K.: An estimate of the rate of convergence of Fourier series in the generalized Hölder metric by deferred cesàro men. JMAA **420**, 563–575 (2014)
15. Nuray, F., Savas, E.: Some new sequence spaces defined by a modulus function. Indian J. Pure. Appl. Math. **24**(11), 657–663 (1993)
16. Ruckle, W.H.: FK Spaces in which the sequences of coordinate vectors is bunded. Can. J. Math. **(25)**, 973–978 (1973)
17. Savaş, E.: On some generalized sequence spaces defined by a modulus. Indian J. Pure Appl. Math. **30**(5), 459–464 (1999)
18. Savaş, E.: Matrix transformations of some generalized sequence spaces. J. Orissa Math. Soc. **4**(1), 37–51 (1985)
19. Savaş, E.: Matrix transformations and absolute almost convergence. Bull. Inst. Math. Acad. Sinica **15**(3), 345–355 (1987)
20. Savaş, E.: Matrix transformations between some new sequence spaces. Tamkang J. Math. **19**(4), 75–80 (1988)

Chapter 18
Certain Classes of Combinatorial Polynomials Involving (Inverse) Binomial Coefficients with Their Applications

Yilmaz Simsek

Dedicated to Academician Gradimir V. Milovanović on the occasion of his 75th birthday

18.1 Introduction

Constructing generating functions for certain finite sums and giving their implementations have been one of the main topics that many researchers have focused on for years. For this reason, these functions are defined using many different branches and techniques of mathematics. Because it is well known that both the sums themselves and their generating functions have been used to build mathematical models that involve many real-world problems. This chapter is written considering this reality. In this chapter, generating functions of previously known finite sums and their computational formulas are given. Applying these, novel formulas are derived. These applicable new results that will have a high potential for researchers to use are also proven using different methods. In this chapter, it is planned to conduct research on many different finite sums and infinite sums, including the powers of the (inverse) binomial coefficients. With the aid of these sums, new classes of polynomials, numbers and their mathematical analysis processes, and their applications in approximation theory are also discussed (*cf.* [3–75]).

In [58, 64] and [66], we have given many computational formulas and generating functions for the sums of finite powers of binomial coefficients in 2019 and for the sums of finite powers of inverse binomial coefficients in 2023. In particular, the formulas found so far for these sums have many useful applications in mathematical models, Franel numbers, Brownian paths, graph theory, and other real-world problems that relate watermelons to more than two chains (*cf.* [4–68]). It is also known that the binomial coefficient also plays a leading role in creating

Y. Simsek (✉)
Department of Mathematics, University of Akdeniz, Faculty of Science, Antalya, Turkey
e-mail: ysimsek@akdeniz.edu.tr

M. P. Stanić et al. (eds.), *Analysis, Approximation, Optimization: Computation and Applications*, Springer Optimization and Its Applications 224,
https://doi.org/10.1007/978-3-031-85743-0_18

mathematical models for cell population dynamics and calculating the growth of the cell number of multicellular organisms (*cf.* [14]).

Throughout our chapter, the following well-known standard notations can be used:

$$\mathbb{N}_0 = \{0, 1, 2, 3, \ldots\}.$$

Here, $\mathbb{N}$ denotes the set of positive integers, $\mathbb{Z}$ denotes the set of integers, $\mathbb{Q}$ denotes the set of rational numbers, $\mathbb{R}$ denotes the set of real numbers, and $\mathbb{C}$ denotes the set of complex numbers. We assume that $\ln z$ denotes the principal branch of the multi-valued function $\ln z$ with the imaginary part $\mathrm{Im}(\ln z)$ constrained by the interval $(-\pi, \pi]$.

18.2 Combinatorial Polynomials and Numbers

In this section, we give various formulas and relations involving certain families of finite sums, combinatorial polynomials and numbers containing sums of powers of (inverse) binomial coefficients with their generating functions, and other special numbers and polynomials.

We [58] defined following the sum:

$$y_6(j, b; \omega, p) = \sum_{k=0}^{b} \binom{b}{k}^p \frac{\omega^k k^j}{b!}, \tag{18.1}$$

where $b, j, p \in \mathbb{N}_0$, $\omega \in \mathbb{R}$ or $\omega \in \mathbb{C}$, and $0^j = 1$ for $j = 0$.

Recently, these numbers have been used by many authors. In [28], Kucukoglu named the sum $y_6(j, b; \omega, p)$ as *combinatorial Simsek numbers of the sixth kind.* In [27], Kucukoglu also analyzed q-analogues of the special case of the numbers $y_6(m, k; \omega, v)$ when $p = 1$, calling them *q-combinatorial Simsek numbers and polynomials of the first kind.*

Let us specifically note here that the sum $y_6(j, b; \omega, p)$ can also be considered as a polynomial with a variable ω. The degree of this polynomial is b.

The sum $y_6(j, b; \omega, p)$ is given by means of the following generating function (for details, see also [58]):

$$\begin{aligned} F_{y6}(u, b; \omega, p) &= \frac{1}{b!}\, {}_pF_{p-1}\left(-b, -b, \ldots, -b; 1, 1, \ldots, 1; (-1)^p \omega e^u\right) \\ &= \sum_{j=0}^{\infty} y_6(j, b; \omega, p) \frac{u^j}{j!}, \end{aligned} \tag{18.2}$$

where ${}_kF_m(x_1, \ldots, x_k; w_1, \ldots, w_m; u)$ denotes the following generalized hypergeometric series:

$$ {}_kF_m(x_1, \ldots, x_k; w_1, \ldots, w_m; u) = \sum_{n=0}^{\infty} \left(\frac{\prod_{j=1}^{k} (x_j)_n}{\prod_{j=1}^{m} (w_j)_n} \right) \frac{u^n}{n!}, \tag{18.3} $$

where the above series converges for all u if $k < m+1$, and for $|u| < 1$ if $k = m+1$. Assuming that all parameters have general values, real or complex, except for the w_j, $j = 1, 2, \ldots, m$ none of which is equal to zero or a negative integer. $(w)_v$ denotes Pochhammer's symbol, defined by

$$ (w)_v = \prod_{j=0}^{v-1} (w + j), $$

and $(w)_0 = 1$ for $w \neq 1$, $v \in \mathbb{N}$, and $w \in \mathbb{C}$ (*cf.* [40, 58]).

Some special values of $y_6(j, b; \omega, p)$ are given as follows.

Substituting $p = 1$ into (18.1), we have

$$ y_6(j, b; \omega, 1) = \sum_{k=0}^{b} \binom{b}{k} \frac{\omega^k k^j}{b!}. $$

Recently, many applications of the numbers $y_6(j, b; \omega, p)$ have been given; for details, see [19–35].

Combining the above equation with Exercise 22 in the book of Riordan [45, p. 41] yields

$$ \begin{aligned} y_6(j, b; \omega, 1) &= \sum_{k=0}^{b} \frac{\binom{b}{k}\binom{b}{m}}{\binom{b}{m}} \cdot \frac{\omega^k k^j}{b!} \\ &= \sum_{k=0}^{b} \frac{\omega^k k^j}{\binom{b}{m} b!} \sum_{v=0}^{b} \binom{k+v}{m} \binom{m}{v} \binom{b}{k+v}. \end{aligned} $$

Substituting $j = 0$ and $p = 3$ into (18.1), we have

$$ y_6(0, b; \omega, 3) = \sum_{k=0}^{b} \binom{b}{k}^3 \frac{\omega^k}{b!}. $$

Combining the above equation with Exercise 22, which was given by Foata (1965), in the book of Riordan [45, p. 41] yields

$$y_6(0, b; \omega, 3) = \frac{1}{b!}\sum_{v=0}^{b} \frac{(b+v)!}{v!^3(b-2v)!} \sum_{d=0}^{b-2v} \binom{b-2v}{d} \omega^{v+d}$$

and

$$y_6(0, b; \omega, 3) = \frac{1}{b!}\sum_{v=0}^{b} \frac{(b+v)!}{v!^3(b-2v)!} \omega^v B_{2v,b}(\omega),$$

where $B_{2v,b}(\omega)$ denotes the beta polynomials of degree b, which are given by

$$B_{k,n}(\omega) = \omega^k (1+\omega)^{n-k},$$

$0 \leq k \leq n$, $n \in \mathbb{N}_0$ (*cf.* [54, Definition 1.2]). When $\omega = -1$ and $b = 2m$, with the aid of the famous Dixon's formula, which was given in 1890 [45, p. 41], the above formulas reduce to

$$y_6(0, 2m; -1, 3) = \frac{(-1)^m (3m)!}{(2m)!(m!)^3}.$$

Integrating $y_6(j, b; \omega, p)$ wrt ω from 0 to y, we get

$$\int_0^y y_6(j, b; \omega, p)\, d\omega = \frac{1}{b!}\sum_{k=0}^{b} \binom{b}{k}^p \frac{k^j y^{k+1}}{k+1}.$$

Substituting $j = 0$ into the above integral formula and assuming that $0^0 = k^0 = 1$, we get

$$\int_0^y y_6(0, b; \omega, 1)\, d\omega = \frac{y}{b!}\sum_{k=0}^{b} \binom{b}{k} \frac{y^k}{k+1}.$$

By combining the left-hand side of the above integral formula with Exercise 11(a) in the book of Riordan [45, p. 35] yields

$$\frac{y}{b!}\sum_{k=0}^{b} \binom{b}{k} \frac{y^k}{k+1} = \frac{1}{(b+1)!}\sum_{k=0}^{b} \binom{b+1}{k+1} y^{k+1} = \frac{(1+y)^{b+1} - 1}{(b+1)!}.$$

Integrating $y_6(j, b; \omega, p)$ wrt ω from 0 to 1, we get

$$\int_0^1 y_6(j, b; \omega, p)\, d\omega = \sum_{k=0}^{b} \binom{b}{k}^p \frac{k^j}{(k+1)\, b!}.$$

Substituting $j = p = 0$ into the above integral formula yields

$$\int_0^1 y_6(0, b; \omega, 0)\, d\omega = \sum_{k=0}^{b} \frac{1}{(k+1)\, b!} = \frac{1}{b!} H_{b+1},$$

where H_b denotes the harmonic numbers, which are defined by means of the following well-known generating function:

$$\ln(1-u) = (u-1) \sum_{v=0}^{\infty} H_v u^v.$$

In [58], using (18.1) and (18.2), we solved Srivastava's [71, p. 416] open problem 1.

In [58], we gave the following open problems for ordinary generating functions for the numbers $y_6(j, b; \omega, p)$.

Let

$$f(t; \omega; p, j) = \sum_{j=0}^{\infty} y_6(j, b; \omega, p) t^j.$$

What is the explicit value of the function $f(t; \omega; p, j)$?

How can we construct ordinary generating function for the numbers $y_6(j, b; \omega, p)$?

We gave the following some special values of the function $f(t; \omega; p, j)$:

$$\sum_{n=0}^{\infty} y_6\,(0, n; 1; 2)\, t^n = \frac{1}{\sqrt{\pi}} \sum_{n=0}^{\infty} \frac{4^n \Gamma\left(\frac{1+2n}{2}\right)}{\Gamma\,(n+1)} \cdot \frac{t^n}{n!},$$

$$f\,(t; 1; 2, 0) = {}_1F_1\left(\frac{1}{2}; 1; 4t\right),$$

$$\sum_{n=0}^{\infty} y_6\,(0, n; \omega; 0)\, t^n = \frac{1}{\omega - 1}\left(\omega e^{\omega t} - e^t\right)$$

and

$$\sum_{n=0}^{\infty} y_6(0, n; \omega; 1)\, t^n = e^{(\omega+1)t}.$$

Recently, Kilar [21] gave the following ordinary generating function for the numbers $y_6(j, b; \omega, p)$:

$$\sum_{j=0}^{\infty} y_6(m, j; \omega, 1)\, t^j = e^{t(\omega+1)} \sum_{k=0}^{m} \omega^k S_2(m, k)\, t^k. \tag{18.4}$$

Here we note that the general solutions to these two problems still remain an open problem.

Substituting $k = p+1$, $m = p$ and $x_1 = x_2 = \cdots = x_{p+1} = 1$, $w_1 = w_2 = \cdots = w_p = -\beta$, $(-\beta \notin \{0, -1, -2, -3, \ldots\})$, and $u = (-1)^p \lambda e^z$ into (18.3), we construct the following generating function involving higher powers of inverse binomial coefficients:

Let $v, p \in \mathbb{N}_0$, $-\beta \notin \{0, -1, -2, -3, \ldots\}$ and $\lambda \in \mathbb{R}$ (or $\mathbb{C}$) with $|\lambda| < 1$. We define the sum $B_v(\beta; \lambda, p)$ by the following generating function:

$$\begin{aligned} F(z, v; \beta, \lambda, p) &= {}_{p+1}F_p\left(1, 1, \ldots, 1; -\beta, \ldots, -\beta; (-1)^p \lambda e^z\right) \\ &= \sum_{v=0}^{\infty} B_v(\beta; \lambda, p) \frac{z^v}{v!} \end{aligned} \tag{18.5}$$

(*cf.* [64]).

The function $F(z, v; \beta, \lambda, p)$ is also related to $\cosh(z)$ and $\sinh(z)$. That is,

$$\begin{aligned} F(z, v; \beta, \lambda, p) &= \sum_{j=0}^{\infty} \frac{\lambda^j}{\binom{\beta}{j}^p} (\cosh(jz) + \sinh(jz)) \\ &= \sum_{v=0}^{\infty} B_v(\beta; \lambda, p) \frac{z^v}{v!}, \end{aligned} \tag{18.6}$$

where $|\lambda| < 1$ (*cf.* [64]).

Let $v, p \in \mathbb{N}_0$ and $\lambda \in \mathbb{R}$ (or $\mathbb{C}$) with $|\lambda| < 1$. By using (18.5), we gave the following explicit formula for $B_v(\beta; \lambda, p)$:

$$B_v(\beta; \lambda, p) = \sum_{m=0}^{\infty} \frac{m^v \lambda^m}{\binom{\beta}{m}^p}, \tag{18.7}$$

where $\binom{\beta}{0} = 1$ and

$$\binom{\beta}{m} = \frac{\prod_{j=0}^{m-1} (\beta - j)}{m!} = \sum_{j=0}^{m} \frac{S_1(m, j)}{m!} \beta^j$$

(*cf.* [64]), where $S_1(m, j)$ denotes the Stirling numbers, which are defined by means of the following well-known generating function:

$$(\ln(1 + w))^j = j! \sum_{v=0}^{\infty} S_1(v, j) \frac{w^v}{v!}.$$

Integrating $\binom{\beta}{m}$ wrt β from 0 to 1, we get

$$\int_0^1 \binom{\beta}{m} \mathrm{d}\beta = \sum_{j=0}^{m} \frac{S_1(m, j)}{(j + 1)\, m!},$$

which gives us formulas for the Bernoulli of the second kind (Cauchy numbers) (*cf.* [8, p. 294]).

By using (18.5), we also have

$$B_v(\beta; \lambda, p) = \sum_{j=0}^{\infty} \frac{\lambda^j}{\binom{\beta}{j}^p} \sum_{k=0}^{j} \binom{j}{k} k! S_2(v, k),$$

where $|\lambda| < 1$ (*cf.* [64]).

The sum $B_v(\beta; \lambda, p)$ is related the Euler-Frobenius polynomial, B-spline, and exponential Euler spline. For example, Gun and Simsek [12] investigated some relations of the exponential Euler splines and B-spline with some special numbers and polynomials associated with the sum $B_v(\beta; \lambda, p)$. Besides, there are also many other applications of the spline and exponential Euler spline for the real-world problems and other mathematical models (*cf.* [37–39, 64]).

Substituting $p = 0$ into (18.7), for $|\lambda| < 1$, we get

$$B_v(\beta; \lambda, 0) = \sum_{m=0}^{\infty} m^v \lambda^m.$$

The above series is related to the analytical continuation of the Lerch transcendent and Lerch zeta function on the curve which is a Hankel contour counterclockwise around the positive real axis. This analytical continuation that gives us $B_v(\beta; \lambda, 0)$

is also related to the Eulerian polynomials and the Apostol-Bernoulli numbers (*cf.* [64]).

With the aid of modification of (18.7), we define the following sequence:

$$K(v) = \sum_{m=v}^{\infty} \frac{1}{\binom{m}{v}},$$

where $v \in \mathbb{N}$ with $v \geq 2$.

We give some special cases of the sequence $K(v)$ as follows:

$$K(2) = \sum_{m=2}^{\infty} \frac{1}{\binom{m}{2}}, \quad K(3) = \sum_{m=3}^{\infty} \frac{1}{\binom{m}{3}}, \quad K(4) = \sum_{m=4}^{\infty} \frac{1}{\binom{m}{4}}, \quad K(5) = \sum_{m=5}^{\infty} \frac{1}{\binom{m}{5}},$$

and so on. In [5, p. 136], explicit formula for the sequence $K_n(v)$ is given by

$$\sum_{m=v}^{\infty} \frac{1}{\binom{m}{v}} = \frac{v}{v-1},$$

where $v \in \mathbb{N}$ with $v \geq 2$. Thus,

$$K(2) = 2, \quad K(3) = \frac{3}{2}, \quad K(4) = \frac{4}{3}, \quad K(5) = \frac{5}{4}, \ldots$$

and so on. We now give a relation between $\sum_{v=2}^{n} K(v)$ and the harmonic numbers:

$$\sum_{v=2}^{n} K(v) = \sum_{v=2}^{n} \sum_{m=v}^{\infty} \frac{1}{\binom{m}{v}}.$$

Thus,

$$\sum_{v=2}^{n} K(v) = \sum_{v=2}^{n} \frac{v}{v-1} = 2 + \frac{3}{2} + \frac{4}{3} + \frac{5}{4} + \cdots + \frac{n}{n-1}.$$

From the above, we have

$$\sum_{v=2}^{n} K(v) = 1 + 1 + \frac{1}{2} + 1 + \frac{1}{3} + 1 + \frac{1}{4} + \cdots + 1 + \frac{1}{n-1} = n - 1 + H_{n-1}.$$

The sum $\sum_{v=2}^{n} K(v)$ is also given by

$$\sum_{v=2}^{n} K(v) = n - \sum_{j=1}^{n-1} (-1)^j \binom{n-1}{j} \frac{1}{j},$$

where

$$H_{n-1} = -\sum_{j=1}^{n-1} (-1)^j \binom{n-1}{j} \frac{1}{j}.$$

The motivation of this chapter derives many new formulas for not only the sums $B_v(\beta; \lambda, p)$ and the numbers $y_6(j, b; \omega, p)$, but also the following finite sum involving higher powers of inverse binomial coefficients:

$$\mathcal{S}_v(n; \lambda, p) := \sum_{j=0}^{n} \frac{j^v}{\binom{n}{j}^p} \lambda^j, \tag{18.8}$$

where $n, v, p \in \mathbb{N}_0$, and $\lambda \in \mathbb{R}$ (or $\mathbb{C}$) (*cf.* [64]).

Putting $v = 0$ and $p = 1$ and $\lambda = -1$ in (18.8) yields

$$\mathcal{S}_0(2n; -1, 1) = 2 + \sum_{j=1}^{2n-1} \frac{(-1)^j}{\binom{2n}{j}},$$

where $n \in \mathbb{N}$. Combining $\mathcal{S}_0(2n; -1, 1)$ with the following identity

$$\sum_{j=1}^{2n-1} \frac{(-1)^{j-1}}{\binom{2n}{j}} = \frac{1}{n+1},$$

which was given in [69], we have

$$\mathcal{S}_0(2n; -1, 1) = \frac{2n+1}{n+1}$$

(*cf.* [64]).

Setting $v = p = 1$ and $\lambda = -1$ in (18.8) yields

$$\mathcal{S}_1(2n; -1, 1) = 2n - \sum_{j=1}^{2n-1} \frac{(-1)^j j}{\binom{2n}{j}}$$

(*cf.* [64]).

Combining $\mathcal{S}_1(2n; -1, 1)$ with

$$\sum_{j=1}^{2n-1} \frac{(-1)^{j-1} j}{\binom{2n}{j}} = \frac{n}{n+1},$$

which was given in [69], we have

$$\mathcal{S}_1(2n; -1, 1) = \frac{2n^2 + n}{n+1}$$

(*cf.* [64]).

Substituting $v = 0$, $p = 1$, and $\lambda = -1$ into (18.8) yields

$$\mathcal{S}_0(n; -1, 1) = \sum_{j=0}^{n} \frac{(-1)^v}{\binom{n}{j}} = (1 + (-1)^n)\frac{n+1}{n+2}$$

and

$$\sum_{j=0}^{n} \frac{(-1)^v}{\binom{x}{j}} = \left(1 + \frac{(-1)^n}{\binom{x+1}{n+1}}\right) \frac{x+1}{x+2},$$

where $x \neq -1, \ -2$ (*cf.* [64, 69]).

For $n \in \mathbb{N}$, we [64] gave the following formulas:

$$2n\mathcal{S}_0(n; 1, 1) - (n+1)\,\mathcal{S}_0(n-1; 1, 1) = 2n \tag{18.9}$$

and

$$\mathcal{S}_0(n; 1, 1) = \sum_{v=0}^{n} \frac{n+1}{v+1} 2^{v-n}, \tag{18.10}$$

where $n \in \mathbb{N}_0$ (*cf.* [10, p. 273]).

Remark 18.1 It is known that formulas such as equations (18.10) and (18.9) and special finite summations containing the sum of inverse binomial coefficients, etc. have been proven by many different researchers with many different techniques and methods (*cf.* [6, 36, 48, 56, 68] and the references cited in each of these earlier works). By the help of (18.9), we [64] also gave solution of Exercise 30* in [10, p. 273]. In [67], we found new formulas involving certain special finite sums, discovering their relations not only with harmonic numbers but also with (inverse) binomial coefficients, some special numbers, and polynomials.

Riordan [45] gave the following finite sum:

$$g_k(n) = \sum_{j=k}^{n} (-1)^j \frac{(j)_{(k)}}{\binom{n}{j}}, \tag{18.11}$$

where $k = 0, 1, \ldots, n$ and $(j)_{(k)} = j(j-1)\cdots(j-k+1)$. Combining (18.11) with the following well-known formula (*cf.* [11, Eq. (2.15)], [68]):

$$\frac{1}{\binom{n}{j}} = (n+1) \sum_{l=0}^{j} (-1)^{j-l} \binom{j}{l} \frac{1}{n-l+1},$$

we get

$$g_k(n) = \sum_{j=k}^{n} \sum_{l=0}^{j} (-1)^l \binom{j}{k} \binom{j}{l} \frac{(n+1)\, k!}{n-l+1}.$$

We now give the following relation between $\mathcal{S}_v(n; \lambda, p)$ and $g_k(n)$:

$$g_k(n) = \sum_{v=0}^{k} S_1(k, v) \sum_{j=k}^{n} (-1)^j \frac{j^v}{\binom{n}{j}}.$$

Thus,

$$g_k(n) = \sum_{v=0}^{k} S_1(k, v) \left(\sum_{j=0}^{n} (-1)^j \frac{j^v}{\binom{n}{j}} - \sum_{j=0}^{k-1} (-1)^j \frac{j^v}{\binom{n}{j}} \right).$$

From the above, we get

$$g_k(n) = \sum_{v=0}^{k} S_1(k, v) \mathcal{S}_v(n; -1, 1) - \sum_{j=0}^{k-1} (-1)^j \frac{1}{\binom{n}{j}} \sum_{v=0}^{k} S_1(k, v) j^v.$$

At this point, asking the following crucial questions can constitute a good research source for this section.

Open Problem 1 What is the relationship between $g_k(n)$ and the following rational sum, $y(m, \phi)$, given by

$$y(m, \phi) = (-1)^m \sum_{k=0}^{m} \frac{1}{(\phi - 1)^{m+1-k}\, \vartheta \left\{ \phi^{k+1} \right\}}, \tag{18.12}$$

where ϑ denotes the following well-known Euler operator [52, p. 80, Eq. (2)]:

$$\vartheta = \phi \frac{\mathrm{d}}{\mathrm{d}\phi},$$

$\phi \neq 0, 1$, $\phi \in \mathbb{R}$, and $m \in \mathbb{N}_0$ (*cf.* [62–64]).

Open Problem 2 Can the relations obtained as answer to open problem 1 be of any relationship with certain family of other special numbers?

In order to solve the above questions, it is useful to provide the following explanatory information. It has been discussed in many of our studies that rational sum $y(m, \phi)$ is related to many different families of special numbers, inverse binomial coefficients, interesting integral formulas, and other special functions involving hypergeometric series, Dirichlet series, and eta-type functions (*cf.* [61, 62]). Now it is time to give the following generating function for this sum:

For $\phi \neq 0, 1$, $\phi \in \mathbb{R}$, $w \in \mathbb{C}$, $0 < |w| < 1$, and $\left|\frac{(\phi-1)w}{\phi}\right| < 1$, this rational sum is given by the following generating function:

$$G(w, \phi) = \frac{\ln\left(1 - \frac{\phi-1}{\phi} w\right)}{(w^2 - w)(1 - \phi)^2} = \sum_{v=0}^{\infty} y(v, \phi)\left((1 - \phi) w\right)^v, \tag{18.13}$$

where the principal branch of the logarithm is assumed to be taken (*cf.* [63]).

18.3 The Numbers $y_6(m, k; \omega, v)$ Related to Arithmetical Functions, k-ary Lyndon Words, and Bell Series

In this section, we give relations among the numbers $y_6(m, k; \omega, v)$ related to arithmetical functions, k-ary Lyndon words, kth powers of the consecutive integers, and Bell series.

Using the Dirichlet convolution of arithmetic functions and the Möbius inversion formula, Kucukoglu [25] gave many novel formulas involving the Möbius function, the Euler's totient function, the numbers of Lyndon words, the numbers of necklaces, the Stirling numbers of the second kind, and the numbers $y_6(m, k; \omega, v)$. Some of her results are given by the following theorems:

Two of the most frequently used arithmetical functions are the Möbius function $\mu(n)$ and Euler's totient function $\varphi(n)$, and these arithmetical functions are respectively defined by (*cf.* [1])

$$\mu(n) = \begin{cases} 1, & \text{if } n = 1, \\ (-1)^m, & \text{if } n \text{ is a square-free integer with } m \text{ distinct prime factors}, \\ 0, & \text{if } n \text{ has a squared prime factor}, \end{cases}$$

and

$$\varphi(n) = \sum_{d|n} \mu(d) \frac{n}{d} = n \prod_{p|n} \frac{p-1}{p} = \sum_{\substack{m=1 \\ (m,n)=1}}^{n} 1,$$

in which the sum runs over all positive integers $m \leq n$ that are relatively prime to the positive integer n, p is a prime number, and $\varphi(1) = 1$ (*cf.* [1])

The Dirichlet convolution $f * g$ of two arithmetical functions f and g is given by

$$(f * g)(n) = \sum_{d|n} f(d) g\left(\frac{n}{d}\right),$$

in which the sum runs over all divisors d of the positive integer n (*cf.* [1]).

The arithmetical function $L_k(n)$ that allows us to count how many k-ary Lyndon words of length n are is given by

$$L_k(n) = \frac{1}{n} \sum_{d|n} \mu\left(\frac{n}{d}\right) k^d \tag{18.14}$$

(*cf* [7, 41, 42]).

The arithmetical function $N_k(n)$ that allows us to count how many necklaces with n beads of k different colors are is given by

$$N_k(n) = \frac{1}{n} \sum_{d|n} \varphi\left(\frac{n}{d}\right) k^d \tag{18.15}$$

(*cf.* [7]).

Kucukoglu [25] gave the following formulas for the Dirichlet convolutions of the numbers $y_6(n, k; \theta, v)$ with the Möbius function $\mu(n)$ and the Euler's totient function $\varphi(n)$ as follows, respectively:

$$(y_6 * \mu)(n) = \sum_{d|n} \mu\left(\frac{n}{d}\right) y_6(d, k; \theta, v),$$

$$(y_6 * \mu)(n) = \frac{1}{k!} \sum_{j=0}^{k} \binom{k}{j}^v \theta^j \sum_{d|n} \mu\left(\frac{n}{d}\right) j^d,$$

$$(y_6 * \mu)(n) = \frac{n}{k!} \sum_{j=0}^{k} \binom{k}{j}^v \theta^j L_j(n),$$

$$(y_6 * \varphi)(n) = \sum_{d|n} \varphi\left(\frac{n}{d}\right) y_6(d, k; \theta, v),$$

$$(y_6 * \varphi)(n) = \frac{n}{k!} \sum_{j=0}^{k} \binom{k}{j}^v \theta^j N_j(n)$$

and

$$(y_6 * \varphi)(n) = \sum_{j=0}^{k} \binom{k}{j}^v N_j(n) \sum_{s=0}^{j} S_1(j, s) \frac{n\theta(\theta - 1) \cdots (\theta - s + 1)}{k!}.$$

Combining the above formulas, Kucukoglu [25] proved the following theorem.

Theorem 18.1 (*cf.* [25]) *Let $n \in \mathbb{N}$. Then we have*

$$\sum_{d|n} \mu\left(\frac{n}{d}\right) y_6(d, k; \theta, v) = \frac{n}{k!} \sum_{j=0}^{k} \binom{k}{j}^v \theta^j L_j(n), \tag{18.16}$$

$$y_6(n, k; \theta, v) = \sum_{d|n} \sum_{j=0}^{k} \binom{k}{j}^v \frac{d}{k!} \theta^j L_j(d) \tag{18.17}$$

and

$$\sum_{d|n} \varphi\left(\frac{n}{d}\right) y_6(d, k; \theta, v) = \frac{n}{k!} \sum_{j=0}^{k} \binom{k}{j}^v \theta^j N_j(n). \tag{18.18}$$

Applying the Euler operator to the numbers $y_6(0, k; \theta, v)$ yields

Theorem 18.2 (*cf.* [25]) *Let $v, k \in \mathbb{N}_0$. Then we have*

$$\sum_{j=0}^{k} \binom{k}{j}^v \theta^j \left(N_j(p) - L_j(p)\right) = k!\theta \frac{d}{d\theta}\{y_6(0, k; \theta, v)\}. \tag{18.19}$$

In [28], Kucukoglu gave the following decomposition of the generating function for the numbers $y_6(n, k; \theta, v)$ with the aid of two sums separately over all divisors and non-divisors of n:

$$n!F_{y6}(t,n;\theta,r) = \sum_{d|n} \binom{n}{d}^r \theta^d \mathrm{e}^{dt} + \sum_{d\nmid n} \binom{n}{d}^r \theta^d \mathrm{e}^{dt} \tag{18.20}$$

and

$$F_{\mathcal{K}}(t,n;\theta,r) := n!F_{y_6}(t,n;\theta,r) - \sum_{d\nmid n} \binom{n}{d}^r \theta^d \mathrm{e}^{dt}, \tag{18.21}$$

which yields

$$F_{\mathcal{K}}(t,n;\theta,r) = \sum_{m=0}^{\infty} \mathcal{K}(n,m;\theta,r) \frac{t^m}{m!} \tag{18.22}$$

and

$$\mathcal{K}(n,m;\theta,r) = \sum_{d|n} \binom{n}{d}^r \theta^d d^m,$$

where $n \in \mathbb{N}$, $m, r \in \mathbb{N} \cup \{0\}$, and $\theta, t \in \mathbb{R}$ (or $\mathbb{C}$).

By using the above generating functions, Kucukoglu [28] gave the following results.

Theorem 18.3 (*cf.* **[28]**) *Let $n \in \mathbb{N}$. Then we have*

$$\sum_{\theta \in \mathcal{A}_n} y_6(m,n;\theta,r) = \frac{1}{n!} \sum_{j=0}^{n} \binom{n}{j}^r j^m \varphi_j(n)$$

and

$$\sum_{\theta \in \mathcal{A}_m} y_6(m,n;\theta,r) = \frac{1}{n!} \sum_{j=0}^{n} \binom{n}{j}^r j^m \varphi_j(m),$$

where $\mathcal{A}_m$ denotes the set of all totatives of the positive integer m,

$$\mathcal{A}_m = \{v \in \mathbb{N} : 1 \leq v < m, \ \gcd(v,m) = 1\}$$

and $\varphi_k(n)$ denotes Thacker's (totient) function:

$$\varphi_k(n) = \sum_{j=0}^{k} (-1)^j \binom{k}{j} n^{k-j} \varphi_j(n), \tag{18.23}$$

(*cf.* [1, p. 48]).

From Apostol's book (*cf.* [1, Exercise 15–16, p. 48], it is easy to see that

$$\varphi_0(n) = \varphi(n),$$
$$\varphi_1(n) = \frac{n\varphi(n)}{2},$$
$$\varphi_2(n) = \frac{2n^2\varphi(n) + n\prod_{p|n}(1-p)}{6}$$

and also a relation between Thacker's (totient) function $\varphi_k(n)$ and the well-known Faulhaber formula is given by

$$\sum_{d|n} \varphi_k(p)\frac{1}{d^k} = \frac{1^k + 2^k + \cdots + n^k}{n^k}; \tag{18.24}$$

see also, for detail, [1]. Here we note that Faulhaber formula, which is known as the kth powers of the consecutive first $(n-1)$ positive integers, traced back to roughly 400 years ago. The explicit values of this formula are given by means of the following well-known formula involving Bernoulli polynomials and numbers:

$$1^k + 2^k + \cdots + (n-1)^k = \frac{B_{k+1}(n) - B_{k+1}}{k+1}$$
$$= \sum_{c=0}^{k} (-1)^c \binom{k+1}{c} \frac{(n-1)^{k+1-c}}{k+1} B_c$$

(*cf.* [9, 24, 28, 32]; see also [53]).

Let f be an arithmetical function and p be a prime number. The Bell series of f modulo p is given by the following formal power series:

$$f_p(t) = \sum_{v=0}^{\infty} f(p)t^v$$

(*cf.* [1, p. 43]). By the aid of formal power series, E. T. Bell (February 7, 1883, to December 21, 1960), who was a Scottish-born mathematician and science fiction, gave many properties of multiplicative arithmetical functions.

If f is completely multiplicative, then its Bell series is given by

$$f_p(t) = \frac{1}{1 - f(p)t}$$

(*cf.* [74]).

Some of the Bell series for well-known arithmetic functions are given as follows: The Bell series of the Möbius function:

$$\mu_p(t) = 1 - t$$

(*cf.* [74]).

The Bell series of Euler's totient function:

$$\varphi_p(t) = \frac{1-t}{1-pt}$$

(*cf.* [74]).

Open Problem 3 Is there any relation between the Bell series of arithmetic functions and the sum $y_6(m, n; \theta, r)$?

18.4 New Classes of Polynomials with Coefficients Given by Higher Powers of Inverses of Binomial Coefficients

We [64] very recently defined new classes of polynomials $\mathcal{P}_v(x, \beta; \lambda, p)$ whose coefficients are the sum $B_j(\beta; \lambda, p)$ by the following generating functions:

$$G(z; \beta, \lambda, p) = F(z; \beta, \lambda, p)e^{xz} = \sum_{v=0}^{\infty} \mathcal{P}_v(x, \beta; \lambda, p)\frac{z^v}{v!}, \tag{18.25}$$

where $-\beta \notin \{0, -1, -2, -3, \ldots\}$, $v, p \in \mathbb{N}_0$ and $x, \lambda \in \mathbb{R}$ (or $\mathbb{C}$) with $|\lambda| < 1$.

Let $n, p \in \mathbb{N}_0$ and $\lambda, \beta \in \mathbb{R}$ (or $\mathbb{C}$) with $|\lambda| < 1$. By using (18.25), we gave the following definition for the polynomial $\mathcal{P}_v(x, \beta; \lambda, p)$:

$$\mathcal{P}_n(x, \beta; \lambda, p) = \sum_{j=0}^{n} \binom{n}{j} B_j(\beta; \lambda, p)x^{n-j} \tag{18.26}$$

(*cf.* [64]).

With the aid of (18.8), we also define the following polynomials:

$$\mathfrak{q}_n(x, n; \lambda, p) = \sum_{j=0}^{n} \binom{n}{j} S_j(n; \lambda, p)x^{n-j}. \tag{18.27}$$

Using the following higher-order partial differential equation

$$\frac{\partial^k}{\partial x^k}\{G(z; \beta, \lambda, p)\} = t^k G(z; \beta, \lambda, p),$$

we arrive at the following higher-order derivative formula for the polynomials $\mathcal{P}_n(x, \beta; \lambda, p)$:

$$\frac{\partial^k}{\partial x^k}\{\mathcal{P}_n(x, \beta; \lambda, p)\} = \left(\prod_{j=0}^{k-1}(n-j)\right)\mathcal{P}_{n-k}(x, \beta; \lambda, p),$$

where we assume that $1 = \prod_{j=0}^{-1}(n-j)$ (*cf.* [64])

$$\frac{\partial}{\partial x}\{\mathcal{P}_n(x, \beta; \lambda, p)\} = n\mathcal{P}_{n-1}(x, \beta; \lambda, p).$$

We also give the following higher-order derivative formula for the polynomials $\mathfrak{q}_n(x, n; \lambda, p)$:

$$\frac{\partial^k}{\partial x^k}\{\mathfrak{q}_n(x, n; \lambda, p)\} = \sum_{j=0}^{n}\binom{n}{j}\left(\prod_{l=0}^{k-1}(n-j-l)\right)\mathcal{S}_j(n; \lambda, p)x^{n-j-k-1}.$$

Equation (18.5) can also be represented as follows:

$$F(z; \beta, \lambda, p) = \sum_{j=0}^{\infty}\frac{(\lambda e^z)^j}{\binom{\beta}{j}^p} = \sum_{v=0}^{\infty}B_v(\beta; \lambda, p)\frac{z^v}{v!}, \tag{18.28}$$

where $|\lambda| < 1$.

By using (18.28), we get

$$\sum_{j=0}^{\infty}\frac{1}{\binom{\beta}{j}^p}\sum_{d=0}^{j}(-1)^{j-d}\binom{j}{d}\left(\lambda e^z + 1\right)^d = \sum_{v=0}^{\infty}B_v(\beta; \lambda, p)\frac{z^v}{v!}.$$

Combining the above equation with the following special numbers $y_1(n, k; \lambda)$, which are given in [57],

$$\left(\lambda e^u + 1\right)^k = \sum_{n=0}^{\infty}k!y_1(n, k; \lambda)\frac{u^n}{n!}, \tag{18.29}$$

where $k \in \mathbb{N}_0$, $\lambda \in \mathbb{C}$, and

$$y_1(n, k; \lambda) = \frac{1}{k!}\sum_{j=0}^{k}\binom{k}{j}j^n\lambda^j,$$

which are called the *combinatorial Simsek numbers of the first kind* (see for detail [16, 28, 75]), we get

$$\sum_{j=0}^{\infty}\frac{1}{\binom{\beta}{j}^{p}}\sum_{d=0}^{j}(-1)^{j-d}\binom{j}{d}d!\sum_{v=0}^{\infty}y_1(v,d;\lambda)\frac{z^v}{v!}=\sum_{v=0}^{\infty}B_v(\beta;\lambda,p)\frac{z^v}{v!}.$$

After some elementary calculations, by equalizing the coefficients of $\frac{z^v}{v!}$ on both sides of the previous equation, we arrive at the following theorem.

Theorem 18.4 *Let $n, p \in \mathbb{N}_0$ and $\lambda, \beta \in \mathbb{R}$ (or $\mathbb{C}$) with $|\lambda| < 1$. Then we have*

$$B_v(\beta;\lambda,p)=\sum_{j=0}^{\infty}\frac{1}{\binom{\beta}{j}^{p}}\sum_{d=0}^{j}(-1)^{j-d}\binom{j}{d}d!y_1(v,d;\lambda).$$

By using (18.28), we also get

$$\sum_{j=0}^{\infty}\frac{1}{\binom{\beta}{j}^{p}}\sum_{d=0}^{j}\binom{j}{d}\left(\lambda e^z-1\right)^d=\sum_{v=0}^{\infty}B_v(\beta;\lambda,p)\frac{z^v}{v!}.$$

Combining the above equation with the generating function for the following λ-Stirling numbers

$$\left(\lambda e^z-1\right)^d=d!\sum_{v=0}^{\infty}S_2(v,d;\lambda)\frac{z^v}{v!},$$

(*cf.* [72]), we get

$$\sum_{j=0}^{\infty}\frac{1}{\binom{\beta}{j}^{p}}\sum_{d=0}^{j}\binom{j}{d}d!\sum_{v=0}^{\infty}S_2(v,d;\lambda)\frac{z^v}{v!}=\sum_{v=0}^{\infty}B_v(\beta;\lambda,p)\frac{z^v}{v!}.$$

After some elementary calculations, by equalizing the coefficients of $\frac{z^v}{v!}$ on both sides of the previous equation, we also arrive at the following theorem.

Theorem 18.5 *Let $n, p \in \mathbb{N}_0$ and $\lambda, \beta \in \mathbb{R}$ (or $\mathbb{C}$) with $|\lambda| < 1$. Then we have*

$$B_v(\beta;\lambda,p)=\sum_{j=0}^{\infty}\frac{1}{\binom{\beta}{j}^{p}}\sum_{d=0}^{j}\binom{j}{d}d!S_2(v,d;\lambda). \tag{18.30}$$

When $\lambda = 1$ in Eq. (18.30), since $S_2(v, d; 1) = 0$ if $d > v$, we get the following corollary.

Corollary 18.1 *Let $n, p \in \mathbb{N}_0$ and $\beta \in \mathbb{R}$ (or $\mathbb{C}$) with $|\lambda| < 1$. Then we have*

$$B_v(\beta; \lambda, p) = \sum_{j=0}^{v} \frac{1}{\binom{\beta}{j}^p} \sum_{d=0}^{j} \binom{j}{d} d! S_2(v, d).$$

In [64], we also gave the following generating function for the polynomial $\mathcal{P}_n(x, \beta; \lambda, p)$:

$$\sum_{j=0}^{\infty} \frac{\lambda^j}{\binom{\beta}{j}^p} e^{(j+x)z} = \sum_{n=0}^{\infty} \mathcal{P}_n(x, \beta; \lambda, p) \frac{z^n}{n!}. \tag{18.31}$$

For $|\lambda| < 1$, the above equation yields

$$\mathcal{P}_n(x, \beta; \lambda, p) = \sum_{j=0}^{\infty} \frac{\lambda^j}{\binom{\beta}{j}^p} (j + x)^n.$$

Integrating $\mathcal{P}_n(x, \beta; \lambda, p)$ from 0 to 1, we get

$$\int_0^1 \mathcal{P}_n(x, \beta; \lambda, p)\, dx = \sum_{j=0}^{\infty} \frac{\lambda^j}{\binom{\beta}{j}^p} \frac{(j + 1)^{n+1} - j^{n+1}}{n + 1}$$

and

$$\int_0^1 \mathcal{P}_n(x, \beta; \lambda, p)\, dx = \sum_{j=0}^{n} \binom{n}{j} \frac{B_j(\beta; \lambda, p)}{n - j + 1}.$$

Combining the above integral formulas, we arrive at the following theorem.

Theorem 18.6 *Let $n, p \in \mathbb{N}_0$ and $\lambda, \beta \in \mathbb{R}$ (or $\mathbb{C}$) with $|\lambda| < 1$. Then we have*

$$\sum_{j=0}^{n} \binom{n}{j} \frac{B_j(\beta; \lambda, p)}{n - j + 1} = \sum_{j=0}^{\infty} \frac{\lambda^j}{\binom{\beta}{j}^p} \cdot \frac{(j + 1)^{n+1} - j^{n+1}}{n + 1}.$$

18.4.1 A New Class of Polynomials with Coefficients $y_6(k, n; \omega, p)$

Let $m, n, p \in \mathbb{N}_0$ and $\omega \in \mathbb{R}$ or $\mathbb{C}$. In [58], we defined a new class of polynomials with coefficients $y_6(k, n; \omega, p)$:

$$P_s(x; m, n; \omega, p) = \sum_{k=0}^{m} \binom{m}{k} x^{m-k} y_6(k, n; \omega, p). \tag{18.32}$$

We gave many properties of the polynomials $P_s(x; m, n; \omega, p)$. Some of them are also given below.

Some relations among the polynomials $P_s(x; m, n; \omega, p)$, the Mirimanoff polynomials, Frobenius-Euler numbers and polynomials, and Bernoulli numbers and polynomials are given as follows:

The Mirimanoff polynomials $f_m(x, n)$ are defined by

$$f_m(x, n) = \sum_{j=0}^{n-1} j^m x^j$$

and

$$f_m(x, n, k) = \sum_{j=0}^{n-1} (j + k)^m x^j$$

(*cf.* [58, 73]).

$$P_s\left(x; m, n; \frac{1}{\omega}, 0\right) = \omega^{n-1} f_n(x, m),$$

$$f_n(x, m) = \frac{\omega H_m\left(x + n; \frac{1}{\omega}\right) - \omega^{-n+1} H_m\left(x + n; \frac{1}{\omega}\right)}{\omega - 1},$$

(*cf.* [58]) and

$$P_s\left(x; m, n; \frac{1}{\omega}, 0\right) = \frac{\omega^n H_m\left(x + n; \frac{1}{\omega}\right) - H_m\left(x + n; \frac{1}{\omega}\right)}{\omega - 1},$$

where $H_m(x; \alpha)$ denotes the Euler-Frobenius polynomials, which are defined by means of the following generating function:

$$\sum_{m=0}^{\infty} \frac{H_m(x; \alpha)}{m!} t^m = \frac{1 - \alpha}{e^t - \alpha} e^{tx}$$

(*cf.* [59, 72]).

We [58] also showed that

$$P_s(0; m - 1, n - 1; 1, 0) = \frac{B_m(n) - B_m}{m}.$$

For other interesting results see [58].

18.4.2 The Michael Vowe Polynomial

In [58], we defined the following polynomials:

$$R_n(x; p) = y_6(0, n; x, p) = \frac{1}{n!}\sum_{k=0}^{n}\binom{n}{k}^p x^k.$$

The Michael Vowe polynomials are given by

$$\mathcal{M}_n(x) = n!R_n(x; 2)$$

(*cf.* [15, 58]). In [58], we showed that the polynomials $\mathcal{M}_n(x)$ are also related to the Legendre polynomial.

18.5 Approximation Formulas for Finite Sums

In this section, we first give some survey on asymptotic expressions for the series with integration. Throughout this chapter, some well-known results of the definitions and notations of approximation theory are given, based on the definitions and notations of Zygmund's book [76] and also see [35, 43].

Let $f(x)$ and $g(x)$ be two functions defined for $x > x_0$, and $g(x) \neq 0$ there. The symbols

$$f(x) = o(g(x)) \quad \text{and} \quad f(x) = O(g(x))$$

mean, respectively, that $\frac{f(x)}{g(x)} \to 0$ as $x \to +\infty$ and that $\frac{f(x)}{g(x)}$ is bounded for x large enough.

An expression is $o(1)$ or $O(1)$ if it tends to 0 or is bounded, respectively.

Two functions, $f(x)$ and $g(x)$ defined in the neighborhood of a point x_0 (finite or infinite), are called *asymptotically equal* if $\frac{f(x)}{g(x)} \to 1$ as $x \to x_0$. It can be written as

$$f(x) \simeq g(x), \quad x \to x_0.$$

If the ratios $\frac{f(x)}{g(x)}$ and $\frac{g(x)}{f(x)}$ are both bounded in the neighborhood of x_0, we say that $f(x)$ and $g(x)$ are of *the same order* as $x \to x_0$. It can also be written as

$$f(x) \sim g(x), \quad x \to x_0.$$

Let $u_0, u_1, u_2, \ldots,$ be a sequence of numbers and let

$$U_n = u_0 + u_1 + u_2 + \cdots + u_n,$$

where $n \in \mathbb{N}_0$.

Let a be finite, and let $f(x)$ be a function defined in a finite or infinite interval $a \le x < b$ and integrable over every interval (a, y), $y < b$. It can be given by

$$F(x) = \int_a^x f(t)\, dt,$$

where $a \le x < b$.

Let $\{u_v\}$ and $\{v_v\}$ be two sequences of numbers. If $u_n = o(v_n)$ and $V_n \to +\infty$, then we have

$$U_n = o(V_n).$$

Suppose that the series $\sum v_v$ converges with $v \in \mathbb{N}$ and that $u_v = o(v_v)$. Then, we have

$$u_n + u_{n-1} + \cdots = o(v_n + v_{n+1} + \cdots)$$

(*cf.* [76, Theorem (8.3)]).

This is obvious.

Theorem 18.7 (*cf.* [76, Theorem (8.4)]) *Let $f(x)$ be a positive, finite, and monotone function defined for $x \ge 0$, and let*

$$F(x) = \int_0^x f(t)\, dt,$$

$$F_n = f(0) + f(1) + \cdots + f(n).$$

Then:

(i) *If $f(x)$ decreases,*

$$F(n) - F_n \to \infty.$$

(ii) *If $f(x)$ decreases,*

$$F(n) \le F_n \le F(n) + f(n).$$

Proof of this theorem was given in [76, Theorem 8.4].

Using (i) and $f(n) < F(k) - F(k-1) < f(k-1)$, for $k \in \mathbb{N}$, we have the following known inequality:

$$0 \le F(k) - F(k-1) - f(k) \le f(k-1) - f(k). \tag{18.33}$$

Since $\sum_{k=1}^{\infty}\{f(k-1)-f(k)\}$ converges, $\sum_{k=1}^{\infty}\{F(k)-F(k-1)-f(k)\}$ also converges.

By using (ii), for $k \in \mathbb{N}$, we have the following known inequality:

$$f(k-1) \le F(k) - F(k-1) \le f(k).$$

Theorem 18.8 (*cf.* [76, Theorem (8.6)]) *Let $f(x)$ be positive, finite, and monotone for $x \ge 0$. If either (i) $f(x)$ decreases and $F(x) \to \infty$ or (ii) $f(x)$ increases and $f(x) = o(F(x))$, then*

$$F(n) \simeq F_n.$$

Proof of this theorem was given in [76, Theorem (8.6)].

Theorem 18.9 (*cf.* [76, Theorem (8.7)]) *Let $f(x)$, $x \ge 0$, be positive, monotone decreasing, and integrable over $(0, \infty)$, and let*

$$F^*(x) = \int_x^{\infty} f\, dt,$$

$$F_n^* = f(n) + f(n+1) + \cdots.$$

Then, $F_{n+1}^* \le F^*(n) \le F_n^*$.

Proof of this theorem was given in [76, Theorem 8.7].

If $f(x) = o(F^*(x))$, then $F_n^* \simeq F^*(n)$. Combining (18.8) with (18.9), we have

$$\sum_{k=1}^{n} k^{\rho} \simeq \frac{n^{\rho+1}}{\rho+1}, \tag{18.34}$$

for $\rho > -1$ (*cf.* [76]).

Open Problem 4 By using (18.34), is it possible to give an approximation formula for the sums involving Faulhaber formula or sums of positive consecutive integers.

We observe that substituting $\rho = 2$ into (18.34) gives the following well-known formula:

$$\sum_{k=1}^{n} k^2 \simeq \frac{n^3}{3}.$$

Using the above formulas, we have

$$b!y_6(2, b; \omega, p) \simeq \frac{b^3}{3}.$$

By using the following Stirling's approximation

$$n! \sim \sqrt{2\pi n}\left(\frac{n}{\mathrm{e}}\right)^n, \tag{18.35}$$

(*cf.* [13, 46, 72]), for $n \to \infty$ and $k \to \infty$, approximation of $\binom{n}{k}$ is given by the following known result:

$$\binom{n}{k} \sim \frac{n^n}{k^k(n-k)^{n-k}}\sqrt{\frac{n}{2\pi k(n-k)}}. \tag{18.36}$$

By using (18.35), Polya and Szego [43, Problem 40, p. 55] gave the following asymptotic expression formula for the finite sum of powers of binomial coefficients:

$$\sum_{k=0}^{n}\binom{n}{k}^p \sim \frac{2^{pn}}{\sqrt{p}}\left(\frac{2}{\pi n}\right)^{\frac{p-1}{2}}.$$

Recently, Kilar [19] gave some asymptotic formulas of the finite sum of powers of binomial coefficients involving the numbers $y_6(m, n; \omega, p)$, one of which is given by the following theorem.

Theorem 18.10 (*cf.* [19]) *Let $n, p \in \mathbb{N}$. Then we have*

$$y_6(0, n; 1, p) \sim \frac{2^{pn-1}}{\sqrt{p}}\left(\frac{\mathrm{e}}{n}\right)^n\left(\frac{2}{\pi n}\right)^{\frac{p}{2}}. \tag{18.37}$$

The following special values of (18.37) are given by Kilar [19]:
Putting $p = 1$, $\omega = 1$, and $m = 0$ in (18.1) yields

$$y_6(0, n; 1, 1) = \frac{2^n}{n!}, \tag{18.38}$$

$$y_6(0, n; 1, 2) = \frac{(2n)!}{(n!)^3} \tag{18.39}$$

(*cf.* [58]), and putting $p = 1, 2, 3$ in (18.37) yields

$$y_6(0, n; 1, 1) \sim \frac{2^{n-\frac{1}{2}}}{\sqrt{n\pi}}\left(\frac{\mathrm{e}}{n}\right)^n, \tag{18.40}$$

$$y_6(0, n; 1, 2) \sim \frac{\mathrm{e}^n 2^{2n-\frac{1}{2}}}{\pi n^{n+1}} \tag{18.41}$$

Table 18.1 Some values of the numbers $y_6(0, n; 1, 1)$ compared with their approximate values

n	$y_6(0, n; 1, 1)$	Approximate values of $y_6(0, n; 1, 1)$
1	2	2.16888
2	2	2.08441
3	1.33333	1.37075
4	0.666667	0.680672
5	0.266667	0.271142
6	0.0888889	0.0901309
7	0.0253968	0.0257008
8	0.00634921	0.00641565
9	0.00141093	0.00142405
10	0.000282187	0.000284548
100	$(1.3583)10^{-128}$	$(1.35943)10^{-128}$
600	$(3.27877)10^{-1228}$	$(3.27922)10^{-1228}$
1000	$(2.66288)10^{-2267}$	$(2.66310)10^{-2267}$
6000	$(5.63886)10^{-18260}$	$(5.63894)10^{-18260}$
10000	$(7.00942)10^{-32650}$	$(7.00948)10^{-32650}$
60000	$(4.03148)\ 10^{-242573}$	$(4.03149)10^{-242573}$

and

$$y_6(0, n; 1, 3) \sim \frac{2^{3n+1}}{\sqrt{6}} \left(\frac{e}{n}\right)^n \left(\frac{1}{\pi n}\right)^{\frac{3}{2}}.$$

By using the same calculation method, which is given by Kilar [19], some numeric and approximate values of the numbers $y_6(0, n; 1, 1)$, $y_6(0, n; 1, 2)$, and $y_6(0, n; 1, 3)$ are given respectively in Tables 18.1, 18.2, and 18.3.

In addition to the approximate values of the sum $y_6(j, b; \omega, p)$, in order to see some inequalities arising from the application of the arithmetic-geometric mean inequality to the sum $y_6(j, b; \omega, p)$, the interested readers may also glance at the recent paper [29] in which Kucukoglu provided some lower bounds, in terms of hyperfactorial and superfactorial, for the combinatorial Simsek numbers.

Recall that approximation formula for the hyperfactorial $H(n) = \prod_{k=1}^{n} k^k$ is given by

$$H(n) \sim (1.2824) n^{\frac{6(n^2+n)+1}{12}} e^{-\frac{n^2}{4}} \tag{18.42}$$

(*cf.* [2]).

Open Problem 5 By considering the approximation formula (18.42), is there any further relationship of the hyperfactorial with the combinatorial Simsek numbers?

Table 18.2 Some values of the numbers $y_6(0, n; 1, 2)$ compared with their approximate values

n	$y_6(0, n; 1, 2)$	Approximate values of $y_6(0, n; 1, 2)$
1	2	2.44731
2	3	3.32624
3	3.33333	3.57202
4	2.91667	3.07223
5	2.1	2.18921
6	1.28333	1.32863
7	0.680952	0.701507
8	0.319196	0.327613
9	0.133984	0.13712
10	0.0509138	0.0519852
100	$(9.70237)10^{-100}$	$(9.7226)10^{-100}$
600	$(3.13305)10^{-1049}$	$(3.13413)10^{-1049}$
1000	$(5.09000)10^{-1968}$	$(5.09106)10^{-1968}$
6000	$(6.21593)10^{-16456}$	$(6.21614)10^{-16456}$
10000	$(7.88966)10^{-29642}$	$(7.88983)10^{-29642}$
60000	$(5.85535)10^{-224514}$	$(5.85538)10^{-224514}$

Table 18.3 Some values of the numbers $y_6(0, n; 1, 3)$ compared with their approximate values

n	$y_6(0, n; 1, 3)$	Approximate values of $y_6(0, n; 1, 3)$
1	2	3.1887
2	5	6.12905
3	9.33333	10.7482
4	14.4167	16.0117
5	18.7667	20.4102
6	21.0889	22.6153
7	20.8254	22.1099
8	18.3324	19.3175
9	14.5528	15.2456
10	10.5173	10.9667
100	$(7.99588)10^{-71}$	$(8.02928)10^{-71}$
600	$(3.45647)10^{-870}$	$(3.45887)10^{-870}$
1000	$(1.12336)10^{-1668}$	$(1.12382)10^{-1668}$
6000	$(7.91195)10^{-14652}$	$(7.91250)10^{-14652}$
10000	$(1.02542)10^{-26633}$	$(1.02546)10^{-26633}$
60000	$(9.81997)10^{-206455}$	$(9.82004)10^{-206455}$

18.6 Implementations and Plots for the Sum $y_6(j, b; \omega, p)$ and the Polynomials $P_s(x; m, n; \omega, p)$

In this section, we present implementations of the sum $y_6(j, b; \omega, p)$ and the polynomials $P_s(x; m, n; \omega, p)$ with some of their plots for various special cases of their parameters (see Figs. 18.1, 18.2, 18.3, and 18.4).

Implementation 1 The following code of the Wolfram Language includes both the procedure **y6Sum** which returns the sum $y_6(j, b; \omega, p)$ and the procedure PsPoly which returns the polynomials $P_s(x; m, n; \omega, p)$

```
Unprotect[Power];
Power[0, 0] = 1;
Protect[Power];
y6Sum[j, b, \begin{displayMath}Omega], p] :=
 Sum[Power[Binomial[b, k], p]
    *((Power[\begin{displayMath}Omega], k]*Power[k, j])/Factorial[b]),
    {k, 0, b}]
PsPoly[x, m, n, \begin{displayMath}Omega], p] :=
 Sum[Binomial[m, k]*Power[x, m = k]
        *y6Sum[k, n, \begin{displayMath}Omega], p],  {k, 0, m}]
```

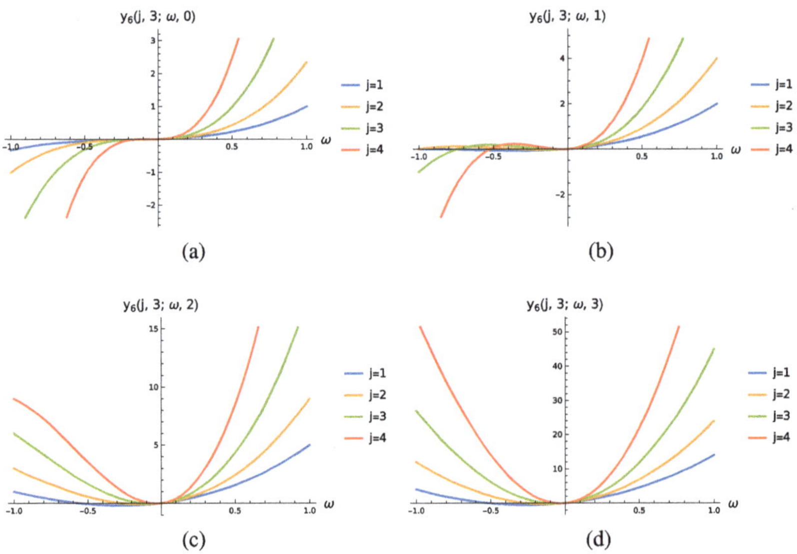

Fig. 18.1 Plots of the sum $y_6(j, b; \omega, p)$ in the special cases when $\omega \in [-1, 1]$, $b = 3$, and $j \in \{1, 2, 3, 4\}$ with (**a**) $p = 0$; (**b**) $p = 1$; (**c**) $p = 2$; (**d**) $p = 3$

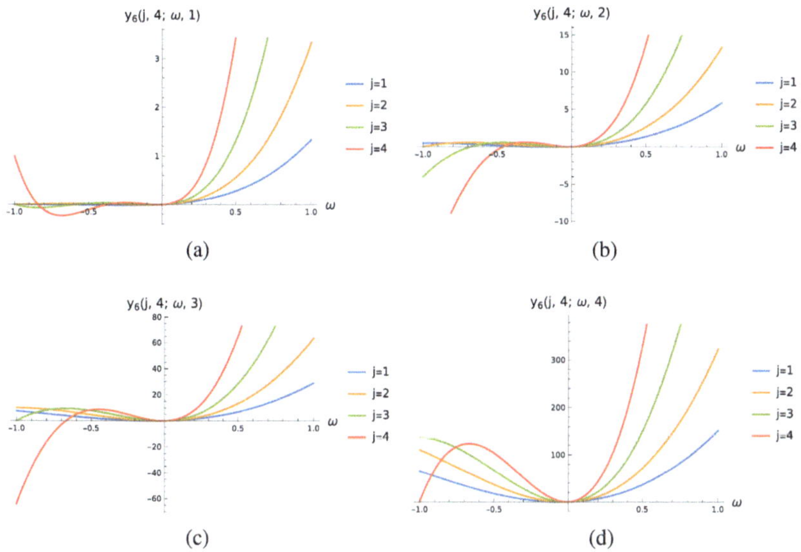

Fig. 18.2 Plots of the sum $y_6(j, b; \omega, p)$ in the special cases when $\omega \in [-1, 1]$, $b = 4$, and $j \in \{1, 2, 3, 4\}$ with (**a**) $p = 1$; (**b**) $p = 2$; (**c**) $p = 3$; (**d**) $p = 4$

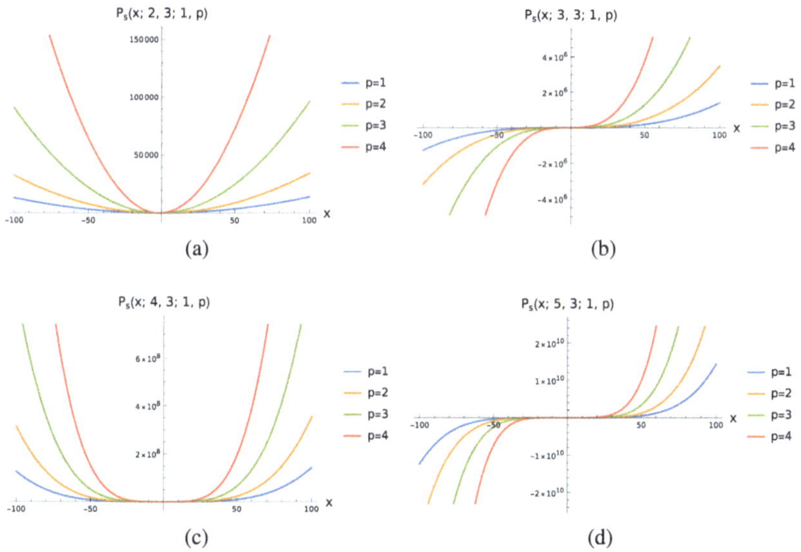

Fig. 18.3 Plots of the polynomials $P_s(x; m, n; \omega, p)$ in the special cases when $x \in [-100, 100]$, $n = 3$, $\omega = 1$, and $p \in \{1, 2, 3, 4\}$ with (**a**) $m = 2$; (**b**) $m = 3$; (**c**) $m = 4$; (**d**) $m = 5$

By using Implementation 1 with Implementation 2, we get Fig. 18.1 that involves the plots of the sum $y_6(j, b; \omega, p)$ in the special cases when $\omega \in [-1, 1]$, $b = 3$, and $j \in \{1, 2, 3, 4\}$ with $p \in \{0, 1, 2, 3\}$.

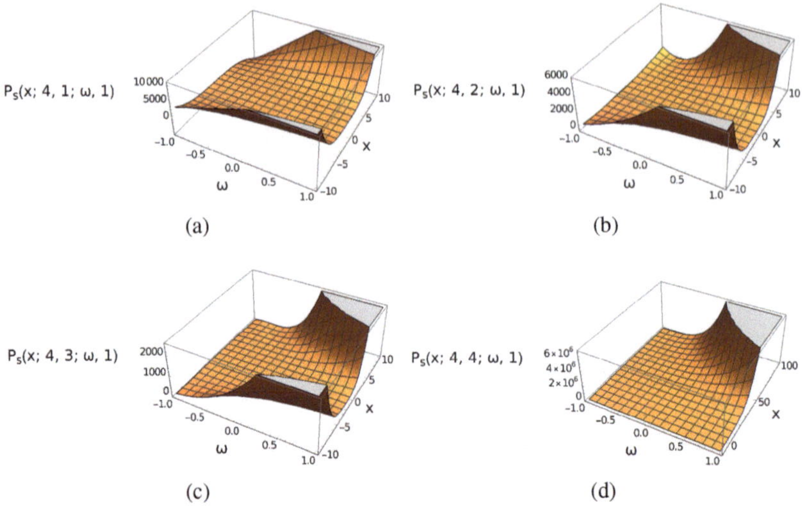

Fig. 18.4 Surface plots of the polynomials $P_s(x; m, n; \omega, p)$ in the special cases when $x \in [-10, 10]$, $m = 4$, $\omega \in [-1, 1]$, and $p = 1$ with **(a)** $n = 1$; **(b)** $n = 2$; **(c)** $n = 3$; **(d)** $n = 4$

Implementation 2 The following code of the Wolfram Language returns Fig. 18.1

```
Table[Plot[Evaluate[Table[y6Sum[j,3,\begin{displayMath}Omega],p], {j, 1, 4}]],
    {\begin{displayMath}Omega], -1, 1}, AxesLabel -> {Style["\begin{displayMath}Omega]",
        Black, FontSize -> 14],
        Style[StringJoin[ToString[Subscript["y", "6"],
            StandardForm],
        ToString["(j, 3; \begin{displayMath}Omega],  ", StandardForm],
        ToString[p, StandardForm],
        ToString[")", StandardForm]],
        Black, FontSize -> 14]},
    PlotLegends -> {"j=1", "j=2", "j=3", "j=4"}], {p,0,3}]
```

By using Implementation 1 with Implementation 3, we get Fig. 18.2 that involves the plots of the sum $y_6(j, b; \omega, p)$ in the special cases when $\omega \in [-1, 1]$, $b = 4$, and $j \in \{1, 2, 3, 4\}$ with $p \in \{1, 2, 3, 4\}$.

Implementation 3 The following code of the Wolfram Language returns Fig. 18.2

```
Table[Plot[Evaluate[Table[y6Sum[j,4,\begin{displayMath}Omega],p],
        {j, 1, 4}]],
    {\begin{displayMath}Omega], -1, 1}, AxesLabel -> {Style["\begin{displayMath}Omega]",
        Black, FontSize -> 14],
    Style[StringJoin[ToString[Subscript["y", "6"],
        StandardForm],
    ToString["(j, 3; \begin{displayMath}Omega],  ", StandardForm],
    ToString[p, StandardForm],ToString[")", StandardForm]],
        Black, FontSize -> 14]},
    PlotLegends -> {"j=1", "j=2", "j=3", "j=4"}], {p,1,4}]
```

By using Implementation 1 with Implementation 4, we get Fig. 18.3 that involves the plots of the polynomials $P_s(x; m, n; \omega, p)$ in the special cases when $x \in [-100, 100]$, $n = 3$, $\omega = 1$, and $p \in \{1, 2, 3, 4\}$ with $m \in \{2, 3, 4, 5\}$.

Implementation 4 The following code of the Wolfram Language returns Fig. 18.3

```
Table[Plot[Evaluate[Table[PsPoly[x,m,3,1,p], {p, 1, 4}]],
    {x, -100,100}, AxesLabel -> {Style["x", Black,
        FontSize -> 14],Style[StringJoin[
        ToString[Subscript["P", "s"], StandardForm],
        ToString["(x; ", StandardForm],
        ToString[m, StandardForm],
        ToString[", 3; 1, p)", StandardForm]], Black,
            FontSize -> 14]},
    PlotLegends -> {"p=1", "p=2", "p=3", "p=4"}], {m,2,5}]
```

By using Implementation 1 with Implementation 5, we get Fig. 18.4 that involves the surface plots of the polynomials $P_s(x; m, n; \omega, p)$ in the special cases when $x \in [-10, 10]$, $m = 4$, $\omega \in [-1, 1]$, and $p = 1$ with $p \in \{1, 2, 3, 4\}$.

Implementation 5 The following code of the Wolfram Language returns Fig. 18.4

```
Table[Plot3D[PsPoly[x,4,n,\begin{displayMath}Omega],1], {\begin{displayMath}Omega],-1,1},
    {x,  -10,10},AxesLabel -> {Style["\begin{displayMath}Omega]", Black,
        FontSize -> 14],Style["x", Black, FontSize -> 14],
        Style[StringJoin[ToString[Subscript["P", "s"],
            StandardForm],
        ToString["(x; 4, ", StandardForm],
        ToString[n, StandardForm],
        ToString["; \begin{displayMath}Omega], 1)    ", StandardForm]],
        Black, FontSize -> 14]}], {n,1,4}]
```

References

1. Apostol, T.M.: Introduction to Analytic Number Theory. Springer, New York (1976)
2. Alabdulmohsin, I.M.: Summability calculus. https://arxiv.org/pdf/1209.5739.pdf
3. Apostol, T.M.: On the Lerch zeta function. Pacific J. Math. **1**(2), 161–167 (1951)
4. Bostan, A., Lairez, P., Salvy, B.: Multiple binomial sums. J. Symbolic Comput. **80**(2), 351–386 (2017)
5. Bonar, D.D., Khoury, Jr M.J.: Real Infinite Series, 1st edn. Mathematical Association of America Textbooks, Cambridge (2006)
6. Belbachir, H., Rahmani, M., Sury, B.: Sums involving moments of reciprocals of binomial coefficients. J. Integer Seq. **14**, 6 (2011) (article 11.6.6.)
7. Berstel, J., Perrin, D.: The origins of combinatorics on words. European J. Combin. **28**, 996–1022 (2007)
8. Comtet, L.: Advanced Combinatorics. D. Reidel, Dordrecht (1974)
9. Faulhaber, J.: Academia Algebrae. Augspurg, bey Johann Ulrich Schönigs (1631)
10. Charalambides, C.A.: Enumerative Combinatorics. Chapman & Hall/CRC, Boca Raton (2002)
11. Choi, J., Srivastava, H.M.: Some summation formulas involving harmonic numbers and generalized harmonic numbers. Math. Comput. Modelling **54**, 2220–2234 (2011)
12. Gun, D., Simsek, Y.: Relations of exponential Euler spline associated with special numbers and polynomials. In: Proceedings Book of 6th Mediterranean International Conference of Pure & Applied Mathematics and Related Areas (MICOPAM 2023), pp. 127–134. Paris-France, August 23–27 (2023)
13. Dutka, J.: The early history of the factorial function. Arch. Hist. Exact Sci. **43**(3), 225–249 (1991)
14. Gavrikov, V.L.: Some properties of binomial coefficients and their application to growth modelling. Arab. J. Basic Appl. Sci. **25**(1), 38–43 (2018)
15. Golombek, R.: Aufgabe 1088. El. Math. **49**, 126–127 (1994)
16. Goubi, M.: An Affirmative answer to two questions concerning special case of Simsek numbers and open problems. Appl. Anal. Discrete Math. **14**, 94–105 (2020)
17. Gould, H.W.: Explicit formulas for Bernoulli numbers. Am. Math. Monthly **79**(1), 44–51 (1972)
18. Graham, R.L., Knuth, D.E., Patashnik, O.: Concrete Mathematics, 2nd edn. Addison-Wesley, New York (1994)
19. Kilar, N.: Asymptotic expressions and formulas for finite sums of powers of binomial coefficients involving special numbers and polynomials. J. Inequal. Spec. Funct. **14**(1), 51–67 (2023)

20. Kilar, N: Remarks on combinatorial numbers containing special numbers and polynomials, AIP Conf. Proc **3094**(1), 090006 (2024). https://doi.org/10.1063/5.0210513
21. Kilar, N.: Observations on asymptotic expressions for combinatorial Simsek numbers. In: Proceedings Book of 6th Mediterranean International Conference of Pure & Applied Mathematics and Related Areas (MICOPAM 2023), pp. 280–284. Paris-France, August 23–27 (2023)
22. Kilar, N., Simsek, Y.: Computational formulas and identities for new classes of Hermite-based Milne-Thomson type polynomials: analysis of generating functions with Euler's formula. Math. Meth. Appl. Sci. **44**(8), 6731–6762 (2021)
23. Kilar, N., Simsek, Y.: Formulas and relations of special numbers and polynomials arising from functional equations of generating functions. Montes Taurus J. Pure Appl. Math. **3**(1), 106–123 (2021)
24. Kim, T., Kim, D.S., Kwon, J.: Analogues of Faulhaber's formula for poly-Bernoulli and type 2 poly-Bernoulli polynomials. Montes Taurus J. Pure Appl. Math. **3**(1), 1–6 (2021)
25. Kucukoglu, I.: Some relationships between the numbers of Lyndon words and a certain class of combinatorial numbers containing powers of binomial coefficients. Adv. Stud. Contemp. Math. **30**(4), 529–538 (2020)
26. Kucukoglu, I.: Identities for the multiparametric higher-order Hermite-based Peters-type Simsek polynomials of the first kind. Montes Taurus J. Pure Appl. Math. **5**(1), 102–123 (2023)
27. Kucukoglu, I.: Formulas for q-combinatorial Simsek numbers and polynomials: Analyzing with computational implementations, Appl. Anal. Discrete Math (2025). https://doi.org/10.2298/AADM240227005K
28. Kucukoglu, I.: Generating functions for divisor sums and totative sums arising from combinatorial Simsek numbers. Filomat **38**(5), 1513–1529 (2024)
29. Kucukoglu, I.: Inequalities obtained by the application of arithmetic-geometric mean inequality to the combinatorial Simsek numbers. Preprint (2024)
30. Kucukoglu, I., Simsek, Y.: Identities and derivative formulas for the combinatorial and Apostol-Euler type numbers by their generating functions. Filomat **32**(20), 6879–6891 (2018)
31. Kucukoglu, I., Simsek, Y.: Observations on identities and relations for interpolation functions and special numbers. Adv. Stud. Contemp. Math. **28**(1), 41–56 (2018)
32. Kucukoglu, I., Simsek, Y.: Identities and relations on the q-Apostol type Frobenius-Euler numbers and polynomials. J. Korean Math. Soc. **56**(1), 265–284 (2019)
33. Kucukoglu, I., Simsek, Y.: New formulas and numbers arising from analyzing combinatorial numbers and polynomials. Montes Taurus J. Pure Appl. Math. **3**(3), 238–259 (2021)
34. Kucukoglu, I., Bayad, A., Simsek, Y.: k-ary Lyndon words and necklaces arising as rational arguments of Hurwitz-Lerch zeta function and Apostol-Bernoulli polynomials. Mediterr. J. Math. **14**, Article Number: 223 (2017)
35. Kucukoglu, I., Milovanović, G.V., Simsek, Y.: Analysis of generating functions for special words and numbers and algorithms for computation. Mediterr. J. Math. **19**(6), 268 (2022). https://doi.org/10.1007/s00009-022-02191-3
36. Mansour, T.: Combinatorial identities and inverse binomial coefficients. Adv. Appl. Math. **28**, 196–202 (2002)
37. Milovanović, G.V., Cvetković, A.S.: Special classes of orthogonal polynomials and corresponding quadratures of Gaussian type. Math. Balkanica **26**, 169–184 (2012)
38. Milovanović, G.V., Udovičić, Z.: Calculation of coefficients of a cardinal B-spline. Appl. Math. Lett. **23**, 1346–1350 (2010)
39. Milovanović, G.V.: Symbolic numeric computation of orthogonal polynomials and Gaussian quadratures with respect to the cardinal B-spline. Numer. Algor. **76**, 333–347 (2017)
40. Milovanović, G.V.: Four unified results for reducibility of Srivastava's triple hypergeometric series H_B. Montes Taurus J. Pure Appl. Math. **3**(3), 155–164 (2021)
41. Lothaire, M.: Combinatorics on Words. Cambridge University, Cambridge (1997)
42. Lyndon, R.: On Burnside problem I. Trans. Am. Math. Soc. **77**, 202–215 (1954)
43. Polya, G., Szegö, G.: Problems and Theorems in Analysis I. Springer, Berlin (1978)
44. Pla, T.J.: The sum of inverses of binomial coefficients revisited. Fibonacci Quart. **35**(4), 342–345 (1997)

45. Riordan, J.: Combinatorial Identities. Wiley, New York (1968)
46. Robbins, H.: A remark on Stirling's formula. Am. Math. Monthly **62**(1), 26–297 (1955)
47. Ripon, S.M.: A generalized inverse binomial summation theorem and some hypergeometric transformation formulas. Int. J. Comb. 2016(1), 4546509 (2016). https://doi.org/10.1155/2016/4546509
48. Rockett, A.M.: Sums of the inverses of binomial coefficients. Fibonacci Quart. **19**, 433–437 (1981)
49. Ruskey, F.: Generating necklaces, J. Algorithms **13**(3), 414–430 (1992)
50. Sándor, J.: A note concerning the Euler totient. https://rgmia.org/papers/v12n3/art51.pdf. Cited 14 Jul 2023
51. Sándor, J., Crstici, B.: Handbook of Number Theory II (Chapter 3: The many facets of Euler's totient). Kluwer Academic Publishers, Dordrecht (2004)
52. Schwatt, I.J.: An Introduction to the Operations with Series (First Edition 1924; Second Edition). Chelsea Publishing Company, New York (2014)
53. Simsek, B., Simsek, Y., Sukruoglu, E.: Identities and approximation formulas for Faulhaber's formula revealing in applications of moment generating function, distribution, and arithmetic functions. ESS Open Archive. October 07 (2024). https://doi.org/10.22541/au.172831149.99500217/v1
54. Simsek, Y.: Beta-type polynomials and their generating functions. Appl. Math. Comput. **254**, 172–182 (2015)
55. Simsek, Y.: Construction of some new families of Apostol-type numbers and polynomials via Dirichlet character and p-adic q-integrals. Turkish J. Math. **42**, 557–577 (2018)
56. Simsek, Y.: Combinatorial sums and binomial identities associated with the Beta-type polynomials. Hacet. J. Math. Stat. **47**(5), 1144–1155 (2018)
57. Simsek, Y.: New families of special numbers for computing negative order Euler numbers and related numbers and polynomials. Appl. Anal. Discrete Math. **12**, 1–35 (2018)
58. Simsek, Y.: Generating functions for finite sums involving higher powers of binomial coefficients: analysis of hypergeometric functions including new families of polynomials and numbers. J. Math. Anal. Appl. **477**, 1328–1352 (2019)
59. Simsek, Y.: Explicit formulas for p-adic integrals: approach to p-adic distributions and some families of special numbers and polynomials. Montes Taurus J. Pure Appl. Math. **1**(1), 1–76 (2019)
60. Simsek, Y.: Identities and relations containing finite sums of powers of binomial coefficients, AIP Conf. Proc **2293**(1), 180002 (2020). https://doi.org/10.1063/5.0026604
61. Simsek, Y.: Interpolation functions for new classes special numbers and polynomials via applications of p-adic integrals and derivative operator. Montes Taurus J. Pure Appl. Math. **3**(1), 1–24 (2021)
62. Simsek, Y.: New integral formulas and identities involving special numbers and functions derived from certain class of special combinatorial sums. Rev. R. Acad. Cienc. Exactas Fis. Nat. Ser. A Mat. RACSAM **115** (2021). https://doi.org/10.1007/s13398-021-01006-6
63. Simsek, Y.: Derivation of computational formulas for certain class of finite sums: Approach to generating functions arising from p-adic integrals and special functions. Math. Methods Appl. Sci. **45**(16), 9520–9544 (2022)
64. Simsek, Y.: Generating functions for series involving higher powers of inverse binomial coefficients and their applications. Math. Methods Appl. Sci. **46**(12), 12591–12617 (2023). https://doi.org/10.1002/mma.9199
65. Simsek, Y.: On combinatorial numbers and polynomials and their applications, AIP Conf. Proc **2849**(1), 060001 (2023). https://doi.org/10.1063/5.0162213
66. Simsek Y: Some families of finite sums associated with interpolation functions for many classes of special numbers and polynomials and Hypergeometric series, Chapter 26: In Series on Computers and Operations Research Analysis, Geometry, Nonlinear Optimization and Applications (2023). https://doi.org/10.1142/9789811261572_0026
67. Simsek, Y.: A note on certain classes of finite sums, AIP Conf. Proc **3094**(1), 090001 (2024). https://doi.org/10.1063/5.0210137

68. Sury, B., Wang, T., Zhao, F.-Z. Identities involving reciprocals of binomial coefficients. J. Integer Seq. **7**, 1–2 (2004) (article 04.2.8)
69. Sprugnoli, R.: Riordan Array Proofs of Identities in Gould's Book. In: Dipartimento di Sistemi e Informatica Viale Morgagni, Firenze (Italy) (2006)
70. Sprugnoli, R.: Alternating weighted sums of inverses of binomial coefficients. J. Integer Seq. **5**, 1–13 (2012) (article 12.6.3)
71. Srivastava, H.M.: Some generalizations and basic (or q-) extensions of the Bernoulli, Euler and Genocchi polynomials. Appl. Math. Inf. Sci. **5**, 390–444 (2011)
72. Srivastava, H.M., Choi, J.: Zeta and q-Zeta Functions and Associated Series and Integrals. Elsevier Science Publishers, Amsterdam (2012)
73. Vandiver, H.S.: An arithmetical theory of the Bernoulli numbers. Trans. Amer. Math. Soc. **51**, 502–531 (1942)
74. Wikipedia: Bell series. The Free Encyclopedia. https://en.wikipedia.org/wiki/Bell_series. Cited 11 Nov 2023
75. Xu, A.: On an open problem of Simsek concerning the computation of a family of special numbers. Appl. Anal. Discrete Math. **13**(1), 61–72 (2019)
76. Zygmund, A.: Trigonometric Series, vols. I & II combined. Cambridge University, Cambridge (2002)

Chapter 19
The Finite Difference Method for Solution of Parabolic Problem with Time-Dependent Coefficients

Bratislav Sredojević, Dejan Bojović, and Zorica Milovanović Jeknić

Dedicated to Academician Gradimir V. Milovanović on the occasion of his 75th birthday

19.1 Introduction

The finite difference method is one of the most commonly used methods for the numerical solution of boundary value problems for partial differential equations. In the context of finite difference method, one of the main problems is proving convergence of difference schemes which approximate boundary problems. Of particular interest is the estimates of the rate of convergence compatible with the smoothness of the coefficients and solution.

When numerically approximating the parabolic initial-boundary value problem with the generalized solution, there are also some additional problems: the coef-

Research supported by the Serbian Ministry of Science, Technological Development and Innovation (Agreement No. 451-03-47/2023-01/200108, 451-03-47/2023-01/200122 and 451-03-47/2023-01/200029)

B. Sredojević (✉)
University of Kragujevac, Faculty of Mechanical and Civil Engineering in Kraljevo, Kraljevo, Serbia
e-mail: sredojevic.b@mfkv.kg.ac.rs

D. Bojović
Department of Mathematics and Informatics, University of Kragujevac, Faculty of Science, Kragujevac, Serbia
e-mail: dejan.bojovic@pmf.kg.ac.rs

Z. M. Jeknić
University "Union-Nikola Tesla", Faculty of Construction Management, Belgrade, Serbia
e-mail: zmilovanovic@unionnikolatesla.edu.rs

M. P. Stanić et al. (eds.), *Analysis, Approximation, Optimization: Computation and Applications*, Springer Optimization and Its Applications 224,
https://doi.org/10.1007/978-3-031-85743-0_19

ficients are not continuous functions, variable coefficients can be time-dependent coefficients, and the solution belongs to nonstandard anisotropic Sobolev spaces.

In the case of linear partial equations of the second order, a complete theory of existence and uniqueness of solutions for the basic initial-boundary problems in the anisotropic Sobolev spaces $W_2^{s,\,s/2}$ is constructed. Therefore, we can use difference analogues of such norms to estimate the rate of convergence.

Let $u = u(x,t)$ be the solution of the initial-boundary problem and v be the solution of the corresponding difference scheme. For the parabolic case, typical estimates are of the form (see [15])

$$\|u - v\|_{W_2^{k,k/2}(Q_{h\tau})} \le C\left(h^{s-k} + \tau^{\frac{s-k}{2}}\right)\|u\|_{W_p^{s,s/2}(Q)}, \quad s \ge k, \tag{19.1}$$

where h is the space step and τ is the time step. If steps h i τ satisfy the natural relationship

$$k_1 h^2 \le \tau \le k_2 h^2, \quad k_1, k_2 = const. > 0,$$

estimate (19.1) is of the form

$$\|u - v\|_{W_2^{k,k/2}(Q_{h\tau})} \le C h^{s-k}\|u\|_{W_2^{s,\,s/2}(Q)}, \quad s \ge k. \tag{19.2}$$

In the case of equations with variable coefficients, constant C depends on the norms of the coefficients (see [18]). If the coefficients are not dependent on t, we get the estimates of the form

$$\|u - v\|_{W_2^{k,k/2}(Q_{h\tau})} \le C h^{s-k}\max_{i,\,j}\|a_{ij}\|_{W_p^{s-1}(\Omega)}\|u\|_{W_2^{s,\,s/2}(Q)}, \quad s \ge k. \tag{19.3}$$

In the case $a_{ij} = a_{ij}(x,t)$, what is especially disturbing is that we have the estimates of the forms

$$\|u - v\|_{W_2^{k,k/2}(Q_{h\tau})} \le C h^{s-k}\max_{i,\,j}\|a_{ij}\|_{W_p^{s-1,\,(s-1)/2}(Q)}\|u\|_{W_2^{s,\,s/2}(Q)}, \quad s \ge k. \tag{19.4}$$

The technique of these estimates is based on the lemma Bramble-Hilbert and its generalizations.

One interesting class of parabolic problems model processes in heat-conducting media with concentrated capacity in which the heat capacity coefficient contains a Dirac delta function as the coefficient with the derivation of the time (see [5, 13, 16]). In a one-dimensional case, such problems are considered in the works of Jovanović, Vulkov, and Bojović (see [1, 2, 6, 10, 12, 13]). Two-dimensional problems are discussed in the works of Jovanović, Bojović, and Sredojević (see [3, 4, 14, 19]).

In Sect. 19.2 we defined the finite differences and Steklov averaging operators.

In Sect. 19.3 one-dimensional parabolic problem with concentrated capacity and time-dependent coefficients is considered (see [21]). Only some of the functionalities are presented in detail, and the rest are analogously derived.

In Sect. 19.4, we discussed two-dimensional parabolic problem with concentrated capacity and time-dependent coefficients (see [22]). An estimate of the rate of convergence in a special discrete $\widetilde{W}_2^{1,1/2}$ Sobolev norm, compatible with the smoothness of the coefficients and solution, is proved.

In Sect. 19.5, we investigate the third initial-boundary value problem for parabolic equation with time-dependent coefficients (see [20]). A finite difference scheme which approximating this problem is derived. A special Sobolev norm $W_2^{1,\,1/2}$ is constructed. In this norm, a convergence rate estimate, compatible with the smoothness of the solution of the boundary value problem, is obtained.

19.2 The Finite Differences and Steklov Averaging Operators

Let $h = (h_1, h_2, \ldots, h_n)$ and $h_i > 0$, $i = 1, 2, \ldots, n$. Note

$$\mathbb{R}_h^n = \{(i_1 h_1, i_2 h_2, \ldots, i_n h_n) \mid i_1, i_2, \ldots, i_n = 0, \pm 1, \pm 2, \ldots\}$$

a set of points that we will call a mesh. In parabolic case in the direction "space variable" $x_1, x_2, \ldots, x_{n-1}$, we use one step

$$h_1 = h_2 = \cdots = h_{n-1} = h,$$

and in the direction "time variable" $x_n = t$ the second

$$h_n = \tau.$$

Further, let v be the function defined on the mesh $\mathbb{R}_h^n$. Let us mark

$$v_{i_1, i_2, \ldots, i_n} = v(i_1 h_1, i_2 h_2, \ldots, i_n h_n).$$

The finite difference operators are defined in the following way (see [17]):

$$\begin{aligned}(v_{x_k})_{i_1,\ldots,i_k,\ldots,x_n} &= (v_{i_1,\ldots,i_k+1,\ldots,i_n} - v_{i_1,\ldots,i_k,\ldots,i_n})/h_k = (v_{\overline{x}_k})_{i_1,\ldots,i_k+1,\ldots,i_n},\\ v_{\dot{x}_k} &= (v_{x_k} + v_{\overline{x}_k})/2.\end{aligned}$$

In further work, we will need the following operators (see [17]):

$$v_{x_i \overline{x}_i}(x, t) = \frac{v^{+i}(x, t) - 2v(x, t) + v^{-i}(x, t)}{h^2}$$

$$v_t(x,t) = \frac{v(x,t+\tau) - v(x,t)}{\tau} = v_{\bar{t}}(x,t+\tau),$$

where $x = (x_1, x_2, \dots, x_n)$, $v^{\pm i}(x,t) = v(x \pm he_i, t)$, and e_i is the unit vector of the axis x_i.

In the theory of partial differential equations with generalized solutions, the coefficients of the equation do not have to be continuous functions. Therefore, the difference schemes, which approximate such problems, must have averaging coefficients. For an arbitrary function $u \in L_{1,loc}(\mathbb{R})$ and $h > 0$ we defined the linear averaging operator (Steklov operator) in the following way (see [7, 23]):

$$(T_h u)(x) = \frac{1}{h} \int\limits_{x-h/2}^{x+h/2} u(\xi)\,\mathrm{d}\xi, \quad x \in \mathbb{R}.$$

Analogously, in the multidimensional case, for function $u = u(x_1, x_2, t)$ we introduce Steklov operators (see [7]):

$$T_1 u(x_1,x_2,t) = T_1^{\pm} u(x_1 \mp h/2, x_2, t) = \frac{1}{h} \int\limits_{x_1-h/2}^{x_1+h/2} u(x_1', x_2, t)\,\mathrm{d}x_1',$$

$$T_2 u(x_1,x_2,t) = T_2^{\pm} u(x_1, x_2 \mp h/2, t) = \frac{1}{h} \int\limits_{x_2-h/2}^{x_2+h/2} u(x_1, x_2', t)\,\mathrm{d}x_2',$$

$$T_1^2 u(x_1,x_2,t) = T_1^{+} T_1^{-} u(x_1,x_2,t) = \frac{1}{h} \int\limits_{x_1-h}^{x_1+h} \left(1 - \frac{|x_1' - x_1|}{h}\right) u(x_1', x_2, t)\,\mathrm{d}x_1',$$

$$T_2^2 u(x_1,x_2,t) = T_2^{+} T_2^{-} u(x_1,x_2,t) = \frac{1}{h} \int\limits_{x_2-h}^{x_2+h} \left(1 - \frac{|x_2' - x_2|}{h}\right) u(x_1, x_2', t)\,\mathrm{d}x_2',$$

$$T_t^{-} u(x_1,x_2,t) = T_t^{+} u(x_1,x_2,t-\tau) = \frac{1}{\tau} \int\limits_{t-\tau}^{t} u(x_1,x_2,t')\,\mathrm{d}t'.$$

These operators commute and transform derivatives into differences, for example:

$$T_i^{+}\left(\frac{\partial u}{\partial x_i}\right) = u_{x_i}, \quad T_i^{-}\left(\frac{\partial u}{\partial x_i}\right) = u_{\bar{x}_i}, \quad T_i^{2}\left(\frac{\partial^2 u}{\partial x_i^2}\right) = u_{\bar{x}_i x_i}, \quad i = 1, 2,$$

$$T_t^+\left(\frac{\partial u}{\partial t}\right) = u_t\,, \quad T_t^-\left(\frac{\partial u}{\partial t}\right) = u_{\bar{t}}\,.$$

19.3 1D Problem with Dirac Delta Function

Let us consider the initial-boundary value problem for the heat equation in the presence of a concentrated capacity at the interior point $x = \xi$:

$$\begin{aligned}
&(1 + K\delta(x-\xi))\frac{\partial u}{\partial t} - \frac{\partial}{\partial x}\left(a(x,t)\frac{\partial u}{\partial x}\right) + b(x,t)u \\
&= f(x,t)\,, \text{ in } Q = \Omega \times (0,T)\,, \\
&u(0,t) = 0\,,\ u(1,t) = 0\,, \quad 0 < t < T, \\
&u(x,0) = u_0(x)\,, \quad x \in (0,1)\,,
\end{aligned} \tag{19.5}$$

where $\delta(x)$ is the Dirac delta function, $K > 0$, and $\Omega = (0,1)$. We shall assume that

$$a \in W_2^{s-1,(s-1)/2}(Q^L) \cap W_2^{s-1,(s-1)/2}(Q^R), \tag{19.6}$$

$$b \in W_2^{s-2,s/2-1}(Q^L) \cap W_2^{s-2,s/2-1}(Q^R)\,, \quad a > 0\,,\ b > 0 \tag{19.7}$$

$$f \in W^{s-2,s/2-1}(Q)\,, \quad u_0(x) \in W_2^{s-1}(\Omega^L) \cap W_2^{s-1}(\Omega^R)\,, \tag{19.8}$$

$$u \in W_2^{s,s/2}(Q^L) \cap W_2^{s,s/2}(Q^R) \tag{19.9}$$

for $5/2 < s \le 3$, where $\Omega^L = (0,\xi)$, $\Omega^R = (\xi,1)$, and $Q^L = \Omega^L \times (0,T)$, $Q^R = \Omega^R \times (0,T)$. Note that conditions (19.6)–(19.8) express the minimal smoothness requirements on the data under which the solution u of (19.5) may belong to the function space stated in (19.9) (see [14]). To guarantee that such u really exists, we also need some additional compatibility conditions at the corners of Q (see [5]).

Finite Difference Approximation Let w_h be the uniform mesh with step size h on $(0,1)$, $\omega_h^- = \omega_h \cup \{0\}$ and $\overline{\omega}_h = \omega_h \cup \{0,1\}$. Suppose that ξ is a rational number. Then one can choose the step h so that $\xi \in \omega_h$. Let ω_τ be a uniform mesh on $(0,T)$ with the step size $\tau = T/m$, $\omega_\tau^+ = \omega_\tau \cup \{T\}$ and $\bar{\omega}_\tau = \omega_\tau \cup \{0,T\}$. Also we assume that the condition $c_1 h^2 \le \tau \le c_2 h^2$ is satisfied.

Problem (19.5) can be approximated on the mesh $\overline{Q}_{h\tau} = \overline{\omega}_h \times \overline{\omega}_\tau$ by the following the finite difference scheme with averaged right-hand side (see [9]):

$$\begin{aligned}
&(1 + K\delta_h(x-\xi))v_{\bar{t}} - \frac{1}{2}((av_x)_{\bar{x}} + (av_{\bar{x}})_x) + (T_x^2 T_t^- b)v \\
&= T_x^2 T_t^- f, \quad (x,t) \in \omega_h \times \omega_\tau^+\,,
\end{aligned}$$

$$v(0,t)=0\,,\ \ v(1,t)=0\,,\quad t\in\omega_\tau^+\,, \tag{19.10}$$
$$v(x,0)=u_0(x)\,,\quad x\in\bar\omega_h\,,$$

where

$$\delta_h(x-\xi)=\begin{cases}0\,, & x\in\omega_h\backslash\{\xi\},\\ 1/h\,, & x=\xi,\end{cases}$$

is the mesh Dirac function, and T_x , T_t^- are Steklov averaging operators defined as in Sect. 19.2.

We also define operators

$$T_x^{2-}f(x,t)=\frac{2}{h}\int\limits_{x-h}^{x}\left(1+\frac{x'-x}{h}\right)f(x',t)\,\mathrm{d}x'\,,$$

$$T_x^{2+}f(x,t)=\frac{2}{h}\int\limits_{x}^{x+h}\left(1-\frac{x'-x}{h}\right)f(x',t)\,\mathrm{d}x'\,.$$

We define the following discrete inner products, norms, and seminorms:

$$(v,u)_{L_2(\omega_h)}=h\sum_{x\in\omega_h}v(x)u(x)\,,\quad \|v\|_{L_2(\omega_h)}=(v,v)^{1/2}_{L_2(\omega_h)}\,,$$

$$(v,u)_{L_2(\omega_h^-)}=h\sum_{x\in\omega_h^-}v(x)u(x)\,,\quad \|v\|_{L_2(\omega_h^-)}=(v,v)^{1/2}_{L_2(\omega_h^-)}\,,$$

$$\|v\|_{W_2^1(\omega_h)}=\|v_x\|^2_{L_2(\omega_h^-)}+\|v\|^2_{L_2(\omega_h)}\,,$$

$$\|v\|^2_{L_2(Q_{h\tau})}=\tau\sum_{t\in\omega_\tau^+}\|v(\cdot,t)\|^2_{L_2(\omega_h)},\quad \|v\|^2_{L_2(\xi\times\omega_\tau)}=\tau\sum_{t\in\omega_\tau^+}|v(\xi,t)|^2,$$

$$|v|^2_{W_2^{1/2}(\omega_\tau;\,L_2(\omega_h))}=\tau\sum_{t\in\overline{\omega}_\tau}\tau\sum_{t'\in\overline{\omega}_\tau,\,t'\neq t}\frac{\|v(\cdot,t)-v(\cdot,t')\|^2_{L_2(\omega_h)}}{|t-t'|^2},$$

$$|v|^2_{W_2^{1/2}(\omega_\tau,\xi)}=\tau\sum_{t\in\overline{\omega}_\tau}\tau\sum_{t'\in\overline{\omega}_\tau,\,t'\neq t}\frac{|v(\xi,t)-v(\xi,t')|^2}{|t-t'|^2},$$

$$\|v\|^2_{\widetilde{W}_2^{1/2}(\omega_\tau;\,L_2(\omega_h))}=|v|^2_{W_2^{1/2}(\omega_\tau;\,L_2(\omega_h))}+\tau\sum_{t\in\overline{\omega}_\tau}\left(\frac{1}{t+\tau}+\frac{1}{T-t+\tau}\right)\|v(\cdot,t)\|^2_{L_2(\omega_h)},$$

$$\|v\|^2_{\widetilde{W}_2^{1/2}(\omega_\tau;\,\xi)} = |v|^2_{W_2^{1/2}(\omega_\tau;\,\xi)} + \tau \sum_{t\in\overline{\omega}_\tau} \left(\frac{1}{t+\tau} + \frac{1}{T-t+\tau}\right) |v(\xi,t)|^2,$$

$$\|v\|^2_{\widetilde{W}_2^{1,\,1/2}(Q_{h\tau})} = \tau \sum_{t\in\omega_\tau^+} \|v(\cdot,t)\|^2_{W_2^1(\omega_h)} + |v|^2_{W_2^{1/2}(\omega_\tau;\,L_2(\omega_h))} + |v|^2_{W_2^{1/2}(\omega_\tau;\,\xi)}.$$

Error Analysis In this section we shall prove the convergence of the difference scheme (19.10) in the $\widetilde{W}_2^{1,\,1/2}(Q_{h\tau})$ norm.

Let u be the solution of the boundary-value problem (19.5) and v the solution of the difference problem (19.10). The error $z = u - v$ satisfies the finite difference scheme:

$$\begin{aligned}
&(1 + K\delta_h(x-\xi))z_{\bar{t}} - \frac{1}{2}((az_x)_{\bar{x}} + (az_{\bar{x}})_x) + (T_x^2 T_t^- b)z \\
&= \eta_{\bar{x}} + \phi_{\bar{t}} + \mu \text{ on } Q_{h\tau}\,, \\
&z(0,t) = z(1,t) = 0\,, \quad t \in \omega_\tau^+\,, \\
&z(x,0) = 0\,, \quad x \in \bar{\omega}_h\,,
\end{aligned} \tag{19.11}$$

where

$$\eta = T_x^+ T_t^- \left(a\frac{\partial u}{\partial x}\right) - \frac{1}{2}(a + a^+)u_x\,,$$

$$\phi = u - T_x^2 u\,, \quad \mu = (T_x^2 T_t^- b)u - T_x^2 T_t^-(bu).$$

and $a^+(x,t) = a(x+h,t)$. Let us set $\phi = \tilde{\phi} + \delta_h\hat{\phi}$, $\mu = \tilde{\mu} + \delta_h\hat{\mu}$ where

$$\hat{\phi} = \frac{h^2}{6}\left[\frac{\partial u}{\partial x}\right]_\xi,$$

$$\hat{\mu} = -\frac{h^2}{3}\left[(T_t^- b)\left(T_t^-\frac{\partial u}{\partial x}\right)\right]_\xi,$$

and $[u]_\xi = u(\xi+0,t) - u(\xi-0,t)$.

The following a priori estimate for the solution of the difference scheme (19.11) is valid (see [14, 19]):

$$\begin{aligned}
\|z\|_{\widetilde{W}_2^{1,1/2}(Q_{h\tau})} \le C(&\|\eta\|_{L_2(Q_{h\tau})} + \|\tilde{\phi}\|_{\widetilde{W}_2^{1/2}(\omega_\tau;\,L_2(\omega_h))} + \|\hat{\phi}\|_{\widetilde{W}_2^{1/2}(\omega_\tau;\,\xi)} \\
&+ \|\tilde{\mu}\|_{L_2(Q_{h\tau})} + \|\hat{\mu}\|_{L_2(\xi\times\omega_\tau)}).
\end{aligned} \tag{19.12}$$

Therefore, in order to estimate the rate of convergence of the difference scheme (19.10), it is sufficient to estimate the right-hand side of inequality (19.12).

By a technique based on the Bramble-Hilbert lemma, we estimate the right-hand side terms in inequality (19.12).

Further, we have the following.

Theorem 19.1 *Suppose that solution and coefficients of the differential problem* (19.5) *satisfy conditions* (19.6)–(19.9). *Then, the solution of the difference scheme* (19.10) *converges in* $\widetilde{W}_2^{1,\,1/2}(Q_{h\tau})$ *to the solution of the differential problem* (19.5) *and, assuming that* $\tau \asymp h^2$, *the following estimate is valid:*

$$\|u - v\|_{\widetilde{W}_2^{1,\,1/2}(Q_{h\tau})} \le Ch^{s-1}\Big(\|a\|_{W_2^{s-1,\,(s-1)/2}(Q^L)} + \|a\|_{W_2^{s-1,\,(s-1)/2}(Q^R)} + \|b\|_{W_2^{s-2,\,s/2-1}(Q^L)} + \|b\|_{W_2^{s-2,\,s/2-1}(Q^R)} + l(h)\Big) \times \Big(\|u\|_{W_2^{s\,/s/2}(Q^L)} + \|u\|_{W_2^{s\,/s/2}(Q^R)}\Big)\,, \quad 5/2 < s \le 3\,, \tag{19.13}$$

where $l(h) = \sqrt{\log 1/h}$.

Numerical Example The test example is problem (19.5), with $a(x,t) = t + 1$, $b(x,t) = 0$, and $T = 0.001$. A concentrated capacity $K = 1$ exists at the interior point $x = \xi = \frac{1}{2}$.

The exact solution of problem (19.5) is

$$u(x,t) = \mathrm{e}^t \sin(2\pi x).$$

Hence,

$$u_0(x) = \sin(2\pi x),$$
$$f(x,t) = \mathrm{e}^t(1 + 4(t+1)\pi^2)\sin(2\pi x).$$

Problem (19.5) is approximated by the finite difference scheme (19.10). Since the input data are smooth functions, Steklov averaging is not applied.

The following table presents the values of error and convergence rate (CR), estimated using the double-mesh principle. Numerical experiment confirms the theoretical results (Table 19.1).

Table 19.1 Error and convergence rate (CR)

Mesh		
h	τ	Error (CR)
0.125,	0.00004	3.3000e-03
0.0625,	0.00001	9.2599e-04 (1.8334)
0.03125,	0.0000025	2.3654e-04 (1.9689)
0.015625,	0.00000062	5.8934e-05 (2.0049)

19.4 2D Problem with Dirac Delta Function

Let us consider the 2D initial-boundary value problem for the heat equation in the presence of a concentrated capacity at the line $x_2 = \xi$:

$$
\begin{aligned}
&(1 + K\delta(x_2 - \xi))\frac{\partial u}{\partial t} - \sum_{i}^{2} \frac{\partial}{\partial x_i}\left(a_i(x,t)\frac{\partial u}{\partial x_i}\right) + a(x,t)u = f\,, \quad \text{in } Q\,,\\
&u = 0\,, \quad \text{on } \partial\Omega \times (0,T)\,, \qquad (19.14)\\
&u(x,0) = u_0(x)\,, \quad \text{on } \Omega\,,
\end{aligned}
$$

where $\delta(x)$ is the Dirac delta function, $K > 0$, $\Omega = (0,1)^2$, and $Q = \Omega \times (0,T)$.

Finite Difference Approximation Let $\bar{\omega}_h$ be the uniform mesh with step size h in $\bar{\Omega}$, $\omega_h = \bar{\omega}_h \cap \Omega$, $\omega_{1h} = \bar{\omega}_h \cap ([0,1) \times (0,1))$, $\omega_{2h} = \bar{\omega}_h \cap ((0,1) \times [0,1))$, $\sigma_h = \omega_h \cap \Sigma$, where $\Sigma = \{(x_1,\xi) | x_1 \in (0,1)\}$. Suppose that ξ is a rational number. Then one can choose step h so that $\sigma_h \neq \emptyset$. Let ω_τ be a uniform mesh on $(0,T)$ with the step size $\tau = T/m$, $\omega_\tau^- = \omega_\tau \cup \{0\}$, $\omega_\tau^+ = \omega_\tau \cup \{T\}$, and $\overline{\omega}_\tau = \omega_\tau \cup \{0,T\}$. Also we assume that the condition $c_1h^2 \le \tau \le c_2h^2$ is satisfied.

Problem (19.14) can be approximated on the mesh $\overline{Q}_{h\tau} = \bar{\omega}_h \times \bar{\omega}_\tau$ by the following finite difference scheme with averaged right-hand side (see [9]):

$$
\begin{aligned}
&(1 + K\delta_h(x_2 - \xi))v_{\bar{t}} + L_h v = T_1^2 T_2^2 T_t^- f\,, \quad \text{in } Q_{h\tau}\,, \qquad (19.15)\\
&v = 0\,, \quad \text{on } \gamma_h \times \omega_\tau^+\,, \quad v(x,0) = u_0(x)\,, \quad \text{on } \omega_h\,,
\end{aligned}
$$

where $L_h v = -\frac{1}{2}\sum_{i=1}^{2}((a_i v_{x_i})_{\bar{x}_i} + (a_i v_{\bar{x}_i})_{x_i}) + (T_1^2 T_2^2 T_t^- a)v$,

$$
\delta_h(x_2 - \xi) = \begin{cases} 0\,, & x \notin \sigma_h,\\ 1/h\,, & x \in \sigma_h, \end{cases}
$$

is the mesh Dirac function, and T_1^2, T_2^2, T_t^- are Steklov averaging operators defined as in Sect. 19.2.

We also define operators

$$
T_2^{2-} f(x_1,x_2) = \frac{2}{h}\int_{x_2-h}^{x_2}\left(1 + \frac{x_2' - x_2}{h}\right) f(x_1,x_2')\,\mathrm{d}x_2'\,,
$$

$$
T_2^{2+} f(x_1,x_2) = \frac{2}{h}\int_{x_2}^{x_2+h}\left(1 - \frac{x_2' - x_2}{h}\right) f(x_1,x_2')\,\mathrm{d}x_2'\,.
$$

We define the following inner products and norms:

$$(v, u)_{L_2(\omega_h)} = h^2 \sum_{x\in\omega_h} v(x)u(x)\,, \quad \|v\|_{L_2(\omega_h)} = (v, v)^{1/2}_{L_2(\omega_h)}\,,$$

$$(v, u)_{L_2(\omega_{ih})} = h^2 \sum_{x\in\omega_{ih}} v(x)u(x)\,, \quad \|v\|_{L_2(\omega_{ih})} = (v, v)^{1/2}_{L_2(\omega_{ih})}\,.$$

Further, we denote $B_h v = (1 + K\delta_h(x_2 - \xi))v$ and define the following norms:

$$\|v\|^2_{B_h} = \|v\|^2_{L_2(\omega_h)} + Kh \sum_{x\in\sigma_h} v^2(x)\,,$$

$$\|v\|^2_{B_h^{-1}} = h^2 \sum_{x\in w_h\setminus\sigma_h} v^2(x) + \frac{h^3}{K+h} \sum_{x\in\sigma_h} v^2(x)\,,$$

$$\|v\|_{\widetilde{W}^2_{2,h}} = \sum_{i=1}^{2} \left(\|v_{x_i\bar{x}_i}\|^2_{B_h^{-1}} + \|v_{x_i}\|^2_{L_2(\omega_{ih})} \right) + \|v\|^2_{B_h}\,,$$

$$\|v\|^2_{\widetilde{W}^{2,1}_2(Q_{h\tau})} = \tau \sum_{t\in\bar{\omega}_\tau} \|v(\cdot,t)\|^2_{\widetilde{W}^2_{2,h}} + \tau \sum_{t\in\omega_\tau^+} \|v_{\bar{t}}(\cdot,t)\|^2_{B_h}\,.$$

Also, we define the $W_2^{1,\,1/2}$ norm as in the previous section.

Convergence in $\widetilde{W}_2^{1,1/2}(Q_{h\tau})$ We shall assume that (see [9, 24])

$$a_i \in W_2^{2+\varepsilon,1+\varepsilon/2}(Q_1) \cap W_2^{2+\varepsilon,1+\varepsilon/2}(Q_2)\,, \tag{19.16}$$

$$a \in W_2^{1+\varepsilon,(1+\varepsilon)/2}(Q_1) \cap W_2^{1+\varepsilon,(1+\varepsilon)/2}(Q_2)\,, \quad a_i > 0\,,\ a > 0\,, \tag{19.17}$$

$$f \in W_2^{1,1/2}(Q_1) \cap W_2^{1,1/2}(Q_2)\,, \tag{19.18}$$

$$u \in W_2^{3,3/2}(Q_1) \cap W_2^{3,3/2}(Q_2) \cap W_2^{3,3/2}(\Sigma \times (0,T))\,, \tag{19.19}$$

where $\Omega_1 = (0,1) \times (0,\xi)\,,\ \Omega_2 = (0,1) \times (\xi,1)\,,\ Q_1 = \Omega_1 \times (0,T)\,,\ Q_2 = \Omega_2 \times (0,T)\,,\ \Sigma = \{(x_1,\xi)|x_1 \in (0,1)\}$, and $\varepsilon > 0$.

Let u be the solution of the initial-boundary value problem (19.14) and v the solution of the difference problem (19.15). The error $z = u - v$ satisfies the finite difference scheme:

$$(1+K\delta_h(x_2-\xi))z_{\bar{t}}+L_h z=\sum_{i=1}^{2}\xi_{i,\bar{x}_i}+\chi_{\bar{t}}+\delta_{\sigma_h}\mu_{\bar{t}}+\eta\,,\quad \text{in } Q_{h\tau}\,, \tag{19.20}$$

$$z=0\,,\quad \text{on } \gamma_h\times\omega_\tau^+\,,\quad z(x,0)=0\,,\quad \text{on } \omega_h\,,$$

where

$$\xi_i=T_i^+T_{3-i}^2T_t^-\left(a_i\frac{\partial u}{\partial x_i}\right)-\frac{1}{2}(a_i+a_i^{+i})u_{x_i}\,,$$

$$\chi=u-T_1^2T_2^2u\,,\quad \mu=ku-T_1^2(ku)\,,\quad \eta=(T_1^2T_2^2T_t^-a)u-T_1^2T_2^2T_t^-(au)\,.$$

Let us set $\xi_1=\widetilde{\xi}_1+\delta_{\sigma_h}\widehat{\xi}_1\,,\ \chi=\widetilde{\chi}+\delta_{\sigma_h}\widehat{\chi}\,,\ \eta=\widetilde{\eta}+\delta_{\sigma_h}\widehat{\eta}$, where

$$\widehat{\xi}_1=\frac{h^2}{6}T_1^+T_t^-\left(\left[a_1\frac{\partial^2 u}{\partial x_1\partial x_2}+\frac{\partial a_1}{\partial x_2}\frac{\partial u}{\partial x_1}\right]_\Sigma\right),$$

$$\widehat{\chi}=\frac{h^2}{6}\left[T_1^2\frac{\partial u}{\partial x_2}\right]_\Sigma\,,$$

$$\widehat{\eta}=-\frac{h^2}{3}\left[(T_1^2T_t^-a)\left(T_1^2T_t^-\frac{\partial u}{\partial x_2}\right)\right]_\Sigma\,,$$

and $[u]_\Sigma=u(x_1,\xi+0,t)-u(x_1,\xi-0,t)$.

The following a priori estimate for the solution of the difference scheme (19.20) is valid (see [14, 19]):

$$\begin{aligned}\|z\|_{\widetilde{W}_2^{1,1/2}(Q_{h\tau})}\le C\Big[&\|\xi_2\|_{L_2(Q_{h\tau})}+\|\widetilde{\xi}_1\|_{L_2(Q_{h\tau})}+|\widehat{\xi}_1|_{L_2(\omega_\tau;\,W_2^{1/2}(\sigma_h))}\\&+\|\widetilde{\eta}\|_{L_2(Q_{h\tau})}+\|\widehat{\eta}\|_{L_2(\sigma_h\times\omega_\tau)}+\|\widetilde{\chi}\|_{\widetilde{W}_2^{1/2}(\omega_\tau,L_2(\omega_h))}\\&+\|\widehat{\chi}\|_{\widetilde{W}_2^{1/2}(\omega_\tau,L_2(\sigma_h))}+\|\mu\|_{\widetilde{W}_2^{1/2}(\omega_\tau,L_2(\sigma_h))}\Big].\end{aligned} \tag{19.21}$$

Therefore, in order to estimate the rate of convergence of the difference scheme (19.15), it is sufficient to estimate the right-hand side of inequality (19.21).

By a technique based on the Bramble-Hilbert lemma we estimate the right-hand side terms in inequality (19.21).

Further, we have the following.

Theorem 19.2 *Suppose that solution and coefficients of the differential problem* (19.14) *satisfy conditions* (19.16)–(19.19). *Then, the solution of the difference scheme* (19.15) *converges in* $\widetilde{W}_2^{1,\,1/2}(Q_{h\tau})$ *to the solution of the differential* *problem* (19.14) *and, assuming that $\tau \asymp h^2$, the following estimate is valid:*

$$\|u - v\|_{\widetilde{W}_2^{1,1/2}(Q_{h\tau})} \le Ch^2\Big(\max_i \|a_i\|_{W_2^{2+\varepsilon,\,1+\varepsilon/2}(Q_1)} + \max_i \|a_i\|_{W_2^{2+\varepsilon,\,1+\varepsilon/2}(Q_2)}$$

$$+\|a\|_{W_2^{1+\varepsilon,\,(1+\varepsilon)/2}(Q_1)} + \|a\|_{W_2^{1+\varepsilon,\,(1+\varepsilon)/2}(Q_2)} + l(h)\Big) \quad (19.22)$$

$$\times \left(\|u\|_{W_2^{3,3/2}(Q_1)} + \|u\|_{W_2^{3,3/2}(Q_2)} + \|u\|_{W_2^{3,3/2}(\Sigma\times(0,T))}\right),$$

where $l(h) = \sqrt{\log 1/h}$.

Convergence in $\widetilde{W}_2^{2,1}(Q_{h\tau})$ In this section we shall prove the convergence of the difference scheme (19.15) in the $\widetilde{W}_2^{2,1}(Q_{h\tau})$ norm.

We shall assume that

$$a_i \in W_2^{3,3/2}(Q_1) \cap W_2^{3,3/2}(Q_2), \tag{19.23}$$

$$a \in W_2^{2,1}(Q_1) \cap W_2^{2,1}(Q_2), \quad a_i > 0,\ a > 0, \tag{19.24}$$

$$f \in W_2^{2,1}(Q_1) \cap W_2^{2,1}(Q_2), \tag{19.25}$$

$$u \in W_2^{4,2}(Q_1) \cap W_2^{4,2}(Q_2) \cap W_2^{4,2}(\Sigma \times (0,T)). \tag{19.26}$$

We also assume that the coefficients $a_i(x,t)$ are decreasing functions in the variable t.

The following a priori estimate for the solution of the difference scheme (19.20) is valid (see [4]):

$$\|z\|_{\widetilde{W}_2^{2,1}(Q_{h\tau})}$$

$$\le C\left(\tau \sum_{t\in w_\tau^+} \left(\|\xi_{1,\bar{x}_1} + \xi_{2,\bar{x}_2}\|^2_{B_h^{-1}} + \|\chi_{\bar{t}} + \delta_{\sigma_h}\mu_{\bar{t}}\|^2_{B_h^{-1}} + \|\eta\|^2_{B_h^{-1}}\right)\right)^{1/2} \tag{19.27}$$

Therefore, in order to estimate the rate of convergence of the difference scheme (19.15), it is sufficient to estimate the right-hand side of inequality (19.27).

By a technique based on the Bramble-Hilbert lemma we estimate the right-hand side terms in inequality (19.27).

Further, we have the following.

Theorem 19.3 *Suppose that solution and coefficients of the differential problem* (19.14) *satisfy conditions* (19.23)–(19.26). *Then, the solution of the difference scheme* (19.15) *converges in $\widetilde{W}_2^{2,1}(Q_{h\tau})$ to the solution of the differential prob-*

Table 19.2 Error and convergence rate (CR)

Mesh		
h	τ	Error (CR)
0.125,	0.00004	3.3000e-03
0.0625,	0.00001	9.2600e-04 (1.8334)
0.03125,	0.0000025	2.3654e-04 (1.9689)
0.015625,	0.0000025	5.8933e-05 (2.0049)

lem (19.14) *and, assuming that* $\tau \asymp h^2$, *the following estimate is valid:*

$$\begin{aligned}\|u-v\|_{\widetilde{W}_2^{2,1}(Q_{h\tau})} \le Ch^2\Big(&\max_i \|a_i\|_{W_2^{3,3/2}(Q_1)} + \max_i \|a_i\|_{W_2^{3,3/2}(Q_2)} \\ &+\|a\|_{W_2^{2,1}(Q_1)} + \|a\|_{W_2^{2,1}(Q_2)} + 1\Big) \\ &\times \Big(\|u\|_{W_2^{4,2}(Q_1)} + \|u\|_{W_2^{4,2}(Q_2)} + \|u\|_{W_2^{4,2}(\Sigma\times(0,T))}\Big)\,.\end{aligned} \tag{19.28}$$

This estimate is compatible with the smoothness of the coefficients and solution of the differential problem (19.14).

Numerical Example The test example is problem (19.14), with $a_i(x,t) = t+1$, $a(x,t) = 0$ and $T = 0.001$. A concentrated capacity $K = 1$ exists at the line $x_2 = \xi = \frac{1}{2}$. The exact solution of problem (19.14) is

$$u(x,t) = \mathrm{e}^t \sin(\pi x_1)\sin(2\pi x_2).$$

Hence,

$$\begin{aligned} u_0(x) &= \sin(\pi x_1)\sin(2\pi x_2), \\ f(x,t) &= \mathrm{e}^t(1+5(t+1)\pi^2)\sin(\pi x_1)\sin(2\pi x_2). \end{aligned}$$

Problem (19.14) is approximated by finite difference scheme (19.15). Since the input data are smooth functions, Steklov averaging is not applied.

The following table presents the values of error and convergence rate (CR), estimated using the double-mesh principle. Numerical experiment confirms the theoretical results (Table 19.2).

19.5 The Third Initial-Boundary Value Problem

As the model problem we consider, in $Q = \Omega \times (0,T) = (0,1)^2 \times (0,T)$, the following initial-boundary value problem for a parabolic equation with variable coefficients:

$$\frac{\partial u}{\partial t} + Lu = f, \quad (x,t) = (x_1, x_2, t) \in Q, \tag{19.29}$$

$$lu = 0, \quad (x,t) \in \Gamma \times (0,T) = \partial\Omega \times (0,T), \tag{19.30}$$

$$u(x,0) = u_0(x), \quad x \in \Omega, \tag{19.31}$$

where

$$Lu := -\sum_{i,j=1}^{2} \frac{\partial}{\partial x_i}\left(a_{ij}\,\frac{\partial u}{\partial x_j}\right),$$

$$lu := \sum_{i,j=1}^{2} a_{ij}\,\frac{\partial u}{\partial x_j}\,\cos(\nu, x_i) + \alpha u,$$

and ν is the unit outward normal to Γ. We assume that the conditions of strong ellipticity are satisfied:

$$a_{ij} = a_{ij}(x,t) = a_{ji}, \quad \alpha = \alpha(x),$$

$$c_0\sum_{i=1}^{2}\xi_i^2 \le \sum_{i,j=1}^{2} a_{ij}\xi_i\xi_j \le c_1\sum_{i=1}^{2}\xi_i^2, \forall x \in \bar{\Omega},\ \forall \xi \in \mathbb{R}^2,\ c_i = \text{const.} > 0, \tag{19.32}$$

$$\alpha_0 \le \alpha(x) \le \alpha_1, \quad \alpha_i = \text{const.} > 0.$$

Let us denote $\Gamma = \cup_{i=1}^{2} \cup_{k=0}^{1} \Gamma_{ik}$, where $\Gamma_{1k} = \{k\} \times [0,1]$, $\Gamma_{2k} = [0,1] \times \{k\}$, and $\Sigma_{ik} = \Gamma_{ik} \times [0,T]$.

We also assume that the generalized solution of problem that's one problem (19.29)–(19.31) belongs to the Sobolev space $W_2^{3,3/2}(Q)$, while the data satisfy the following smoothness conditions: $a_{ij} \in W_2^{2+\epsilon,1+\epsilon/2}(Q)$, $\alpha \in W_2^{3/2}(\Gamma_{ik})$, $\alpha \in C(\Gamma)$, $f \in W_2^{1,1/2}(Q)$ and $u_0 \in W_2^2(\Omega)$.

Finite Difference Approximation Let $n, m \in \mathbb{N}$, $n \ge 2$, $m \ge 1$, $h = 1/n$, and $\tau = T/m$. We consider the uniform spatial mesh $\bar{\omega}$ with mesh size h on $\bar{\Omega}$ and the uniform temporal mesh $\bar{\omega}_\tau$ with mesh size τ on $[0,T]$. We also denote $\omega = \bar{\omega} \cap \Omega$, $\omega_\tau = \bar{\omega}_\tau \cap (0,T)$, $\omega_\tau^- = \bar{\omega}_\tau \cap [0,T)$, $\omega_\tau^+ = \bar{\omega}_\tau \cap (0,T]$, $\gamma = \bar{\omega} \cap \Gamma$, $\bar{\gamma}_{ik} = \bar{\omega} \cap \Gamma_{ik}$, $\gamma_{ik} = \{x \in \bar{\gamma}_{ik} : 0 < x_{3-i} < 1\}$, $\gamma_{ik}^- = \{x \in \bar{\gamma}_{ik} : 0 \le x_{3-i} < 1\}$, $\gamma_{ik}^+ = \{x \in \bar{\gamma}_{ik} : 0 < x_{3-i} \le 1\}$, $\gamma_{ik}^\star = \bar{\gamma}_{ik} \setminus \gamma_{ik}$, $\gamma^\star = \gamma \setminus \{\cup_{i,k} \gamma_{ik}\}$, $\sigma_{ik} = \gamma_{ik} \times \omega_\tau^+$, $\bar{\sigma}_{ik} = \bar{\gamma}_{ik} \times \omega_\tau^+$, $i = 1,2$, $k = 0,1$, and $\bar{Q}_{h\tau} = \bar{\omega} \times \bar{\omega}_\tau$.

In addition to the standard Steklov averaging operators defined in Section 1.2, we also need operators

$$T_i^{2\pm} f(x,t) = 2\int_0^1 (1-x')f(x \pm hx'e_i)\,\mathrm{d}x', \quad i = 1,2. \tag{19.33}$$

We approximate the initial-boundary value problem (19.29)–(19.31) with the following implicit finite difference scheme:

$$v_{\bar t} + L_h v = \tilde f, \quad x \in \bar\omega, \quad t \in \omega_\tau^+, \tag{19.34}$$
$$v(x,0) = u_0(x), \quad x \in \bar\omega,$$

where

$$L_h v = \begin{cases} -\frac{1}{2}\sum\limits_{i,j=1}^{2}\Big[\big(a_{ij}\, v_{x_j}\big)_{\bar x_i} + \big(a_{ij}\, v_{\bar x_j}\big)_{x_i}\Big], & x \in \omega, \\ \frac{2}{h}\Big(-\frac{a_{11}+a_{11}^{+1}}{2}\, v_{x_1} - a_{12}\,\frac{v_{x_2}+v_{\bar x_2}}{2} + \tilde\alpha v\Big) - \big(a_{12}\, v_{\bar x_2}\big)_{x_1} & \\ \quad -\big(a_{21}\, v_{x_1}\big)_{\bar x_2} - \frac{1}{2}\big(a_{22}\, v_{x_2}\big)_{\bar x_2} - \frac{1}{2}\big(a_{22}\, v_{\bar x_2}\big)_{x_2}, & x \in \gamma_{10}, \\ \frac{2}{h}\Big[-\frac{a_{11}+a_{11}^{+1}}{2}\, v_{x_1} - a_{12} v_{x_2} - a_{21} v_{x_1} - \frac{a_{22}+a_{22}^{+2}}{2}\, v_{x_2} & \\ \quad +(\tilde\alpha_1+\tilde\alpha_2)v\Big], & x = (0,0), \\ \frac{2}{h}\Big[-\frac{a_{11}+a_{11}^{+1}}{2}\, v_{x_1} - a_{12} v_{\bar x_2} + a_{21} v_{x_1} + \frac{a_{22}+a_{22}^{-2}}{2}\, v_{\bar x_2} & \\ \quad +(\tilde\alpha_1+\tilde\alpha_2)v\Big] - 2\big(a_{12}\, v_{\bar x_2}\big)_{x_1} - 2\big(a_{21}\, v_{x_1}\big)_{\bar x_2}, & x = (0,1), \\ \text{and analogously at the other boundary nodes,} & x \in \gamma \setminus \bar\gamma_{10}, \end{cases}$$

$$\tilde f = \begin{cases} T_1^2 T_2^2 T_t^- f, & x \in \omega, \\ T_i^{2\pm} T_{3-i}^2 T_t^- f, & x \in \gamma_{i,\,0.5\mp 0.5}, \\ T_1^{2\pm} T_2^{2\pm} T_t^- f, & x = (0.5\mp 0.5,\; 0.5\mp 0.5), \in \gamma^\star \end{cases}$$

and

$$\tilde\alpha = T_{3-i}^2 \alpha\,, \quad x \in \gamma_{i0}\cup\gamma_{i1}\,, \quad i = 1,\, 2,$$
$$\tilde\alpha_i = T_i^{2\pm}\alpha\,, \quad x \in \gamma^\star, \quad x_i = 0.5\mp 0.5\,, \quad i = 1,\, 2\,.$$

Error Analysis Let u be the solution of the initial-boundary value problem (19.29)–(19.31), and let v denote the solution of the finite difference scheme (19.34). The error $z = u - v$ is defined on $\bar Q_{h\tau}$ and satisfies the following conditions:

$$z_{\bar t} + L_h z = \psi, \quad x \in \bar\omega, \quad t \in \omega_\tau^+,, \quad z(x,0) = 0, \quad x \in \bar\omega, \tag{19.35}$$

where

$$\psi = \begin{cases} \xi_{\bar{t}} + \sum\limits_{i,j=1}^{2} \eta_{ij,\bar{x}_i}, & x \in \omega, \\ \tilde{\xi}_{\bar{t}} + \frac{2}{h}\eta_{11} + \frac{2}{h}\eta_{12} + \tilde{\eta}_{21,\bar{x}_2} + \tilde{\eta}_{22,\bar{x}_2} + \frac{2}{h}\zeta, & x \in \gamma_{10}, \\ \tilde{\tilde{\xi}}_{\bar{t}} + \frac{2}{h}\tilde{\eta}_{11} + \frac{2}{h}\tilde{\eta}_{12} + \frac{2}{h}\tilde{\eta}_{21} + \frac{2}{h}\tilde{\eta}_{22} + \frac{2}{h}(\zeta_1 + \zeta_2), & x = (0,0), \\ \text{and analogously at the other boundary nodes,} & x \in \gamma \setminus \gamma_{10}^{-}, \end{cases}$$

$$\xi = u - T_1^2 T_2^2 u, \quad x \in \omega,$$

$$\tilde{\xi} = u - T_i^2 T_{3-i}^{2\pm} u, \quad x \in \gamma_{3-i,\,0.5\mp 0.5},$$

$$\tilde{\tilde{\xi}} = u - T_1^{2\pm} T_2^{2\pm} u, \quad x = (0.5 \mp 0.5,\ 0.5 \mp 0.5) \in \gamma^\star,$$

$$\eta_{ij} = T_i^+ T_{3-i}^2 T_t^- \Big(a_{ij}\frac{\partial u}{\partial x_j}\Big) - \frac{1}{2}\big(a_{ij} u_{x_j} + a_{ij}^{+i} u_{\bar{x}_j}^{+i}\big), \quad x \in \omega,$$

$$\tilde{\eta}_{ii} = T_i^+ T_{3-i}^{2\pm} T_t^- \Big(a_{ii}\frac{\partial u}{\partial x_i}\Big) - \frac{a_{ii} + a_{ii}^{+i}}{2} u_{x_i}, \quad x \in \gamma_{3-i,\,0.5\mp 0.5}^{-},$$

$$\tilde{\eta}_{i,3-i} = \begin{cases} T_i^+ T_{3-i}^{2+} T_t^- \Big(a_{i,3-i}\frac{\partial u}{\partial x_{3-i}}\Big) - a_{i,3-i}\, u_{x_{3-i}}, & x \in \gamma_{3-i,0}^{-}, \\ T_i^+ T_{3-i}^{2-} T_t^- \Big(a_{i,3-i}\frac{\partial u}{\partial x_{3-i}}\Big) - a_{i,3-i}^{+i}\, u_{\bar{x}_{3-i}}^{+i}, & x \in \gamma_{3-i,1}^{-}, \end{cases}$$

$$\zeta = (T_i^2 \alpha) u - T_i^2 T_t^- (\alpha u), \quad x \in \gamma_{3-i,0} \cup \gamma_{3-i,1},$$

$$\zeta_i = (T_i^{2\pm}\alpha) u - T_i^{2\pm} T_t^- (\alpha u), \quad x \in \gamma^\star, \quad x_i = 0.5 \mp 0.5.$$

We define the following discrete inner products and norms:

$$[v, w] = h^2 \sum_{x\in\omega} v(x)w(x) + \frac{h^2}{2} \sum_{x\in\gamma\setminus\gamma^\star} v(x)w(x) + \frac{h^2}{4} \sum_{x\in\gamma^\star} v(x)w(x),$$

$$|[v]|^2 = [v, v],$$

$$[v, w)_i = h^2 \sum_{x\in\omega\cup\gamma_{i0}} v(x)w(x) + \frac{h^2}{2} \sum_{x\in\gamma_{3-i,0}^{-}\cup\gamma_{3-i,1}^{-}} v(x)w(x),$$

$$|[v\|_i^2 = [v, v)_i,$$

$$(v, w]_i = h^2 \sum_{x \in \omega \cup \gamma_{i1}} v(x)w(x) + \frac{h^2}{2} \sum_{x \in \gamma_{3-i,0}^+ \cup \gamma_{3-i,1}^+} v(x)w(x),$$

$$\|v]|_i^2 = (v, v]_i, \quad [v, w) = h^2 \sum_{x \in \omega \cup \gamma_{10}^- \cup \gamma_{20}^-} v(x)w(x), \quad |[v\|^2 = [v, v),$$

$$(v, w] = h^2 \sum_{x \in \omega \cup \gamma_{11}^+ \cup \gamma_{21}^+} v(x)w(x), \quad \|v]|^2 = (v, v],$$

$$|[v]|^2_{W_2^1(\bar{\omega})} = |[v]|^2 + |[v_{x_1}|]_1^2 + |[v_{x_2}\|_2^2, \quad \|[v]|_{C(\bar{\omega})} = \max_{x \in \bar{\omega}} |v(x)|,$$

$$[v, w]_{\bar{\gamma}_{ik}} = h \sum_{x \in \gamma_{ik}} v(x)w(x) + \frac{h}{2} \sum_{x \in \gamma_{ik}^\star} v(x)w(x), \quad |[v]|^2_{\bar{\gamma}_{ik}} = [v, v]_{\bar{\gamma}_{ik}},$$

$$[v, w)_{\gamma_{ik}^-} = h \sum_{x \in \gamma_{ik}^-} v(x)w(x), \quad |[v||^2_{\gamma_{ik}^-} = [v, v)_{\gamma_{ik}^-},$$

$$\|v\|^2_{\gamma_{ik}} = h \sum_{x \in \gamma_{ik}} v^2(x),$$

$$|v|^2_{W_2^{1/2}(\gamma_{ik}^-)} = h^2 \sum_{x, x' \in \gamma_{ik}^-,\ x' \neq x} \left[\frac{v(x) - v(x')}{x_{3-i} - x'_{3-i}} \right]^2,$$

$$|[v\|^2_{W_2^{1/2}(\gamma_{ik}^-)} = |v|^2_{W_2^{1/2}(\gamma_{ik}^-)} + |[v\|^2_{\gamma_{ik}^-},$$

$$|[v\|^2_{\ddot{W}_2^{1/2}(\gamma_{ik}^-)} = |v|^2_{W_2^{1/2}(\gamma_{ik}^-)} + h \sum_{x \in \gamma_{ik}^-} \left(\frac{1}{x_{3-i} + h/2} + \frac{1}{1 - x_{3-i} - h/2} \right) v^2(x),$$

$$\|v\|^2_\tau = \tau \sum_{t \in \omega_\tau^+} v^2(t), \quad |[v\|^2_{i,h\tau} = \tau \sum_{t \in \omega_\tau^+} |[v(\cdot, t)\|_i^2,$$

$$\|v\|^2_{\sigma_{ik}} = \tau \sum_{t \in \omega_\tau^+} \|v(\cdot, t)\|^2_{\gamma_{ik}}, \quad |[v]|^2_{\bar{\sigma}_{ik}} = \tau \sum_{t \in \omega_\tau^+} |[v(\cdot, t)]|^2_{\bar{\gamma}_{ik}},$$

$$|[v\|^2_{L_2(\omega_\tau^+,\, \ddot{W}_2^{1/2}(\gamma_{ik}^-))} = \tau \sum_{t \in \omega_\tau^+} |[v(\cdot, t)\|^2_{\ddot{W}_2^{1/2}(\gamma_{ik}^-)},$$

$$|v|^2_{W_2^{1/2}(\bar{\omega}_\tau, L_2(\bar{\omega}))} = \tau^2 \sum_{t, t' \in \bar{\omega}_\tau,\ t' \neq t} \frac{|[v(\cdot, t) - v(\cdot, t')]|^2}{(t - t')^2},$$

$$|[v]|^2_{\ddot{W}_2^{1/2}(\bar{\omega}_\tau, L_2(\bar{\omega}))} = |v|^2_{W_2^{1/2}(\bar{\omega}_\tau, L_2(\bar{\omega}))}$$

$$+\tau \sum_{t\in\bar{\omega}_\tau} \left(\frac{1}{t+\tau/2} + \frac{1}{T-t+\tau/2}\right) \|[v(\cdot,t)]\|^2,$$

$$\|[v]\|^2_{L_2(\omega_\tau^+,\, W_2^1(\bar{\omega}))} = \tau \sum_{t\in\omega_\tau^+} \|[v(\cdot,t)]\|^2_{W_2^1(\bar{\omega})},$$

$$\|[v]\|^2_{W_2^{1,\,1/2}(Q_{h\tau})} = \|[v]\|^2_{L_2(\omega_\tau^+,\, W_2^1(\bar{\omega}))} + |v|^2_{W_2^{1/2}(\bar{\omega}_\tau, L_2(\bar{\omega}))}.$$

We shall prove a suitable a priori estimate for the finite difference scheme (19.35) which will be used to estimate its convergence rate.

The following proposition holds (see [8]).

Lemma 19.1 *Let $w \in W_2^r(\Gamma_{ik})$, $0 < r \le 0.5$. Then,*

$$|T_{3-i}^+ w|_{W_2^{1/2}(\gamma_{ik}^-)} \le C(r) h^{r-1/2} |w|_{W_2^r(\Gamma_{ik})}.$$

Let us rearrange the summands in truncation error ψ in the following manner:

$$\tilde{\eta}_{ij} = \eta_{ij} + \eta'_{ij}, \quad \tilde{\xi} = \xi + \xi', \quad \tilde{\tilde{\xi}} = \xi + \xi^\star,$$

where

$$\eta'_{ii} = \pm\frac{h}{3}\, T_i^+ T_t^- \left(\frac{\partial}{\partial x_{3-i}}\left(a_{ii}\,\frac{\partial u}{\partial x_i}\right)\right), \qquad x \in \gamma_{3-i,\,0.5\mp0.5},$$

$$\eta'_{i,3-i} = \pm\frac{h}{3}\, T_i^+ T_t^- \left(\frac{\partial}{\partial x_{3-i}}\left(a_{i,3-i}\,\frac{\partial u}{\partial x_{3-i}}\right)\right) \mp \frac{h}{2}\, T_i^+ T_t^- \left(a_{i,3-i}\,\frac{\partial^2 u}{\partial x_{3-i}^2}\right)$$
$$+\frac{h}{2}\, T_i^+ T_t^- \left(\frac{\partial}{\partial x_i}\left(a_{i,3-i}\,\frac{\partial u}{\partial x_{3-i}}\right)\right), \quad x \in \gamma_{3-i,\,0.5\mp0.5},$$

$$\xi' = \mp\frac{h}{3}\, T_{3-i}^2\left(\frac{\partial u}{\partial x_i}\right), \quad x \in \gamma_{i,\,0.5\mp0.5},$$

$$\xi^\star = \mp\frac{h}{3}\, T_2^{2\pm}\left(\frac{\partial u}{\partial x_1}\right) \mp \frac{h}{3}\, T_1^{2\pm}\left(\frac{\partial u}{\partial x_2}\right), \quad x = (0.5\mp0.5,\ 0.5\mp0.5) \in \gamma^\star.$$

Using the boundary condition (19.30) we further obtain

$$\xi'_{\bar{t}} = \lambda_{i,\,\bar{x}_{3-i}} + \mu_i + \nu_i, \quad x \in \gamma_{i,\,0.5\mp0.5},$$

where

$$\lambda_i = \pm\frac{h}{3}\, T_{3-i}^+ T_t^- \left(\frac{a_{i,3-i}}{a_{ii}}\,\frac{\partial u}{\partial t}\right),$$

$$\mu_i = \mp\frac{h}{3}\, T_{3-i}^2 T_t^- \left(\frac{\partial}{\partial x_{3-i}} \left(\frac{a_{i,3-i}}{a_{ii}} \right) \frac{\partial u}{\partial t} \right),$$

$$\nu_i = -\frac{h}{3}\, T_{3-i}^2 T_t^- \left(\frac{\alpha}{a_{ii}} \frac{\partial u}{\partial t} \right).$$

Similarly, for $x = (0, 0)$, we obtain

$$\xi_t^\star = \frac{2}{h}\lambda_1 - \frac{2}{h}\lambda_1^\star + \mu_1 + \nu_1 + \frac{2}{h}\lambda_2 - \frac{2}{h}\lambda_2^\star + \mu_2 + \nu_2,$$

where λ_1 and λ_2 are the same as before and

$$\lambda_i^\star = \pm\frac{h}{3}\, T_t^- \left(\frac{a_{i,3-i}}{a_{ii}} \frac{\partial u}{\partial t} \right),$$

$$\mu_i = -\frac{h}{3}\, T_{3-i}^{2+} T_t^- \left(\frac{\partial}{\partial x_{3-i}} \left(\frac{a_{i,3-i}}{a_{ii}} \right) \frac{\partial u}{\partial t} \right),$$

$$\nu_i = -\frac{h}{3}\, T_{3-i}^{2+} T_t^- \left(\frac{\alpha}{a_{ii}} \frac{\partial u}{\partial t} \right),$$

with an analogous representation at other nodes from $\gamma^\star$.

The following a priori estimate is valid (see [8]).

Theorem 19.4 *The finite difference scheme* (19.35) *is stable in the sense of the a priori estimate:*

$$\begin{aligned}
|[z]|_{W_2^{1,1/2}(Q_{h\tau})} \le\ & C\Bigg\{ |[\xi]|_{\ddot{W}_2^{1/2}(\bar\omega_\tau,\, L_2(\bar\omega))} + \sum_{i,j=1}^{2} |[\eta_{ij}\|_{i,h\tau} + \sum_{k=0}^{1}\sum_{i=1}^{2} \|\zeta\|_{\sigma_{ik}} \\
& + h \sum_{k=0}^{1}\sum_{i,j=1}^{2} |[\eta'_{ij}\|_{L_2(\omega_\tau^+,\, \ddot{W}_2^{1/2}(\gamma_{3-i,k}^-))} \\
& + h \sum_{k=0}^{1}\sum_{i=1}^{2} |[\lambda_i\|_{L_2(\omega_\tau^+,\, \ddot{W}_2^{1/2}(\gamma_{ik}^-))} \\
& + h \sum_{k=0}^{1}\sum_{i=1}^{2} \Big(|[\mu_i]|_{\bar\sigma_{ik}} + |[\nu_i]|_{\bar\sigma_{ik}} \Big) \\
& + h\sqrt{\log\tfrac{1}{h}} \sum_{i=1}^{2}\sum_{x\in\gamma^\star} \Big(\|\zeta_i(x,\cdot)\|_\tau + \|\lambda_i^\star(x,\cdot)\|_\tau \Big) \Bigg\}.
\end{aligned} \tag{19.36}$$

In accordance with Theorem 19.4, the problem of deriving the convergence rate estimate for the finite difference scheme (19.34) is reduced to estimating the right-hand side terms in inequality (19.36).

Let us assume that $\tau \asymp h^2$, i.e., $c_2h^2 \le \tau \le c_3h^2$ for some positive constants c_2 and c_3.

By a technique based on the Bramble-Hilbert lemma we estimate the right-hand side terms in inequality (19.36).

Further, we have

Theorem 19.5 *Let the assumptions from the beginning of Sect. 19.5 hold and let $\tau \asymp h^2$. Then the finite difference scheme* (19.34) *converges in the norm $W_2^{1,1/2}(Q_{h\tau})$ and the convergence rate estimates*

$$\|[z]\|_{W_2^{1,1/2}(Q_{h\tau})} \le Ch^2 \left(\log \tfrac{1}{h}\right)^{\frac{3}{2}} \left(1 + \max_{i,j} \|a_{ij}\|_{W_2^{2+\varepsilon,\,1+\varepsilon/2}(Q)} + \max_{i,k} \|\alpha\|_{W_2^{3/2}(\Gamma_{ik})}\right) \|u\|_{W_2^{3,3/2}(Q)}$$

hold.

Remark 19.1 Note that the obtained error bounds are "almost" compatible with the smoothness of the data (up to additional logarithmic factors).

Remark 19.2 The obtained error bounds have been derived under the assumption $\tau \asymp h^2$, which links the temporal mesh size τ to the spatial mesh size h, despite the fact that the implicit finite difference scheme is unconditionally stable, and therefore, from the point of view of stability, there should be no limitation on the choice of mesh sizes. This limitation can be avoided in certain cases by careful study of truncation error functionals (see, e.g., [11]).

Remark 19.3 Assuming that the generalized solution of the problem belongs to the Sobolev space $W_2^{s,s/2}$, $2.5 < s \le 3$, the estimate of the convergence rate of the finite difference schemes for parabolic problem with variable coefficients (but not time-dependent) is obtained (see [8]).

Numerical Example The test example in problem (19.29)–(19.31), with $a_{ij}(x,t) = t + 1$, $T = 0.001$, $Lu = -(t+1)\Delta u$, where $\alpha(x)$, is defined in such a manner that u,

$$u(x_1, x_2, t) = (x_1 - 1/4)(9/4 - x_1)(x_2 + 1/4)(7/4 - x_2)\mathrm{e}^{-t}$$

is the exact solution of problem (19.29)–(19.31). Hence,

$$\alpha(0, x_2) = -\frac{40}{9}, \quad \alpha(x_1, 0) = \frac{24}{7}, \quad \alpha(x_1, 1) = \frac{8}{15}, \quad \alpha(1, x_2) = -\frac{8}{15},$$
$$u_0(x_1, x_2) = (-x_1 - 1/4)(9/4 + x_1)(x_2 + 1/4)(7/4 - x_2),$$

Table 19.3 Error and convergence rate (CR)

Mesh		
h	τ	Error (CR)
0.125,	0.00004	1.1059e-08
0.0625,	0.00001	2.0449e-09 (2.4351)
0.03125,	0.0000025	5.1073e-10 (2.0014)
0.015625,	0.000000625	1.2768e-10 (2.0000)

$$\begin{aligned} f(x_1, x_2, t) = & \; 2(-x_1 - 1/4)(9/4 + x_1)(1 + t)\mathrm{e}^{-t} \\ & +2(x_2 + 1/4)(7/4 - x_2)(1 + t)\mathrm{e}^{-t} \\ & -(-x_1 - 1/4)(9/4 + x_1)(x_2 + 1/4)(7/4 - x_2)\mathrm{e}^{-t}. \end{aligned}$$

Problem (19.29)–(19.31) are approximated by the finite difference scheme (19.34). Since the input data are smooth functions, Steklov averaging is not applied.

The following table presents values of error and convergence rate (CR), estimated using the double-mesh principle. Numerical experiment confirms the theoretical results (Table 19.3).

References

1. Bojović, D.R., Jovanović, B.S.: Convergence of finite difference method for the parabolic problem with concentrated capacity and variable operator. J. Comp. Appl. Math. **189**, 286–303 (2006)
2. Bojović, D.R., Jovanović, B.S.: Finite difference method for a parabolic problem with concentrated capacity and time-dependent operator. In: Approximation and Computation. Series: Springer Optimization and its Applications, pp. 286–296 (2011)
3. Bojović, D.R., Jovanović, B.S.: Convergence of a finite difference method for solving 2D parabolic interface problems. J. Comp. Appl. Math. **236**, 3605–3612 (2012)
4. Bojović, D.R., Sredojević, B.V., Jovanović, B.S.: Numerical approximation of a two-dimensional parabolic time-dependent problem containing a delta function. J. Comp. Appl. Math. **259**, 129–137 (2014)
5. Grisvard, P.: Singularities in Boundary Value Problems. In: Research in Applied Mathematics, vol. 22. Masson, Paris (1992)
6. Jovanović, B.S.: On the convergence of finite-difference schemes for parabolic equations with variable coefficients. Numer. Math. **54**, 395–404 (1989)
7. Jovanović, B.S.: Finite difference method for boundary value problems with weak solutions. In: Posebna izdanja Matematičkog Instituta 16, Beograd (1993)
8. Jovanović, B.S., Milovanović, Z.D.: Finite difference approximation of a parabolic problem with variable coefficients. Publ. Inst. Math. **95**(109), 49–62 (2014)
9. Jovanović, B.S., Süli, E.: Analysis of Finite Difference Schemes. In: Springer Series in Computational Mathematics (2014)
10. Jovanović, B.S., Vulkov, L.G.: Operator approach to the problems with concentrated factors. In: Numerical analysis and its applications (Rousse, 2000). Lecture Notes in Computer Science 1988, pp. 439–450. Springer, Berlin (2001)
11. Jovanović, B.S., Vulkov, L.G.: Finite difference approximation of strong solutions of a parabolic interface problem on disconnected domains. Publ. Inst. Math. **84**(98), 37–48 (2008)

12. Jovanović, B.S., Vulkov, L.G.: On the convergence of difference schemes for hyperbolic problems with concentrated data. SIAM J. Numer. Anal. **41**(2), 516–538 (2003)
13. Jovanović, B.S., Vulkov, L.G.: On the convergence of finite difference schemes for the heat equation with concentrated capacity. Numer. Math. **89**(4), 715–734 (2001)
14. Jovanović, B.S., Vulkov, L.G.: Finite difference approximation for some interface problems with variable coefficients. Appl. Numer. Math. **59**, 349–372 (2009)
15. Lazarov, R.D., Makarov, V.L., Samarskii, A.A.: Applications of exact difference schemes for construction and studies of difference schemes on generalized solutions. Math. Sbornik **117**, 469–480 (1982) (Russian)
16. Lykov, A.V.: Heat and Mass Transfer. Nauka, Moscow (1989) (Russian)
17. Samarskii, A.A.: Theory of Difference Schemes. Nauka, Moscow (1989) (Russian; English edition: Pure and Appl. Math., vol. 240. Marcel Dekker, New York 2001)
18. Samarskii, A.A., Lazarov, R.D., Makarov, V.L.: Difference Schemes for Differential Equations with Generalized Solutions. Vysshaya Shkola, Moscow (1987) (Russian)
19. Sredojević, B.V., Bojović, D.R.: Finite difference approximation for parabolic interface problem with time-dependent coefficients. Publ. Inst. Math. **99**(113), 67–76 (2016)
20. Sredojević, B.S., Milovanović Jeknić, Z.D., Bojović, D.R.: On the estimate of the convergence rate of the finite difference scheme for the approximation the third initial-boundary-value parabolic problem. ETNA (2023) (in publication)
21. Sredojević, B.V., Bojović, D.R.: Fractional order convergence rate estimate of finite-difference method for the heat equation with concentrated capacity. Filomat **35**(1), 331–338 (2021)
22. Sredojević, B.V., Bojović, D.R.: Finite difference approximation for the 2D heat equation with concentrated capacity. Filomat **32**(20), 6979–6987 (2018)
23. Vladimirov, V.S.: Generalized Functions in Mathematical Physics. (English ed.) Mir Publishers, Moscow (1979)
24. Wloka, J.: Partial Differential Equations. Cambridge University, Cambridge (1987)

Chapter 20
Quadrature Rules for Trigonometric Polynomials

Tatjana V. Tomović Mladenović and Marija P. Stanić

Dedicated to Academician Gradimir V. Milovanović on the occasion of his 75th birthday

20.1 Introduction

First, results about quadrature rules of Gaussian type with maximal even trigonometric degree of exactness are given in [50]. Turetzkii's approach is similar to the approach which is used in the construction of the famous Gaussian quadrature rule with maximal algebraic degree of exactness, i.e., nodes of the quadrature rule are zeros of the corresponding orthogonal polynomial. Results given in this chapter are based on Turetzkii's approach. For some different approaches one can see [4, 5, 10, 13, 16, 17, 19, 20, 30, 31, 48, 49].

Let us denote by $\mathcal{T}_m^\gamma$ the linear space $\mathcal{L}\{\sin(k+\gamma)x, \cos(k+\gamma)x \mid k = 0, 1, \dots, m\}$, where m is nonnegative integer and $\gamma \in \{0, 1/2\}$. Thus, $\mathcal{T}_m = \mathcal{T}_m^0$ is the linear space of all trigonometric polynomials of degree less than or equal to m, and $\mathcal{T}_m^{1/2}$ is the linear space of all trigonometric polynomials of semi-integer degree less than or equal to $m + 1/2$. Trigonometric polynomials of semi-integer

The authors were supported in part by the Serbian Ministry of Science, Technological Development, and Innovations, Serbia, contract numbers 451-03-136/2025-03/ 200122.

T. V. Tomović Mladenović (✉) · M. P. Stanić
Department of Mathematics and Informatics, Faculty of Science, University of Kragujevac, Kragujevac, Serbia
e-mail: tatjana.tomovic@pmf.kg.ac.rs; marija.stanic@pmf.kg.ac.rs

M. P. Stanić et al. (eds.), *Analysis, Approximation, Optimization: Computation and Applications*, Springer Optimization and Its Applications 224,
https://doi.org/10.1007/978-3-031-85743-0_20

degree $n + 1/2$ and integer degree n are of the following form:

$$T_n^{1/2}(x) = \sum_{k=0}^{n} \left(p_k^{(n),1/2} \cos\left(k + \frac{1}{2}\right) x + q_k^{(n),1/2} \sin\left(k + \frac{1}{2}\right) x \right), \tag{20.1}$$

$$T_n(x) = 1 + \sum_{k=1}^{n} \left(p_k^{(n),0} \cos kx + q_k^{(n),0} \sin kx \right),$$

respectively, where $p_k^{(n),\gamma}, q_k^{(n),\gamma} \in \mathbb{R}$, $\left|p_n^{(n),\gamma}\right| + \left|q_n^{(n),\gamma}\right| \neq 0$, $\gamma \in \{0, 1/2\}$, and $p_n^{(n),\gamma}$, and $q_n^{(n),\gamma}$ are called the leading coefficients.

Every trigonometric polynomial T_n^{γ} can be represented in the form

$$T_n^{\gamma}(x) = \prod_{k=1}^{2(n+\gamma)} \sin \frac{x - x_k}{2}. \tag{20.2}$$

It is obvious that every T_n^{γ} given by (20.2) is a trigonometric polynomial of degree $n + \gamma$.

Conversely, a function T_n^{γ} has exactly $2(n + \gamma)$ zeros provided that we count them as usual with their multiplicities, with restriction to the strip $-\pi \leq \Re z < \pi$. Furthermore, the non-real zeros appear in conjugate pairs (see [6]).

Let us emphasize that when we say trigonometric polynomial of degree $n + \gamma$, $n \in \mathbb{N}$, $\gamma \in \{0, 1/2\}$, it means trigonometric polynomial of semi-integer degree (for $\gamma = 1/2$), or trigonometric polynomials of integer degree (for $\gamma = 0$).

In [50] quadrature rules with maximal trigonometric degree of exactness and an odd number of nodes were considered on interval $[0, 2\pi)$ and in [22], this was generalized to any interval $E = [a, a + 2\pi)$, $a \in \mathbb{R}$, of length 2π. Also, a case with an even number of nodes was considered in [5]. In the sequel of this chapter we will consider interval $[-\pi, \pi)$.

Let $n \in \mathbb{N}$ and $\gamma \in \{0, 1/2\}$. For an integrable and nonnegative weight function w on the interval $[-\pi, \pi)$, vanishing there only on a set of a measure zero, the quadrature rule

$$\int_{-\pi}^{\pi} f(x)w(x)\,dx = \sum_{i=1}^{2(n+\gamma)} A_i f(x_i) + R_n(f), \tag{20.3}$$

where $-\pi \leq x_1 < x_2 < \cdots < x_n < \pi$, has trigonometric degree of exactness equal to d if $R_n(f) = 0$ for all $f \in \mathcal{T}_d$ and there exists $g \in \mathcal{T}_{d+1}$ such that $R_n(g) \neq 0$ (see [22, 50]). We are interested in quadrature rules with maximal trigonometric degree of exactness.

The chapter is organized as follows. In Sect. 20.2 we consider quadrature rules with maximal trigonometric degree of exactness, i.e., quadrature rules of Gaussian type for trigonometric polynomials. In Sect. 20.3 results about anti-

Gaussian quadrature rules for trigonometric polynomials are given. Section 20.4 is devoted to the optimal set of quadrature rules with an odd number of nodes for trigonometric polynomials in the sense of Borges. In Sect. 20.5 we consider quadrature rules for trigonometric polynomials with an even number of preassigned nodes. Finally, in Sect. 20.6 numerical examples are included.

20.2 Quadrature Rules of Gaussian Type for Trigonometric Polynomials

One way of generalization of the famous Gaussian quadrature rule for algebraic polynomials is quadrature rule with maximal degree of exactness in some other linear space, for example, in space of trigonometric polynomials. Such quadrature rules are important in the numerical integration of 2π periodic functions.

The maximal trigonometric degree of exactness of quadrature rules (20.3) with n nodes is $n-1$. Such quadrature rules are known as quadrature rules of Gaussian type for trigonometric polynomials. Since $\dim(\mathcal{T}_{2n+2\gamma-1}) = 2(2n+2\gamma-1)+1 = 4n+4\gamma-1$, from the conditions of maximal degree of exactness of quadrature rule (20.3), we have a nonlinear system of $4n+4\gamma-1$ equations for $4n+4\gamma$ unknowns $x_1, x_2, \ldots x_{2n+2\gamma}, A_1, A_2, \ldots, A_{2n+2\gamma}$, so one unknown should be fixed in advance. In order to simulate the development of Gaussian quadrature rules for algebraic polynomials, where nodes are not fixed in advance, but they are the zeros of the corresponding orthogonal polynomials, one has to consider quadrature rules with odd or even number of nodes. Namely, if quadrature rule has an odd or even number of nodes, then nodes are zeros of the corresponding orthogonal trigonometric polynomial of semi-integer degree or integer degree, respectively.

Some basic properties about orthogonal trigonometric polynomials of semi-integer and integer degree which are important for the quadrature rules for trigonometric polynomials will be given in the sequel, and a detailed study can be found in [8, 9, 22–25, 50] for the semi-integer case and [5, 6, 43] for the integer case.

For the given weight function w, on $[-\pi,\pi)$, we introduce the following inner product:

$$\langle f, g\rangle_w = I(fg), \tag{20.4}$$

where

$$I(f) = \int_{-\pi}^{\pi} f(x)w(x)\,\mathrm{d}x. \tag{20.5}$$

Definition 20.1 The trigonometric polynomial $T_n^\gamma \in \mathcal{T}_n^\gamma$, $n \in \mathbb{N}$, $\gamma \in \{0, 1/2\}$, is orthogonal trigonometric polynomial with respect to the inner product (20.4), i.e., with respect to the weight function w on interval $[-\pi,\pi)$, if and only if $\langle T_n^\gamma, T_k^\gamma\rangle = 0$ for $k = 0, 1, \ldots, n-1$.

Since $\dim(\mathcal{T}_k^\gamma) = 2(k+\gamma)+1$, $k = 0, 1, \ldots, n-1$, $\gamma \in \{0, 1/2\}$, from the Definition 20.1 we have $2(n+\gamma)-1$ conditions for the $2(n+\gamma)+1$ unknown coefficients $p_k^{(n),\gamma}$, $q_k^{(n),\gamma}$ of the polynomials (20.1), so two coefficients have to be fixed in advance, as it is stated in the following theorem.

Theorem 20.1 *Trigonometric polynomial $T_n^\gamma \in \mathcal{T}_n^\gamma$, $n \in \mathbb{N}$, $\gamma \in \{0, 1/2\}$ which is orthogonal to all trigonometric polynomials from $\mathcal{T}_{n-1}^\gamma$ with respect to the weight function w on interval $[-\pi, \pi)$ is unique if two coefficients are fixed in advance.*

We choose to fix the leading coefficients $p_n^{(n),\gamma}$ and $q_n^{(n),\gamma}$, in (20.1), with the restriction $\left|p_n^{(n),\gamma}\right| + \left|q_n^{(n),\gamma}\right| \neq 0$. Specially, for the selection $(p_n^{(n),\gamma}, q_n^{(n),\gamma}) = (1, 0)$ and $(p_n^{(n),\gamma}, q_n^{(n),\gamma}) = (0, 1)$, we use the notation $T_n^{C,\gamma}$ and $T_n^{S,\gamma}$, respectively, i.e.,

$$T_n^{C,\gamma}(x) = \cos(n+\gamma)x + \sum_{k=0}^{n-1}\left(a_k^{(n),\gamma}\cos(k+\gamma)x + b_k^{(n),\gamma}\sin(k+\gamma)x\right),$$
$$T_n^{S,\gamma}(x) = \sin(n+\gamma)x + \sum_{k=0}^{n-1}\left(c_k^{(n),\gamma}\cos(k+\gamma)x + d_k^{(n),\gamma}\sin(k+\gamma)x\right). \tag{20.6}$$

Orthogonal trigonometric polynomials $T_n^{C,\gamma}$ and $T_n^{S,\gamma}$ satisfy five-term recurrence relations which is important for their construction.

Let us introduce the following notation:

$$I_k^{C,\gamma} = \left\langle T_k^{C,\gamma}, T_k^{C,\gamma}\right\rangle_w, \quad I_k^{S,\gamma} = \left\langle T_k^{S,\gamma}, T_k^{S,\gamma}\right\rangle_w, \quad I_k^{\gamma} = \left\langle T_k^{C,\gamma}, T_k^{S,\gamma}\right\rangle_w, \tag{20.7}$$
$$J_{k,\ell}^{C,\gamma} = \left\langle 2\cos x\, T_k^{C,\gamma}, T_\ell^{C,\gamma}\right\rangle_w, \quad J_{k,\ell}^{S,\gamma} = \left\langle 2\cos x\, T_k^{S,\gamma}, T_\ell^{S,\gamma}\right\rangle_w,$$
$$I_{k,\ell}^{\gamma} = \left\langle 2\cos x\, T_k^{C,\gamma}, T_\ell^{S,\gamma}\right\rangle_w.$$

Theorem 20.2 *Trigonometric polynomials $T_k^{C,\gamma}$ and $T_k^{S,\gamma}$, $k \geq 1$, $\gamma \in \{0, 1/2\}$, orthogonal with respect to the inner product* (20.4), *satisfy the following recurrence relations:*

$$T_k^{C,\gamma}(x) = \left(2\cos x - \alpha_k^{(1),\gamma}\right)T_{k-1}^{C,\gamma}(x) - \beta_k^{(1),\gamma}T_{k-1}^{S,\gamma}(x) - \alpha_k^{(2),\gamma}T_{k-2}^{C,\gamma}(x)$$
$$-\beta_k^{(2),\gamma}T_{k-2}^{S,\gamma}(x),$$
$$T_k^{S,\gamma}(x) = \left(2\cos x - \eta_k^{(1),\gamma}\right)T_{k-1}^{S,\gamma}(x) - \delta_k^{(1),\gamma}T_{k-1}^{C,\gamma}(x) - \eta_k^{(2),\gamma}T_{k-2}^{S,\gamma}(x)$$
$$-\delta_k^{(2),\gamma}T_{k-2}^{C,\gamma}(x),$$

where the coefficients $\alpha_k^{(j),\gamma}$, $\beta_k^{(j),\gamma}$, $\eta_k^{(j),\gamma}$, $\delta_k^{(j),\gamma}$, $k \geq 1$, $j \in 1, 2$, *are given by*

$$\alpha_k^{(j),\gamma} I_{k-j}^{C,\gamma} + \beta_k^{(j),\gamma} I_{k-j}^{\gamma} = J_{k-1,k-j}^{C,\gamma}, \quad \alpha_k^{(j),\gamma} I_{k-j}^{\gamma} + \beta_k^{(j),\gamma} I_{k-j}^{S,\gamma} = J_{k-1,k-j}^{\gamma}, \tag{20.8}$$

$$\eta_k^{(j),\gamma} I_{k-j}^{C,\gamma} + \delta_k^{(j),\gamma} I_{k-j}^{\gamma} = J_{k-1,k-j}^{\gamma}, \quad \eta_k^{(j),\gamma} I_{k-j}^{\gamma} + \delta_k^{(j),\gamma} I_{k-j}^{S,\gamma} = J_{k-1,k-j}^{S,\gamma},$$

with $\alpha_1^{(2),\gamma} = \beta_1^{(2),\gamma} = \eta_1^{(2),\gamma} = \delta_1^{(2),\gamma} = 0.$

It is easy to prove that $I_k^{C,\gamma} = J_{k,k-1}^{C,\gamma}$, $I_k^{S,\gamma} = J_{k,k-1}^{S,\gamma}$, $I_k^{\gamma} = J_{k,k-1}^{\gamma}$, $k \geq 1$. Using the notation $J_k^{C,\gamma} = J_{k,k}^{C,\gamma}$, $J_k^{S,\gamma} = J_{k,k}^{S,\gamma}$, and $J_k^{\gamma} = J_{k,k}^{\gamma}$, from (20.8) we obtain the recurrence coefficients:

$$\alpha_k^{(1),\gamma} = \frac{I_{k-1}^{S,\gamma} J_{k-1}^{C,\gamma} - I_{k-1}^{\gamma} J_{k-1}^{\gamma}}{D_{k-1}^{\gamma}}, \quad \alpha_k^{(2),\gamma} = \frac{I_{k-1}^{C,\gamma} I_{k-2}^{S,\gamma} - I_{k-1}^{\gamma} I_{k-2}^{\gamma}}{D_{k-2}^{\gamma}},$$

$$\beta_k^{(1),\gamma} = \frac{I_{k-1}^{C,\gamma} J_{k-1}^{\gamma} - I_{k-1}^{\gamma} J_{k-1}^{C,\gamma}}{D_{k-1}^{\gamma}}, \quad \beta_k^{(2),\gamma} = \frac{I_{k-1}^{\gamma} I_{k-2}^{C,\gamma} - I_{k-1}^{C,\gamma} I_{k-2}^{\gamma}}{D_{k-2}^{\gamma}},$$

$$\eta_k^{(1),\gamma} = \frac{I_{k-1}^{S,\gamma} J_{k-1}^{\gamma} - I_{k-1}^{\gamma} J_{k-1}^{S,\gamma}}{D_{k-1}^{\gamma}}, \quad \eta_k^{(2),\gamma} = \frac{I_{k-1}^{\gamma} I_{k-2}^{S,\gamma} - I_{k-1}^{S,\gamma} I_{k-2}^{\gamma}}{D_{k-2}^{\gamma}},$$

$$\delta_k^{(1),\gamma} = \frac{I_{k-1}^{C,\gamma} J_{k-1}^{S,\gamma} - I_{k-1}^{\gamma} J_{k-1}^{\gamma}}{D_{k-1}^{\gamma}}, \quad \delta_k^{(2),\gamma} = \frac{I_{k-1}^{S,\gamma} I_{k-2}^{C,\gamma} - I_{k-1}^{\gamma} I_{k-2}^{\gamma}}{D_{k-2}^{\gamma}},$$

where $D_{k-j}^{\gamma} = I_{k-j}^{C,\gamma} I_{k-j}^{S,\gamma} - (I_{k-j}^{\gamma})^2$, $j = 1, 2$, for $k \geq 1$.

Next, the important property of the zeros of orthogonal trigonometric polynomials is given in the following theorem (for the proof see [50, Theorem 3] for the case $\gamma = 1/2$ and [5, Corollary 4.12] for the case $\gamma = 0$). This property is very important for the construction of quadrature rules of Gaussian type for trigonometric polynomials.

Theorem 20.3 *The orthogonal trigonometric polynomial* $T_n^{\gamma} \in \mathcal{T}_n^{\gamma}$, $n \in \mathbb{N}$, $\gamma \in \{0, 1/2\}$ *with respect to the weight function* w *on interval* $[-\pi, \pi)$ *has exactly* $2(n+\gamma)$ *simple zeros on interval* $[-\pi, \pi)$.

Let us consider the interpolation problem:

$$\widetilde{T}_n(x_i) = y_i, \quad i = 1, 2, \ldots, 2n, \tag{20.9}$$

where $x_i \in [-\pi, \pi)$, $i = 1, 2, \ldots, 2n$, are $2n$ distinct nodes and y_i, $i = 1, 2, \ldots, 2n$, are real numbers.

Then we have (see [5, Theorem 3.3.]) the following.

1° If $\sum_{i=1}^{2n} x_i \neq k\pi$ for all $k \in \mathbb{Z}$, then there exists a unique solution of interpolation problem (20.9) in $\mathcal{T}_n \backslash \mathcal{L}\{\cos nx\}$ and $\mathcal{T}_n \backslash \mathcal{L}\{\sin nx\}$.
2° If $\sum_{i=1}^{2n} x_i = k\pi$, where k is an odd integer, then there exists a unique solution of interpolation problem (20.9) in $\mathcal{T}_n \backslash \mathcal{L}\{\cos nx\}$.
3° If $\sum_{i=1}^{2n} x_i = k\pi$, where k is an odd integer, then there exists a unique solution of interpolation problem (20.9) in $\mathcal{T}_n \backslash \mathcal{L}\{\sin nx\}$.

Also, in [5] an explicit representation for the unique solution is given:

$$\widetilde{T}_n(x) = \sum_{i=1}^{2n} f(x_i)\widetilde{s}_i(x), \tag{20.10}$$

where

$$\widetilde{s}_i(x) = \frac{1}{2T_n'(x_i)\sin\eta_n} \sin\left(\frac{x+\alpha_i}{2}\right) \frac{T_n(x)}{\sin\left(\frac{x-x_i}{2}\right)} \tag{20.11}$$

or

$$\widetilde{s}_i(x) = \frac{1}{2T_n'(x_i)\cos\eta_n} \cos\left(\frac{x+\alpha_i}{2}\right) \frac{T_n(x)}{\sin\left(\frac{x-x_i}{2}\right)}, \tag{20.12}$$

if $\widetilde{\mathcal{T}}_n$ is the space $\mathcal{T}_n \backslash \mathcal{L}\{\sin nx\}$ or $\mathcal{T}_n \backslash \mathcal{L}\{\cos nx\}$, respectively, and

$$T_n(x) = A\prod_{i=1}^{2n} \sin\frac{x-x_i}{2} \quad \text{const. } A \neq 0, \quad \eta_n = \frac{1}{2}\sum_{i=1}^{2n} x_i, \quad \alpha_i = \prod_{\substack{k=1\\k\neq i}}^{2n} x_k.$$

Now, we can give the following theorem which is the counterpart to the famous theorem about the Gaussian quadrature rules for algebraic polynomials (see [5, 8, 22, 50]).

Theorem 20.4 *Quadrature rule* (20.3) *is of Gaussian type for trigonometric polynomials, i.e., is exact for* $t \in \mathcal{T}_{2(n+\gamma)-1}$, *if and only if the nodes* $x_i \in [-\pi, \pi)$, $i = 1, 2, \ldots, 2(n+\gamma)$, *are the zeros of trigonometric polynomial* $T_n^\gamma \in \mathcal{T}_n^\gamma$, *which is orthogonal with respect to the weight function* w *on* $[-\pi, \pi)$ *to all trigonometric polynomials from* $\mathcal{T}_{n-1}^\gamma$. *Weight coefficients* A_i, $i = 1, \ldots, 2(n+\gamma)$, *are given as follows.*

- If $\gamma = 1/2$, *then (see [22])*

$$A_i = \int_{-\pi}^{\pi} \ell_i(x)w(x)\,\mathrm{d}x, \quad \ell_i(x) = \prod_{\substack{j=1\\j\neq\nu}}^{2n+1} \frac{\sin\left(\frac{x-x_j}{2}\right)}{\sin\left(\frac{x_\nu-x_j}{2}\right)}, \qquad i = 1, 2, \ldots, 2n+1.$$

- *If* $\gamma = 0$, *then (see [5])*

$$A_i = \int_{-\pi}^{\pi} \widetilde{s}_i(x)w(x)\,\mathrm{d}x, \quad i = 1, 2, \ldots, 2n,$$

where $\widetilde{s}_i(x)$, $i = 1, 2, \ldots, 2n$, *are given by* (20.11) *or* (20.12).

Let us notice that the quadrature rule of Gaussian type for trigonometric polynomials (20.3) integrates exactly trigonometric polynomials of integer degree at most $2(n+\gamma)-1$, i.e., $2n$ for $\gamma = 1/2$ and $2n-1$ for $\gamma = 0$. But such quadrature rule has an odd number of nodes, for $\gamma = 1/2$, or an even number of nodes for $\gamma = 0$. In order to achieve the maximal trigonometric degree of exactness, the nodes of the quadrature rule of Gaussian type for trigonometric polynomials are zeros of the corresponding orthogonal trigonometric polynomials of semi-integer degree in the case $\gamma = 1/2$ or integer degree in the case $\gamma = 0$.

The remainder term of quadrature rules of Gaussian types for trigonometric polynomials was analyzed in [43–46].

20.2.1 Even Weight Function

In this section we will consider the case when weight function w is even on interval $(-\pi, \pi)$. It is an important special case, because trigonometric polynomials (20.6) reduce to polynomials that consist of just cosine and sine functions, respectively, i.e.,

$$T_n^{C,\gamma}(x) = \cos(n+\gamma)x + \sum_{k=0}^{n-1} a_k^{(n),\gamma} \cos(k+\gamma)x, \tag{20.13}$$

$$T_n^{S,\gamma}(x) = \sin(n+\gamma)x + \sum_{k=0}^{n-1} d_k^{(n),\gamma} \sin(k+\gamma)x.$$

According to [22, Lemma 4.1] and [47, Lemma 2.5] five-term recurrence relations from Theorem 20.2 reduces to the following tree-term recurrence relation:

$$T_k^{C,\gamma}(x) = (2\cos x - \alpha_k^{(1),\gamma})T_{k-1}^{C,\gamma}(x) - \alpha_k^{(2),\gamma}T_{k-2}^{C,\gamma}(x), \tag{20.14}$$

$$T_k^{S,\gamma}(x) = (2\cos x - \eta_k^{(1),\gamma})T_{k-1}^{S,\gamma}(x) - \eta_k^{(2),\gamma}T_{k-2}^{S,\gamma}(x).$$

In order to simplify the construction of the quadrature rules of Gaussian type for trigonometric polynomials in the case with even weight function, we consider connection with certain Gaussian quadrature rule for algebraic polynomials. The following lemmas, which give the mentioned connections for the case $\gamma = 0$ and $\gamma = 1/2$, were proved in [22, 43].

Lemma 20.1 *Let t_ν, ω_ν, $\nu = 1, 2, \ldots, n$, be the nodes and the weights of the n-point Gaussian quadrature rule for algebraic polynomials with respect to the weight function:*

$$\widetilde{w}_1(x) = w(\arccos x)\sqrt{\frac{1+x}{1-x}}, \quad x \in (-1, 1). \tag{20.15}$$

Then, for the quadrature rule of Gaussian type for trigonometric polynomials (20.3) *with respect to even weight function w on $(-\pi, \pi)$ and $\gamma = 1/2$, we have*

$$A_{2n-\nu+1} = A_\nu = \frac{\omega_\nu}{1+t_\nu}, \quad \nu = 1, 2, \ldots, n, \quad A_{2n+1} = \int_{-\pi}^{\pi} w(x)\,\mathrm{d}x - \sum_{\nu=1}^{2n} A_\nu,$$

$$x_{2n-\nu+1} = -x_\nu = \arccos t_\nu, \quad \nu = 1, 2, \ldots, n, \quad x_{2n+1} = \pi.$$

Lemma 20.2 *Let t_ν, ω_ν, $\nu = 1, 2, \ldots, n$, be the nodes and the weights of the n-point Gaussian quadrature rule for algebraic polynomials with respect to the weight function:*

$$\widetilde{w}_2(x) = w(\arccos x)\sqrt{\frac{1-x}{1+x}}, \quad x \in (-1, 1). \tag{20.16}$$

Then, for the quadrature rule of Gaussian type for trigonometric polynomials (20.3) *with respect to even weight function w on $(-\pi, \pi)$ and $\gamma = 1/2$, we have*

$$A_{2n-\nu+2} = A_\nu = \frac{\omega_\nu}{1-t_\nu}, \quad \nu = 1, 2, \ldots, n, \quad A_{n+1} = \int_{-\pi}^{\pi} w(x)\,\mathrm{d}x - \sum_{\substack{\nu=1 \\ \nu \neq n+1}}^{2n+1} A_\nu,$$

$$x_{2n-\nu+2} = -x_\nu = \arccos t_\nu, \quad \nu = 1, 2, \ldots, n, \quad x_{n+1} = 0.$$

Lemma 20.3 *Let t_ν, ω_ν, $\nu = 1, 2, \ldots, n$, be the nodes and the weights of the n-point Gaussian quadrature rule for algebraic polynomials with respect to the weight function:*

$$\widetilde{w}_3(x) = \frac{w(\arccos x)}{\sqrt{1-x^2}}, \quad x \in (-1, 1). \tag{20.17}$$

Then, for the quadrature rule of Gaussian type for trigonometric polynomials (20.3) *with respect to even weight function w on $(-\pi, \pi)$ and $\gamma = 0$, we have*

$$A_{2n+1-\nu} = A_\nu = \omega_\nu, \quad \nu = 1, 2, \ldots, n,$$

$$x_{2n+1-\nu} = -x_\nu = \arccos t_\nu, \quad \nu = 1, 2, \ldots, n.$$

Lemma 20.4 *Let t_ν, ω_ν, $\nu = 1, 2, \ldots, n$, be the nodes and the weights of the n-point Gaussian quadrature rule for algebraic polynomials with respect to the weight function:*

$$\widetilde{w}_4(x) = w(\arccos x)\sqrt{1-x^2}, \quad x \in (-1, 1). \tag{20.18}$$

Then, for the quadrature rule of Gaussian type for trigonometric polynomials (20.3) *with respect to even weight function w on $(-\pi, \pi)$ and $\gamma = 0$, we have*

$$A_{2n+1-\nu} = A_\nu = \frac{\omega_\nu}{1-t_\nu^2}, \quad \nu = 1, 2, \ldots, n,$$

$$x_{2n+1-\nu} = -x_\nu = \arccos t_\nu, \quad \nu = 1, 2, \ldots, n.$$

20.3 Anti-Gaussian Quadrature Rules for Trigonometric Polynomials

Motivation for the anti-Gaussian quadrature rules lies in the difficulty to obtain an accurate error estimate for famous Gaussian quadrature rules for an arbitrary function f. Anti-Gaussian quadrature rules for algebraic polynomials, that is, quadrature rules that give an error equal in magnitude but of opposite sign to that of the corresponding Gaussian quadrature rule, were introduced by Laurie [21].

Different generalizations of the anti-Gaussian quadrature rules for real-valued measures were investigated in [3, 34], generalizations to matrix-valued measures were given in [1, 11], and applications for the estimates of the error of Gaussian quadrature rules can be found in [38]. In order to obtain more accurate estimates of error of Gaussian quadrature rule, Laurie in [21] introduced an averaged Gaussian rule, based on anti-Gaussian quadrature rule. Different averaged Gaussian quadrature rules and its applications can be found in [35, 37, 39, 40].

Anti-Szegő quadrature rules were introduced in [18], and by using Szegő and anti-Szegő quadrature rules, generalized averaged Szegő quadrature rules were defined in [20].

In this section, we consider an anti-Gaussian quadrature rule for trigonometric polynomials, i.e., quadrature rules that give an error equal in magnitude but of opposite sign to that of the corresponding Gaussian quadrature rules for trigonometric polynomials. Such quadrature rules are generalizations to anti-Gaussian quadrature rules for algebraic polynomials. Detailed analysis and numerical construction are restricted to the anti-Gaussian quadrature rules for trigonometric polynomials with respect to an even weight function on $(-\pi, \pi)$. Also, we introduce the averaged Gaussian quadrature rule for trigonometric polynomials. More details about this topic can be found in [33].

For the weight function w on the interval $[-\pi, \pi)$ and $\gamma \in \{0, 1/2\}$, we consider approximation of the integral (20.5) by Gaussian quadrature rule for trigonometric polynomials:

$$G_n(f) = \sum_{i=1}^{2(n+\gamma)} A_i f(x_i). \tag{20.19}$$

The anti-Gaussian quadrature rule

$$H_{n+2}(f) = \sum_{k=1}^{2(n+1+\gamma)} \lambda_k f(\xi_k) \tag{20.20}$$

is an $2(n+1+\gamma)$-point rule such that $I(t) - H_{n+2}(t) = -(I(t) - G_n(t))$ for all $t \in \mathcal{T}_{2(n+1+\gamma)-1}$, $\gamma \in \{0, 1/2\}$. Here, we see that $H_{n+2}(t) = (2I - G_n)(t)$ for each $t \in \mathcal{T}_{2(n+\gamma)+1}$. Because of that, for the fixed positive integer n, we introduce the following bilinear form:

$$(f, g)_w = (2I - G_n)(fg). \tag{20.21}$$

It is obvious that error $I(f) - G_n(f)$ can be estimated by $(H_{n+2}(f) - G_n(f))/2$, whereas for the approximation of the integral $I(f)$ one can use the following formula:

$$A_{2n+2}f = \frac{H_{n+2}(f) + G_n(f)}{2}, \tag{20.22}$$

which is the averaged Gaussian quadrature rule for trigonometric polynomials.

Since the bilinear form (20.21) depends on the number n, in what follows we consider the bilinear form (20.21) with respect to the positive integer n fixed in advance.

Let positive integer n be fixed in advance. By (V_k^γ), $V_k^\gamma \in \mathcal{T}_k^\gamma$, $k \in \mathbb{N}_0$, we denote the sequence of trigonometric polynomials orthogonal with respect to the bilinear form (20.21). The corresponding Gaussian rule for trigonometric polynomials has $2(n+\gamma)$ nodes, while the corresponding anti-Gaussian rule for trigonometric polynomials (where $n \mapsto n+1$) has $2(n+\gamma+1)$ nodes (so we denote it by H_{n+2}, because of two additional nodes, and it is exact for all $t \in \mathcal{T}_{2n+2\gamma+1}$). The nodes of anti-Gaussian rule for trigonometric polynomials are the zeros of the trigonometric polynomial:

$$V_{n+1}^\gamma(x) = \sum_{k=0}^{n+1} \left(p_k^{(n+1),\gamma} \cos{(k+\gamma)x} + q_k^{(n+1),\gamma} \sin{(k+\gamma)x} \right),$$

where $\left|p_{n+1}^{(n+1),\gamma}\right| + \left|q_{n+1}^{(n+1),\gamma}\right| \neq 0$, which is orthogonal with respect to the bilinear form (20.21) to all trigonometric polynomials from $\mathcal{T}_k^{\gamma}$, $k \leq n$. Specially, if we choose $(p_{n+1}^{(n+1),\gamma}, q_{n+1}^{(n+1),\gamma}) = (1, 0)$ and $(p_{n+1}^{(n+1),\gamma}, q_{n+1}^{(n+1),\gamma})$, we get

$$V_{n+1}^{C,\gamma}(x) = \cos(n+1+\gamma)x + \sum_{k=0}^{n}\Big(p_k^{(n+1),\gamma}\cos(k+\gamma)x \\ +q_k^{(n+1),\gamma}\sin(k+\gamma)x\Big)$$

and

$$V_{n+1}^{S,\gamma}(x) = \sin(n+1+\gamma)x + \sum_{k=0}^{n}\Big(r_k^{(n+1),\gamma}\cos(k+\gamma)x \\ +s_k^{(n+1),\gamma}\sin(k+\gamma)x\Big),$$

respectively.

Similar to notation (20.7) introduced with respect to the inner product $\langle\cdot,\cdot\rangle_w$, we introduce the following notation:

$$\widehat{I}_k^{C,\gamma} = \left(V_{k,\gamma}^C, V_{k,\gamma}^C\right)_w, \quad \widehat{I}_k^{S,\gamma} = \left(V_{k,\gamma}^S, V_{k,\gamma}^S\right)_w, \quad \widehat{I}_k^{\gamma} = \left(V_{k,\gamma}^C, V_{k,\gamma}^S\right)_w,$$
$$\widehat{J}_{k,\ell}^{C,\gamma} = \left(2\cos x\, V_{k,\gamma}^C, V_{\ell,\gamma}^C\right)_w, \quad \widehat{J}_{k,\ell}^{S,\gamma} = \left(2\cos x\, V_{k,\gamma}^S, V_{\ell,\gamma}^S\right)_w,$$
$$\widehat{J}_{k,\ell}^{\gamma} = \left(2\cos x\, V_{k,\gamma}^C, V_{\ell,\gamma}^S\right)_w,$$

for $k, \ell \in \mathbb{N}_0$, where $(\cdot,\cdot)_w$ is given by (20.21).

Theorem 20.5 *Let positive integer n be fixed in advance. Trigonometric polynomials $V_k^{C,\gamma}(x)$ and $V_k^{S,\gamma}(x)$, $k \geqslant 1$, orthogonal with respect to the bilinear form* (20.21)*, satisfy the following recurrence relations:*

$$V_k^{C,\gamma}(x) = \left(2\cos x - a_k^{(1),\gamma}\right)V_{k-1}^{C,\gamma}(x) - b_k^{(1),\gamma}V_{k-1}^{S,\gamma}(x) - a_k^{(2),\gamma}V_{k-2}^{C,\gamma}(x) \\ -b_k^{(2),\gamma}V_{k-2}^{S,\gamma}(x), \tag{20.23}$$

$$V_k^{S,\gamma}(x) = \left(2\cos x - d_k^{(1),\gamma}\right)V_{k-1}^{S,\gamma}(x) - c_k^{(1),\gamma}V_{k-1}^{C,\gamma}(x) - d_k^{(2),\gamma}V_{k-2}^{S,\gamma}(x) \\ -c_k^{(2),\gamma}V_{k-2}^{C,\gamma}(x), \tag{20.24}$$

where the coefficients are the solutions of the following systems for $j = 1, 2$:

$$\widehat{J}^{C,\gamma}_{k-1,\,k-j} = a_k^{(j),\gamma}\widehat{I}^{C,\gamma}_{k-j} + b_k^{(j,\gamma)}\widehat{I}^{\gamma}_{k-j}, \quad \widehat{J}^{\gamma}_{k-1,\,k-j} = a_k^{(j),\gamma}\widehat{I}^{\gamma}_{k-j} + b_k^{(j),\gamma}\widehat{I}^{S,\gamma}_{k-j}, \tag{20.25}$$

$$\widehat{J}^{\gamma}_{k-1,\,k-j} = c_k^{(j),\gamma}\widehat{I}^{C,\gamma}_{k-j} + d_k^{(j),\gamma}\widehat{I}^{\gamma}_{k-j}, \quad \widehat{J}^{S,\gamma}_{k-1,\,k-j} = c_k^{(j),\gamma}\widehat{I}^{\gamma}_{k-j} + d_k^{(j),\gamma}\widehat{I}^{S,\gamma}_{k-j},$$

with $a_1^{(2),\gamma} = b_1^{(2),\gamma} = c_1^{(2),\gamma} = d_1^{(2),\gamma} = 0$.

Proof Since the polynomials $V_j^{C,\gamma}(x)$ and $V_j^{S,\gamma}(x)$, for $j = 0, \dots, k$, are linearly independent, we have

$$2\cos x V_{k-1}^{C,\gamma}(x) = V_k^{C,\gamma}(x) + \sum_{j=0}^{k-1}\left[a_k^{(k-j),\gamma}V_j^{C,\gamma}(x) + b_k^{(k-j),\gamma}V_j^{S,\gamma}(x)\right]. \tag{20.26}$$

Multiplying by $V_i^{C,\gamma}(x)$ and $V_i^{C,\gamma}(x)$, for $i = 0, 1, \dots, k-1$, and applying the bilinear form (20.21), we get

$$\begin{aligned}\left(2\cos x V_{k-1}^{C,\gamma}(x), V_i^{C,\gamma}(x)\right)_w &= \left(V_k^{C,\gamma}(x), V_i^{C,\gamma}(x)\right)_w \\ &+ \sum_{j=0}^{k-1} a_k^{(k-j),\gamma}\left(V_j^{C,\gamma}(x), V_i^{C,\gamma}(x)\right)_w \\ &+ \sum_{j=0}^{k-1} b_k^{(k-j),\gamma}\left(V_j^{S,\gamma}(x), V_i^{C,\gamma}(x)\right)_w\end{aligned}$$

and

$$\begin{aligned}\left(2\cos x V_{k-1}^{C,\gamma}(x), V_i^{S,\gamma}(x)\right)_w &= \left(V_k^{C,\gamma}(x), V_i^{S,\gamma}(x)\right)_w \\ &+ \sum_{j=0}^{k-1} a_k^{(k-j),\gamma}\left(V_j^{C,\gamma}(x), V_i^{S,\gamma}(x)\right)_w \\ &+ \sum_{j=0}^{k-1} b_k^{(k-j),\gamma}\left(V_j^{S,\gamma}(x), V_i^{S,\gamma}(x)\right)_w,\end{aligned}$$

respectively. By using orthogonality conditions, we get homogeneous systems of linear equations:

$$\begin{aligned} \widehat{I}_i^{C,\gamma} \cdot a_k^{(k-i),\gamma} + \widehat{I}_i^{\gamma} \cdot b_k^{(k-i),\gamma} &= 0, \\ \widehat{I}_i^{\gamma} \cdot a_k^{(k-i),\gamma} + \widehat{I}_i^{S,\gamma} \cdot b_k^{(k-i),\gamma} &= 0, \end{aligned} \qquad i = 0, \ldots, k-3, \tag{20.27}$$

where the corresponding determinants of these systems are given by

$$\widehat{D}_i^{\gamma} = \begin{vmatrix} \widehat{I}_i^{C,\gamma} & \widehat{I}_i^{\gamma} \\ \widehat{I}_i^{\gamma} & \widehat{I}_i^{S,\gamma} \end{vmatrix}.$$

Since $I(f) = G_n(f)$ for all $f \in \mathcal{T}_{2n+2\gamma-1}$, we have

$$H_{n+2}(f) = (2I - G_n)(f) = I(f) + (I - G_n)(f) = I(f),$$

for each $f \in \mathcal{T}_{2n+2\gamma-1}$. Further, $V_k^{C,\gamma}(x) = T_k^{C,\gamma}(x)$ and $V_k^{S,\gamma}(x) = T_k^{S,\gamma}(x)$, for every $k = 0, \ldots, n$, where $T_k^{C,\gamma}(x)$ and $T_k^{S,\gamma}(x)$ are orthogonal trigonometric polynomials with respect to the inner product (20.4). For $2k + 1 < 2n$ it follows:

$$\widehat{I}_k^{C,\gamma} = \left(V_k^{C,\gamma}, V_k^{C,\gamma}\right)_w = (2I - G_n)\left(V_k^{C,\gamma} \cdot V_k^{C,\gamma}\right) = I\left(V_k^{C,\gamma} \cdot V_k^{C,\gamma}\right) = I_k^{C,\gamma}.$$

Similarly, $\widehat{I}_k^{S,\gamma} = I_k^{S,\gamma}$ and $\widehat{I}_k^{\gamma} = I_k^{\gamma}$. Now, using Cauchy–Schwarz–Bunyakovsky integral inequality, for $i = 0, \ldots, k-3$, we have

$$\begin{aligned} \widehat{D}_i^{\gamma} &= \widehat{I}_i^{C,\gamma} \widehat{I}_i^{S,\gamma} - (\widehat{I}_i^{\gamma})^2 = I_i^{C,\gamma} I_i^{S,\gamma} - (I_i^{\gamma})^2 \\ &= \left(\int_{-\pi}^{\pi} \left(T_i^{C,,\gamma}(x)\right)^2 w(x)\,\mathrm{d}x\right) \cdot \left(\int_{-\pi}^{\pi} \left(T_i^{S,\gamma}(x)\right)^2 w(x)\,\mathrm{d}x\right) \\ &\quad - \left(\int_{-\pi}^{\pi} T_i^{C,\gamma}(x) T_i^{S,\gamma}(x) w(x)\,\mathrm{d}x\right)^2 \\ &> \left(\int_{-\pi}^{\pi} \left(T_i^{C,\gamma}(x)\right)^2 w(x)\,\mathrm{d}x\right) \cdot \left(\int_{-\pi}^{\pi} \left(T_i^{S,\gamma}(x)\right)^2 w(x)\,\mathrm{d}x\right) \\ &\quad - \left(\int_{-\pi}^{\pi} \left(T_i^{C,\gamma}(x)\right)^2 w(x)\,\mathrm{d}x\right) \cdot \left(\int_{-\pi}^{\pi} \left(T_i^{S,\gamma}(x)\right)^2 w(x)\,\mathrm{d}x\right) \\ &= 0. \end{aligned}$$

Therefore, for all $i = 0, \ldots, k-3$ system (20.27) has only the trivial solutions $a_k^{(k-i),\gamma} = b_k^{(k-i),\gamma} = 0$. Thus, equality (20.23) follows from (20.26).

Analogously, one can obtain the recurrence relation (20.24) for $V_k^{S,\gamma}(x)$.
Using recurrence relation (20.23), for $j = 1, 2$, we have

$$\begin{aligned}
&\left(2\cos x V_{k-1}^{C,\gamma}(x) V_{k-j}^{C,\gamma}(x)\right)_w \\
&= a_k^{(1),\gamma}\left(V_{k-1}^{C,\gamma}(x), V_{k-j}^{C,\gamma}(x)\right)_w + b_k^{(1),\gamma}\left(V_{k-1}^{S,\gamma}(x), V_{k-j}^{C,\gamma}(x)\right)_w \\
&+ a_k^{(2),\gamma}\left(V_{k-2}^{C,\gamma}(x), V_{k-j}^{C,\gamma}(x)\right)_w + b_k^{(2),\gamma}\left(V_{k-2}^{S,\gamma}(x), V_{k-j}^{C,\gamma}(x)\right)_w,
\end{aligned}$$

i.e., $\widehat{J}_{k-1,k-j}^{C,\gamma} = a_{k,\gamma}^{(j)}\widehat{I}_{k-j}^{C,\gamma} + b_{k,\gamma}^{(j)}\widehat{I}_{k-j}^{\gamma}$, $j = 1, 2$.

Analogously, one gets $\widehat{J}_{k-1,k-j}^{\gamma} = a_k^{(j),\gamma}\widehat{I}_{k-j}^{\gamma} + b_k^{(j),\gamma}\widehat{I}_{k-j}^{S,\gamma}$, $j = 1, 2$.

Similarly, starting from (20.24) one can obtain $\widehat{J}_{k-1,k-j}^{\gamma} = c_k^{(j),\gamma}\widehat{I}_{k-j}^{C,\gamma} + d_k^{(j),\gamma}\widehat{I}_{k-j}^{\gamma}$, $\widehat{J}_{k-1,k-j}^{S,\gamma} = c_k^{(j),\gamma}\widehat{I}_{k-j}^{\gamma} + d_k^{(j),\gamma}\widehat{I}_{k-j}^{S,\gamma}$, $j = 1, 2$. □

Using the recurrence relations (20.23), (20.24) and orthogonality conditions, for $k \geqslant 1$, the following equations can be proved:

$$\widehat{I}_k^{C,\gamma} = \widehat{J}_{k,k-1}^{C,\gamma}, \quad \widehat{I}_k^{S,\gamma} = \widehat{J}_{k,k-1}^{S,\gamma}, \quad \widehat{I}_k^{\gamma} = \widehat{J}_{k,k-1}^{\gamma}.$$

Corollary 20.1 *For the recurrence coefficients in* (20.23) *and* (20.24) *the following formulas:*

$$\begin{aligned}
a_k^{(1),\gamma} &= \frac{\widehat{I}_{k-1}^{S,\gamma}\widehat{J}_{k-1}^{C,\gamma} - \widehat{I}_{k-1}^{\gamma}\widehat{J}_{k-1}^{\gamma}}{\widehat{D}_{k-1}^{\gamma}}, & a_k^{(2),\gamma} &= \frac{\widehat{I}_{k-1}^{C,\gamma}\widehat{I}_{k-2}^{S,\gamma} - \widehat{I}_{k-1}^{\gamma}\widehat{I}_{k-2}^{\gamma}}{\widehat{D}_{k-2}^{\gamma}}, \\
b_k^{(1),\gamma} &= \frac{\widehat{I}_{k-1}^{C,\gamma}\widehat{J}_{k-1}^{\gamma} - \widehat{I}_{k-1}^{\gamma}\widehat{J}_{k-1}^{C,\gamma}}{\widehat{D}_{k-1}^{\gamma}}, & b_k^{(2),\gamma} &= \frac{\widehat{I}_{k-1}^{\gamma}\widehat{I}_{k-2}^{C,\gamma} - \widehat{I}_{k-1}^{C,\gamma}\widehat{I}_{k-2}^{\gamma}}{\widehat{D}_{k-2}^{\gamma}}, \\
c_k^{(1),\gamma} &= \frac{\widehat{I}_{k-1}^{S,\gamma}\widehat{J}_{k-1}^{\gamma} - \widehat{I}_{k-1}^{\gamma}\widehat{J}_{k-1}^{S,\gamma}}{\widehat{D}_{k-1}^{\gamma}}, & c_k^{(2),\gamma} &= \frac{\widehat{I}_{k-1}^{\gamma}\widehat{I}_{k-2}^{S,\gamma} - \widehat{I}_{k-1}^{S,\gamma}\widehat{I}_{k-2}^{\gamma}}{\widehat{D}_{k-2}^{\gamma}}, \\
d_k^{(1),\gamma} &= \frac{\widehat{I}_{k-1}^{C,\gamma}\widehat{J}_{k-1}^{S,\gamma} - \widehat{I}_{k-1}^{\gamma}\widehat{J}_{k-1}^{\gamma}}{\widehat{D}_{k-1}^{\gamma}}, & d_k^{(2),\gamma} &= \frac{\widehat{I}_{k-1}^{S,\gamma}\widehat{I}_{k-2}^{C,\gamma} - \widehat{I}_{k-1}^{\gamma}\widehat{I}_{k-2}^{\gamma}}{\widehat{D}_{k-2}^{\gamma}},
\end{aligned}$$

hold, where $\widehat{D}_{k-j}^{\gamma} = \widehat{I}_{k-j}^{C,\gamma}\widehat{I}_{k-j}^{S,\gamma} - (\widehat{I}_{k-j}^{\gamma})^2$, $j = 1, 2$, *for* $k \geqslant 1$, *and* $\widehat{J}_k^{C,\gamma} = \widehat{J}_{k,k}^{C,\gamma}$, $\widehat{J}_k^{S,\gamma} = \widehat{J}_{k,k}^{S,\gamma}$, $\widehat{J}_k^{\gamma} = \widehat{J}_{k,k}^{\gamma}$.

Proof Formulas can be obtained by solving systems (20.25). □

Similar to the algebraic case, we can obtain the relations between coefficients in the five-term recurrent relations given in Theorem 20.2 and Theorem 20.5.

Since $V_k^{C,\gamma}(x) = T_k^{C,\gamma}(x)$ and $V_k^{S,\gamma}(x) = T_k^{S,\gamma}(x)$, for every $k = 0, \ldots, n$, we have

$$a_{k,\gamma}^{(j)} = \alpha_{k,\gamma}^{(j)}, \quad b_{k,\gamma}^{(j)} = \beta_{k,\gamma}^{(j)}, \quad c_{k,\gamma}^{(j)} = \gamma_{k,\gamma}^{(j)}, \quad d_{k,\gamma}^{(j)} = \delta_{k,\gamma}^{(j)}, \quad j = 1, 2,$$

where $\alpha_{k,\gamma}^{(j)}, \beta_{k,\gamma}^{(j)}, \gamma_{k,\gamma}^{(j)}$ and $\delta_{k,\gamma}^{(j)}$ are the coefficients in the recurrence relations for polynomials $T_k^{C,\gamma}$ and $T_k^{S,\gamma}$ given in Theorem 20.2.

Lemma 20.5 *Let positive integer n be fixed in advance. For $k = 0, \ldots, n-1$ the following equations*

$$\widehat{I}_k^{S,\gamma} = I_k^{S,\gamma}, \quad \widehat{J}_k^{S,\gamma} = J_k^{S,\gamma}, \quad \widehat{I}_k^{C,\gamma} = I_k^{C,\gamma}, \quad \widehat{J}_k^{C,\gamma} = J_k^{C,\gamma}, \quad \widehat{I}_k^{\gamma} = I_k^{\gamma},$$
$$\widehat{J}_k^{\gamma} = J_k^{\gamma},$$

hold, while

$$\widehat{I}_n^{S,\gamma} = 2I_n^{S,\gamma}, \quad \widehat{J}_n^{S,\gamma} = 2J_n^{S,\gamma}, \quad \widehat{I}_n^{C,\gamma} = 2I_n^{C,\gamma}, \quad \widehat{J}_n^{C,\gamma} = 2J_n^{C,\gamma}, \quad \widehat{I}_n^{\gamma} = 2I_n^{\gamma},$$
$$\widehat{J}_n^{\gamma} = 2J_n^{\gamma},$$

and for $k = n+1$ coefficients in the five-term recurrence relations (20.23) *and* (20.24) *satisfy*

$$a_{n+1}^{(1),\gamma} = \alpha_{n+1}^{(1),\gamma}, \quad a_{n+1}^{(2),\gamma} = 2\alpha_{n+1}^{(2),\gamma}, \quad b_{n+1}^{(1),\gamma} = \beta_{n+1}^{(1),\gamma}, \quad b_{n+1}^{(2),\gamma} = 2\beta_{n+1}^{(2),\gamma},$$
$$c_{n+1}^{(1),\gamma} = \gamma_{n+1}^{(1),\gamma}, \quad c_{n+1}^{(2),\gamma} = 2\gamma_{n+1}^{(2),\gamma}, \quad d_{n+1}^{(1),\gamma} = \delta_{n+1}^{(1),\gamma}, \quad d_{n+1}^{(2),\gamma} = 2\delta_{n+1}^{(2),\gamma}. \tag{20.28}$$

Proof For $k = 0, \ldots, n-1$, we have

$$\begin{aligned}\widehat{I}_k^{\gamma} &= \left(V_k^{C,\gamma}, V_k^{S,\gamma}\right)_w = (2I - G_n)\left(V_k^{C,\gamma} V_k^{S,\gamma}\right) = (2I - G_n)\left(T_k^{C,\gamma} T_k^{S,\gamma}\right)\\ &= I\left(T_k^{C,\gamma} T_k^{S,\gamma}\right) + (I - G_n)\left(T_k^{C,\gamma} T_k^{S,\gamma}\right) = \left\langle T_k^{C,\gamma}, T_k^{S,\gamma}\right\rangle_w = I_k^{\gamma}.\end{aligned}$$

By using similar arguments, other equalities for $k \leqslant n-1$ can be proved.

Since the nodes of the Gaussian quadrature rule for trigonometric polynomials are the zeros of the corresponding orthogonal trigonometric polynomial $T_n^{;\gamma}(x)$, for $k = n$, we have

$$\widehat{I}_n^{\gamma} = \left(V_n^{C,\gamma}, V_n^{S,\gamma}\right)_w = (2I - G_n)\left(V_n^{C,\gamma} V_n^{S,\gamma}\right) = (2I - G_n)\left(T_n^{C,\gamma} T_n^{S,\gamma}\right) = 2I_n^{\gamma}.$$

The other equalities can be proved in a similar way.

Using the obtained equalities, we have

$$\widehat{D}_k^{\gamma} = \widehat{I}_k^{C,\gamma}\widehat{I}_k^{S,\gamma} - (\widehat{I}_k^{\gamma})^2 = I_k^{C,\gamma} I_k^{S,\gamma} - (I_k^{\gamma})^2 = D_k^{\gamma}, \quad k = 0, \dots, n-1,$$

$$\widehat{D}_n^{\gamma} = \widehat{I}_n^{C,\gamma}\widehat{I}_n^{S,\gamma} - (\widehat{I}_n^{\gamma})^2 = 2I_n^{C,\gamma} \cdot 2I_n^{S,\gamma} - (2I_n^{\gamma})^2 = 4D_n^{\gamma},$$

where $D_k^{\gamma} = I_k^{C,\gamma} I_k^{S,\gamma} - (I_k^{\gamma})^2$, $k = 0, 1, \dots, n$.

Now, for the coefficients $a_{n+1}^{(1),\gamma}$ and $a_{n+1}^{(2),\gamma}$ we have

$$a_{n+1}^{(1),\gamma} = \frac{\widehat{I}_n^{S,\gamma}\widehat{J}_n^{C,\gamma} - \widehat{I}_n^{\gamma}\widehat{J}_n^{\gamma}}{\widehat{D}_n^{\gamma}} = \frac{2I_n^{S,\gamma} \cdot 2J_n^{C,\gamma} - 2I_n^{\gamma} \cdot 2J_n^{\gamma}}{4 \cdot D_n^{\gamma}} = \alpha_{n+1}^{(1),\gamma}$$

and

$$a_{n+1}^{(2),\gamma} = \frac{\widehat{I}_n^{C,\gamma}\widehat{I}_{n-1}^{S,\gamma} - \widehat{I}_n^{\gamma}\widehat{I}_{n-1}^{\gamma}}{\widehat{D}_{n-1}^{\gamma}} = \frac{2I_n^{C,\gamma} I_{n-1}^{S,\gamma} - 2I_n^{\gamma} I_{n-1}^{\gamma}}{D_{n-1}^{\gamma}} = 2\alpha_{n+1}^{(2),\gamma}.$$

The other equalities in (20.28) can be proved in a similar way. □

20.3.1 Even Weight Function

In this subsection we consider only even weight functions w on $(-\pi, \pi)$. Let positive integer n be fixed in advance. Since $V_k^{C,\gamma}(x) = T_k^{C,\gamma}(x)$, $V_k^{S,\gamma}(x) = T_k^{S,\gamma}(x)$, for $k = 0, \dots, n$, and (20.28), we have

$$\begin{aligned} V_{n+1,\gamma}^{C}(x) &= \left(2\cos x - a_{n+1}^{(1),\gamma}\right) V_n^{C,\gamma}(x) - b_{n+1}^{(1),\gamma} V_n^{S,\gamma}(x) - a_{n+1}^{(2),\gamma} V_{n-1}^{C,\gamma}(x) \\ &\quad - b_{n+1}^{(2),,\gamma} V_{n-1}^{S,\gamma}(x) \\ &= \left(2\cos x - \alpha_{n+1}^{(1),\gamma}\right) T_n^{C,\gamma}(x) - \beta_{n+1}^{(1),\gamma} T_n^{S,\gamma}(x) - 2\alpha_{n+1}^{(2),\gamma} T_{n-1}^{C,\gamma}(x) \\ &\quad - 2\beta_{n+1}^{(2),\gamma} T_{n-1}^{S,\gamma}(x) \\ &= T_{n+1}^{C,\gamma}(x) - \alpha_{n+1}^{(2),\gamma} T_{n-1}^{C,\gamma}(x) - \beta_{n+1}^{(2),\gamma} T_{n-1}^{S,\gamma}(x). \end{aligned}$$

By using similar arguments as in Sect. 20.2.1, we obtain that $\beta_{k,\gamma}^{(j)} = 0$, $j = 1, 2$, for each $k \in \mathbb{N}$, so $\beta_{n+1,\gamma}^{(2)} = 0$. Also, because of the form $T_k^{C,\gamma}$ and $T_k^{C,\gamma}$, given in (20.13), we have

$$V_{n+1}^{C,\gamma}(x) = \sum_{k=0}^{n+1} p_{k,\gamma}^{(n+1)} \cos(k+\gamma)x, \quad p_{n+1,\gamma}^{(n+1)} = 1,$$

$$V_{n+1}^{S,\gamma}(x) = \sum_{k=0}^{n+1} s_{k,\gamma}^{(n+1)} \sin(k+\gamma)x, \quad s_{n+1,\gamma}^{(n+1)} = 1.$$

Now, it is obvious, that for every $k \in \mathbb{N}_0$ we have

$$V_{k+1}^{C,\gamma}(\pi) = 0, \quad V_{k+1}^{S,\gamma}(0) = 0. \tag{20.29}$$

Also, the five-term recurrence relations (20.23), (20.24) reduce to the following three-term recurrence relations:

$$V_{n+1}^{C,\gamma}(x) = \left(2\cos x - a_{n+1}^{(1),\gamma}\right) V_n^{C,\gamma}(x) - a_{n+1}^{(2),\gamma} V_{n-1}^{C,\gamma}(x), \tag{20.30}$$

$$V_{n+1}^{S,\gamma}(x) = \left(2\cos x - d_{n+1}^{(1),\gamma}\right) V_n^{S,\gamma}(x) - d_{n+1}^{(2),\gamma} V_{n-1}^{S,\gamma}(x). \tag{20.31}$$

For the numerical construction of anti-Gaussian quadrature rules for trigonometric polynomials in the case of even weight function, an important feature is that the nodes and the weights can be connected with the nodes and the weights of anti-Gaussian quadrature rules for algebraic polynomials. In order to obtain such formulas we have to consider cases when $\gamma = 0$ and $\gamma = 1/2$ separately.

20.3.1.1 Case $\gamma = 0$

Let us introduce the following notation for $k = 0, 1, \ldots$:

$$C_{k,0}(x) = \sum_{j=0}^{k} p_{j,0}^{(k)} T_j(x), \quad S_{k,0}(x) = \sum_{j=0}^{k} s_{j,0}^{(k)} U_{j-1}(x),$$

where $T_k(x) = \cos(k \cdot \arccos x)$ and $U_k(x) = \sin((k+1)\arccos x)/\sqrt{1-x^2}$, $x \in (-1, 1)$, are Chebyshev polynomials of the first and second kind, respectively.

The connections of quadrature rules of Gaussian type for trigonometric polynomials (20.19), in the case $\gamma = 0$, with certain Gaussian quadrature rules for algebraic polynomials were considered in [43]. Using these connections we get the following results.

Theorem 20.6 *Let positive integer n be fixed in advance. For an even weight function $w(x)$ on $(-\pi,\pi)$, the following equalities*

$$\big(C_{n,0}(x), C_{k,0}(x)\big)_{\widetilde{w}_3}=0 \quad \text{and} \quad \big(S_{n,0}(x), S_{k,0}(x)\big)_{\widetilde{w}_4}=0$$

hold for all $k=0,1,\ldots,n-1$, where $\widetilde{w}_3$ and $\widetilde{w}_4$ are given by (20.17) *and* (20.18), *respectively.*

Proof Let t_k and ω_k, $k=1,\ldots,n$, denote the nodes and weights of the Gaussian quadrature rule constructed for the algebraic polynomials with respect to the weight function (20.17). By using [43, Lemma 2.4.], i.e., Lemma 20.3, orthogonality of the polynomials $V_k^{C,0}(x)$ and substitution $x=\arccos t$, for each $k=0,\ldots,n-1$, we have

$$\begin{aligned}
0 &= \left(V_n^{C,0}(x), V_k^{C,0}(x)\right)_w = (2I-G_n)\left(V_n^{C,0}(x)V_k^{C,0}(x)\right)\\
&= 4\int_0^{\pi} V_n^{C,0}(x)V_k^{C,0}(x)w(x)\,\mathrm{d}x - \sum_{j=1}^{2n} A_j V_n^{C,0}(x_j)V_k^{C,0}(x_j)\\
&= 4\int_{-1}^{1} V_n^{C,0}(\arccos t)V_k^{C,0}(\arccos t)\frac{w(\arccos t)}{\sqrt{1-t^2}}\,\mathrm{d}t\\
&\quad -2\sum_{j=1}^{n}\omega_j V_n^{C,0}\left(\arccos t_j\right) V_k^{C,0}\left(\arccos t_j\right)\\
&= 2\,(2I-G_n)\left(V_n^{C,0}(\arccos x)V_k^{C,0}(\arccos x)\right)\\
&= 2\left(V_n^{C,0}(\arccos x), V_k^{C,0}(\arccos x)\right)_{\widetilde{w}_3}.
\end{aligned}$$

Now, using $V_k^{C,0}(\arccos x)=\sum_{j=0}^{k} p_{j,0}^{(k)}T_j(x)$, for all $k=0,\ldots,n-1$ we get

$$\begin{aligned}
\left(V_n^{C,0}(\arccos x), V_k^{C,0}(\arccos x)\right)_{\widetilde{w}_3} &= \left(\sum_{j=0}^{n} p_{j,0}^{(n)}T_j(x), \sum_{j=0}^{k} p_{j,0}^{(k)}T_j(x)\right)_{\widetilde{w}_3}\\
&= \big(C_{n,0}(x), C_{k,0}(x)\big)_{\widetilde{w}_3},
\end{aligned}$$

i.e., $\big(C_{n,0}(x), C_{k,0}(x)\big)_{\widetilde{w}_3}=0$.

The equality $\big(S_{n,0}(x), S_{k,0}(x)\big)_{\widetilde{w}_4}=0$ can be proved in a similar way, using Lemma 20.4, i.e., [43, Lemma 2.5.] (see [33]). □

If we put $x := \arccos x$ in (20.30), using equality $V_n^{C,0}(\arccos x) = C_{n,0}(x)$, we get

$$C_{n,0}(x) = \left(2x - a_n^{(1),0}\right) C_{n-1,0}(x) - a_n^{(2),0} C_{n-2,0}(x), \quad a_1^{(2),0} = 0, \quad C_{0,0}(x) = 1.$$

Therefore, the zeros of the polynomial $C_{n+1,0}(x)$ (and hence the zeros of the trigonometric polynomial $V_{n+1}^{C,0}(x)$) can be calculated using the QR-algorithm.

Lemma 20.6 *Let w be the even weight function on $(-\pi, \pi)$ and let τ_k and σ_k, $k = 1, \ldots, n+1$, be the nodes and weights of $(n+1)$-point anti-Gaussian quadrature rule constructed for algebraic polynomials with respect to the weight function* (20.17) *on $(-1, 1)$. Then the weights λ_k and the nodes ξ_k, $k = 1, 2, \ldots, 2n+2$, of the $(2n+2)$-point anti-Gaussian quadrature rule* (20.20) *($\gamma = 0$) with respect to w are given as follows:*

$$\lambda_k = \lambda_{2n+3-k} = \sigma_k, \quad k = 1, \ldots, n+1,$$

$$\xi_k = -\xi_{2n+3-k} = -\arccos \tau_k, \quad k = 1, \ldots, n+1.$$

Proof Due to the equality $V_{n+1}^{C,0}(\arccos x) = C_{n+1,0}(x)$ it is easy to see that the nodes of the anti-Gaussian quadrature rule for trigonometric polynomials are given by

$$\xi_k = -\xi_{2n+3-k} = -\arccos \tau_k, \quad k = 1, \ldots, n+1.$$

Weights can be constructed using the Shohat formula (see [22, 36]):

$$\sigma_k = \mu_0 \left[\sum_{j=0}^{n} \left(\frac{C_{j,0}(\tau_k)}{\prod_{i=2}^{j} a_i^{(2),0}} \right)^2 \right]^{-1} = \frac{\mu_0}{\sum_{j=0}^{n} \left(\frac{V_j^{C,0}(\arccos \tau_k)}{\prod_{i=2}^{j} a_i^{(2),0}} \right)^2}, \quad k = 1, \ldots, n+1,$$

where

$$\mu_0 = \int_{-1}^{1} \frac{w(\arccos x)}{\sqrt{1-x^2}}\, \mathrm{d}x.$$

Since the function $w(x)$ is even on $(-\pi, \pi)$, we have

$$\lambda_{2n+3-k} = \frac{\widehat{\mu}_0}{2 \cdot \sum_{j=0}^{n} \left(\frac{V_j^{C,0}(\xi_{2n+3-k})}{\prod_{i=2}^{j} a_{i,0}^{(2)}} \right)^2}, \quad k = 1, \ldots, n+1,$$

where

$$\widehat{\mu}_0 = \int_{-\pi}^{\pi} w(x)\, \mathrm{d}x = 2 \int_{-1}^{1} w(\arccos t) \frac{\mathrm{d}t}{\sqrt{1-t^2}} = 2\mu_0.$$

Therefore, we get $\lambda_{2n+3-k} = \lambda_k = \sigma_k$, for $k = 1, \ldots, n+1$. □

The following result could be proved by using similar arguments.

Lemma 20.7 *Let w be the even weight function on $(-\pi, \pi)$ and let τ_k and σ_k, $k = 1, \ldots, n+1$, be the nodes and weights of $(n+1)$-point anti-Gaussian quadrature rule constructed for algebraic polynomials with respect to the weight function* (20.18) *on $(-1, 1)$. Then the weights λ_k and the nodes ξ_k, $k = 1, 2, \ldots, 2n+2$, of the $(2n+2)$-point anti-Gaussian quadrature rule* (20.20) *($\gamma = 0$) with respect to w are given as follows:*

$$\lambda_k = \lambda_{2n+3-k} = \frac{\sigma_k}{1-\tau_k^2}, \quad k = 1, \ldots, n+1,$$

$$\xi_k = -\xi_{2n+3-k} = -\arccos \tau_k, \quad k = 1, \ldots, n+1.$$

20.3.1.2 Case $\gamma = 1/2$

Let us introduce the following notation for $k = 0, 1, \ldots$:

$$C_{k,1/2}(x) = \sum_{j=0}^{k} p_{j,1/2}^{(k)} \big(T_j(x) - (1-x)U_{j-1}(x)\big),$$

$$S_{k,1/2}(x) = \sum_{j=0}^{k} s_{j,1/2}^{(k)} \big(T_j(x) + (1+x)U_{j-1}(x)\big).$$

The following theorem can be proved in a similar way as Theorem 20.6 (see [33]).

Theorem 20.7 *Let positive integer n be fixed in advance. For an even weight function $w(x)$ on $(-\pi, \pi)$, the following equalities*

$$\big(C_{n,1/2}(x), C_{k,1/2}(x)\big)_{\widetilde{w}_1} = 0 \quad \text{and} \quad \big(S_{n,1/2}(x), S_{k,1/2}(x)\big)_{\widetilde{w}_2} = 0$$

hold for all $k = 0, 1, \ldots, n-1$, where $\widetilde{w}_1$ and $\widetilde{w}_2$ are given by (20.15) *and* (20.16), *respectively.*

If we put $x := \arccos x$ in (20.30), using equality

$$V_k^{C,1/2}(\arccos x) = \sqrt{\frac{1+x}{2}}\, C_{k,1/2}(x), \tag{20.32}$$

and dividing both sides by $\sqrt{(1+x)/2}$, we get

$$C_{n+1,1/2}(x) = \left(2x - a_{n+1}^{(1),1/2}\right) C_{n,1/2}(x) - a_{n+1}^{(2),1/2} C_{n-1,1/2}(x), \tag{20.33}$$

where $a_1^{(2),1/2} = 0$, $C_{0,1/2}(x) = 1$, which means that the zeros of the trigonometric polynomial $V_{n+1}^{C,1/2}(x)$ can be calculated by using the QR-algorithm.

Lemma 20.8 *Let w be the even weight function on $(-\pi, \pi)$ and let τ_k and σ_k, $k = 1, \ldots, n+1$, be the nodes and weights of $(n+1)$-point anti-Gaussian quadrature rule constructed for algebraic polynomials with respect to the weight function* (20.15) *on $(-1, 1)$. Then the weights λ_k and the nodes ξ_k, $k = 1, 2, \ldots, 2n+3$, of the $(2n+3)$-point anti-Gaussian quadrature rule* (20.20) *$(\gamma = 1/2)$ with respect to w are given as follows:*

$$\lambda_k = \lambda_{2n+3-k} = \frac{\sigma_k}{1+\tau_k}, \quad k = 1, \ldots, n+1, \quad \lambda_{2n+3} = \int_{-\pi}^{\pi} w(x)\,\mathrm{d}x - \sum_{k=1}^{2n+2} \lambda_k,$$

$$\xi_k = -\xi_{2n+3-k} = -\arccos \tau_k, \quad k = 1, \ldots, n+1, \quad \xi_{2n+3} = \pi.$$

Proof Using the three-term recurrence relation (20.33), we can construct the sequence of the polynomials $(C_{k,1/2}(x))$, and using the QR-algorithm we can obtain the zeros of the polynomial $C_{n+1,1/2}(x)$, i.e., nodes τ_k, $k = 1, \ldots, n+1$. According to (20.32) it is easy to see that $\xi_k = -\xi_{2n+3-k} = -\arccos \tau_k$, $k = 1, \ldots, n+1$, and by (20.29) we have $\xi_{2n+2} = \pi$.

The weights can be constructed using Shohat formula (see [22, 36]):

$$\sigma_k = \mu_0 \left[\sum_{j=0}^{n} \left(\frac{C_{j,1/2}(\tau_k)}{\prod_{i=2}^{j} a_{i,1/2}^{(2)}} \right)^2 \right]^{-1} = \frac{\mu_0(1+\tau_k)}{2 \sum_{j=0}^{n} \left(\frac{V_j^{C,1/2}(\xi_{2n+3-k})}{\prod_{i=2}^{j} a_i^{(2),1/2}} \right)^2}, \quad k = 1, \ldots, n+1,$$

where

$$\mu_0 = \int_{-1}^{1} \sqrt{\frac{1+x}{1-x}}\, w(\arccos x)\, \mathrm{d}x.$$

Since the function $w(x)$ is even on $(-\pi, \pi)$, we have

$$\lambda_{2n+3-k} = \frac{\widehat{\mu}_0}{2 \sum_{j=0}^{n} \left(\frac{V_j^{C,1/2}(\xi_{2n+3-k})}{\prod_{i=2}^{j} a_i^{(2),1/2}} \right)^2}, \quad k = 1, \ldots, n+1,$$

where

$$\widehat{\mu}_0 = \int_{-\pi}^{\pi} \cos^2 \frac{x}{2}\, w(x)\, \mathrm{d}x = 2 \int_{-1}^{1} \frac{1+t}{2}\, w(\arccos t) \frac{\mathrm{d}t}{\sqrt{1-t^2}} = \mu_0.$$

So, we obtain

$$\lambda_{2n+3-k} = \lambda_k = \frac{\sigma_k}{1+\tau_k}, \quad k = 1, \ldots, n+1.$$

Finally, using equality $\sum_{k=1}^{2n+3} \lambda_k = \int_{-\pi}^{\pi} w(x)\, \mathrm{d}x$, we get the last weight

$$\lambda_{2n+3} = \int_{-\pi}^{\pi} w(x)\, \mathrm{d}x - \sum_{k=1}^{2n+2} \lambda_k.$$

□

Lemma 20.9 *Let w be the even weight function on $(-\pi, \pi)$ and let τ_k and σ_k, $k = 1, \ldots, n+1$, be the nodes and weights of $(n+1)$-point anti-Gaussian quadrature rule constructed for algebraic polynomials with respect to the weight function* (20.16) *on* $(-1, 1)$. *Then the weights λ_k and the nodes ξ_k, $k = 1, 2, \ldots, 2n+3$,* *of the* $(2n+3)$*-point anti-Gaussian quadrature rule* (20.20) *(*$\gamma = 1/2$*) with respect to* w *are given as follows:*

$$\lambda_k = \lambda_{2n+4-k} = \frac{\sigma_k}{1-\tau_k}, \quad k = 1, \ldots, n+1, \quad \lambda_{n+2} = \int_{-\pi}^{\pi} w(x)\,dx - \sum_{\substack{k=1\\ k\neq n+2}}^{2n+3} \lambda_k,$$

$$\xi_k = -\xi_{2n+4-k} = -\arccos \tau_k, \quad k = 1, \ldots, n+1, \quad \xi_{n+2} = 0.$$

Proof This lemma can be proved in a similar way as the previous one, using polynomials $V_j^{S,1/2}(x)$, relation

$$V_k^{S,1/2}(\arccos x) = \sqrt{\frac{1-x}{2}}\, S_{n,1/2}(x),$$

and $\widehat{\mu}_0 = \int\limits_{-\pi}^{\pi} \sin^2 \frac{x}{2}\, w(x)\,dx$. □

20.4 Optimal Set of Quadrature Rules for Trigonometric Polynomials in the Sense of Borges

Motivation for introducing the optimal set of quadrature rules for trigonometric polynomials in the sense of Borges appeared in the evaluation of computer graphics illumination models (see [2]). Namely, this problem reduces to the problem of numerically evaluating a set of r definite integrals taken with respect to r distinct weight functions, but related to a common integrand and the same interval of integration, i.e., the set of integrals of the following form:

$$\int_{-\pi}^{\pi} f(x) w_k(x)\,dx, \quad k = 1, 2, \ldots, r.$$

We will restrict our attention to the quadrature rules with odd number of nodes. One solution is to use r Gaussian quadrature rules for trigonometric polynomials, given in Sect. 20.2 for $\gamma = 1/2$, but valuable information will be wasted. Analogously with [2], we introduce the performance ratio with respect to trigonometric degree of exactness as follows:

$$R_T = \frac{\text{Overall trigonometric degree of exactness} + 1}{\text{Number of integrand evaluation}}.$$

If we take distinct nodes for each of r quadrature rule, it is obvious that R_T will be maximal when we have the set of r Gaussian quadrature rule, i.e.,

$$R_T = \frac{2n+1}{r(2n+1)} = \frac{1}{r},$$

so, $R_T < 1/2$ for all $r > 2$.

More efficient solution is that distinct nodes are common for all quadrature rules. If we select a set of $2n+1$ distinct nodes, common for all quadrature rules, weight coefficients for each of r quadrature rules can be chosen in such a way that $R_T > 1/2$. Of course, the arbitrary selection of common nodes is not the best choice. But, if we choose common nodes as zeros of corresponding trigonometric polynomial of semi-integer degree, we obtain an optimal set of quadrature rules for trigonometric polynomials in the sense of Borges. So, such trigonometric polynomial of semi-integer degree is orthogonal with respect to r distinct weight functions on the same interval, i.e., it is the type II multiple orthogonal trigonometric polynomial of semi-integer degree.

Let r be a positive integer and $\mathbf{n} = (n_1, n_2, \ldots, n_r)$ be a multi-index, where $n_1, n_2, \ldots, n_r$ are nonnegative integers, and $|\mathbf{n}| = n_1 + n + 2 + \cdots + n_r$ is its length. We introduce a partial order on multi-indices $\mathbf{m} \preceq \mathbf{n}$ if $m_k \leq n_k$ for every $k = 1, 2, \ldots, r$. Let $W = (w_1, w_2, \ldots, w_r)$ be a system of weight functions, which are integrable and nonnegative on the interval $[-\pi, \pi)$, vanishing there only on a set of measure zero.

We introduce the following inner products:

$$\langle f, g\rangle_k = \int_{-\pi}^{\pi} f(x)g(x)\, w_\nu(x)\, \mathrm{d}x, \quad k = 1, 2, \ldots, r, \quad f, g \in \mathcal{T}^{1/2}. \tag{20.34}$$

Two types of multiple orthogonal trigonometric polynomials of semi-integer degree are introduced in [29, 41].

1° Type I multiple orthogonal trigonometric polynomials of semi-integer degree with respect to $(W, \mathbf{n})$ are given in a vector of trigonometric polynomials of semi-integer degree $(T_{\mathbf{n},1}^{1/2}, T_{\mathbf{n},2}^{1/2}, \ldots, T_{\mathbf{n},r}^{1/2})$, where $T_{\mathbf{n},k}^{1/2}$ has semi-integer degree $n_k - 1/2$, $k = 1, 2, \ldots, r$, such that the following orthogonality conditions hold:

$$\sum_{k=1}^{r}\left\langle T_{\mathbf{n},k}^{1/2}, \cos\left(\nu+\frac{1}{2}\right)x\right\rangle_k = 0, \quad \nu = 0, 1, 2, \ldots, |\mathbf{n}|-2, \tag{20.35}$$

$$\sum_{k=1}^{r}\left\langle T_{\mathbf{n},k}^{1/2}, \sin\left(\nu+\frac{1}{2}\right)x\right\rangle_k = 0, \quad \nu = 0, 1, 2, \ldots, |\mathbf{n}|-2,$$

with the normalizations:

$$\sum_{k=1}^{r}\left\langle T_{\mathbf{n},k}^{1/2}, \cos\left(|\mathbf{n}|-\frac{1}{2}\right)x\right\rangle_k = 1, \tag{20.36}$$

$$\sum_{k=1}^{r}\left\langle T_{\mathbf{n},k}^{1/2}, \sin\left(|\mathbf{n}|-\frac{1}{2}\right)x\right\rangle_k = 0,$$

or

$$\sum_{k=1}^{r}\left\langle T_{\mathbf{n},k}^{1/2}, \cos\left(|\mathbf{n}|-\frac{1}{2}\right)x\right\rangle_k = 0, \tag{20.37}$$

$$\sum_{k=1}^{r}\left\langle T_{\mathbf{n},k}^{1/2}, \sin\left(|\mathbf{n}|-\frac{1}{2}\right)x\right\rangle_k = 1.$$

The conditions (20.35)–(20.36) or (20.35)–(20.37) give a linear system of $2|\mathbf{n}|$ equations, for $2|\mathbf{n}|$ unknown coefficients of the trigonometric polynomials of semi-integer degree $(T_{\mathbf{n},1}^{1/2}, T_{\mathbf{n},2}^{1/2}, \ldots, T_{\mathbf{n},r}^{1/2})$. The multi-index $\mathbf{n}$ is normal for type I if the system (20.35)–(20.36) or (20.35)–(20.37) has a unique solution.

2° Type II multiple orthogonal trigonometric polynomial of semi-integer degree with respect to $(W, \mathbf{n})$ is trigonometric polynomial $T_{\mathbf{n}}^{1/2}$ of semi-integer degree $|\mathbf{n}|+1/2$ which satisfies the following orthogonality conditions:

$$\begin{aligned}&\left\langle T_{\mathbf{n}}^{1/2}(x), \cos\left(\nu_k+\frac{1}{2}\right)x\right\rangle_\nu = 0,\\&\left\langle T_{\mathbf{n}}^{1/2}(x), \sin\left(\nu_k+\frac{1}{2}\right)x\right\rangle_\nu = 0, \quad \nu_k = 0, 1, \ldots, n_k-1, \quad k = 1, 2, \ldots, r.\end{aligned} \tag{20.38}$$

Let us notice that if some $n_k = 0$, we do not have orthogonality conditions (20.38) for the corresponding weight w_k and if $r = 1$ we have the case of ordinary orthogonal trigonometric polynomials of semi-integer degree (see Sect. 20.2 for $\gamma = 1/2$).

Since $T_{\mathbf{n}}^{1/2} \in \mathcal{T}_{\mathbf{n}}^{1/2}$ and $\dim(\mathcal{T}_{\mathbf{n}}^{1/2}) = 2|\mathbf{n}|+2$, conditions (20.38) give $2|\mathbf{n}|$ equations for the $2|\mathbf{n}|+2$ unknown coefficients of polynomials $T_{\mathbf{n}}^{1/2}$, we have to fix two coefficients. Similar to the case of orthogonal trigonometric polynomials of semi-integer degree, we choose to fix the leading coefficients $(c_{|\mathbf{n}|}, d_{|\mathbf{n}|}) \in \{(1,0),(0,1)\}$ and we use notation $T_{|\mathbf{n}|}^{C,1/2}$ and $T_{|\mathbf{n}|}^{S,1/2}$, respectively, i.e.,

$$T_{|\mathbf{n}|}^{C,1/2}(x) = \cos\left(|\mathbf{n}| + \frac{1}{2}\right)x + \sum_{k=0}^{|\mathbf{n}|-1}\left(c_k \cos\left(k+\frac{1}{2}\right)x + d_k \sin\left(k+\frac{1}{2}\right)x\right), \tag{20.39}$$

$$T_{|\mathbf{n}|}^{S,1/2}(x) = \sin\left(|\mathbf{n}| + \frac{1}{2}\right)x + \sum_{k=0}^{|\mathbf{n}|-1}\left(f_k \cos\left(k+\frac{1}{2}\right)x + g_k \sin\left(k+\frac{1}{2}\right)x\right).$$

The uniqueness of such polynomials can only be guaranteed under additional assumptions on r weights, i.e., if the set of functions

$$\left\{w_k(x)\cos\left(\nu_k+\frac{1}{2}\right)x,\, w_k(x)\sin\left(\nu_k+\frac{1}{2}\right)x \mid \nu_k = 0,1,\ldots,n_k-1,\right.$$
$$\left. k=1,2,\ldots,r\right\}$$

form a Chebyshev system on $[-\pi,\pi)$ for the multi-index $\mathbf{n}$. Such system of weight functions $W = (w_1, w_2, \ldots, w_r)$ is called trigonometric T system (TT system) for multi-index $\mathbf{n}$ (see [29]). If all multi-indices are normal, then we have a perfect system.

The properties of zeros of type II multiple orthogonal trigonometric polynomials of semi-integer degree, which are important for the construction of the optimal set of quadrature rules for trigonometric polynomials in the sense of Borges, are given in the following theorem.

Theorem 20.8 *Suppose that* $\mathbf{n}$ *is a multi-index such that* $W = (w_1, w_2, \ldots, w_r)$ *is a TT system of weight functions for all multi-indices* $\mathbf{m}$ *such that* $\mathbf{m} \preceq \mathbf{n}$. *Type II multiple trigonometric orthogonal polynomial of semi-integer degree* $T_{\mathbf{n}}^{1/2}$ *with respect to* $(W, \mathbf{n})$ *has exactly* $2|\mathbf{n}|+1$ *simple zeros on* $[-\pi,\pi)$.

Now we can define optimal set of quadrature rules for trigonometric polynomials in the sense of Borges (see [29]).

Definition 20.2 Let $\mathbf{n}$ be a multi-index and let $W = (w_1, w_2, \ldots, w_r)$ be a TT system for $\mathbf{n}$ on interval $[-\pi,\pi)$. A set of quadrature rules of the form

$$\int_{-\pi}^{\pi} f(x)w_k(x)\,\mathrm{d}x \approx \sum_{i=1}^{2|\mathbf{n}|+1} A_{k,i} f(x_i), \quad k=1,2,\ldots,r, \tag{20.40}$$

is an optimal set with respect to $(W, \mathbf{n})$ if and only if the weight coefficients $A_{k,i}$, $k=1,2,\ldots,r$, $i=1,2,\ldots,2|\mathbf{n}|+1$, and the nodes x_i, $i=1,2,\ldots,2|\mathbf{n}|+1$, satisfy the following equations:

$$\sum_{k=1}^{2|\mathbf{n}|+1} A_{k,i} = \int_{-\pi}^{\pi} w_k(x)\,\mathrm{d}x, \tag{20.41}$$

$$\sum_{k=1}^{2|\mathbf{n}|+1} A_{k,i}\cos(\nu_k x_i) = \int_{-\pi}^{\pi} \cos(\nu_k x)\, w_k(x)\,\mathrm{d}x, \quad \nu_k = 1, 2, \ldots, |\mathbf{n}| + n_k,$$

$$\sum_{k=1}^{2|\mathbf{n}|+1} A_{k,i}\sin(\nu_k x_i) = \int_{-\pi}^{\pi} \sin(\nu_k x)\, w_k(x)\,\mathrm{d}x, \quad \nu_k = 1, 2, \ldots, |\mathbf{n}| + n_k,$$

for $k = 1, 2, \ldots, r$.

Let us notice that

$$R_T = \frac{|\mathbf{n}| + m + 1}{2|\mathbf{n}| + 1} = \frac{1}{2} + \frac{m + 1/2}{2|\mathbf{n}| + 1} > \frac{1}{2},$$

since overall trigonometric degree of exactness is equal to $|\mathbf{n}| + m$, where $m = \min\{n_1, n_2, \ldots, n_r\}$ and an optimal set of quadrature rules has $2|\mathbf{n}| + 1$ distinct nodes, common for all quadrature rules.

The following theorem gives the characterization for the optimal set of quadrature rules for trigonometric polynomials in the sense of Borges, which is a generalization of Theorem 20.4, for $\gamma = 1/2$, and a counterpart to the fundamental theorem of Gaussian quadrature.

Theorem 20.9 *Let* $\mathbf{n}$ *be a multi-index and let* $W = (w_1, w_2, \ldots, w_r)$ *be a TT system for* $\mathbf{n}$ *on interval* $[-\pi, \pi)$. *A set of quadrature rules* (20.40) *is an optimal set with respect to* $(W, \mathbf{n})$ *if and only if:*

1° *All rules are exact for all trigonometric polynomials from* $\mathcal{T}_{|\mathbf{n}|}$;
2° $T_{\mathbf{n}}^{1/2}(x) = \prod_{i=1}^{2|\mathbf{n}|+1} \sin \frac{x - x_i}{2}$ *is the type II multiple orthogonal trigonometric polynomial of semi-integer degree* $|\mathbf{n}| + 1/2$ *with respect to* $(W, \mathbf{n})$.

20.4.1 System of Even Weight Functions

Similar as in the case of quadrature rules for trigonometric polynomials with respect to the one weight function given in Sect. 20.2.1, we will consider even system of weight functions.

Let $W = (w_1, w_2, \ldots, w_r)$ be a TT system for multi-index $\mathbf{n}$, of even weight functions w_k, $k = 1, 2, \ldots, r$, on interval $(-\pi, \pi)$. Construction of the corresponding type II multiple orthogonal trigonometric polynomials of semi-integer degree is easier in the case of a system of even weight functions, i.e., nodes of the corresponding optimal set of quadrature rules for trigonometric polynomials in the sense of Borges can be obtained in the easier way.

In what follows we simple say that the TT system $W = (w_1, w_2, \ldots, w_r)$ is a TT system of even weight functions on the interval $[-\pi, \pi)$ with respect to $\mathbf{n}$, when all weight functions from W are even on the interval $(-\pi, \pi)$.

Theorem 20.10 *Let* $\mathbf{n}$ *be a multi-index and* $W = (w_1, w_2, \ldots, w_r)$ *be a TT system of even weight functions on the interval* $[-\pi, \pi)$ *with respect to* $\mathbf{n}$. *The type II multiple orthogonal trigonometric polynomials of semi-integer degree* (20.39) *reduce to*

$$T_{\mathbf{n}}^{C,1/2}(x) = \cos\left(|\mathbf{n}| + \frac{1}{2}\right)x + \sum_{k=0}^{|\mathbf{n}|-1} c_k \cos\left(k + \frac{1}{2}\right)x \tag{20.42}$$

and

$$T_{\mathbf{n}}^{S,1/2}(x) = \sin\left(|\mathbf{n}| + \frac{1}{2}\right)x + \sum_{k=0}^{|\mathbf{n}|-1} g_k \sin\left(k + \frac{1}{2}\right)x. \tag{20.43}$$

Proof Let us consider the type II multiple orthogonal trigonometric polynomial of semi-integer degree $T_{\mathbf{n}}^{C,1/2}(x)$ with respect to the system of even weight functions W. Then, the system of Eqs. (20.38) reduces to the following two independent systems:

$$\sum_{k=0}^{|\mathbf{n}|-1} c_k \left\langle \cos\left(k + \frac{1}{2}\right)x, \cos\left(k_\nu + \frac{1}{2}\right)x \right\rangle_\nu = -\left\langle \cos\left(|\mathbf{n}| + \frac{1}{2}\right)x, \cos\left(k_\nu + \frac{1}{2}\right)x \right\rangle_\nu$$

and

$$\sum_{k=0}^{|\mathbf{n}|-1} d_k \left\langle \sin\left(k + \frac{1}{2}\right)x, \sin\left(k_\nu + \frac{1}{2}\right)x \right\rangle_\nu = 0, \tag{20.44}$$

where $k_\nu = 0, 1, \ldots, n_\nu - 1$, $\nu = 1, 2, \ldots, r$, and unknown coefficients c_k and d_k, $k = 0, 1, \ldots, |\mathbf{n}| - 1$, respectively. One obvious solution of system (20.44) is the trivial solution $d_k = 0$, $k = 0, 1, \ldots, |\mathbf{n}| - 1$. This system has a unique solution, since W is a TT system with respect to $\mathbf{n}$ on the interval $[-\pi, \pi)$, so we conclude that system (20.44) has only the trivial solution $d_k = 0$, $k = 0, 1, \ldots, |\mathbf{n}| - 1$, i.e., $T_{\mathbf{n}}^{C,1/2}(x)$ depends only on cosine functions.

The statement for $T_{\mathbf{n}}^{S,1/2}(x)$ can be proved in a similar way. □

The following corollary follows immediately from (20.42) and (20.43).

Corollary 20.2 *Let* $\mathbf{n}$ *be a multi-index and* $W = (w_1, w_2, \ldots, w_r)$ *be a TT system of even weight functions on the interval* $[-\pi, \pi)$ *with respect to* $\mathbf{n}$. *Then*

$$T_{\mathbf{n}}^{C,1/2}(-\pi) = 0 \quad \text{and} \quad T_{\mathbf{n}}^{S,1/2}(0) = 0.$$

Theorem 20.11 *Let* $\mathbf{n}$ *be a multi-index and* $W = (w_1, w_2, \ldots, w_r)$ *be a TT system of even weight functions on the interval* $[-\pi, \pi)$ *with respect to* $\mathbf{n}$. *Then we have*

$$\int_{-1}^{1} C_{\mathbf{n}}(x) C_{k_\nu}(x) \sqrt{\frac{1+x}{1-x}} w_\nu(\arccos x) \mathrm{d}x = 0$$

and

$$\int_{-1}^{1} S_{\mathbf{n}}(x) S_{k_\nu}(x) \sqrt{\frac{1-x}{1+x}} w_\nu(\arccos x) \mathrm{d}x = 0,$$

where $k_\nu = 0, 1, \ldots, n_\nu - 1$, $\nu = 1, 2, \ldots, r$, *and* C_{k_ν}, S_{k_ν} *are algebraic polynomials of degree less or equal to* k_ν; $C_{\mathbf{n}}$, $S_{\mathbf{n}}$, *are algebraic polynomials of degree less or equal to* $|\mathbf{n}|$, *given by*

$$C_{\mathbf{n}}(x) = \sum_{k=0}^{|\mathbf{n}|} c_k \big(T_k(x) - (1-x) U_{k-1}(x)\big)$$

and

$$S_{\mathbf{n}}(x) = \sum_{k=0}^{|\mathbf{n}|} g_k \big(T_k(x) + (1+x) U_{k-1}(x)\big),$$

T_k *and* U_k, $k \in \mathbb{N}_0$ *are Chebyshev polynomials of the first and second kind, respectively.*

Proof Since $T_{\mathbf{n}}^{C,1/2}$ is the type II multiple orthogonal trigonometric polynomial of semi-integer degree with respect to the system of even weight functions W on the interval $(-\pi, \pi)$, from the orthogonality conditions, we have that

$$\int_{0}^{\pi} T_{\mathbf{n}}^{C,1/2}(x) T_{k_\nu}^{C,1/2}(x)\, w_\nu(x)\, \mathrm{d}x = 0, \quad k_\nu = 0, 1, \ldots, n_\nu - 1, \; \nu = 1, 2, \ldots, r.$$

Applying the substitution $x := \arccos x$ in the previous integral, we get

$$\int_{-1}^{1} T_{\mathbf{n}}^{C,1/2}(\arccos x) T_{k_\nu}^{C,1/2}(\arccos x) \frac{w_\nu(\arccos x)}{\sqrt{1-x^2}} \mathrm{d}x = 0. \tag{20.45}$$

By using elementary trigonometric transformations we get

$$T_{\mathbf{n}}^{C,1/2}(\arccos x) = \sqrt{\frac{1+x}{2}} \sum_{k=0}^{|\mathbf{n}|} c_k^{(\mathbf{n})} \big(T_k(x) - (1-x)U_{k-1}(x)\big),$$

where $T_k(x)$ and $U_{k-1}(x)$, are Chebyshev polynomials of the first and second kind, respectively. By substituting the obtained formulas in (20.45), after some elementary transformations, we get the first assertion.

The second assertion can be proved similarly, by using orthogonality conditions for $T_{\mathbf{n}}^{S,1/2}$ and equality

$$\sin\left(\left(k+\frac{1}{2}\right)\arccos x\right) = \sqrt{\frac{1-x}{2}}T_k(x) + \sqrt{\frac{1+x}{2}}\sqrt{1-x^2}U_{k-1}(x).$$

□

From the proof of Theorem 20.11 we conclude that

$$T_{\mathbf{n}}^{C,1/2}(\arccos x) = \sqrt{\frac{1+x}{2}}C_{\mathbf{n}}(x), \tag{20.46}$$

where $C_{\mathbf{n}}$ is the type II multiple orthogonal algebraic polynomial with respect to the multi-index $\mathbf{n}$ and weight functions

$$\left(\sqrt{\frac{1+x}{1-x}}w_1(\arccos x), \sqrt{\frac{1+x}{1-x}}w_2(\arccos x), \ldots, \sqrt{\frac{1+x}{1-x}}w_r(\arccos x)\right) \tag{20.47}$$

on the interval $[-1, 1)$. Accordingly, the zeros of the corresponding type II multiple orthogonal algebraic polynomial (see [27, 28]) can be used for the calculation of zeros the type II multiple orthogonal trigonometric polynomial of semi-integer degree $T_{\mathbf{n}}^{C,1/2}$. Formulas for this calculation are given in the following two lemmas.

Lemma 20.10 *Let* $\mathbf{n}$ *be a multi-index and let* $W = (w_1, w_2, \ldots, w_r)$ *be a TT system of even weight functions on the interval* $[-\pi, \pi)$ *with respect to* $\mathbf{n}$. *If* τ_k, $k = 1, 2, \ldots, |\mathbf{n}|$, *are the zeros of the type II multiple orthogonal algebraic polynomial of degree* $|\mathbf{n}|$ *with respect to the vector of weight functions* (20.47) *on the interval* $[-1, 1)$, *then the zeros of the type II multiple orthogonal trigonometric polynomial of semi-integer degree* $|\mathbf{n}| + 1/2$ *with respect to* $(W, \mathbf{n})$ *on the interval* $[-\pi, \pi)$ *are given by*

$$x_1 = -\pi, \quad x_{2|\mathbf{n}|-k+2} = -x_{k+1} = \arccos \tau_k, \quad k = 1, 2, \ldots, |\mathbf{n}|.$$

Proof First, we have that $x_1 = -\pi$ (see Corollary 20.2). From (20.46) it follows that

$$x_{2|\mathbf{n}|-k+2} = -x_{k+1} = \arccos \tau_k, \quad k = 1, 2, \ldots, |\mathbf{n}|,$$

where τ_k, $k = 1, 2, \ldots, |\mathbf{n}|$, are the zeros of the type II multiple orthogonal algebraic polynomial with respect to the multi-index $\mathbf{n}$ and the vector of weight functions (20.47) on the interval $[-1, 1)$. □

The proof of the following lemma is similar.

Lemma 20.11 *Let* $\mathbf{n}$ *be a multi-index and let* $W = (w_1, w_2, \ldots, w_r)$ *be a TT system of even weight functions on the interval* $(-\pi, \pi)$ *with respect to* $\mathbf{n}$. *If* τ_k, $k = 1, 2, \ldots, |\mathbf{n}|$, *are the zeros of the type II multiple orthogonal algebraic polynomial of degree* $|\mathbf{n}|$ *with respect to the vector of weight functions*

$$\left(\sqrt{\frac{1-x}{1+x}}w_1(\arccos x), \sqrt{\frac{1-x}{1+x}}w_2(\arccos x), \ldots, \sqrt{\frac{1-x}{1+x}}w_r(\arccos x)\right)$$

on the interval $[-1, 1)$, *then the zeros of the type II multiple orthogonal trigonometric polynomial of semi-integer degree* $|\mathbf{n}| + 1/2$ *with respect to* $(W, \mathbf{n})$ *on the interval* $[-\pi, \pi)$ *are given by*

$$x_{|\mathbf{n}|+1} = 0, \quad x_{2|\mathbf{n}|-k+2} = -x_k = \arccos \tau_k, \quad k = 1, 2, \ldots, |\mathbf{n}|.$$

20.4.2 Recurrence Relations in the Case of a System of Even Weight Functions and Nearly Diagonal Multi-Index

Recurrence relations are fundamental for the general theory of multiple orthogonal trigonometric polynomials. In this subsection we will give recurrence relations for the type II multiple orthogonal trigonometric polynomials of semi-integer degree for nearly diagonal multi-index and system of even weight functions on $(-\pi, \pi)$. These recurrence relations are generalization of the recurrence relations for multiple orthogonal algebraic polynomials for nearly diagonal multi-index (see [27, 51]). Also, such recurrence relations reduce to (20.14) if we have one even weight function.

Let m be a nonnegative integer, which can be written as $m = \ell r + j$, for $\ell = [m/r]$ and $j \in \{0, 1, \ldots, r-1\}$. The nearly diagonal multi-index $\mathbf{d}(m)$ corresponding to m is multi-index of the form

$$\mathbf{d}(m) = (\underbrace{\ell+1, \ell+1, \ldots, \ell+1}_{j \text{ times}}, \underbrace{\ell, \ell, \ldots, \ell}_{r-j \text{ times}}).$$

The corresponding type II monic multiple orthogonal trigonometric polynomials of semi-integer degree with respect to the system of even weight functions $W = (w_1, w_2, \ldots, w_r)$ will be denoted as

$$T_m^{C,1/2} = T_{\mathbf{d}(m)}^{C,1/2}, \quad T_m^{S,1/2} = T_{\mathbf{d}(m)}^{S,1/2}.$$

Theorem 20.12 *Let m be a nonnegative integer and let $W = (w_1, w_2, \ldots, w_r)$ be a TT system of even weight functions on the interval $[-\pi, \pi)$ with respect to all nearly diagonal multi-indices $\mathbf{d}(k) \preceq \mathbf{d}(m+1)$, $k \in \mathbb{N}$. The type II multiple orthogonal trigonometric polynomials of semi-integer degree with nearly diagonal multi-indices $T_m^{C,1/2}$ and $T_m^{S,1/2}$ satisfy the following recurrence relations:*

$$2\cos x T_m^{C,1/2}(x) = T_{m+1}^{C,1/2}(x) + \sum_{k=0}^{r} \alpha_{m,r-k} T_{m-k}^{C,1/2}(x), \tag{20.48}$$

$$2\cos x T_m^{S,1/2}(x) = T_{m+1}^{S,1/2}(x) + \sum_{k=0}^{r} \beta_{m,r-k} T_{m-k}^{S,1/2}(x), \tag{20.49}$$

with the initial conditions $T_0^{C,1/2}(x) = \cos(x/2)$, $T_i^{C,1/2}(x) = 0$, $i = -1, -2, \ldots, -r$ and $T_0^{S,1/2}(x) = \sin(x/2)$, $T_i^{S,1/2}(x) = 0$, $i = -1, -2, \ldots, -r$, for the relations (20.48) *and* (20.49), *respectively.*

Proof Multiplying (20.42) by $2\cos x$ we obtain

$$2\cos x T_m^{C,1/2}(x) = T_{m+1}^{C,1/2}(x) + \sum_{k=0}^{m} \alpha_{m,k} T_k^{C,1/2}(x). \tag{20.50}$$

Let us assume that $m = \ell r + j$. If $m \leq r$, then equality (20.50) is the same as equality (20.48) with the given initial conditions.

Let us now suppose that $m > r$. If we prove that $\alpha_{m,pr+i} = 0$ for $p = 0, 1, \ldots, \ell - 2$ and $i = 0, 1, \ldots, r-1$, and for $p = \ell - 1$ and $i = 0, 1, \ldots, j-1$, then the right-hand side of (20.50) reduces to

$$T_{m+1}^{C,1/2}(x) + \sum_{k=m-r}^{m} \alpha_{m,k} T_k^{C,1/2}(x),$$

i.e., to the right-hand side of (20.48).

First, by multiplying both sides of (20.50) by $T_p^{C,1/2}(x)w_1(x)$ and integrate on the interval $[-\pi,\pi)$, for all $p=0,1,\ldots,\ell-2$, we obtain

$$\int_{-\pi}^{\pi} 2\cos x T_m^{C,1/2}(x)T_p^{C,1/2}(x)w_1(x)\,\mathrm{d}x \tag{20.51}$$

$$= \int_{-\pi}^{\pi} T_{m+1}^{C,1/2}(x)T_p^{C,1/2}(x)w_1(x)\,\mathrm{d}x + \sum_{k=0}^{m}\alpha_{m,k}\int_{-\pi}^{\pi} T_k^{C,1/2}(x)T_p^{C,1/2}(x)w_1(x)\,\mathrm{d}x.$$

By using the orthogonality conditions (20.38) (the first coordinate of the multi-index $\mathbf{d}(m)$ is equal to $\ell+1$, $\ell\geq p+2$) we have that

$$\int_{-\pi}^{\pi} \cos x T_p^{C,1/2}(x)T_m^{C,1/2}(x)\,w_1(x)\,\mathrm{d}x$$

is equal to zero whenever $m>r$, i.e., the left-hand side of (20.51) is equal to zero. The right-hand side of (20.51) reduces to

$$\alpha_{m,pr}\int_{-\pi}^{\pi} T_p^{C,1/2}(x)T_{pr}^{C,1/2}(x)\,w_1(x)\,\mathrm{d}x.$$

If we assume that the previous integral is equal to zero, then we have one orthogonality condition more which implies that $T_{pr}^{C,1/2}(x)\equiv 0$. So, it has to be $\alpha_{m,pr}=0$. Now, we multiply both sides of equality (20.50) by $T_p^{C,1/2}(x)w_2(x)$ and integrate on $[-\pi,\pi)$, so we prove that $\alpha_{m,pr+1}=0$. Generally, multiplying (20.50) with $T_p^{C,1/2}(x)w_{i+1}(x)$ and integrating on the interval $[-\pi,\pi)$, from orthogonality conditions (20.38) we prove that $\alpha_{m,pr+i}=0$, $i=0,1,2,\ldots,r-1$.

Finally, suppose that $p=\ell-1$. Multiplying (20.50) by $T_{\ell-1}^{C,1/2}(x)w_{i+1}(x)$ and integrating on $[-\pi,\pi)$, by using the orthogonality, we prove that $\alpha_{m,(\ell-1)r+i}=0$ for $i=0,1,\ldots,j-1$.

In a similar way one can obtain the recurrence relation (20.49) for $T_m^{S,1/2}(x)$. □

According to the previous theorem, if we have coefficients of the recurrence relations, then the type II multiple orthogonal trigonometric polynomials of semi-integer degree $T_m^{C,1/2}$ and $T_m^{C,1/2}$ can be obtained.

In order to simplify the notation, first, we consider the simplest case with two weight functions, i.e., $r=2$ for trigonometric polynomials $T_m^{C,1/2}$ (in a similar way one can obtain the corresponding coefficients of the recurrence relations for trigonometric polynomials $T_m^{S,1/2}$). In this case we have the multi-indices $\mathbf{d}(m)=(m_1,m_2)$, where $m_1=[(m+1)/2]$ and $m_2=[m/2]$ (obviously, $m_1+m_2=m$). Recurrence relation (20.48) reduces to

$$T_{m+1}^{C,1/2}(x)=(2\cos x-\alpha_{m,2})T_m^{C,1/2}-\alpha_{m,1}T_{m-1}^{C,1/2}(x)-\alpha_{m,0}T_{m-2}^{C,1/2}(x), \tag{20.52}$$

for $m = 0, 1, 2, \ldots$, with the initial conditions $T_0^{C,1/2}(x) = \cos(x/2)$, $T_{-1}^{C,1/2}(x) = T_{-2}^{C,1/2}(x) = 0$. In order to determine the recurrence coefficients we use recurrence relations (20.52) and the orthogonality conditions

$$\left\langle T_m^{C,1/2}, T_i^{C,1/2} \right\rangle_1 = 0, \quad \text{for } i \leq \left[\frac{m-1}{2}\right],$$

$$\left\langle T_m^{C,1/2}, T_i^{C,1/2} \right\rangle_2 = 0, \quad \text{for } i \leq \left[\frac{m-2}{2}\right],$$

where $\langle \cdot, \cdot \rangle_i$, $i = 1, 2$, are the inner products given by (20.34).

Since $\left\langle T_1^{C,1/2}, T_0^{C,1/2} \right\rangle_1 = 0$, from (20.52) for $m = 0$ we get

$$\alpha_{02} = \frac{\left\langle 2\cos x T_0^{C,1/2}, T_0^{C,1/2} \right\rangle_1}{\left\langle T_0^{C,1/2}, T_0^{C,1/2} \right\rangle_1}. \tag{20.53}$$

By using (20.52) for $m = 1$, $\left\langle T_2^{C,1/2}, T_0^{C,1/2} \right\rangle_1 = 0$ and $\left\langle T_2^{C,1/2}, T_0^{C,1/2} \right\rangle_2 = 0$, we have

$$\alpha_{11} = \frac{\left\langle 2\cos x T_1^{C,1/2}, T_0^{C,1/2} \right\rangle_1}{\left\langle T_0^{C,1/2}, T_0^{C,1/2} \right\rangle_1}, \quad \alpha_{12} = \frac{\left\langle 2\cos x T_1^{C,1/2} - \alpha_{11} T_0^{C,1/2}, T_0^{C,1/2} \right\rangle_2}{\left\langle T_1^{C,1/2}, T_0^{C,1/2} \right\rangle_2}. \tag{20.54}$$

Similarly, (20.52) for $m = 2$ and the orthogonality conditions $\left\langle T_3^{C,1/2}, T_0^{C,1/2} \right\rangle_1 = 0$, $\left\langle T_3^{C,1/2}, T_0^{C,1/2} \right\rangle_2 = 0$, and $\left\langle T_3^{C,1/2}, T_1^{C,1/2} \right\rangle_1 = 0$ give

$$\begin{aligned}
\alpha_{20} &= \frac{\left\langle 2\cos x T_2^{C,1/2}, T_0^{C,1/2} \right\rangle_1}{\left\langle T_0^{C,1/2}, T_0^{C,1/2} \right\rangle_1}, \\
\alpha_{21} &= \frac{\left\langle 2\cos x T_2^{C,1/2} - \alpha_{20} T_0^{C,1/2}, T_0^{C,1/2} \right\rangle_2}{\left\langle T_1^{C,1/2}, T_0^{C,1/2} \right\rangle_2}, \\
\alpha_{22} &= \frac{\left\langle 2\cos x T_2^{C,1/2} - \alpha_{20} T_0^{C,1/2} - \alpha_{21} T_1^{C,1/2}, T_1^{C,1/2} \right\rangle_1}{\left\langle T_2^{C,1/2}, T_1^{C,1/2} \right\rangle_1}.
\end{aligned} \tag{20.55}$$

In general, this procedure can be generalized, and one can prove the following result.

Theorem 20.13 *Let $m = 2\ell + \nu$, where $\ell = [m/2]$ and $\nu \in \{0, 1\}$. The recurrence coefficients in* (20.52) *can be expressed in the form*

$$\alpha_{m,0} = \frac{\left\langle 2\cos x\, T_m^{C,1/2}, T_{[(m-2)/2]}^{C,1/2} \right\rangle_{\nu+1}}{\left\langle T_{m-2}^{C,1/2}, T_{[(m-2)/2]}^{C,1/2} \right\rangle_{\nu+1}}, \tag{20.56}$$

$$\alpha_{m,1} = \frac{\left\langle 2\cos x\, T_m^{C,1/2} - \alpha_{m,0} T_{m-2}^{C,1/2}, T_{[(m-1)/2]}^{C,1/2} \right\rangle_{\nu}}{\left\langle T_{m-1}^{C,1/2}, T_{[(m-1)/2]}^{C,1/2} \right\rangle_{\nu}},$$

$$\alpha_{m,2} = \frac{\left\langle 2\cos x\, T_m^{C,1/2} - \alpha_{m,0} T_{m-2}^{C,1/2} - \alpha_{m,1} T_{m-1}^{C,1/2}, T_{[m/2]}^{C,1/2} \right\rangle_{\nu+1}}{\left\langle T_m^{C,1/2}, T_{[m/2]}^{C,1/2} \right\rangle_{\nu+1}},$$

where we put $\langle\cdot,\cdot\rangle_{j+2k} = \langle\cdot,\cdot\rangle_j$, $j = 1, 2$, for each $k \in \mathbb{Z}$.

Now, the previous theorem can be generalized to the case $r \in \mathbb{N}$, $r \geq 3$, for the even weight functions w_j, $j = 1, 2, \ldots, r$. By using $\langle\cdot,\cdot\rangle_{j+rk} = \langle\cdot,\cdot\rangle_j$, $k \in \mathbb{Z}$, the following result can be proved.

Theorem 20.14 *Let m be a nonnegative integer and $W = (w_1, w_2, \ldots, w_r)$ be a TT system of even weight functions on the interval $[-\pi, \pi)$ with respect to all nearly diagonal multi-indices $\mathbf{d}(k) \preceq \mathbf{d}(m+1)$, $k \in \mathbb{N}$. The type II multiple orthogonal trigonometric polynomials of semi-integer degree with nearly diagonal multi-indices $T_m^{C,1/2}$ and $T_m^{S,1/2}$ satisfy the recurrence relations:*

$$T_{m+1}^{C,1/2}(x) = (2\cos x - \alpha_{m,r}) T_m^{C,1/2}(x) - \sum_{k=0}^{r-1} \alpha_{m,k} T_{m-r+k}^{C,1/2}(x), \tag{20.57}$$

$$T_{m+1}^{S,1/2}(x) = (2\cos x - \beta_{m,r}) T_m^{S,1/2}(x) - \sum_{k=0}^{r-1} \beta_{m,k} T_{m-r+k}^{S,1/2}(x), \tag{20.58}$$

with coefficients given by

$$\alpha_{m,0} = \frac{\left\langle 2\cos x\, T_m^{C,1/2}, T_{[(m-r)/p]}^{C,1/2} \right\rangle_{j+1}}{\left\langle T_{m-r}^{C,1/2}, T_{[(m-r)/r]}^{C,1/2} \right\rangle_{j+1}},$$

$$\alpha_{m,k} = \frac{\left\langle 2\cos x\, T_m^{C,1/2} - \sum_{i=0}^{k-1} \alpha_{m,i} T_{m-r+i}^{C,1/2}, T_{[(m-r+k)/p]}^{C,1/2} \right\rangle_{j+k+1}}{\left\langle T_{m-r+k}^{C,1/2}, T_{[(m-r+k)/p]}^{C,1/2} \right\rangle_{j+k+1}},$$

and

$$\beta_{m,0} = \frac{\left\langle 2\cos x T_m^{S,1/2}, T_{[(m-r)/r]}^{S,1/2} \right\rangle_{j+1}}{\left\langle T_{m-r}^{S,1/2}, T_{[(m-r)/r]}^{S,1/2} \right\rangle_{j+1}},$$

$$\beta_{m,k} = \frac{\left\langle 2\cos x T_m^{S,1/2} - \sum_{i=0}^{k-1} \beta_{m,i} T_{m-r+i}^{S,1/2}, T_{[(m-r+k)/r]}^{S,1/2} \right\rangle_{j+k+1}}{\left\langle T_{m-r+k}^{S,1/2}, T_{[(m-r+k)/r]}^{S,1/2} \right\rangle_{j+k+1}},$$

where $k = 1, 2, \dots, r$, *for* $\ell = [m/r]$ *and* $j = m - \ell r \in \{0, 1, \dots, r-1\}$.

The inner products $\langle \cdot, \cdot \rangle_\nu$, $\nu = 1, 2, \dots, r$, which appear in the previous theorem, can be computed exactly, except for rounding errors, by using quadrature rules of Gaussian type for trigonometric polynomials with respect to the corresponding weight functions w_ν. Finally, in order to numerically construct the type II multiple orthogonal trigonometric polynomials of semi-integer degree $T_m^{C,1/2}$ and $T_m^{S,1/2}$, $m \geq 0$, for nearly diagonal multi-indices with respect to even weight functions, we only need recurrence relations (20.57) and (20.58) and quadrature rules of Gaussian type for trigonometric polynomials, given in Sect. 20.2. Such procedure is similar to the discretized Stieltjes–Gautschi procedure (see [15]).

20.5 Quadrature Rules for Trigonometric Polynomials with Preassigned Nodes

An optimal set of quadrature rules for algebraic polynomials with preassigned nodes was considered in [26]). In this section we consider quadrature rules for trigonometric polynomials with preassigned nodes, i.e., quadrature rule considered in Sects. 20.2 and 20.4 with preassigned nodes. In the case of quadrature rules for trigonometric polynomials, a number of preassigned nodes have to be even in order to stay in the linear space of trigonometric polynomials (see [42]).

Definition 20.3 Let w be a weight function on $[-\pi, \pi)$ and $\gamma \in \{0, 1/2\}$. Suppose that $t_i \in [-\pi, \pi)$, $i = 1, 2, \dots, 2k$, are fixed in advance. The quadrature rule

$$\int_{-\pi}^{\pi} f(x) w(x) \mathrm{d}x = \sum_{i=1}^{2k} a_i f(t_i) + \sum_{i=1}^{2(n+\gamma)} A_i f(x_i) + R_n(f) \tag{20.59}$$

is the quadrature rule of Gaussian type for trigonometric polynomials with preassigned nodes $t_i \in [-\pi, \pi)$, $i = 1, 2, \dots, 2k$, if it is exact for all $f \in \mathcal{T}_{2(n+\gamma)+k-1}$, i.e., $R_n(f) = 0$ for all $f \in \mathcal{T}_{2(n+\gamma)+k-1}$.

The method for construction of Gaussian quadrature rules for trigonometric polynomials with preassigned nodes, which is again a counterpart to the fundamental theorem of Gaussian quadrature, is given in the following theorem.

Theorem 20.15 *Let $\gamma \in \{0, 1/2\}$ and w be a weight function on $[-\pi, \pi)$ and $t_i \in [-\pi, \pi)$, $i = 1, 2, \ldots, 2k$, are fixed in advance. Quadrature rule* (20.59) *is of Gaussian type for trigonometric polynomials with preassigned nodes $t_i \in [-\pi, \pi)$, $i = 1, 2, \ldots, 2k$, if and only if the nodes $x_i \in [-\pi, \pi)$, $i = 1, 2, \ldots, 2(n + \gamma)$, are the zeros of $T_n^\gamma \in \mathcal{T}_n^\gamma$, which is orthogonal with respect to the weight function $\widetilde{w}(x) = \prod_{j=1}^{2k} \sin\left(\frac{x-t_j}{2}\right) \cdot w(x)$ on interval $[-\pi, \pi)$.*

The existence of Gaussian-type quadrature rules for trigonometric polynomials with preassigned nodes (20.59) and internity of the nodes is not guaranteed since the weight function $\widetilde{w}$ is a variable sign function on the interval of integration $[-\pi, \pi)$. The standard construction of quadrature rule of Gaussian type for trigonometric polynomials, given in Theorem 20.4, cannot be applied always. A similar situation appeared in the case of Gauss–Kronrod quadrature rules for algebraic polynomials (see [32, 48]), where results for the existence of quadrature rule and internity of the nodes were obtained for the specific weight functions. For example, let us consider integral $\int_{-\pi}^{\pi} f(x)w(x)\mathrm{d}x$, where $w(x) = (\sin^2 x)^\alpha$, $0 < \alpha \le 2$, $x \in [-\pi, \pi)$. This weight function is even on interval $(-\pi, \pi)$. Now, by using $x := \arccos x$, we obtain integral $\int_{-1}^{1} f_1(x)(1 - x^2)^{\alpha-1/2}\mathrm{d}x$, $0 < \alpha \le 2$, $f_1(x) = f(-\arccos x) + f(\arccos x)$ (see [43, Lemma 2.3 and 2.4]). The obtained weight function is Gegenbauer weight function, for which the existence and internity of the nodes of Gauss–Kronrod quadrature rules for algebraic polynomials were proved (see [32]).

The Gaussian-type quadrature rules for trigonometric polynomials with preassigned nodes do not always exist, but still they are useful for estimating the error of some classes of quadrature rules. Such quadrature rules give higher trigonometric degree of exactness, with a smaller number of calculations, so they are worth of further research and application.

In the sequel we give a generalization of the optimal set of quadrature rules for trigonometric polynomials with an odd number of nodes (given in Definition 20.2) by adding an even number of preassigned nodes.

Definition 20.4 Let $\mathbf{n}$ be a multi-index, $W = (w_1, w_2, \ldots, w_r)$ be a TT system for $\mathbf{n}$ on interval $[-\pi, \pi)$, and nodes $t_i \in [-\pi, \pi)$, $i = 1, 2, \ldots, 2k$, be fixed and prescribed in advance. A set of quadrature rules of the form

$$\int_{-\pi}^{\pi} f(x)w_m(x)\mathrm{d}x \approx \sum_{i=1}^{2k} a_{m,i} f(t_i) + \sum_{i=1}^{2|\mathbf{n}|+1} A_{m,i} f(x_i), \qquad m = 1, 2, \ldots, r, \tag{20.60}$$

are called an optimal set of quadrature rules for trigonometric polynomials with preassigned nodes $t_i \in [-\pi, \pi)$, $i = 1, 2, \ldots, 2k$, with respect to $(W, \mathbf{n})$ if and only if the weight coefficients, $a_{m,i}$, $A_{m,i}$, $m = 1, 2, \ldots, r$, $i = 1, 2, \ldots, 2|\mathbf{n}| + 1$, and the nodes x_i, $i = 1, 2, \ldots, 2|\mathbf{n}| + 1$, satisfy the following equations:

$$\sum_{i=1}^{2k} a_{m,i} + \sum_{i=1}^{2|\mathbf{n}|+1} A_{m,i} = \int_{-\pi}^{\pi} w_m(x)\mathrm{d}x,$$

$$\sum_{i=1}^{2k} a_{m,i} \cos(\ell_m t_i) + \sum_{i=1}^{2|\mathbf{n}|+1} A_{m,i} \cos(\ell_m x_i) = \int_{-\pi}^{\pi} \cos(\ell_m x) w_m(x)\mathrm{d}x,$$

$$\sum_{i=1}^{2k} a_{m,i} \sin(\ell_m t_i) + \sum_{i=1}^{2|\mathbf{n}|+1} A_{m,i} \sin(\ell_m x_i) = \int_{-\pi}^{\pi} \sin(\ell_m x) w_m(x)\mathrm{d}x,$$

for $\ell_m = 1, 2, \ldots, |\mathbf{n}| + n_m + k$ and $m = 1, 2, \ldots, r$.

Theorem 20.16 *Let $W = (w_1, w_2, \ldots, w_r)$ be a TT system for $\mathbf{n}$ on interval $[-\pi, \pi)$ and let $t_i \in [-\pi, \pi)$, $i = 1, 2, \ldots, 2k$, are fixed and prescribed in advance. Suppose that $\widetilde{W} = (\widetilde{w}_1, \widetilde{w}_2, \ldots, \widetilde{w}_r)$, $\widetilde{w}_m(x) = \prod_{j=1}^{2k} \sin\left(\frac{x-t_j}{2}\right) \cdot w_m(x)$, $m = 1, 2, \ldots, r$, is also a TT system for $\mathbf{n}$ on interval $[-\pi, \pi)$. The set of quadrature rules (20.60) is the optimal set of quadrature rules for trigonometric polynomials with preassigned nodes $t_i \in [-\pi, \pi)$, $i = 1, 2, \ldots, 2k$, with respect to $(W, \mathbf{n})$ if and only if:*

(a) They are exact for all trigonometric polynomials $t \in \mathcal{T}_{|\mathbf{n}|+k}$;
(b) $T_{\mathbf{n}}^{1/2}(x) = \prod_{k=1}^{2|\mathbf{n}|+1} \sin\left(\frac{x-x_k}{2}\right)$ is the type II multiple orthogonal trigonometric polynomial of semi-integer degree $|\mathbf{n}| + 1/2$ with respect to $(\widetilde{W}, \mathbf{n})$.

Since $\widetilde{W}$ is a system of variable sign functions on interval $[-\pi, \pi)$ for arbitrary preassigned nodes $t_i \in [-\pi, \pi)$, $i = 1, 2, \ldots, 2k$, we need additional condition in order to ensure the existence of the type II multiple orthogonal trigonometric polynomial of semi-integer degree $T_{\mathbf{n}}^{1/2}(x)$ with respect to $(\widetilde{W}, \mathbf{n})$ on interval $[-\pi, \pi)$. By adding the condition that $\widetilde{W}$ is a TT system for $\mathbf{n}$ on interval $[-\pi, \pi)$, we obtain the existence of the type II multiple orthogonal trigonometric polynomials of semi-integer degree and the corresponding optimal set of quadrature rules for trigonometric polynomials with an odd number of nodes.

20.6 Numerical Examples

In this section we give numerical examples related to the presented quadrature rules of Gaussian type for trigonometric polynomials. In Example 20.1 and 20.2 we give error estimates by using Gaussian, anti-Gaussian, and corresponding averaged quadrature rules for trigonometric polynomials for the calculation of certain integrals. Example 20.3 gives comparison of our method and other methods.

Examples 20.4, 20.5, and 20.6 are devoted to the quadrature rules for trigonometric polynomials with preassigned nodes.

In the numerical constructions we use the software package OrthogonalPolynomials [7]. All computations were performed in Wolfram Mathematica, Version 13.1.

Example 20.1 In this example we consider case $\gamma = 0$, i.e., quadrature rules with an even number of nodes. The compared errors Gaussian (G_n) (20.19), anti-Gaussian (H_{n+2}) (20.20), and averaged (A_{2n+2}) (20.22) quadrature rules for trigonometric polynomials for the integral

$$\int_{-\pi}^{\pi} f(x)w(x)\,\mathrm{d}x,$$

where $w(x) = 1-\cos^2 x$ and $f(x) = (1+\cos x)(\mathrm{e}^{-x}+4/3)$ are given in Table 20.1.

Example 20.2 In this example we consider case $\gamma = 1/2$, i.e., quadrature rules with an odd number of nodes. The compared errors of Gaussian (G_n) (20.19), anti-Gaussian (H_{n+2}) (20.20), and averaged (A_{2n+2}) (20.22) quadrature rules for trigonometric polynomials with respect to the weight function $w(x) = 1 + \cos x$, $x \in (-\pi, \pi)$, and integrand $f(x) = (1+\cos x)(\mathrm{e}^{-x}+4/3)$ are given in Table 20.2.

As we can see from Tables 20.1 and 20.2 errors obtained by using Gaussian and anti-Gaussian quadrature rule are similar, but the averaged Gaussian quadrature rules for trigonometric polynomials give a significant improvement, as expected.

Example 20.3 In this example we will compare our method with method given in [18].

The nodes of the generalized averaged Szegő quadrature rule are the eigenvalues of corresponding $(2n-2) \times (2n-2)$ unitary upper Hessenberg matrix $\breve{H}_{2n-2}(\tau)$,

Table 20.1 Errors of Gaussian (G_n), anti-Gaussian (H_{n+2}), and averaged (A_{2n+2}) formula for $w(x) = 1 - \cos^2 x$, $x \in (-\pi, \pi)$, and $f(x) = (1 + \cos x)(\mathrm{e}^{-x} + 4/3)$

n	$I(f) - G_n(f)$	$I(f) - H_{n+2}(f)$	$I(f) - A_{2n+2}(f)$
40	$-9.30463 \cdot 10^{-9}$	$8.99386 \cdot 10^{-9}$	$-1.55389 \cdot 10^{-10}$
30	$-4.97942 \cdot 10^{-8}$	$4.82213 \cdot 10^{-8}$	$-7.86464 \cdot 10^{-10}$
20	$-5.16734 \cdot 10^{-7}$	$5.00653 \cdot 10^{-7}$	$-8.04024 \cdot 10^{-9}$
10	-0.0000254069	0.0000246255	$-3.90685 \cdot 10^{-7}$

Table 20.2 Errors of Gaussian (G_n), anti-Gaussian (H_{n+2}), and averaged (A_{2n+2}) formula for $w(x) = 1 + \cos x$, $x \in (-\pi, \pi)$, and $f(x) = (1 + \cos x)(\mathrm{e}^{-x} + 4/3)$

n	$I(f) - G_n(f)$	$I(f) - H_{n+2}(f)$	$I(f) - A_{2n+2}(f)$
40	$-4.63804 \cdot 10^{-9}$	$4.49229 \cdot 10^{-9}$	$-7.28786 \cdot 10^{-11}$
30	$-2.48222 \cdot 10^{-8}$	$2.40457 \cdot 10^{-8}$	$-3.88281 \cdot 10^{-10}$
20	$-2.56852 \cdot 10^{-7}$	$2.48826 \cdot 10^{-7}$	$-4.01318 \cdot 10^{-9}$
10	-0.0000124339	0.0000120453	$-1.94297 \cdot 10^{-7}$

Table 20.3 Errors for several quadrature rules for $w(x) = 2\sin^2(x/2)$, $x \in (-\pi, \pi)$, and $f(x) = 1/2\log(5 + 4\cos x)$

Rule	$n = 12$	$n = 15$	$n = 18$
$S_1^n(f)$	$-2.2 \cdot 10^{-5}$	$2.2 \cdot 10^{-6}$	$-2.3 \cdot 10^{-7}$
$A_1^n(f)$	$2.3 \cdot 10^{-5}$	$-2.3 \cdot 10^{-6}$	$2.4 \cdot 10^{-7}$
$\widehat{S}_1^{(2n-2)}(f)$	$-1.5 \cdot 10^{-7}$	$9.2 \cdot 10^{-9}$	$-6.7 \cdot 10^{-10}$
$G_n(f)$	$-1.98 \cdot 10^{-5}$	$1.38 \cdot 10^{-5}$	$-2 \cdot 10^{-7}$
$H_{n+2}(f)$	$1.98 \cdot 10^{-5}$	$-1.38 \cdot 10^{-5}$	$2 \cdot 10^{-7}$
A_{2n+2}	$-5.31 \cdot 10^{-10}$	$1.04 \cdot 10^{-10}$	$-8.75 \cdot 10^{-14}$

and weights are determined by the squares of the first components of associated unit eigenvectors (see [18, 20]). There are several algorithms for the computation of the eigensystem by using compact representation of Hessenberg matrices (see, e.g., [12, 14]).

In our method in order to escape the numerical non-stability which is characteristic for the Gram–Schmidt method, we have used recurrence relations. Also, we demonstrated how, in the case of symmetric weight function, the anti-Gaussian quadrature rules can be constructed by using already known orthogonal polynomials on the real line.

Using averaged Gaussian quadrature rules for trigonometric polynomials introduced in this article, we were able to achieve much greater accuracy in comparison with averaged Szegő quadrature rules on the class of symmetric weight functions, which is shown in Table 20.3. Also, introducing the anti-Gaussian quadrature rules with an odd number of nodes (the case $\gamma = 1/2$), we achieved accuracy for the trigonometric polynomials of higher degree, for all $t \in \mathcal{T}_{2n+2}$. We have compared one example from [18] with our results.

Let us consider the weight function $w(x) = 2\sin^2(x/2)$ on the interval $(-\pi, \pi)$ and the integrand $f(x) = 1/2\log(5 + 4\cos x)$. In Table 20.3 we give errors in Szegő ($S_1^n(f)$), anti-Szegő ($A_1^n(f)$), generalized averaged Szegő ($\widehat{S}_1^{(2n-2)}(f)$), Gaussian ($G_n(f)$), anti-Gaussian ($H_{n+2}(f)$), and averaged Gaussian (A_{2n+2}) formula.

It is obvious that errors obtained by using Gaussian and anti-Gaussian quadrature rule are similar with those obtained by using Szegő and anti-Szegő quadrature rules, respectively. However, the averaged Gaussian quadrature rules for trigonometric polynomials give a significant improvement in comparison with the generalized averaged Szegő quadrature rules.

Example 20.4 Let us construct the quadrature rule of Gaussian type for trigonometric polynomials (20.59) on $E = [-\pi, \pi)$, for $n = 2$, $\gamma = 1/2$, with respect to the weight function $w(x) = 2 - \sin 2x$ and the preassigned nodes $\{y_1, y_2\} = \{-\pi/5, \pi/4\}$.

Table 20.4 The nodes y_i, $i = 1, 2$, x_i, $i = 1, 2, \ldots, 5$, and the weight coefficients a_i, $i = 1, 2$, A_i, $i = 1, 2, \ldots, 5$, of the quadrature rule of Gaussian type for trigonometric polynomials (20.59) for $\gamma = 1/2$ with respect to $w(x) = 2 - \sin 2x$, $x \in [-\pi, \pi)$, and the preassigned nodes $\{y_1, y_2\} = \{-\pi/5, \pi/4\}$

i	y_i	a_i
1	$-\pi/5$	2.79281984766475
2	$\pi/4$	– 1.417342815770321
i	x_i	A_i
1	– 2.8242640557319621	1.631353294992642
2	– 1.585174275925089	2.01044202848654
3	0.497212029865876	2.54917211861256
4	1.452427960918445	2.19469675845722
5	2.459798340872731	2.80522938191579

It is easy to check that the orthogonal trigonometric polynomial of semi-integer degree $T_2^{1/2}(x)$, with respect to the weight function

$$w^G(x) = \sin\left(\frac{x + \pi/3}{2}\right) \sin\left(\frac{x - \pi/3}{2}\right) (2 - \sin 2x)$$

exists. By using procedure given in [22, Section 5], we obtain the nodes x_i, $i = 1, 2, \ldots, 5$, and the weight coefficients a_i, $i = 1, 2$, A_i, $i = 1, 2, \ldots, 5$, which are given in Table 20.4.

Example 20.5 Let us construct the quadrature rule of Gaussian type for trigonometric polynomials (20.59) on $E = [-\pi, \pi)$, for $n = 3$, $\gamma = 1/2$, with respect to the weight function $w(x) = \sqrt{1 - (x/\pi)^2}$ (the Chebyshev weight function of the second kind scaled to the interval E), and the preassigned nodes $\{y_1, y_2\} = \{-\pi/4, \pi/2\}$.

By using the same procedure as in Example 20.4 we obtain the nodes x_i, $i = 1, 2, \ldots, 7$, and the weight coefficients a_i, $i = 1, 2$, A_i, $i = 1, 2, \ldots, 7$, which are given in Table 20.5.

The obtained quadrature rule integrates exactly all trigonometric polynomials from $\mathcal{T}_7$, as it is stated in Theorem 20.15. For the integrand $f(x) = (1 + \cos x)/\sqrt{5 + \cos x}$, the approximation error is $-3.4168555 \times 10^{-8}$.

Example 20.6 Let us construct the optimal set of quadrature rules for trigonometric polynomials (20.60) on $E = [-\pi, \pi)$, for $r = 2$, multi-index $\mathbf{n} = (2, 1)$, with respect to the weight functions $w_1(x) = 1$ and $w_2(x) = 2 - \sin 2x$ and the preassigned nodes $\{y_1, y_2\} = \{-\pi/5, \pi/4\}$.

It is easy to see that $W = (w_1, w_2)$ is TT system for multi-index $\mathbf{n}$, as well as $\widetilde{W} = (\widetilde{w}_1, \widetilde{w}_2)$, where

$$\widetilde{w}_1(x) = \sin\left(\frac{x + \pi/5}{2}\right) \sin\left(\frac{x - \pi/4}{2}\right),$$

Table 20.5 The nodes y_i, $i = 1, 2$, x_i, $i = 1, 2, \ldots, 7$, and the weight coefficients a_i, $i = 1, 2$, A_i, $i = 1, 2, \ldots, 7$, of the quadrature rule of Gaussian type for trigonometric polynomials (20.59) for $\gamma = 1/2$ with respect to the weight function $w(x) = \sqrt{1-(x/\pi)^2}$, $x \in [-\pi, \pi)$, and the preassigned nodes $\{y_1, y_2\} = \{-\pi/4, \pi/2\}$

i	y_i	a_i
1	$-\pi/4$	0.7302004920494538
2	$\pi/2$	0.6273402200502432
i	x_i	A_i
1	− 3.135385513084978	0.2905157247063272
2	− 2.303421145210531	0.5194617256444344
3	− 1.542485834689003	0.6613144018791357
4	− 0.0393589763330550	0.7312150005887421
5	0.614350240503910	0.4706384077889477
6	0.949619073471995	0.3924742651577427
7	2.315089501751868	0.5116419626796524

Table 20.6 The nodes y_i, $i = 1, 2$, x_i, $i = 1, 2, \ldots, 7$, and the weight coefficients $a_{m,i}$, $i = 1, 2$, $m = 1, 2$, $A_{m,i}$, $i = 1, 2, \ldots, 7$, $m = 1, 2$, of the optimal set of quadrature rules (20.60) with respect to $W = (1, 2 - \sin 2x)$, $\mathbf{n} = (2, 1)$ and the preassigned nodes $\{y_1, y_2\} = \{-\pi/5, \pi/4\}$

i	y_i	$a_{1,i}$	$a_{2,i}$
1	$-\pi/5$	0.800736958704361	2.36302001982287
2	$\pi/4$	2.36665196898271	2.36665196898271
i	x_i	$A_{1,i}$	$A_{2,i}$
1	− 3.005941674854158	0.790889182235539	1.369831136882345
2	− 2.215286690772217	0.790702456658771	0.821894005267554
3	− 1.423895857757946	0.792535990680536	1.81458436226368
4	0.199490138020088	0.917236116124360	1.478145343771460
5	0.701822851834176	− 1.77450396222122	− 1.79923552635939
6	1.687633606477333	0.805514084705667	1.79754799358853
7	2.485381300257828	0.793422511308864	2.35393131013942

$$\widetilde{w}_2(x) = \sin\left(\frac{x+\pi/5}{2}\right)\sin\left(\frac{x-\pi/4}{2}\right)(2-\sin 2x).$$

First, we obtain the zeros x_i, $i = 1, 2, \ldots, 7$, of the type II multiple orthogonal trigonometric polynomial of semi-integer degree $T_3^{1/2}(x)$, with respect to $\widetilde{W}$, i.e., the nodes of the optimal set of quadrature rules (20.60). By requiring that quadrature rule integrates exactly trigonometric polynomials from $\mathcal{T}_4$, with respect to the weight functions w_m, $m = 1, 2$, we obtain the weight coefficients. The obtained nodes x_i, $i = 1, 2, \ldots, 7$, and the weight coefficients $a_{m,i}$, $i = 1, 2$, $m = 1, 2$, $A_{m,i}$, $i = 1, 2, \ldots, 7$, $m = 1, 2$, are given in Table 20.6.

Let us notice that the obtained optimal set of quadrature rules integrates exactly all trigonometric polynomials from $\mathcal{T}_5$ and all trigonometric polynomials from

$\mathcal{T}_4$, with respect to the weight function w_1 and w_2, respectively, as it is stated in Theorem 20.16.

References

1. Alqahtani, H., Reichel, L.: Simplified anti-Gauss quadrature rules with applications in linear algebra. Numer. Algorithms **77**, 577–602 (2018)
2. Borges, C.F.: On a class of Gauss-like quadrature rules. Numer. Math. **67**, 271–288 (1994)
3. Calvetti, D., Reichel, L.: Symmetric Gauss-Lobatto and modified anti-Gauss rules. BIT Numer. Math. **43**, 541–554 (2003)
4. Cantero, M.J., Cruz-Barroso, R., Gonzalez-Vera, P.: A matrix approach to the computation of quadrature formulas on the unit circle. Appl. Numer. Math. **58**, 296–318 (2008)
5. Cruz-Barroso, R., Darius, L., Gonzáles-Vera, P., Njåstad, O., Quadrature rules for periodic integrands. Bi-orthogonality and para-orthogonality. Annals of Mathematical et Informancae **32**, 5–44 (2005)
6. Cruz-Barroso, R., Gonzáles-Vera, P., Njåstad, O.: On bi-orthogonal systems of trigonometric functions and quadrature formulas for periodic integrands. Numer. Algorithm **44**(4), 309–333 (2007)
7. Cvetković, A.S., Milovanović, G.V.: The Mathematica Package "OrthogonalPolynomials". Facta Univ. Ser. Math. Inform. **19**, 17–36 (2004)
8. Cvetković, A.S., Stanić, M.P.: Trigonometric orthogonal systems. In: (Gautschi, W., Mastroianni, V.G., Rassias, Th.M. (eds.) Approximation and Computation - In Honor of Gradimir V. Milovanović, pp. 103–116. Springer, Berlin-Heidelberg-New York (2010)
9. Cvetković, A.S., Stanić, M.P., Marjanović, Z.M., Tomović, T.V.: Asymptotic behavior of orthogonal trigonometric polynomials of semi-integer degree, Appl. Math. Comput. **218**(23), 11528–11533 (2012)
10. Davis, P.J., Rabinowitz, P.: Methods of Numerical Integration. Academic Press, New York, San Francisco, London (1975)
11. Fenu, C., Martin, D., Reichel, L., Rodriguez, G.: Block Gauss and anti-Gauss quadrature with application to networks. SIAM J. Matrix Anal. Appl. **34**(4), 1655–1684 (2013)
12. Gragg, W.B.: The QR algorithm for unitary Hessenberg matrices. J. Comput. Appl. Math. **16**, 1–8 (1986)
13. Gragg, W.B.: Positive definite Toeplitz matrices, the Arnoldi process for isometric operators and Gaussian quadrature on the unit circle. J. Comput. Appl. Math. **46**, 183–198 (1993)
14. Gu, M., Guzzo, R., Chi, X-B., Cao, X-Q.: A stable divide and conquer algorithm for the unitary eigenproblem. SIAM J. Matrix Anal. Appl. **25**, 385–404 (2003)
15. Gautschi, W.: Orthogonal polynomials: applications and computation. Acta Numer. **5**, 45–119 (1996)
16. Jagels, C., Reichel, L.: Szegö-Lobatto quadrature rules. J. Comput. Appl. Math. **200**, 116–126 (2007)
17. Jagels, C., Reichel, L.: On the construction of Szegö polynomials. J. Comput. Appl. Math. **46**, 241–254 (1993)
18. Jagels, C., Reichel, L., Tang, T.: Generalized averaged Szegő quadrature rules. J. Comput. Appl. Math. **311**, 645–654 (2017)
19. Jones, W.B., Njåstad, O., Thron, W.J.: Moment theory, orthogonal polynomials, quadrature, and continued fractions associated with the unit circle. Bull. London Math. Soc. **21**, 113–152 (1989)
20. Kim, S.-M., Reichel, L.: Anti-Szegö quadrature rules. Math. Comp. **76**, 795–810 (2007)
21. Laurie, D.: Anti-Gaussian quadrature formulas. Math. Comp. **65**(214), 739–747 (1996)

22. Milovanović, G.V., Cvetković, A.S., Stanić, M.P.: Trigonometric orthogonal systems and quadrature formulae. Comput. Math. Appl. **56(11)**, 2915–2931 (2008)
23. Milovanović, G.V., Cvetković, A.S., Stanić, M.P.: Explicit formulas for five-term recurrence coefficients of orthogonal trigonometric polynomials of semi-integer degree. Appl. Math. Comput. **198**(2), 559–573 (2008)
24. Milovanović, G.V., Cvetković, A.S., Stanić, M.P.: Christofell-Darboux formula for orthogonal trigonometric polynomials of semi-integer degree. Facta Univ. Ser. Math. Inform. **23**, 29–37 (2008)
25. Milovanović, G.V., Cvetković, A.S., Stanić, M.P.: A trigonometric orthogonality with respect to a nonnegative Borel measure. Filomat **26**(4), 689–696 (2012)
26. Milovanović, G.V., Stanić, M.: Multiple orthogonality and quadratures of Gaussian type. Rend. Circ. Mat. Palermo, Serie II, Suppl. **76**, 75–90 (2005)
27. Milovanović, G.V., Stanić, M.: Construction of multiple orthogonal polynomials by discretized Stieltjes–Gautschi procedure and corresponding Gaussian quadratures. Facta Univ. Ser. Math. Inform. **18**, 9–29 (2003)
28. Milovanović, G.V., Stanić, M.: Multiple orthogonality and applications in numerical integration. In: Georgiev, P., Pardalos, P.M., Srivastava, H.M. (eds.) Nonlinear Analysis: Stability, Approximation, and Inequalities, Series: Springer Optimization and Its Applications, Vol. 68 Chapter 26, pp. 431–455, Springer, Berlin, Heidelberg, New York (2012)
29. Milovanović, G.V., Stanić, M.P., Tomović, T.V.: Trigonometric multiple orthogonal polynomials of semi-integer degree and the corresponding quadrature formulas. Publ. Inst. Math. (Beograd) (N.S.) **96**(110), 211–226 (2014)
30. Mysovskikh, I.P.: Quadrature formulae of the highest trigonometric degree of accuracy. Ž. Vučisl. Mat. i Mat. Fiz. **25**(8), 1246–1252 (1985) (in Russian); U.S.S.R. Computational Mathematics and Mathematical Physics. **25**, 180–184 (1985) (English)
31. Mysovskikh, I.P.: Algorithms to construct quadrature formulae of highest trigonometric degree of precision. Metody Vychisl. **16**, 5–16 (1991) (in Russian)
32. Notaris, S.E.: Gauss-Kronrod quadrature formulae—a survey of fifty years research. Electron. Trans. Numer. Anal. **45**, 371–404 (2016)
33. Petrović, N.Z., Stanić, M.P., Tomović Mladenović, T.V.: Anti-Gaussian quadrature rule for trigonometric polynomials. Filomat **36**(3), 1005–1019 (2022)
34. Pranić, M.S., Reichel, L.: Generalized anti-Gauss quadrature rules. J. Comput. Appl. Math. **284**, 235–243 (2015)
35. Reichel, L., Spalević, M.M., Tang, T.: Generalized averaged Gauss quadrature rules for the approximation of matrix functionals. BIT Numeri. Math. **56**, 1045–1067 (2016)
36. Shohat, J.A.: On a certain formula of mechanical quadratures with non-equidistant ordinates. Trans. Amer. Math. Soc. **31**, 448–463 (1929)
37. Spalević, M.M.: A note on generalized averaged Gaussian formulas. Numer. Algorithms **46**(3), 253–264 (2007)
38. Spalević, M.M.: Error estimates of anti-Gaussian quadrature formulae. J. Comput. Appl. Math. **236**, 3542–3555 (2012)
39. Spalević, M.M.: On generalized averaged Gaussian formulas. Math. Comp. **76**(259), 1483–1492 (2007)
40. Spalević, M.M.: On generalized averaged Gaussian formulas. II, Math. Comp. **86**(306), 1877–1885 (2017)
41. Stanić, M.P., Tomović, T.V.: Multiple orthogonality in the space of trigonometric polynomials of semi-integer degree. Filomat **29**(10), 2227–2237 (2015)
42. Stanić, M.P., Tomović Mladenović, T.V., Jovanovicć, A.Ne.: Quadrature rules of Gaussian type for trigonometric polynomials with preassigned nodes. Appl. Numer. Math. **200**, 399–408 (2024)
43. Stanić, M.P., Cvetković, A.S., Tomović, T.V.: Error estimates for some quadrature rules with maximal trigonometric degree of exactness. Math. Methods Appl. Sci. **37**(11), 1687–1699 (2014)

44. Stanić, M.P., Cvetković, A.S., Tomović, T.V.: Error bound of certain Gaussian quadrature rules for trigonometric polynomials. Kragujevac J. Math. **36**(1), 63–72 (2012)
45. Stanić, M.P., Cvetković, A.S., Tomović, T.V.: Error estimates for quadrature rules with maximal even trigonometric degree of exactness. Rev. R. Acad. Cienc. Exactas, Fis. Nat. Ser. A. Mat. RACSAM **108**, 603–615 (2014)
46. Stanić, M.P., Cvetković, A.S., Tomović, T.V.: Error bounds for some quadrature rules with maximal trigonometric degree of exactness. AIP Conf. Proc. **1479**, 1042–1045 (2012)
47. Tomović, T.V.: Analyses and applications of quadrature formulas of Gaussian type for trigonometric polynomials. Ph.D Thesis, Faculty of Science, University of Kragujevac (2014)
48. Szegö, G.: Ober gewisse orthogonale polynome, die zu einer oszillierenden Belegungsfunktion gehřen. Math. Ann. 501–513 (1935). In: Askey, R. (ed.) Collected Papers, vol. 2, pp. 545–557. Birkhäuser, Boston (1982)
49. Szegö, G.: Orthogonal Polynomials. 4th edn. American Mathematical Society, Providence (1975)
50. Turetzkii, A.H.: On quadrature formulae that are exact for trigonometric polynomials. East J. Approx. **11(3)**, 337–359 (2005), Translation in English from Uchenye Zapiski, Vypusk 1 (149). Seria Math. Theory of Functions, Collection of papers, Izdatel'stvo Belgosuniversiteta imeni V.I. Lenina, Minsk, 31–54 (1959)
51. Van Assche, W.: Nonsymmetric linear difference equations for multiple orthogonal polynomials. CRM Proceedings and Lecture Notes, vol. 25, 391–405. American Mathematical Society, Providence (2000)

Chapter 21
A Boubaker Collocation Method to Solve the Single Degree of Freedom System

Şuayip Yüzbaşı and Beyza Çetin

Dedicated to Academician Gradimir V. Milovanović on the occasion of his 75th birthday

21.1 Introduction

The research of the free vibration of damped spring-mass systems having single degree of freedom illuminates on numerous developed problems, as a complex system can be designed as a single degree of freedom spring-mass system. For this purpose, many researchers have developed various methods for linear differential equations [1–4]. In this research, a collocation method based on Boubaker polynomials is presented for the approximate solution of single degree of freedom system [5] which is a class of 2th order linear differential equations as

$$My''(\eta) + Cy'(\eta) + Ky(\eta) = F_0 \cos(w\eta), \tag{21.1}$$

with initial conditions

$$y(0) = \lambda_0, \quad y'(0) = \lambda_1. \tag{21.2}$$

In this research, it aims to find approximate solutions of single degree of freedom system by means of a collocation method constructed by using Boubaker polynomials. Among the most regularly applied bases, Boubaker polynomials are one of

Ş. Yüzbaşı (✉)
Department of Mathematics, Bartın University, Bartın, Türkiye
e-mail: suayipyuzbasi@bartin.edu.tr

B. Çetin
Department of Mathematics, Missouri University Science and Technology, Rolla, MO, USA

M. P. Stanić et al. (eds.), *Analysis, Approximation, Optimization: Computation and Applications*, Springer Optimization and Its Applications 224,
https://doi.org/10.1007/978-3-031-85743-0_21

the orthogonal polynomials defined by Yalçınbaş and Akkaya [6]

$$B_w(\eta) = \sum_{j=0}^{[|w/2|]} (-1)^j \binom{w-4j}{i-j} \binom{w-j}{w} \eta^{w-2j}, \tag{21.3}$$

and reiteration relation of the Boubaker polynomials defined as

$$B_w(\eta) = tB_{w-1}(\eta) - B_{w-2}(\eta), \quad w \geq 3.$$

The features of these polynomials are exploited to construct the new matrices. Then these matrices and numerical approximation technique are utilized to transform the problem to a system of linear algebraic equations.

21.2 The Operational Matrices

21.2.1 Matrix Representation for Single Degree of Freedom System

The approximate solution based on the Boubaker polynomials can be acquired in the form [7]

$$y(\eta) \approxeq y_N(\eta) = \sum_{n=0}^{N} a_n B_n(\eta). \tag{21.4}$$

Initially, the approximate solution can be typed in matrix form [8]

$$y(\eta) \approxeq y_N(\eta) = \sum_{n=0}^{N} a_n B_n(\eta) = \mathbf{B}(\eta)\mathbf{A}, \tag{21.5}$$

where

$$\mathbf{B}(\eta) = \begin{bmatrix} B_0(\eta) \; B_1(\eta) \; B_2(\eta) \; \ldots \; B_N(\eta) \end{bmatrix} \quad \text{and} \quad \mathbf{A} = \begin{bmatrix} a_0 \; a_1 \; a_2 \; \ldots \; a_N \end{bmatrix}.$$

Then the vector $\mathbf{B}(\eta)$ composed Boubaker polynomials can be expressed as [9]

$$\mathbf{B}(\eta) = \mathbf{X}(\eta)\mathbf{R}, \tag{21.6}$$

where

$$\mathbf{X}(\eta) = \begin{bmatrix} 1 \; \eta \; \eta^2 \; \ldots \; \eta^N \end{bmatrix}$$

and if N is odd

$$\mathbf{R}^T = \begin{bmatrix} \xi_{0,0} & 0 & 0 & \dots & 0 & 0 \\ 0 & \xi_{1,0} & 0 & \dots & 0 & 0 \\ \xi_{2,1} & 0 & \xi_{2,0} & \dots & 0 & 0 \\ \vdots & \vdots & \vdots & \ddots & \vdots & \vdots \\ \xi_{N-1,\frac{N-1}{2}} & 0 & \xi_{N-1,\frac{N-3}{2}} & \dots & \xi_{N-1,0} & 0 \\ 0 & \xi_{N,\frac{N-1}{2}} & 0 & \dots & 0 & \xi_{N,0} \end{bmatrix}$$

and if N is even

$$\mathbf{R}^T = \begin{bmatrix} \xi_{0,0} & 0 & 0 & \dots & 0 & 0 \\ 0 & \xi_{1,0} & 0 & \dots & 0 & 0 \\ \xi_{2,1} & 0 & \xi_{2,0} & \dots & 0 & 0 \\ \vdots & \vdots & \vdots & \ddots & \vdots & \vdots \\ 0 & \xi_{N-1,\frac{N-2}{2}} & 0 & \dots & \xi_{N-1,0} & 0 \\ \xi_{N,\frac{N}{2}} & 0 & \xi_{N,\frac{N-2}{2}} & \dots & 0 & \xi_{N,0} \end{bmatrix}.$$

The relation between the terms of the Boubaker polynomials is as follows:

$$\xi_{n,w} = (-1)^w \left(\frac{n - 4w}{w!} \right) \prod_{j=w+1}^{2w-1} (n - j), \quad n, w = 0, 1, \dots, N.$$

By putting (21.6) into the (21.5), we get

$$y(\eta) \approxeq y_N(\eta) = \mathbf{X}(\eta)\mathbf{R}\mathbf{A}. \tag{21.7}$$

Using (21.6), we can derive derivative relations of the Boubaker polynomials as [10]

$$\mathbf{B}'(\eta) = \mathbf{X}(\eta)\mathbf{S}\mathbf{R}, \tag{21.8}$$

and

$$\mathbf{B}''(\eta) = \mathbf{X}(\eta)\mathbf{S}^2\mathbf{R}, \tag{21.9}$$

where

$$\mathbf{S} = \begin{bmatrix} 0\,1\,0\,0 \ldots 0 \\ 0\,0\,2\,0 \ldots 0 \\ 0\,0\,0\,3 \ldots 0 \\ \vdots\,\vdots\,\vdots\,\vdots \ddots \vdots \\ 0\,0\,0\,0 \ldots N \\ 0\,0\,0\,0 \ldots 0 \end{bmatrix}.$$

By making use of Eqs. (21.8)–(21.9), we arrive at the matrix form of the derivatives of (21.7) as

$$y'(\eta) \approxeq y'_N(\eta) = \mathbf{X}(\eta)\mathbf{SRA}, \tag{21.10}$$

$$y''(\eta) \approxeq y''_N(\eta) = \mathbf{X}(\eta)\mathbf{S}^2\mathbf{RA}. \tag{21.11}$$

21.2.2 Matrix Representation for Initial Conditions

The initial conditions of (21.1), $y(0) = \lambda_0$ and $y'(0) = \lambda_1$, can be generalized as

$$\sum_{k=0}^{1} a_{ik} y^{(k)}(0) = \lambda_i, \quad i = 0, 1. \tag{21.12}$$

By means of the relations (21.10),(21.11) can be written in matrix form as

$$\sum_{k=0}^{1} a_{ik}\mathbf{X}(0)\mathbf{S}^k\mathbf{R} = [\lambda_i], \quad i = 0, 1. \tag{21.13}$$

21.3 Determination of Boubaker Polynomial Method

In this study, we analyze the viscously damped single degree of freedom system subjected to harmonic excitation gives as

$$My''(\eta) + Cy'(\eta) + Ky = F_0 \cos(w\eta), \tag{21.14}$$

with initial conditions

$$y(0) = \lambda_0, \quad y'(0) = \lambda_1. \tag{21.15}$$

This equation is a second-order linear differential equation with inhomogeneous. If Eqs. (21.7), (21.10), and (21.11) obtained in the previous section are substituted in Eq. (21.1), we get

$$M\mathbf{X}(\eta)\mathbf{S}^2\mathbf{RA} + C\mathbf{X}(\eta)\mathbf{S}^1\mathbf{RA} + K\mathbf{X}(\eta)\mathbf{RA} = f(\eta), \tag{21.16}$$

where

$$f(\eta) = F_0 \cos(w\eta).$$

Now we can define collocation point as

$$\eta_j = \frac{b}{N} j, \quad j = 0, 1, \ldots, N, \quad 0 \leq \eta \leq b, \tag{21.17}$$

and they are written in Eq. (21.16), and then we obtain

$$M\mathbf{X}(\eta_j)\mathbf{S}^2\mathbf{RA} + C\mathbf{X}(\eta_j)\mathbf{S}^1\mathbf{RA} + K\mathbf{X}(\eta_j)\mathbf{RA} = f(\eta_j) \tag{21.18}$$

or briefly the fundamental matrix equation is

$$\left\{ M\mathbf{XS}^2\mathbf{R} + C\mathbf{XS}^1\mathbf{R} + K\mathbf{XR} \right\} \mathbf{A} = \mathbf{F}, \tag{21.19}$$

where

$$\mathbf{X} = \begin{bmatrix} \mathbf{X}(\eta_0) \\ \mathbf{X}(\eta_1) \\ \mathbf{X}(\eta_2) \\ \vdots \\ X(\eta_N) \end{bmatrix}, \quad \mathbf{F} = \begin{bmatrix} f(\eta_0) \\ f(\eta_1) \\ f(\eta_2) \\ \vdots \\ f(\eta_N) \end{bmatrix}.$$

This leads us to a linear algebraic system. In order to find matrix **A** of unknown coefficients, we solve the linear algebraic system:

$$[\mathbf{W}; \mathbf{F}],$$

where

$$\mathbf{W} = M\mathbf{S}^2\mathbf{R} + C\mathbf{SR} + K\mathbf{R}. \tag{21.20}$$

The matrix form of the initial conditions of (21.1) is

$$\mathbf{U}_i\mathbf{A} = [\lambda_i] \quad \text{or} \quad [\mathbf{U}_i; \lambda_i], \quad i = 0, 1. \tag{21.21}$$

instead of

$$\mathbf{U}_i = \sum_{k=0}^{1} a_{ik}\mathbf{X}(a)\mathbf{S}^k\mathbf{R} = [u_{i0} \quad u_{i1} \cdots u_{iN}], \quad i = 0, 1. \tag{21.22}$$

Substituting the given initial conditions into the (21.21), we get

$$[u_{00} \quad u_{01} \cdots u_{0N}] = \lambda_0 \quad \text{and} \quad [u_{10} \quad u_{11} \cdots u_{1N}] = \lambda_1. \tag{21.23}$$

If the 2 rows we found from the conditions are replaced in the last 2 rows of the augmented matrix, we get

$$\left[\tilde{\mathbf{W}}; \tilde{\mathbf{F}}\right] = \begin{bmatrix} \omega_{00} & \omega_{01} & \dots & \omega_{0N} & ; & f(\eta_0) \\ \omega_{10} & \omega_{11} & \dots & \omega_{1N} & ; & f(\eta_1) \\ \omega_{20} & \omega_{21} & \dots & \omega_{2N} & ; & f(\eta_2) \\ \vdots & \vdots & \vdots & \vdots & \ddots & \vdots \\ \omega_{N-2,0} & \omega_{N-2,1} & \dots & \omega_{N-2,N} & ; & f(\eta_{N-2}) \\ u_{00} & u_{01} & \dots & u_{0N} & ; & \lambda_0 \\ u_{10} & u_{11} & \dots & u_{1N} & ; & \lambda_1 \end{bmatrix}.$$

By solving the augmented matrix as $\mathbf{A} = \tilde{\mathbf{W}}^{-1}\tilde{\mathbf{F}}$, we obtain the matrix $\mathbf{A}$. This matrix is replaced by the following equation:

$$y(\eta) \approxeq y_N(\eta) = \mathbf{X}(\eta)\mathbf{R}\mathbf{A}. \tag{21.24}$$

In this way, the approximate solution is found. We note that rows from conditions can be written on the rows of $\left[\tilde{\mathbf{W}}; \tilde{\mathbf{F}}\right]$. For convenience, we select the last rows. If this matrix is written in (21.7), the required approximate solution is found.

21.4 Applications

In the spring-mass-damper system (21.14)–(21.15), taking the mass $M = 10$ kg, the damping coefficient $C = 20$ kg/s, spring stiffness $K = 4000$ N/m subject to the excitation force of amplitude is 100, and the frequency $w = 10$ rad/s. The initial conditions are $y(0) = 0.01$, $y'(0) = 0$. If these parameters are substituted in Eqs. (21.14)–(21.15), the problem becomes as follows:

$$\begin{cases} 10y''(\eta) + 20y'(\eta) + 4000y = 100\cos(10\eta), \\ y(0) = 0,01, \quad y'(0) = 0. \end{cases} \tag{21.25}$$

The exact solution is

$$y(\eta) = e^{-\eta}\left(-\frac{131}{5650}\cos(\sqrt{399}\eta) - \frac{48}{21161}\sin(\sqrt{366}\eta)\right) + \frac{1}{452}\sin(10\eta) + \frac{15}{452}\cos(10\eta).$$

Now, applying the method we developed in the previous section, an approximate solution for $N = 3$. Hence, the collocation points are calculated as

$$\eta_0 = 0, \quad \eta_1 = \frac{1}{3}, \quad \eta_3 = \frac{2}{3}, \quad \eta_4 = 1.$$

For this purpose, we can determine the matrices **X** and **F** as follows:

$$\mathbf{X} = \begin{bmatrix} 0 & 0 & 0 & 0 \\ 1 & 1/3 & (1/3)^2 & (1/3)^3 \\ 1 & 2/3 & (2/3)^2 & (2/3)^3 \\ 1 & 1 & 1 & 1 \end{bmatrix}, \quad \mathbf{F} = \begin{bmatrix} 100 \\ -22284/227 \\ 1762/19 \\ -11747/140 \end{bmatrix}.$$

The fundamental matrix equation of the (21.25) is

$$\left\{10\mathbf{XS}^2\mathbf{R} + 20\mathbf{XS}^1\mathbf{R} + 4000\mathbf{XR}\right\}\mathbf{A} = \mathbf{F}. \tag{21.26}$$

If these calculations are made, we get the augmented matrix

$$[\mathbf{W}; \mathbf{F}] = \begin{bmatrix} 4040 & 20 & 20 & 0 & ; & 100 \\ 44600/9 & 41260/27 & 4300/9 & 4720/27 & ; & -22284/227 \\ 68840/9 & 106340/27 & 16420/9 & 33800/27 & ; & 1762/19 \\ 12120 & 8140 & 4060 & 4120 & ; & -11747/140 \end{bmatrix}.$$

The matrix form of the initial conditions from (21.21) is

$$[\mathbf{U_0}; \lambda_\mathbf{0}] = [1 \quad 0 \quad 0 \quad 0 \quad ; \quad 1/100]$$

and

$$[\mathbf{U_1}; \lambda_\mathbf{1}] = [0 \quad 1 \quad 0 \quad 0 \quad ; \quad 0],$$

Table 21.1 Numerical results of the solution $y_5(\eta)$ and other methods' result for $N = 5$ of Eq. (21.25)

x	Present method	Operational matrix method [11]	Taylor matrix method [12]
0.0	4.4525e-05	1.8150e-02	0
0.1	1.7107e-03	1.1962e-02	9.3475e-03
0.2	5.6281e-03	4.9328e-03	8.2347e-02
0.3	7.8290e-03	2.6896e-02	6.3098e-02
0.4	1.3982e-02	4.5657e-03	2.2306e+00
0.5	6.1395e-03	8.8396e-03	5.2935e+00
0.6	6.5531e-02	2.0145e-02	1.0108e+01
0.7	1.5162e-01	2.2662e-02	1.6825e+01
0.8	1.1275e+00	6.0898e-03	2.5175e+01
0.9	3.6891e+00	1.4639e-03	3.4432e+01
1.0	9.1345e+00	1.7488e-03	4.3360e+01

If the augmented matrix based on the conditions is calculated, we get

$$[\mathbf{W}; \mathbf{F}] = \begin{bmatrix} 4040 & 20 & 20 & 0 & ; & 100 \\ 44600/9 & 41260/27 & 4300/9 & 4720/27 & ; & -22284/227 \\ 1 & 0 & 0 & 0 & ; & 1/100 \\ 0 & 1 & 0 & 0 & ; & 0 \end{bmatrix}.$$

By solving this system, matrix **A** is found as

$$\mathbf{A} = [1/100 \quad 0 \quad 149/50 \quad -2571/286]^T.$$

If this matrix is written in (21.7), the approximate solution sought is found as

$$y_3(\eta) - 2571/286\eta^3 + 3\eta^2 + 1/100.$$

This problem has been solved using different methods in the literature. To demonstrate the effectiveness of the method we have presented, we can compare our results with the results obtained from different methods. Table 21.1 was created for this purpose.

It can be seen from the values in Table 21.1 that the method performs better than the Taylor method.

The numerical values here are shown graphically in Fig. 21.1. For small values of η, our method works better.

In addition, the error values obtained from the proposed method for different values of N are shown in Fig. 21.2. For larger N values, the error gradually decreases. In other words, it is closer to the exact solution.

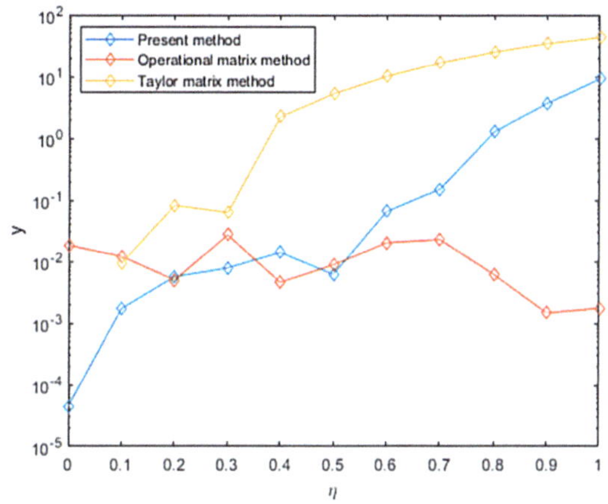

Fig. 21.1 Comparison of our results with other methods for $N = 5$ of Eq. (21.25)

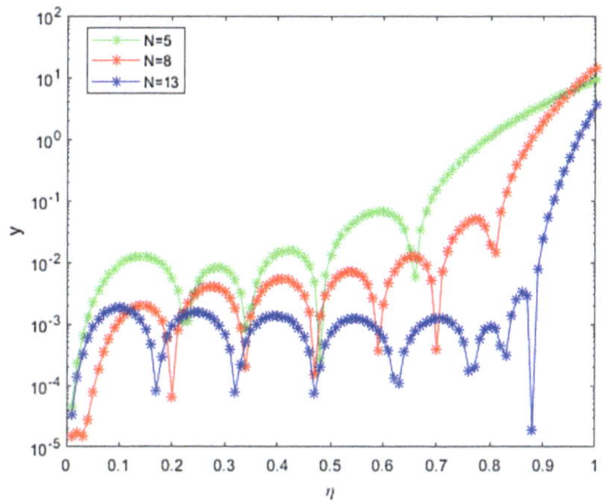

Fig. 21.2 Comparison of the actual error functions for $N = 5$, 8 and 13 of Eq. (21.25)

21.5 Conclusion

In this article, a collocation method based on the Boubaker polynomials for the single degree of freedom system is presented. The matrix relationships created for this method are discussed in Sect. 21.2. Also, in Sect. 21.3, the solution method consisting of these matrix relations is explained. The application of the method is discussed in the fourth section. The results of this application are shown by the data in Table 21.1 that the method gives better results. This practice allowed the method to be better explained and commented on its effectiveness. Also the absolute error functions are compared in Fig. 21.2 for different values of N. As a result, it is obtained that if N is selected larger, the exact solution will be closer. This is presented more clearly with tables and graphs.

References

1. Yüzbaşı, Ş.: Numerical solutions of fractional Riccati type differential equations by means of the Bernstein polynomials. Appl. Math. Comput. **219**(11), 6328–6343 (2013)
2. Yüzbaşı, Ş.: A new Bell function approach to solve linear fractional differential equations. Appl. Numer. Math. **174**, 221–235 (2022)
3. Yalçınbaş, S., Özsoy, N., Sezer, M.: Approximate solution of higher order linear differential equations by means of a new rational Chebyshev collocation method. Math. Comput. Appl. **15**(1), 45–56 (2010)
4. Çetin, M., Sezer, M., Kocayiğit, H.: An efficient method based on Lucas polynomials for solving high-order linear boundary value problems. Gazi Univ. J. Sci. **28**(3), 483–496 (2015)
5. Wang, Z.Q., Guo, B.Y.: Legendre-Gauss-Radau collocation method for solving initial value problems of first order ordinary differential equations. J. Sci. Comput. **52**(1), 226–255 (2012)
6. Yalçınbaş, S., Akkaya, T.: A numerical approach for solving linear integro-differential-difference equations with Boubaker polynomial bases. Ain Shams Eng. J. **3**(2), 153–161 (2012)
7. Yüzbaşı, Ş., Gök, E., Sezer, M.: Laguerre matrix method with the residual error estimation for solutions of a class of delay differential equations. Math. Methods Appl. Sci. **37**(4), 453–463 (2014)
8. Yüzbaşı, Ş., Sezer, M.: Shifted Legendre approximation with the residual correction to solve pantograph-delay type differential equations. Appl. Math. Model. **39**(21), 6529–6542 (2015)
9. Yüzbaşı, Ş., Yıldırım, G.: Approximate solution of the Duffing equation by using first Boubaker polynomials. In: 3rd International Conference on Applied Engineering and Natural Sciences, Konya, Turkey, pp. 369–378 (2022)
10. Akkaya, T., Yalçınbaş, S., Sezer, M.: Numeric solutions for the pantograph type delay differential equation using First Boubaker polynomials. Appl. Math. Comput. **219**(17), 9484–9492 (2013)
11. Yüzbaşı, Ş., Yıldırım, G.: An operational matrix method for the approximate solutions of the viscously damped single degree of freedom system subjected to harmonic excitation. In: 2nd International Congress on Mathematics and Geometry, pp. 48–58 (2021)
12. Kurt, N., Cevik, M.: Polynomial solution of the single degree of freedom system by Taylor matrix method. Mech. Res. Commun. **35**(8), 530–536 (2008)

Chapter 22
New Multiplicative Method of Multicriteria Analysis

Mališa R. Žižović, Dragan Pamučar, Miodrag M. Žižović, and Miloljub Albijanić

Dedicated to Academician Gradimir V. Milovanović on the occasion of his 75th birthday

22.1 Introduction

The multiplicative concept of evaluation in multicriteria analysis emerges at the onset of the scientific approach to this issue. In applications, it has not been particularly prevalent; in fact, it has been quite rare. The reason for this is the fact that the additive concept of evaluation was more convenient. Specifically, when solving multicriteria tasks (problems), normalizations across all criteria must be performed to bring various criterion requirements to the same scale, resulting in small positive numbers [4, 7, 9]. The clear summation of these values is much more convenient in applications than multiplication, as summation increases the result, whereas multiplication decreases it. It is more practical to work with larger numbers than smaller ones!

The idea that the expected result of multicriteria analysis for any criterion should be number one is promoted in the paper [15].

Clearly, better values than the expected ones, for any criterion, had a value greater than one, while weaker values than expected had a positive value less than

M. R. Žižović
Faculty of Technical Sciences in Čačak, University of Kragujevac, Čačak, Serbia

D. Pamučar (✉)
Faculty of Organizational Sciences, University of Belgrade, Belgrade, Serbia
e-mail: dragan.pamucar@fon.bg.ac.rs

M. M. Žižović
Economics Institute, Belgrade, Serbia

M. Albijanić
FEFA Faculty, University Metropolitan, Belgrade, Serbia
e-mail: malbijanic@fefa.edu.rs

M. P. Stanić et al. (eds.), *Analysis, Approximation, Optimization: Computation and Applications*, Springer Optimization and Its Applications 224,
https://doi.org/10.1007/978-3-031-85743-0_22

one. In this way, a method was obtained in which evaluation can function well in a multiplicative form, i.e., a multiplicative method of multicriteria analysis was derived. Further explanation of the functioning of this method is provided in the paper [18].

Since we are presenting a new method of multicriteria analysis here, we will point out a change compared to the previously mentioned method, before presenting the new method. In the previously mentioned method, increases in factors (by criteria) were performed by adding appropriate values to one, while decreases were achieved by subtracting corresponding values (factors are always positive). In the new method, increases are simply accomplished by multiplication and decreases by division with appropriate numbers greater than one.

22.2 New Concept

Step 22.2.1 We are solving a multicriteria ranking problem of m alternatives $A_1, A_2, \ldots, A_m$ with n criteria $C_1, C_2, \ldots, C_n$ specified by the decision matrix (Table 22.1).

The values x_{ij} in the decision matrix represent the performance of each alternative A_i according to criterion C_j, $i = 1, 2, \ldots, m$, $j = 1, 2, \ldots, n$. These values, for each criterion C_j, range between the "anti-ideal" or the least acceptable values (there are no values below them for max criteria or above them for min criteria – alternatives with such values are discarded) and those that are "ideal" for the decision-maker (if in one of the alternatives we have a value better than the ideal, we would record n_{ji} as the ideal value in our matrix).

In addition to these two values, a characteristic value for this multicriteria analysis method within the range of values for an arbitrary criterion is identified as the "normally expected" value for that criterion.

So, for an arbitrary maximization-type criterion C_j we observe the anti-ideal value X_{jA}, normally expected value X_{jN}, and ideal value X_{jI}, arranged in this order $X_{jA} < X_{jN} < X_{jI}$.

For a minimization-type criterion, this order is $X_{jI} < X_{jN} < X_{jA}$.

Note that these values do not necessarily have to be present in the decision matrix of the given multicriteria problem.

Table 22.1 Decision matrix

	C_1	C_2	$\cdots$	C_n
A_1	x_{11}	x_{12}	$\cdots$	x_{1n}
A_2	x_{21}	x_{22}	$\cdots$	x_{2n}
$\vdots$	$\vdots$	$\vdots$	$\vdots$	$\ldots$
A_m	x_{m1}	x_{m2}	$\cdots$	x_{mn}

Step 22.2.2 We form a new matrix from the previously given matrix, consisting of real numbers within the interval $[-1, 1]$ in the following way.

For a maximization-type criterion C_j the decision-maker chooses a nondecreasing function f_j that maps the interval $[X_{jA}, X_{jI}]$ to a closed interval $[-1, 1]$, satisfying the condition $f_j(X_{jA}) = -1$, $f_j(X_{jN}) = 0$ and $f_j(X_{jI}) = 1$.

For a minimization-type criterion C_j the decision-maker chooses a nonincreasing function f_j that maps the interval $[X_{jI}, X_{jA}]$ to a closed interval $[-1, 1]$, satisfying the condition $f_j(X_{jA}) = -1$, $f_j(X_{jN}) = 0$ and $f_j(X_{jI}) = 1$.

By choosing these functions, the decision-maker expresses their preferences and their attitude towards the values within the range for the given criterion.

Later on, the procedure for determining the weighting coefficients for each criterion will be explained. Thus, in this method, the decision-maker has two ways of expressing their preferences, which can further influence the quality of decisions.

Now, using the previously chosen functions, we transform the initial decision matrix into a new decision matrix, which is the subject of further work. More precisely, for each $j = 1, \ldots, n$, we map the values from column j of the original matrix to column j of the new numerical matrix using the function f_j.

The numbers a_{ij}, $i = 1, 2, \ldots, m$, $j = 1, 2, \ldots, n$ appearing in this matrix are between minus one and one and may eventually include these values.

Step 22.2.3 For each criterion C_j, $j = 1, 2, \ldots, n$ the weighting coefficient is defined $k_j > 0$ with the following meaning:

i. The ideal value x_{jI} of the criterion C_j is better than the normal value X_{jN} for that criterion exactly $1 + k_j$ times.
ii. The anti-ideal value x_{jA} of the criterion C_j is weaker than the normal value X_{jN} for that criterion exactly $\frac{1}{1+k_j}$ times.

So, if the criteria $C_1, C_2, \ldots, C_n$ are arranged in increasing order of importance, then their weighting coefficients are positive numbers $k_1, k_2, \ldots, k_n$ in such a way that each subsequent one is less than or equal to the previous one.

Step 22.2.4 In Section STEP 22.2.3, the role of weighting coefficients in this concept of solving the multicriteria problem is explained. However, the technique for obtaining these weighting coefficients is not proposed.

Considering that there is a well-developed methodology for obtaining additive weighting coefficients in multicriteria analysis (positive numbers $w_1, w_2, \ldots, w_n$, assigned to criteria $C_1, C_2, \ldots, C_n$ such that their sum is one), as seen in numerous works (see reference books [3, 5, 11, 14] and papers [1, 2, 6, 8, 10, 12, 13, 16] and [17]), a natural question arises as to how this extensive experience and knowledge can be utilized to obtain multiplicative weighting coefficients for solving our task.

Here, we present one methodology.

Let the relationship between the ideal alternative and the normal alternative be given by the relation

$$\frac{V(A_I)}{V(A_N)} = R, \quad R > 0.$$

Then,

$$V(A_I) = (1 + k_1)(1 + k_2) \cdots (1 + k_n) = RV(A_N) = R.$$

So, it is

$$\ln(1 + k_1) + \ln(1 + k_2) + \cdots + \ln(1 + k_n) = \ln R.$$

Now follows the assumption about the connection between multiplicative weighting coefficients k_j, $j = 1, 2, \ldots, n$ and additive weighting coefficients w_j, $j = 1, 2, \ldots, n$ or the solution for multiplicative weighting coefficients through the relation

$$w_j = \frac{\ln(1 + k_j)}{\ln R}, \quad j = 1, 2, \ldots, n,$$

from which $k_j = R^{w_j} - 1$.

Example of calculating multiplicative weighting coefficients from additive ones:

If the additive weighting coefficients are given for four criteria $w_1 = 0.4$, $w_2 = 0.3$, $w_3 = 0.2$, and $w_4 = 0.1$ and if the relationship between the ideal and normal alternatives is given as $R = 5$, then by the previously defined procedure, it is easily obtained that

$$k_1 = 0.904, \quad k_2 = 0.621, \quad k_3 = 0.380, \quad k_4 = 0.175,$$

Step 22.2.5 We define the function for evaluating alternatives in the following way.

For the alternative Ai, whose values according to criteria are given in matrix Table 22.2, the function for its evaluation is given using the following formula:

$$V(A_i) = \left((1 + |a_{i1}|k_1)^{\mathrm{sgn}(a_{i1})}\right)\left((1 + |a_{i2}|k_2)^{\mathrm{sgn}(a_{i2})}\right) \cdots \left((1 + |a_{in}|k_n)^{\mathrm{sgn}(a_{in})}\right),$$

i.e.,

$$V(A_i) = \prod_{j=1}^{n} \left((1 + |a_{i1}|k_1)^{\mathrm{sgn}(a_{i1})}\right)$$

Table 22.2 Numerical decision matrix

	C_1	C_2	$\cdots$	C_n
A_1	a_{11}	a_{12}	$\cdots$	a_{1n}
A_2	a_{21}	a_{22}	$\cdots$	a_{2n}
$\vdots$	$\vdots$	$\vdots$	$\vdots$	$\cdots$
A_m	a_{m1}	a_{m2}	$\cdots$	a_{mn}

where

$$\operatorname{sgn} a_{ij} = \begin{cases} -1, & \text{for} \quad -1 \le a_{ij} < 0, \\ 0, & \text{for} \quad a_{ij} = 0, \\ 1, & \text{for} \quad 0 < a_{ij} \le 1. \end{cases} \tag{22.1}$$

The correctness of this chosen function for evaluating alternatives arises from the fact that it satisfies:

Statement 22.2.1 If a_{xj} and a_{yj} are elements from column j of the decision matrix Table 22.2 (the column corresponding to criterion C_j, such that $a_{xj} < a_{yj}$, then the factor corresponding to criterion C_j in the evaluation of alternative A_y is greater than the factor corresponding to criterion C_j in the evaluation of alternative A_x).

Proof Firstly, if $a_{yj} > a_{xj}$, then $1 + a_{yj} > 1 + a_{xj}$.

Secondly, if $a_{yj} > 0 > a_{xj}$, then

$$1 + a_{yj} > 1 > \frac{1}{1 + |a_{xj}|}.$$

Thirdly, if $a_{xj} < a_{yj} < 0$, then $|a_{yj}| < |a_{xj}|$, so it is

$$1 < 1 + |a_{yj}|k_j < 1 + |a_{xj}|k_j,$$

i.e.,

$$\frac{1}{1 + |a_{yj}|k_j} > \frac{1}{1 + |a_{xj}|k_j}.$$

This defined function has the maximum value when all factors are maximum, i.e., in the case when the alternative is ideal:

$$V(A_I) = (1 + k_1)(1 + k_2) \cdots (1 + k_n).$$

The minimum value of this defined function is when all factors are minimum (reciprocal values of the maximum), i.e., in the case of the anti-ideal alternative:

$$V(A_A) = \frac{1}{(1+k_1)(1+k_2)\cdots(1+k_n)}.$$

It is also evident that the value of the normal alternative is

$$V(A_N) = 1.$$

□

Step 22.2.6 (Evaluation of Alternatives) If A_p and A_q are alternatives from the set $\{A_1, A_2, \ldots, A_m\}$, then we say that alternative A_p is better than alternative A_q if and only if $V(A_p) > V(A_q)$.

In the case where $V(A_p) = V(A_q)$, we consider the values of these alternatives according to the strongest criterion. The alternative with the higher value is considered better. If these values are equal, we move on to the next criterion in strength and apply the same reasoning, and so on.

Finally, it can be said that for alternatives A_p and A_q, we cannot determine which one is better unless they have the same corresponding values.

It can be relatively easily proven the following highly important:

Theorem If we solve the multicriteria model with alternatives $A_1, A_2, \ldots, A_m$ and criteria $C_1, C_2, \ldots, C_n$ using the proposed method and then expand it by adding new alternatives $B_1, \ldots, B_s$ and solve it again with m+s alternatives and n criteria, we obtain an order of these m+s alternatives in which the order of the initial alternatives $A_1, A_2, \ldots, A_m$ remains unchanged.

This fact has one very important consequence.

Consequence Introducing new alternatives into the model does not affect the ranking of existing alternatives in the model. In other words, introducing new alternatives will not favor the existing alternatives in the model.

It is also important to note that introducing a possible new criterion into an existing multicriteria model (with clearly defined elements: ideal, normal, and anti-ideal values, corresponding weighting coefficient for the newly introduced criterion, and a function mapping the interval of values of that criterion to an $[-1, 1]$ interval) does not require any change in the previously performed calculations. In other words, it is only necessary to multiply the already obtained products with the new factor and perform a new ranking.

22.3 Example

For illustrating the method, we have chosen an example presented in the book [5], which was solved using the AHP method and was also solved in the paper [18] using

Table 22.3 Decision matrix of the example

	C_1	C_2	C_3	C_4	C_5	C_6
A_1	4	45	50	90	40	30
A_2	3	30	50	70	40	50
A_3	2	20	30	10	20	60
	min	min	max	max	max	max

a modified multiplicative method. The example consists of three alternatives and six criteria. The goal is to make a comparison of this new multiplicative method with the previously mentioned methods (Table 22.3).

To calculate, firstly, define the ideal, normal, and anti-ideal alternatives:

$$A_I(2, 20, 50, 90, 40, 60),$$

$$A_N(3, 30, 30, 40, 20, 40),$$

$$A_A(4, 45, 10, 10, 10, 20).$$

Next, we assume that the relationship between the ideal alternative and the normal alternatives is given by the relation

$$\frac{V(A_I)}{V(A_N)} = 5.$$

Now, knowing that the additive weighting coefficients are given

$$w_1 = 0.18182, \quad w_2 = 0.13636, \quad w_3 = 0.09091,$$

$$w_4 = 0.22727, \quad w_5 = 0.04545 \quad \text{and} \quad w_6 = 0.31818.$$

Using the previously defined procedure, we obtain the multiplicative weighting coefficients:

$$k_1 = 0.33994, \quad k_2 = 0.24542, \quad k_3 = 0.15756,$$

$$k_4 = 0.44163, \quad k_5 = 0.07589, \quad k_6 = 0.66878.$$

If, as defined in step 22.2.2, we opt for linear functions for each criterion, then we easily propose them and obtain

For the criterion C_1

$$f_1 = -x + 3, \quad \text{for } 2 \le x \le 4.$$

For the criterion C_2

$$f_2 = \left\{ -\frac{x}{10} + 3, \quad \text{for } 20 \leq x \leq 30 - \frac{x}{25} + 2 \text{ for } 30 \leq x \leq 4.5. \right.$$

For the criterion C_3

$$f_3 = \frac{x}{20} - \frac{3}{2}, \quad \text{for } 10 \leq x \leq 50.$$

For the criterion C_4

$$f_4 = \left(\left\{ \frac{x}{30} - \frac{4}{3}, \quad \text{for } 10 \leq x \leq 40 \frac{x}{50} - \frac{4}{5} \text{ for } 40 \leq x \leq 90. \right. \right.$$

For the criterion C_5

$$f_5 = \left\{ \frac{x}{10} - 2, \quad \text{for } 10 \leq x \leq 20 \frac{x}{20} - 1 \text{ for } 20 \leq x \leq 40. \right.$$

For the criterion C_6

$$f_6 = \frac{x}{20} - 2, \quad \text{for } 20 \leq x \leq 60.$$

Now, we simply recalculate the values from the initial decision matrix for our example using the defined functions and enter the values into the normalized decision matrix (Table 22.4):

We evaluate alternatives after a simple calculation, determined by the established order:

$$V(A_1) = 0.806276,$$
$$V(A_2) = 2.102217,$$
$$V(A_3) = 1.795448.$$

So, the best alternative is A_2 and then A_3 and the worst is A_1 which is consistent with the results obtained in the previously mentioned places.

Table 22.4 Normalized decision matrix of the example

	C_1	C_2	C_3	C_4	C_5	C_6
A_1	-1	-1	1	1	1	$-\frac{1}{2}$
A_2	0	0	1	$\frac{3}{5}$	1	$\frac{1}{2}$
A_3	1	1	0	-1	0	1

22.4 Conclusion

This chapter is written under the assumption that for each criterion, the normal (expected) value is between the ideal and anti-ideal values. At the same time, a logically expected requirement is set that the ideal value is exactly as much better than the expected (normal) value by how much the expected value is better than the anti-ideal value. Certainly, other assumptions are possible and will be presented in future papers. Other aspects of the application of this method are expected in future works.

References

1. Edwards, W., Barron, F.H.: Smarts and smarter: Improved simple methods for multiattribute utility measurement. Organ. Behav. Hum. Decis. Process. **60**(3), 306–325 (1994)
2. Graham, A.: Nonnegative Matrices and Applicable Topics in Linear Algebra. Ellis Horward, Chichester (1987)
3. Keeney, R.L., Raiffa, H.: Decisions with Multiple Objectives: Preferences and Value Trade-Offs. Wiley, New York (1976)
4. Mukhametzyanov, I.: On the conformity of scales of multidimensional normalization: an application for the problems of decision making. Decis. Mak. Appl. Manag. Eng. **6**(1), 399–341 (2023)
5. Nikolic, I., Borovic, S.: Višekriterijumska optimizacija, metode, primena u logistici, softver. CVŠ VJ (1996)
6. Pamučar, D., Stević, Ž., Sremac, S.:: A new model for determining weight coefficients of criteria in mcdm models: Full consistency method (fucom). Symmetry **10**(9), 393 (2018)
7. Pamučar, D., Žižović, M., and Djuričić, D.: Modification of the CRITIC method using fuzzy rough numbers. Decis. Mak. Appl. Manag. Eng. **5**(2), 362–371 (2022)
8. Pöyhönen, M., Hämäläinen, R.P.: On the convergence of multiattribute weighting methods. Eur. J. Oper. Res. **129**(3), 569–585 (2001)
9. Puska, A., Stilić, A., Stojanović, I.: Approach for multi-criteria ranking of balkan countries based on the index of economic freedom. J. Decis. Analytics Intell. Comput. **3**(1), 1–14 (2016)
10. Rezai, J.: Best-worst multi-criteria decision-method. Omega **53**, 49–57 (2015)
11. Saaty, T.L.: Analytic Hierarchy Process. McGraw-Hill, New York (1980)
12. Valipour, A., Yahaya, N., Md Noor, N., Antuchevičienė, J., Tamošaitienė, J.: Hybrid swara-copras method for risk assessment in deep foundation excavation project: an iranian case study. J. Civil Eng. Manag. **23**(4), 524–532 (2017)
13. Zavadskas, E.K., Govindan, K., Antucheviciene, J., Turskis, Z.: Hybrid multiple criteria decision-making methods: A review of applications for sustainability issues. Econom. Res.-Ekonomska istraživanja **29**(1), 857–887 (2016)
14. Žižović, M., Fatić, D., Pamučar, D., Žižović, M.M.: Determining criteria weight coefficients assessment method – ideally change expected. In: SYM-OP-IS 2023 Proceedings Conference, pp. 483–487. Belgrade, Serbia (2023)
15. Žizović, M., Damljanović, N.: Multiplicative multi-criteria analysis method for decision-making. Maejo Int. J. Sci. Technol. **10**(2), 233–241 (2016)
16. Žižović, M., Pamucar, D.: New model for determining criteria weights: level based weight assessment (lbwa) model. Decisi. Mak. Appl. Manag. Eng. **2**(2), 126–137 (2019)

17. Žižović, M., Pamučar, D., Ćirović, G., Žižović, M.M., Miljković, B.D.: A model for determining weight coefficients by forming a non-decreasing series at criteria significance levels (ndsl). Mathematics **8**(5), 745 (2020)
18. Žižović, M.M., Žižović, M., Damljanović, N., Pavlović, K.: Multiplicative method based on expected criteria values. Rep. Mech. Eng. **4**(1), 317–325 (2023)

MIX
Papier aus verantwortungsvollen Quellen
Paper from responsible sources
FSC® C105338

If you have any concerns about our products,
you can contact us on
ProductSafety@springernature.com

In case Publisher is established outside the EU,
the EU authorized representative is:
Springer Nature Customer Service Center GmbH
Europaplatz 3, 69115 Heidelberg, Germany

Printed by Libri Plureos GmbH
in Hamburg, Germany